Environmental Chemistry for a Sustainable World

Other Publications by the Editors

Books

Environmental Chemistry
http://www.springer.com/978-3-540-22860-8

Organic Contaminants in Riverine and Groundwater Systems
http://www.springer.com/978-3-540-31169-0

Sustainable Agriculture
Volume 1: http://www.springer.com/978-90-481-2665-1
Volume 2: http://www.springer.com/978-94-007-0393-3

Book series

Environmental Chemistry for a Sustainable World
http://www.springer.com/series/11480

Sustainable Agriculture Reviews
http://www.springer.com/series/8380

Journals

Environmental Chemistry Letters
http://www.springer.com/10311

Agronomy for Sustainable Development
http://www.springer.com/13593

More information about this series at http://www.springer.com/series/11480

Grégorio Crini • Eric Lichtfouse
Editors

Green Adsorbents for Pollutant Removal

Fundamentals and Design

Editors
Grégorio Crini
Laboratoire Chrono-environnement,
UMR 6249, UFR Sciences et Techniques
Université Bourgogne Franche-Comté
Besançon, France

Eric Lichtfouse
CEREGE, Aix Marseille Univ, CNRS, IRD,
INRA, Coll France
Aix-en-Provence, France

ISSN 2213-7114 ISSN 2213-7122 (electronic)
Environmental Chemistry for a Sustainable World
ISBN 978-3-030-06365-8 ISBN 978-3-319-92111-2 (eBook)
https://doi.org/10.1007/978-3-319-92111-2

Softcover re-print of the Hardcover 1st edition 2018

Printed on acid-free paper

This Springer imprint is published by the registered company Springer International Publishing AG part of Springer Nature.
The registered company address is: Gewerbestrasse 11, 6330 Cham, Switzerland

Preface

Biosorbents have a potential application in both environmental control and metal recovery operations. Bohumil Volesky (1990)

Over the past three decades, there has been an increasing interest in the phenomenon of contaminants sequestration by nonconventional green adsorbents using oriented-adsorption processes. This phenomenon has a high potential for applications in water and wastewater treatments. Liquid-solid adsorption using green adsorbents can be simply defined as the removal of contaminants from solutions using products and by-products of biological, agricultural, and industrial origin. Green adsorbents represent cheap filter materials often with high affinity, capacity, and selectivity to interact with contaminants. They are abundant and already available in most places in large quantities. The list of green adsorbents is extremely extensive, including carbons from agricultural solid wastes and industrial by-products; agro-food wastes; industrial by-products; natural products, e.g., clays, hemp, flax, and cotton; and biological materials such as dead biomass, living plants, algae, and biopolymers. This book, *Green Adsorbents for Pollutant Removal*, is the first volume of two volumes published in the series *Environmental Chemistry for a Sustainable World*. Written by 80 international contributors coming from 23 different countries who are leading experts in the adsorption field, these two volumes show a typical selection of green materials studied and used in wastewater treatment, with emphasis on industrial effluents.

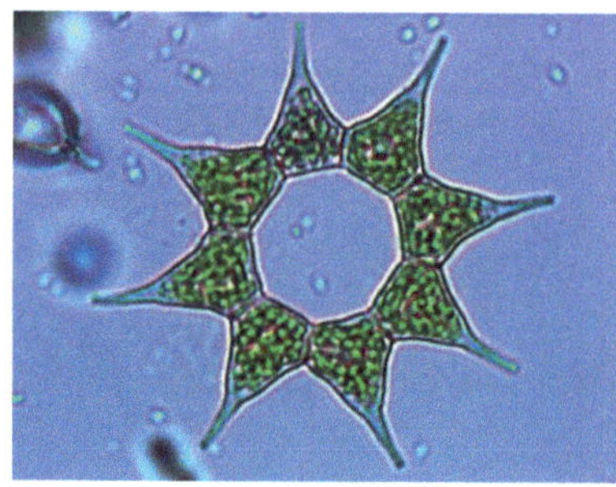

Pediastrum simplex, a green algae living in freshwater environments. Copyright INRA, Jean-Claude DRUART, 2018

The first volume provides an overview of fundamentals and design of adsorption processes. For people who are new to the field, the book starts by two overview chapters presenting the principles and properties of wastewater treatment in Chap. 1 by Grégorio Crini and Eric Lichtfouse, and adsorption processes in Chap. 2 by Grégorio Crini et al. The book also provides a comprehensive source of knowledge on acid-base properties of biosorbents in Chap. 3 by Pablo Lodeiro et al. In Chap. 4, Marco Balsamo and Fabio Montagnaro discuss fractal-like kinetic models for fluid-solid adsorption. Paola Rodríguez-Estupiñán et al. report in Chap. 5 the chemical characterization of oxidized activated carbons for metal removal. In Chap. 6, Mohamed Nasser Sahmoune summarizes the thermodynamic properties of metal adsorption by green adsorbents. Biosorption of polycyclic aromatic hydrocarbons and organic pollutants is reviewed by Bhawana Pathak et al. in Chap. 7 and by Mustapha Mohammed Bello and Abdul Aziz Abdul Raman in Chap. 8. Then, Sofia F. Soares et al. discuss the use of magnetic biosorbents in water treatment in Chap. 9. The final chapter by Manviri Rani and Uma Shanker (Chap. 10) summarizes recent trends and impact of nanomaterials as green adsorbent and potential catalysts for environmental applications. The second volume "Green Adsorbents for Pollutant Removal Innovative Materials" covers innovative materials proposed as green adsorbents.

The audience for this book includes students, environmentalists, engineers, water scientists, and civil and industrial personnel who wish to specialize in adsorption technology. Academically, this book will be of use to students in chemical and environmental engineering who wish to learn about adsorption and its fundamentals. It has also been compiled for practicing engineers who wish to know about recent developments on adsorbent materials in order to promote further research toward improving and developing newer adsorbents and processes for the efficient removal of pollutants from industrial effluents. However, the book is not meant to be an extensive treatise on adsorption and adsorbents. For example, particular aspects on modeling or biosorption are not considered because the reader can find abundant information on these topics in the literature. It is hoped that the book will serve two main functions: (1) a readable and useful presentation not only for undergraduate and postgraduate students but also for the water scientists and engineers; and (2) a convenient reference handbook in the form of numerous recent examples and appended information.

The editors extend their thanks to all the authors who contributed to this book for their efforts in producing timely and high-quality chapters. The creation of this book would not have been possible without the assistance of several friends deserving acknowledgment. They have helped by choosing contributors, reviewing chapters, and in many other ways. Finally, we would like to thank the staff at Springer Nature for their highly professional editing of the publication.

Besançon Cedex, France Grégorio Crini
Aix-en-Provence Cedex, France Eric Lichtfouse

Contents

About the Editors

Grégorio Crini 52, is a researcher at University of Bourgogne Franche-Comté, UMR Chrono-environnement, Besançon, France. His current interests focus on the design of novel polymer networks and the environmental aspects of polysaccharide chemistry. He has published books and over 180 papers in international journals, and he is a highly cited researcher. The total citation of his publications is over 7500 according to ISI Web of Science, h-index: 32.

Eric Lichtfouse 58, is a biogeochemist at the University of Aix-Marseille, CEREGE, Aix-en-Provence, France. He obtained his PhD in Organic Geochemistry from the University of Strasbourg in 1989 for the discovery of new fossil steroids in sediments and petroleum. He has invented the ^{13}C-dating method allowing to measure the dynamics of soil organic molecules. He is Chief Editor of the journal *Environmental Chemistry Letters* and former Chief Editor of the journal *Agronomy for Sustainable Development*. He has published the book *Scientific Writing for Impact Factor Journal*, describing the micro-article, a new tool to identify the novelty of experimental results. He has published 85 research papers, with an h index of 23, and edited more than 50 books. He received the Analytical Chemistry Prize from the French Chemical Society, the Grand University Prize from Nancy University, and a Journal Citation Award by the Essential Science Indicators.

Contributors

Marco Balsamo Dipartimento di Ingegneria Chimica, dei Materiali e della Produzione Industriale, Università degli Studi di Napoli Federico II, Naples, Italy

José L. Barriada Departamento de Química, Universidade da Coruña, A Coruña, Spain

Mustapha Mohammed Bello Department of Chemical Engineering, Faculty of Engineering, University of Malaya, Kuala Lumpur, Malaysia

Grégorio Crini Laboratoire Chrono-environnement, UMR 6249, UFR Sciences et Techniques, Université Bourgogne Franche-Comté, Besançon, France

Ana L. Daniel-da-Silva Department of Chemistry and CICECO-Aveiro Institute of Materials, University of Aveiro, Aveiro, Portugal

Tiago Fernandes Department of Chemistry and CICECO-Aveiro Institute of Materials, University of Aveiro, Aveiro, Portugal

Liliana Giraldo Departamento de Química, Facultad de Ciencias, Universidad Nacional de Colombia, Bogotá, Colombia

Shalini Gupta School of Environment and Sustainable Development, Central University of Gujarat, Gandhinagar, Gujarat, India

Roberto Herrero Departamento de Química, Universidade da Coruña, A Coruña, Spain

Eric Lichtfouse CEREGE, Aix Marseille Univ, CNRS, IRD, INRA, Coll France, Aix-en-Provence, France

Pablo Lodeiro GEOMAR Helmholtz Centre for Ocean Research Kiel, Kiel, Germany

María Martínez-Cabanas Departamento de Química, Universidade da Coruña, A Coruña, Spain

Fabio Montagnaro Dipartimento di Scienze Chimiche, Università degli Studi di Napoli Federico II, Complesso Universitario di Monte Sant'Angelo, Naples, Italy

Juan Carlos Moreno-Piraján Facultad de Ciencias, Departamento de Química, Grupo de Investigación en Sólidos Porosos y Calorimetría, Universidad de los Andes, Bogotá, Colombia

Nadia Morin-Crini Laboratoire Chrono-environnement, UMR 6249, UFR Sciences et Techniques,, Université Bourgogne Franche-Comté, Besançon, France

Bhawana Pathak School of Environment and Sustainable Development, Central University of Gujarat, Gandhinagar, Gujarat, India

Abdul Aziz Abdul Raman Department of Chemical Engineering, Faculty of Engineering, University of Malaya, Kuala Lumpur, Malaysia

Manviri Rani Department of Chemistry, Dr. B. R. Ambedkar National Institute of Technology, Jalandhar, Punjab, India

Pilar Rodríguez-Barro Departamento de Química, Universidade da Coruña, A Coruña, Spain

Paola Rodríguez-Estupiñán Facultad de Ciencias, Departamento de Química, Grupo de Investigación en Sólidos Porosos y Calorimetría, Universidad de los Andes, Bogotá, Colombia

Mohamed Nasser Sahmoune Department of Process Engineering, Faculty of Engineering Sciences, University of Boumerdes, Boumerdes, Algeria

Laboratory of Coatings, Materials and Environment, University of Boumerdes, Boumerdes, Algeria

Manuel E. Sastre de Vicente Departamento de Química, Universidade da Coruña, A Coruña, Spain

Uma Shanker Department of Chemistry, Dr. B. R. Ambedkar National Institute of Technology, Jalandhar, Punjab, India

Sofia F. Soares Department of Chemistry and CICECO-Aveiro Institute of Materials, University of Aveiro, Aveiro, Portugal

Tito Trindade Department of Chemistry and CICECO-Aveiro Institute of Materials, University of Aveiro, Aveiro, Portugal

Reeta Verma School of Environment and Sustainable Development, Central University of Gujarat, Gandhinagar, Gujarat, India

Teresa Vilariño Departamento de Química, Universidade da Coruña, A Coruña, Spain

Lee D. Wilson Department of Chemistry, University of Saskatchewan, Saskatoon, SK, Canada

Chapter 1
Wastewater Treatment: An Overview

Grégorio Crini and Eric Lichtfouse

Contents

Abstract During the last 30 years, environmental issues, especially concerning the chemical and biological contamination of water, have become a major concern for both society and public authorities, but more importantly, for the whole industrial world. Any activities whether domestic or agricultural but also industrial produce wastewaters or effluents containing undesirable contaminants which can also be toxic. In this context, a constant effort must be made to protect water resources. In general, conventional wastewater treatment consists of a combination of physical, chemical, and biological processes and operations to remove insoluble particles and soluble contaminants from effluents. This chapter briefly discusses the different types of effluents, gives a general scheme of wastewater treatment, and describes the advantages and disadvantages of technologies available.

G. Crini (✉)
Laboratoire Chrono-environnement, UMR 6249, UFR Sciences et Techniques, Université Bourgogne Franche-Comté, Besançon, France
e-mail: gregorio.crini@univ-fcomte.fr

E. Lichtfouse
CEREGE, Aix Marseille Univ, CNRS, IRD, INRA, Coll France, Aix-en-Provence, France
e-mail: eric.lichtfouse@inra.fr

G. Crini, E. Lichtfouse (eds.), *Green Adsorbents for Pollutant Removal*, Environmental Chemistry for a Sustainable World 18,
https://doi.org/10.1007/978-3-319-92111-2_1

Abbreviations

AC	Activated carbons
AOP	Advanced oxidation processes
AOX	Adsorbable organic halogen
BAS	Biological activated sludge
BOD	Biochemical oxygen demand
CAA	Commercial activated alumina
CAC	Commercial activated carbons
COD	Chemical oxygen demand
CW	Constructed wetlands
CWAO	Catalytic wet air oxidation
D	Dialysis
DPS	Dangerous priority substances
E	Electrolysis
EC	Electro-coagulation
ED	Electrodialysis
EED	Electroelectro-dialysis
EF	Electro-flocculation
ELM	Emulsion liquid membranes
ICP	Inductively coupled plasma
MF	Microfiltration
MVP	Membrane pervaporation
NF	Nanofiltration
PAH	Polycyclic aromatic hydrocarbons
PCB	Polychlorobiphenyls
PS	Priority substances
RS	Reverse osmosis
SLM	Supported liquid membranes
SS	Suspended solids
TOC	Total organic carbon
TOD	Total oxygen demand
UF	Ultrafiltration
VOC	Volatile organic compounds
WAO	Non catalytic wet air oxidation
WFD	Water Framework Directive

1.1 Introduction

Actually, water pollution by chemicals has become a major source of concern and a priority for both society and public authorities, but more importantly, for the whole industrial world (Sonune and Ghate 2004; Crini 2005; Cox et al. 2007;

Sharma 2015; Rathoure and Dhatwalia 2016). What is water pollution? Water pollution can be defined in many ways. Pollution of water occurs when one or more substances that will modify the water in negative fashion are discharged in it. These substances can cause problems for people, animals and their habitats and also for the environment. There are various classifications of water pollution (Morin-Crini and Crini 2017). The two chief sources can be seen as point and non-point. The first refers to the pollutants that belong to a single source such as emissions from industries into the water, and the second on the other hand means pollutants emitted from multiple sources.

The causes of water pollution are multiple: industrial wastes, mining activities, sewage and waste water, pesticides and chemical fertilizers, energy use, radioactive waste, urban development, etc. The very fact that water is used means that it will become polluted: any activities whether domestic or agricultural but also industrial produce effluent containing undesirable pollutants which can also be toxic. In this context, a constant effort must be made to protect water resources (Khalaf 2016; Rathoure and Dhatwalia 2016; Morin-Crini and Crini 2017).

The legislation covering liquid industrial effluent is becoming stricter, especially in the more developed countries, and imposes the treatment of any wastewater before it is released into the environment. Since the end of the 1970s, in Europe, the directives are increasingly severe and zero rejection is being sought by 2020. Currently, the European policy on water results from the Water Framework Directive (WFD) of 2000 which establishes guidelines for the protection of surface water, underground water, and coastal water in Europe (Morin-Crini and Crini 2017).

The WFD also classified chemicals into two main lists of priority substances. The first, the "Black List", involves dangerous priority substances (DPS) considered to be persistent, highly toxic or to lead to bioaccumulation. The second list, the "Grey List", gathers priority substances (PS) presenting a significant risk for the environment. The selection of these substances can either be based on individual substances of families of substances (e.g. metals, chlorobenzenes, alkylphenols, etc.) or on the basis of the industrial sector (e.g. agro-food industry, chemicals industry, metal-finishing sector, etc.). Currently, Europe is now asking industrials to innovate to reduce and/or eliminate the release of DPS and PS chemicals in their wastewaters.

Moreover, recycling wastewater is starting to receive active attention from industry in the context of sustainable development (e.g. protection of the environment, developing concepts of "green chemistry", use of renewable resources), improved water management (recycling of waste water) and also health concerns (Kentish and Stevens 2001; Cox et al. 2007; Sharma and Sanghi 2012; Khalaf 2016; Rathoure and Dhatwalia 2016; Morin-Crini and Crini 2017). Thus, for the industrial world, the treatment of effluents has become a priority.

During the past three decades, several physical, chemical and biological technologies have been reported such as flotation, precipitation, oxidation, solvent extraction, evaporation, carbon adsorption, ion-exchange, membrane filtration, electrochemistry, biodegradation, and phytoremediation (Berefield et al. 1982; Liu and Liptak 2000; Henze 2001; Harvey et al. 2002; Chen 2004; Forgacs et al. 2004; Anjaneyulu et al. 2005; Crini and Badot 2007; Cox et al. 2007; Hai et al. 2007;

Barakat 2011; Rathoure and Dhatwalia 2016; Morin-Crini and Crini 2017). Which is the best method? There is no direct answer to this question because each treatment has its own advantages and constraints not only in terms of cost but also in terms of efficiency, feasibility, and environmental impact. In general, elimination of pollutants is done by both physical, chemical and biological means. At the present time, there is no single method capable of adequate treatment, mainly due to the complex nature of industrial effluents. In practice, a combination of different methods is often used to achieve the desired water quality in the most economical way.

After a brief discussion on the main contaminants/pollutants and the different types of effluents, this chapter proposes a general scheme of wastewater treatment and presents the advantages and disadvantages of different individual techniques used.

1.2 Water Pollution

1.2.1 Contamination and Contaminants

Contamination/Pollution arises from all sectors of human activity (i.e. domestic, industrial and agricultural), and is not only due to natural (petroleum, minerals, etc.) and anthropogenic causes (e.g. sewage treatment sludge or persistent organic pollutants produced by the incineration of waste) but also, and especially, to synthetic substances produced by chemical industries (e.g. dyes, fertilizers, pesticides, and so on). The terms contamination/pollution and contaminant/pollutant are often used in relation to subjects like environment, food and medicine (Amiard 2011; Rathoure and Dhatwalia 2016). Both contaminant and pollutant refer to undesirable or unwanted substances. Pollutant refers to a harmful substance but contaminant is not necessarily harmful since contamination refers simply to the presence of a chemical substance where it should not be. This means that all pollutants are contaminants, but not all contaminants are pollutants. In this chapter, both these terms were used.

A chemical pollutant is a substance toxic for flora and fauna, and for humanity, and present at concentrations such that, in nature, it has repercussions on the environment and on health in general. Pollutants can be categorized according to the sources they are derived from, such as water pollutants, soil pollutants, air pollutants or noise pollutants (Crini and Badot 2007). Examples of pollutants known to the public and found in waters are numerous and various. The list includes nitrates, phosphates, detergents, pesticides and other crop sprays, chlorinated solvents but also metals (e.g. lead, mercury, chromium, cadmium, arsenic), dyes, organics (benzene, bisphenol A…), mineral derivatives (especially arsenic and cyanides) and microorganisms (e.g. bacteria, virus). Others are less well known but are considered to be high on the list of dangerous substances: volatile organic compounds (VOC), polycyclic aromatic hydrocarbons (PAH), polychlorobiphenyls (PCB), bromine-containing flame-retardants, phthalates, and many more (Liu and Liptak 2000; Sonune and Ghate 2004; Sharma and Sanghi 2012).

One way of measuring the quality of water is to take samples of this water and measure the concentrations of different substances that it contains, using analytical techniques such as, for instance, inductively coupled plasma (ICP) for metals, and/or determine chemical indicators or global parameters (Morin-Crini and Crini 2017). Typically, water quality is determined by comparing the physical and chemical characteristics of a sample with water quality guidelines or standards based on scientifically assessed acceptable levels of toxicity to either humans or aquatic organisms. Biological indicators using living organism such as fish can be also used. From the wastewater treatment point of view, it is also important to list the exact chemical composition of the effluents to be treated (Liu and Liptak 2000; Lacorte et al. 2003; Pokhrel and Viraraghavan 2004; Sharma 2015; Druart et al. 2016). Indeed, before any actions can be taken to reduce and/or eliminate any chemicals, it is necessary to identify all the dissolved substances in the effluents qualitatively and quantitatively using. However, a real effluent can be also a non-uniform mixture, colored and/or smelly, contain suspended solids (SS), immiscible liquids (e.g. oils, fats, hydrocarbons), soluble and/or biodegradable molecules, substances that can give waters redox potential, acidity, or pathogenicity problems (Anjaneyulu et al. 2005; Crini and Badot 2010; Sharma and Sanghi 2012; Sharma 2015; Morin-Crini and Crini 2017). In this case, wastewater quality can then be defined by physical, chemical and biological characteristics or general parameters (Cooper 1993; Liu and Liptak 2000; Crini and Badot 2007). Physical parameters include color, temperature, solids, turbidity, odor, oil and grease. Solids can be further classified into suspended and dissolved substances as well as organic and inorganic fractions. Chemical parameters associated with the organic content of industrial wastewater include the chemical oxygen demand (COD), biochemical oxygen demand (BOD), total organic carbon (TOC), and total oxygen demand (TOD). Inorganic chemical parameters include salinity, pH (acidity, alkalinity), metals, chlorides, sulfates, nitrogen, phosphorus, etc. Bacteriological parameters include coliforms, fecal coliforms, specific pathogens, and viruses. Recent books can be consulted on these topics (Sharma and Sanghi 2012; Sharma 2015; Khalaf 2016; Rathoure and Dhatwalia 2016; Morin-Crini and Crini 2017).

Recently, Druart et al. (2016) investigated the chemical composition of discharge waters from a metal-finishing industry sampled over a three-month period. All these samples respected the regulatory standards. Twenty-on water parameters and 164 substances were monitored, among them organic and metallic compounds. The results indicated, that, on average, 52 substances were found with a high variability, both qualitative and quantitative. Inorganics such as calcium, sodium and chloride were present at concentrations close to g/L and metals higher than mg/L. Organics were detected at trace levels (ng/L of μg/L).

1.2.2 Different Types of Effluents

There are various sources of water contamination (e.g. households, industry, mines, infiltration) but one of the greatest remains its large scale use by industry (Anjaneyulu et al. 2005; Hai et al. 2007). Four categories of water are generally distinguished: (1) rainwater (runoff from impermeable surfaces), (2) domestic wastewater, (3) agricultural water and (4) industrial wastewaters (Crini and Badot 2007). The last group can be subdivided into cooling water, washing effluent (of variable composition), and manufacturing or process water (biodegradable and/or potentially toxic). In general, process waters (i.e. wastewaters or effluents) pose the greatest problems. Wastewaters differ significantly from drinking water sources (usually rivers, lakes, or reservoirs) in one important way: the contaminant levels in most drinking water sources are quite low as compared with contaminant levels in wastewaters derived from industrial-type activities (Cooney 1999). However, their toxicity depends, of course, on their composition, which in turn depends on their industrial origin.

It can be noted that some effluents such as from surface treatment or coke-production plants are serious polluters whereas the effluent from other sectors such as the agro-food industry (including dairies, sugar mills and fruit and vegetable processing units) may be heavily loaded but the substances it contains are easily biodegradable and even recyclable. Pollution issues have a strong impact on the population. Colored effluent, for instance from pulp and paper mills or from textile mills, has a strong visual impact due to its color and is perceived by the public as an indication of the presence of dangerous pollution – however toxic the coloring actually is (Lacorte et al. 2003; Pokhrel and Viraraghavan 2004; Forgacs et al. 2004; Rana et al. 2004; Anjaneyulu et al. 2005; Crini 2005; Hai et al. 2007; Wojnárovits and Takács 2008). Colored effluent can lead to nature protection associations or other stakeholders in the water bodies suing the parties responsible. In addition, it is known that paper-mill wastewater contains nutrient elements that can lead to eutrophication and thus to a heavy organic load for the aquatic environment due to the proliferation of algae at the expense of other aquatic species (Lacorte et al. 2003; Rana et al. 2004). Effluent with high levels of heavy metals from surface treatment industries is also a serious source of toxicity for aquatic ecosystems, again creating worries for the population (Rana et al. 2004; Anjaneyulu et al. 2005; Morin-Crini and Crini 2017).

The industrial sectors of agro-food, textiles, pulp industry and surface-treatment industries are today considered to be the four largest consumers of water and the most polluting, in spite of the considerable effort made to clean up the processes over the last 30 years. These activities are all energy- and water-consuming as well as highly chemically polluting. The problems encountered during wastewater treatment are generally very complex as the effluent contains pollutants of various types depending on its origin. So, there are different types of effluent to treat, each with its own characteristics requiring specific treatment processes.

1.3 Wastewater Treatment

1.3.1 General Scheme of Wastewater Treatment

When water is polluted and decontamination becomes necessary, the best purification approach should be chosen to reach the decontamination objectives (as established by legislation). A purification process generally consists of five successive steps as described in Fig. 1.1: (1) preliminary treatment or pre-treatment (physical and mechanical); (2) primary treatment (physicochemical and chemical); (3) secondary treatment or purification (chemical & biological); (4) tertiary or final treatment (physical and chemical); and (5) treatment of the sludge formed (supervised tipping, recycling or incineration). In general, the first two steps are gathered under the notion of pre-treatment or preliminary step, depending on the situation (Anjaneyulu et al. 2005; Crini and Badot 2007, 2010).

Pre-treatment consists of eliminating the (floating) solid particles and all suspended substances from the effluent. This pre-treatment stage, which can be carried out using mechanical or physical means is indispensable, before envisaging secondary treatment because particulate pollution (e.g. SS, colloids, fats, etc.) will hinder later treatment, make it less efficient or damage the decontamination equipment. Primary chemical treatment such as oxidation for cyanide destruction and Cr(VI) reduction, pH adjustment, pre-reduction of a high organic load may also be required. For instance effluent from paper mills contains abundant SS such as fibres, fillers and other solids (Pokhrel and Viraraghavan 2004; Anjaneyulu et al. 2005; Sharma 2015). Effluents from textile mills have a very variable pH although it is often alkaline, containing a high organic

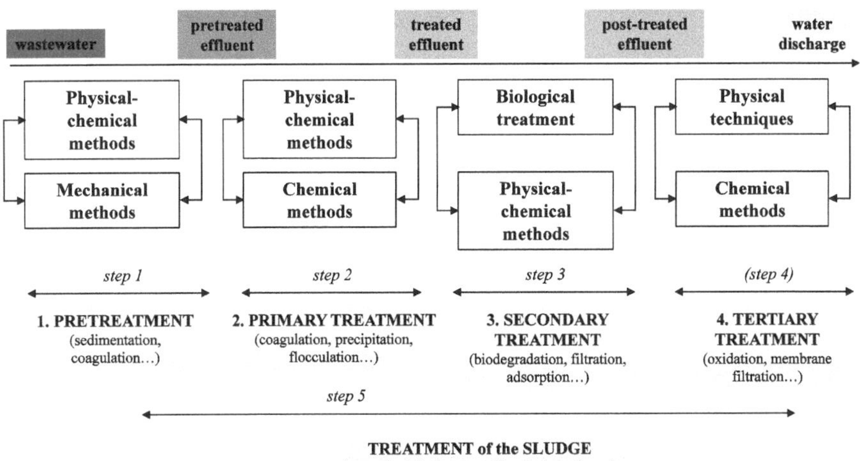

Fig. 1.1 Overview of the main processes for the decontamination of contaminated industrial wastewaters

load. It is therefore indispensable to pre-treat these effluents before considering secondary treatment. However, these treatments alone are, in many cases, incapable of meeting the legislation requirements.

Before its discharge into the environment or its reuse, the pre-treated effluent must undergo secondary purification treatment using the most appropriate of the biological, physical or chemical techniques available to remove the chemical pollution. In certain cases, a final or tertiary treatment (step 4 in Fig. 1.1) can also be required to remove the remaining pollutants or the molecules produced during the secondary purification (e.g. the removal of salts produced by the mineralization of organic matter). However, the use of tertiary treatment in Europe is limited, though it may be necessary in the future if new restrictions are applied. The main tertiary treatments employed to date at a few industrial sites are adsorption using activated carbons (AC), ion-exchange, membrane filtration (ultrafiltration, reverse osmosis), advanced oxidation, and constructed wetlands (CW). In Europe, most of the CW are applied for domestic sewage and municipal wastewater treatment. However, the diversity of CW configurations makes them versatile for implementation to treat industrial effluents (e.g. tannery wastewater, pulp and paper post-treated effluents).

1.3.2 Technologies Available for Contaminant Removal

In general, conventional wastewater treatment consists of a combination of physical, chemical, and biological processes and operations to remove solids including colloids, organic matter, nutrients, soluble contaminants (metals, organics…) etc. from effluents. A multitude of techniques classified in conventional methods, established recovery processes and emerging removal methods can be used (Fig. 1.2). Table 1.1 lists the advantages and disadvantages of different individual techniques (Berefield et al. 1982; Henze 2001; Sonune and Ghate 2004; Chen 2004; Pokhrel and

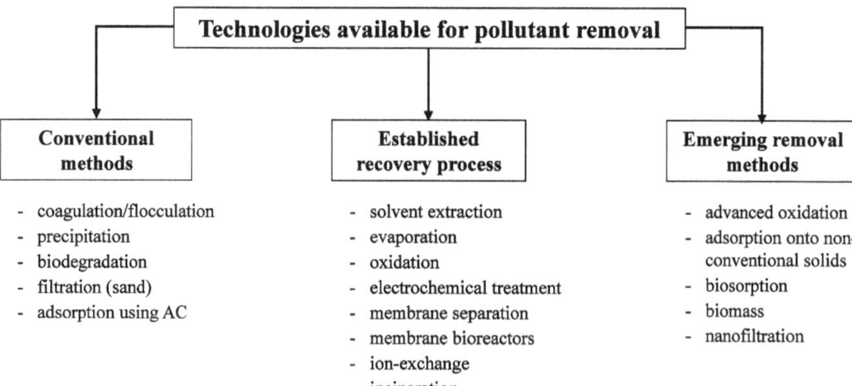

Fig. 1.2 Classification of technologies available for pollutant removal and examples of techniques

Table 1.1 Advantages and disadvantages of the main conventional methods used for the treatment of polluted industrial wastewater

Process	Main characteristic(s)	Advantages	Disadvantages
Chemical precipitation	Uptake of the pollutants and separation of the products formed	Technological simple (simple equipment)	Chemical consumption (lime, oxidants, H_2S...)
		Integrated physicochemical process	Physicochemical monitoring of the effluent (pH)
		Both economically advantageous and efficient	Ineffective in removal of the metal ions at low concentration
		Adapted to high pollutant loads	Requires an oxidation step if the metals are complexed
		Very efficient for metals and fluoride elimination	High sludge production, handling and disposal problems (management, treatment, cost)
		Not metal selective	
		Significant reduction of the COD	
Coagulation/ flocculation	Uptake of the pollutants and separation of the products formed	Process simplicity	Requires adjunction of non-reusable chemicals (coagulants, flocculants, aid-chemicals)
		Integrated physicochemical process	Physicochemical monitoring of the effluent (pH)
		A large range of chemicals is available commercially	Increased sludge volume generation (management, treatment, cost)
		Inexpensive capital cost	Low removal of arsenic
		Very efficient for SS and colloidal particles	
		Good sludge settling and dewatering characteristics	
		Significant reduction of the COD and BOD	
		Interesting reduction of TOX and AOX (pulp and paper industry)	
		Bacterial inactivation capability	
		Rapid and efficient for insoluble contaminants (pigments...) removal	
Flotation	Separation process	Integrated physicochemical process	High initial capital cost
Froth flotation		Different types of collectors (non-ionic or ionic)	Energy costs
		Efficient for small particles removal and can remove	Maintenance and operation costs no negligible

(continued)

Table 1.1 (continued)

Process	Main characteristic(s)	Advantages	Disadvantages
		low density particles which would require long settling periods	
		Useful for primary clarification	Chemicals required (to control the relative hydrophobicities between the particles and to maintain proper froth characteristics)
		Metal selective	Selectivity is pH-dependent
		Low retention time	
		Used as an efficient tertiary treatment in the pulp and paper industry	
		Mechanisms: true flotation, entrainment, and aggregation	
Chemical oxidation	Use of an oxidant (e.g. O_3, Cl_2, ClO_2, H_2O_2, $KMnO_4$)	Integrated physicochemical process	Chemicals required
Simple oxidation		Simple, rapid and efficient process	Production, transport and management of the oxidants (other than ozone)
Ozone			
Hypochlorite treatment		Generation of ozone *on-site* (no storage-associated dangers)	Pre-treatment indispensable
Hydrogen peroxide			
		Quality of the outflow (effective destruction of the pollutants, efficient reduction of color)	Efficiency strongly influenced by the type of oxidant
		Good elimination of color and odor (ozone)	Short half-life (ozone)
		Efficient treatment for cyanide and sulfide removal	A few dyes are more resistant to treatment and necessitate high ozone doses
		Initiates and accelerates azo-bond cleavage (hypochlorite treatment)	Formation of (unknown) intermediates
		Increases biodegradability of product	No diminution of COD values or limited effect (ozone)
		High throughput	No effect on salinity (ozone)
		No sludge production	Release of volatile compounds and aromatic

(continued)

Table 1.1 (continued)

Process	Main characteristic(s)	Advantages	Disadvantages
			amines (hypochlorite treatment)
		Possibility of water recycle	No effect on the COD
		Disinfection (bacteria, viruses)	Generates sludge
Biological methods Bioreactors Biological activated sludge (BAS) Microbiological treatments Enzymatic decomposition Lagoon	Use of biological (pure or mixed) cultures	The application of microorganisms for the biodegradation of organic contaminants is simple, economically attractive and well accepted by the public	Necessary to create an optimally favorable environment
		Large number of species used in mixed cultures (consortiums) or pure cultures (white-rot fungus)	Requires management and maintenance of the microorganisms and/or physicochemical pretreatment (inefficient on non-degradable compounds or when toxic compounds are present)
		White-rot fungi produce a wide variety of extracellular enzymes with high biodegradability capacity	Slow process (problems of kinetics)
		Efficiently eliminates biodegradable organic matter, NH_3, NH_4^+, iron	Low biodegradability of certain molecules (dyes)
		Attenuates color well	Poor decolorization (BAS)
		High removal of BOD and SS (BAS)	Possible sludge bulking and foaming (BAS)
		Decisive role of microbiological processes in the future technologies used for the removal of emergent contaminants from waters	Generation of biological sludge and uncontrolled degradation products
			The composition of mixed cultures may change during the decomposition process
			Complexity of the microbiological mechanisms
			Necessity to have a good knowledge of the enzymatic processes

(continued)

Table 1.1 (continued)

Process	Main characteristic(s)	Advantages	Disadvantages
			governing the decomposition of the substances
Adsorption/ filtration	Non-destructive process	Technological simple (simple equipment) and adaptable to many treatment formats	Relatively high investment (CAC)
CAC	Use of a solid material	Large range of commercial products	Cost of materials (CAC, CAA)
CAA			
Sand		Wide variety of target contaminants (adsorption)	Non-destructive processes
Mixed materials			
Silica gel		Highly effective process (adsorption) with fast kinetics	Non-selective methods
		Excellent quality of the treated effluent	Performance depends on the type of materials (CAC)
		Global elimination (CAC) but possibly selective depending on adsorbent	Requirement for several types of adsorbent
		Excellent ability to separate a large range of pollutants, in particular refractory molecules (CAC is the most effective material)	Chemical derivatization to improve their adsorption capacity
		CAC: efficient for COD removal; highly efficient treatment when coupled to coagulation to reduce both SS, COD and color	Rapid saturation and clogging of the reactors (regeneration costly)
		Sand: efficient for turbidity and SS removal	Not efficient with certain types of dyestuffs and some metals (CAC)
		Alumina: efficient for fluoride removal	Elimination of the adsorbent (requires incineration, regeneration, or replacement of the material)
			Regeneration is expensive and results in loss of material (CAC)
			Economically non-viable for certain industries

(continued)

Table 1.1 (continued)

Process	Main characteristic(s)	Advantages	Disadvantages
			(pulp and paper, textile. . .)
Ion-exchange Chelating resins Selective resins Macroporous resins Polymeric adsorbents Polymer-based hybrid adsorbents	Non-destructive process	Large range of commercial products available from several manufacturers	Economic constraints (initial cost of the selective resin, maintenance costs, regeneration time-consuming. . .)
		Technological simple (simple equipment)	Large volume requires large columns
		Well established and tested procedures; easy control and maintenance	Rapid saturation and clogging of the reactors
		Easy to use with other techniques (e.g. precipitation and filtration in an integrated wastewater process)	Saturation of the cationic exchanger before the anionic resin (precipitation of metals and blocking of reactor)
		Can be applied to different flow regimes (continuous, batch)	Beads easily fouled by particulates and organic matter (organics, oils); requires a physicochemical pretreatment (e.g. sand filtration or carbon adsorption) to remove these contaminants
		High regeneration with possibility of external regeneration of resin	Matrix degrades with time and with certain waste materials (radioactive, strong oxidants. . .)
		Rapid and efficient process	Performance sensitive to pH of effluent
		Produce a high-quality treated effluent	Conventional resins not selective
		Concentrates all types of pollutants, particularly minerals	Selective resins have limited commercial use
		Relatively inexpensive and efficient for metal removal; clean-up to ppb levels (to ppt levels for selective resins) Can be selective for certain metals (with suitable resins)	Not effective for certain target pollutants (disperse dyes, drugs. . .)

(continued)

Table 1.1 (continued)

Process	Main characteristic(s)	Advantages	Disadvantages
		Interesting and efficient technology for the recovery of valuable metals	Elimination of the resin
Incineration	Destruction by combustion	Simple process	Initial investment costs
Thermal oxidation		Useful for concentrated effluents or sludges	Transport and storage of the effluents
Catalytic oxidation		Highly efficient	High running costs
Photocatalytic destruction		Eliminates all types of organics	Formation of dioxins and others pollutants (metals...)
		Production of energy	Local communities always have opposed the presence of incinerating plant in the locality
Electrochemistry	Electrolysis (E)	Efficient technology for the recovery/recycling of valuable metals (E)	High initial cost of the equipment
Electrodeposition			
Electro-coagulation (EC)			
Electro-flocculation (EF)		Adaptation to different pollutant loads and different flow rates (E)	Cost of the maintenance (sacrificial anodes...)
Electro-flotation			
Electrooxidation			
Electrochemical oxidation			
Electrochemical reduction		More effective and rapid organic matter separation than in traditional coagulation (EC)	Requires addition of chemicals (coagulants, flocculants, salts)
Cementation			
Indirect electro-oxidation with strong oxidants		Efficient elimination of SS, oils, greases, color and metals (EC, EF)	Anode passivation and sludge deposition on the electrodes that can inhibit the electrolytic process in continuous operation
Photo-assisted electrochemical methods		EC: pH control is not necessary; generation of coagulants *in situ*; economically feasible and very effective in removing suspended solids, dissolved metals, tannins and dyes (effluents from textile, catering, petroleum, municipal sewage,	Requires post-treatment to remove high concentrations of iron and aluminum ions

(continued)

Table 1.1 (continued)

Process	Main characteristic(s)	Advantages	Disadvantages
		oil-water emulsion, dye-stuff, clay suspension...)	
		EF: Widely used in the miming industries	EF: Separation efficiency depends strongly on bubble sizes
		Effective in treatment of drinking water supplies for small or medium sized communities (EC)	Filtration process for flocs
		Interesting method for the recovery of gold and silver from rinse baths (E)	Formation of sludge (filtering problems)
		Very effective treatment for the reduction, coagulation and separation of copper (EC)	Cost of sludge treatment (electro-coagulation)
		Increases biodegradability (E)	
		Cementation: Efficient for copper removal	
Membrane filtration Microfiltration (MF) Ultrafiltration (UF)	Non-destructive separation	Large range of commercial membrane available from several manufacturers; large number of applications and module configurations	Investment costs are often too high for small and medium industries
Nanofiltration (NF) Reverse osmosis (RO) Dialysis (D) Electrodialysis (ED) Electroelectrodialysis (EED) Emulsion liquid membranes (ELM)	Semi-permeable barrier	Small space requirement	High energy requirements
Supported liquid membranes (SLM)		Simple, rapid and efficient, even at high concentrations	The design of membrane filtration systems can differ significantly
		Produces a high-quality treated effluent	High maintenance and operation costs
		No chemicals required	

(continued)

Table 1.1 (continued)

Process	Main characteristic(s)	Advantages	Disadvantages
			Rapid membrane clogging (fouling with high concentrations)
		Low solid waste generation	Low throughput
		Eliminates all types of dyes, salts and mineral derivatives	Limited flow rates
		Efficient elimination of particles, SS and micro-organisms (MF, UF, NF, RO), volatile and non-volatile organics (NF, RO), dissolved inorganic matter (ED, EED), and phenols, cyanide and zinc (ELM)	Not interesting at low solute feed concentrations
		Possible to be metal selective	The choice of the membrane is determined by the specific application (hardness reduction, particulate or TOC removal, potable water production…)
		A wide range of real applications: clarification or sterile filtration (MF), separation of polymers (UF), multivalents ions (NF), salts from polymer solutions (D) and non-ionic solutes (ED), desalination and production of pure water (RO)	Specific processes
		Well-known separation mechanisms: Size-exclusion (NF, UF, MF), solubility/diffusivity (RO, pervaporation), charge (electrodialysis)	Elimination of the concentrate
Evaporation	Concentration technique	Several types of evaporators exist on the market	Expensive costs for high volumes of wastewater (energy consumption, volume of the concentrate and costs of disposal)
	Thermal process	Versatile technique (the number of cells can be	Investment costs are often too high for small and medium industries

(continued)

Table 1.1 (continued)

Process	Main characteristic(s)	Advantages	Disadvantages
		adapted to the required evaporation capacity)	
Membrane pervaporation (MPV)	Separation process	The energy-costs are well-known for the different configurations	High pollution load in the concentrates
		Efficient processes	Crystallization due to the concentration of the wastewater and corrosion of the heating elements in the evaporator due to the chemical aggressiveness of the concentrated effluent
		Interesting for the production of water for rinsing operations (recycling of distillates), the concentration of rinsing effluents for re-introduction into the process and for the purification of treatment baths (to maintain their nominal concentration)	Problem with the evaporation of effluents containing free cyanide
		Also interesting for the separation of phenol by steam distillation	Requires the installation of a cleaning circuit (to prevent atmospheric pollution)
		MPV: a quite recent technology applied to the removal of organics from water	Potential contamination of the distillate preventing reuse (due to the presence of some VOC or hydrocarbons in the effluent)
Liquid-liquid (solvent) extraction	Separation technology	A well-known established separation technology for wastewater recycling	High investment (equipment)
Membrane-based solvent extraction	Solvent extraction	Principally used for large-scale operations where the load of contaminants are high	Uneconomic when contaminant concentrations are low (< 0.5 g/L)
		Extraction/stripping operations easy to perform	Use of large volumes of organic extractants
		Simple control and monitoring of process	Use of potential toxic solvents
		Economically viable when both solute concentrations and wastewater flowrates are high	Not interesting at low solute feed concentrations

(continued)

Table 1.1 (continued)

Process	Main characteristic(s)	Advantages	Disadvantages
		Relatively low operating costs	Hydrodynamic constraints (flooding, entrainment)
		Recyclability of extractants	Entrainment of phases giving poor effluent quality
		Selectivity of the exchangers for metals efficient for metal removal (cations, anions, ion pairs)	Possible cross-contamination of the aqueous stream
		Efficient for the separation of phenol	Emulsification of phase with poor separation
		A good alternative to classical lime precipitation for phosphoric acid recuperation	Fire risk from use of organic solvents and VOC emissions
Advanced oxidation processes (AOP)	Emerging processes	*In situ* production of reactive radicals	Laboratory scale
Photolysis Heterogeneous and homogeneous photocatalytic reactions Non catalytic wet air oxidation (WAO) Catalytic wet air oxidation (CWAO)	Destructive techniques	Little or no consumption of chemicals	Economically non-viable for small and medium industries
Supercritical water gasification		Mineralization of the pollutants	Technical constraints
		No production of sludge	Formation of by-products
		Rapid degradation	Low throughput
		Efficient for recalcitrant molecules (dyes, drugs...)	High-pressure and energy-intensive conditions (WAO)
		Very good abatement of COD and TOC	pH-dependence (in particular for WAO)
		WAO: technology suitable for effluent too dilute for incineration and too toxic and/or concentrated for biological treatment	WAO: completed mineralization not achieved
		Destruction of phenol in water solution: WAO, CWAO	

(continued)

Table 1.1 (continued)

Process	Main characteristic(s)	Advantages	Disadvantages
		Insoluble organic matter is converted to simpler soluble compounds without emissions of dangerous substances (WAO)	

Viraraghavan 2004; Parsons 2004; Forgacs et al. 2004; Anjaneyulu et al. 2005; Chuah et al. 2005; Crini 2005, 2006; Bratby 2006; Crini and Badot 2007, 2010; Cox et al. 2007; Mohan and Pittman 2007; Hai et al. 2007; Wojnárovits and Takács 2008; Barakat 2011; Sharma and Sanghi 2012; Rathoure and Dhatwalia 2016; Morin-Crini and Crini 2017).

Selection of the method to be used will thus depend on the wastewater characteristics (Anjaneyulu et al. 2005; Crini 2005; Crini and Badot 2007; Cox et al. 2007). Each treatment has its own constraints not only in terms of cost, but also in terms of feasibility, efficiency, practicability, reliability, environmental impact, sludge production, operation difficulty, pre-treatment requirements and the formation of potentially toxic byproducts. However, among the various treatment processes currently cited for wastewater treatment, only a few are commonly employed by the industrial sector for technological and economic reasons. In general, removal of pollutants from effluents is done by physicochemical and/or biological means, with research concentrating on cheaper effective combinations of systems or new alternatives.

1.4 Conclusion

The development of cheaper, effective and novel methods of decontamination is currently an active field of research, as shown by the numerous publications appearing each year. Preserving the environment, and in particular the problem of water pollution, has become a major preoccupation for everyone – the public, industry, scientists and researchers as well as decision-makers on a national, European, or international level. The public demand for pollutant-free waste discharge to receiving waters has made decontamination of industrial wastewaters a top priority. However, this is a difficult and challenging task (Sonune and Ghate 2004; Anjaneyulu et al. 2005; Barakat 2011; Sharma and Sanghi 2012). It is also difficult to define a universal method that could be used for the elimination of all pollutants from wastewaters. This chapter described the advantages and disadvantages of technologies available. A multitude of techniques classified in conventional methods, established recovery processes and emerging removal methods can be used. However, among the numerous and various treatment processes currently cited for wastewater treatment, only a few are commonly used by the industrial sector for economic and technological reasons. Adsorption onto activated carbons is nevertheless often cited as the procedure of choice to remove many different types of pollutants because it gives the best results in terms of efficiency and technical feasibility at the industrial scale.

References

Amiard JC (2011) Les risques chimiques environnementaux. Editions TEC & DOC Lavoisier, Paris, 782 p (in French)

Anjaneyulu Y, Sreedhara Chary N, Samuel Suman Raj D (2005) Decolourization of industrial effluents – available methods and emerging technologies – a review. Rev Environ Sci Biotechnol 4:245–273. https://doi.org/10.1007/s11157-005-1246-z

Barakat MA (2011) New trends in removing heavy metals from industrial wastewater. Arab J Chem 4:361–377. https://doi.org/10.1016/j.arabjc.2010.07.019

Berefield LD, Judkins JF, Weand BL (1982) Process chemistry for water and wastewater treatment. Prentice-Hall, Englewood Cliffs, 510 p

Bratby J (2006) Coagulation and flocculation in water and wastewater treatment. IWA Publishing, London, p 407

Chen G (2004) Electrochemical technologies in wastewater treatment. Sep Purif Technol 38:11–41. https://doi.org/10.1016/j.seppur.2003.10.006

Chuah TG, Jumasiah A, Azni I, Katayon S, Choong SYT (2005) Rice husk as a potentially low-cost biosorbent for heavy metal and dye removal: an overview. Desalination 175:305–316. https://doi.org/10.1016/j.desal.2004.10.014

Cooney DO (1999) Adsorption design for wastewater treatment. Lewis Publishers, Boca Raton, 208 p

Cooper P (1993) Removing colour from dye house waste waters – a critical review of technology available. J Soc Dyers Colour 109:97–100. https://doi.org/10.1111/j.1478-4408.1993.tb01536.x

Cox M, Négré P, Yurramendi L (2007) Industrial liquid effluents. INASMET Tecnalia, San Sebastian, p 283

Crini G (2005) Recent developments in polysaccharide-based materials used as adsorbents in wastewater treatment. Prog Polym Sci 30:38–70. https://doi.org/10.1016/j.progpolymsci.2004.11.002

Crini G (2006) Non-conventional low-cost adsorbents for dye removal. Bioresour Technol 97:1061–1085. https://doi.org/10.1016/j.biortech.2005.05.001

Crini G, Badot PM (2007) Traitement et épuration des eaux industrielles polluées. PUFC, Besançon, 353 p (in French)

Crini G, Badot PM (eds) (2010) Sorption processes and pollution. PUFC, Besançon, 489 p

Druart C, Morin-Crini N, Euvrard E, Crini G (2016) Chemical and ecotoxicological monitoring of discharge water from a metal-finishing factory. Environ Process 3:59–72. https://doi.org/10.1007/s40710-016-0125.7

Forgacs E, Cserhati T, Oros G (2004) Removal of synthetic dyes from wastewaters: a review. Environ Int 30:953–971. https://doi.org/10.1016/j.envint.2004.02.001

Hai FI, Yamamoto K, Fukushi K (2007) Hybrid treatment systems for dye wastewater. Crit Rev Environ Sci Technol 37:315–377. https://doi.org/10.1080/10643380601174723

Harvey PJ, Campanella BF, Castro PM, Harms H, Lichtfouse E, Schäffner AR, Smrcek S, Werck-Reichhart D (2002) Phytoremediation of polyaromatic hydrocarbons, anilines and phenols. Environ Sci Pollut Res Int 9:29–47

Henze M (ed) (2001) Wastewater treatment – biological and chemical processes. Springer, Berlin/New York

Kentish SE, Stevens GW (2001) Innovations in separations technology for the recycling and re-use of liquid waste streams. Chem Eng J 84:149–159

Khalaf MN (2016) Green polymers and environmental pollution control. CRC Press; Apple Academic Press, Inc, Oakville, 436 p

Lacorte S, Latorre A, Barcelo D, Rigol A, Malqvist A, Welander T (2003) Organic compounds in paper-mill process waters and effluents. Trends Anal Chem 22:725–737

Liu DHF, Liptak BG (eds) (2000) Wastewater treatment. CRC Press, Boca Raton

Mohan D, Pittman CU (2007) Arsenic removal from waste/wastewater using adsorbents – a critical review. J Hazard Mater 142:1–53. https://doi.org/10.1016/j.jhazmat.2007.01.006

Morin-Crini N, Crini G (eds) (2017) Eaux industrielles contaminées. PUFC, Besançon, 513 p (in French)

Parsons S (ed) (2004) Advanced oxidation process for water and wastewater treatment. Editions IWA Publishing, London

Pokhrel D, Viraraghavan T (2004) Treatment of pulp and paper mill wastewater – a review. Sci Total Technol 333:37–58. https://doi.org/10.1016/j.scitotenv.2004.05.017

Rana T, Gupta S, Kumar D, Sharma S, Rana M, Rathore VS, Pereira BMJ (2004) Toxic effects of pulp and paper-mill effluents on male reproductive organs and some systemic parameters in rats. Environ Toxicol Pharmacol 18:1–7. https://doi.org/10.1016/j.etap.2004.04.005

Rathoure AK, Dhatwalia VK (2016) Toxicity and waste management using bioremediation. IGI Global, Hershey, 421 p

Sharma SK (2015) Green chemistry for dyes removal from wastewater. Scrivener Publishing LLC Wiley, Beverley, 496 p

Sharma SK, Sanghi R (2012) Advances in water treatment and pollution prevention. Springer, Dordrecht, 457 p

Sonune A, Ghate R (2004) Developments in wastewater treatment methods. Desalination 167:55–63. https://doi.org/10.1016/j.desal.2004.06.113

Wojnárovits L, Takács E (2008) Irradiation treatment of azo dye containing wastewater: an overview. Rad Phys Chem 77:225–244. https://doi.org/10.1016/j.radphyschem.2007.05.003

Chapter 2
Adsorption-Oriented Processes Using Conventional and Non-conventional Adsorbents for Wastewater Treatment

Grégorio Crini, Eric Lichtfouse, Lee D. Wilson, and Nadia Morin-Crini

Contents

G. Crini (✉) · N. Morin-Crini
Laboratoire Chrono-environnement, UMR 6249, UFR Sciences et Techniques, Université Bourgogne Franche-Comté, Besançon, France
e-mail: gregorio.crini@univ-fcomte.fr

E. Lichtfouse
CEREGE, Aix Marseille Univ, CNRS, IRD, INRA, Coll France, Aix-en-Provence, France

L. D. Wilson
Department of Chemistry, University of Saskatchewan, Saskatoon, SK, Canada

G. Crini, E. Lichtfouse (eds.), *Green Adsorbents for Pollutant Removal*, Environmental Chemistry for a Sustainable World 18,
https://doi.org/10.1007/978-3-319-92111-2_2

Abstract The removal of contaminants from wastewaters is a matter of great interest in the field of water pollution. Amongst the numerous techniques of contaminant removal, adsorption using solid materials (named adsorbents) is a simple, useful and effective process. The adsorbent may be of mineral, organic or biological origin. Activated carbon is the preferred material at industrial scale and is extensively used not only for removing pollutants from wastewater streams but also for adsorbing contaminants from drinking water sources (e.g. rivers, lakes or reservoirs). However, its widespread use is restricted due to high cost. In the last three decades, numerous approaches have been studied for the development of cheaper and more effective adsorbents capable to eliminate pollutants at trace levels. This chapter gives a general overview of liquid-solid adsorption processes using conventional and non-conventional materials for pollutant removal. It outlines some of the principles of adsorption and proposes a classification for the different types of materials. Finally, the chapter discusses different mechanisms involved in the adsorption phenomena.

Abbreviations

AAS Atomic absorption spectroscopy
AC Activated carbons
BOD Biochemical oxygen demand
CAA Commercial activated alumina
CAC Commercial activated carbons
COD Chemical oxygen demand
DSC Differential scanning calorimetry
ESR Electron spin resonance spectroscopy
FT-IR Fourier transform infrared spectroscopy
GAC Granular activated carbon
ICP Inductively coupled plasma
ISE Ion selective electrode
LC-MS Liquid chromatography mass spectrometry
NMR Nuclear magnetic resonance
OM Organic matter
PAC Powder activated carbon
PAH Polycyclic aromatic hydrocarbons
PCB Polychlorobiphenyls
PET Polyethylene terephthalate
PZC Point of zero charge
SEM Surface electron microscopy
SMEs Small and medium-size enterprises
SS Suspended solids
TGA Thermogravimetric analysis
TOC Total organic carbon

VOC	Volatile organic compounds
XAS	X-ray absorption spectroscopy
XPS	X-ray photoelectron spectroscopy

2.1 Introduction

Man's use of chemical substances, in particular metals, began to affect the environment during the "*Industrial Revolution*". Although some metal ions are disseminated into the environment naturally by both geological and biological activity, human activity today produces a greater input. The toxicity of many of these pollutants/contaminants is well known. Today, we are in the "*Pollutant Removal Age*" and, it is, therefore, not surprising that there has been considerable effort to develop technologies to reduce contaminant emissions (Morin-Crini and Crini 2017). A significant proportion of these emissions are in the form of industrial wastewaters. Indeed, the industrial sector consumes significant volumes of water, and consequently generates considerable amounts of wastewater discharge containing both mineral and organic contamination. This sector is today considered to be one of the most polluting in spite of the considerable effort made to clean up the processes over the last 30 years (Berefield et al. 1982; Liu and Liptak 2000; Landy et al. 2012a; Khalaf 2016; Morin-Crini and Crini 2017).

Wastewater treatment is becoming ever more critical due to diminishing water resources, increasing wastewater disposal costs, and stricter discharge regulations that have lowered permissible contaminant levels in waste streams. The diversity of water pollutants calls for a wide range of treatment methods that are not only effective, but also technologically and economically feasible. The most common methods for the removal of contaminants from industrial effluents include biodegradation, precipitation, chemical oxidation, solvent extraction, evaporation, electrochemical approaches, cementation, membrane filtration, phytoremediation, ion-exchange, and carbon adsorption (Berefield et al. 1982; Volesky 1990; Liu and Liptak 2000; Harvey et al. 2002; Crini and Badot 2007; Cox et al. 2007; Sharma 2015; Morin-Crini and Crini 2017).

Over the last few decades, adsorption has gained importance as a separation, purification and/or detoxification process on an industrial scale (Table 2.1). Adsorption is used to purify, decolorize, detoxify, deodorize, separate, and concentrate to allow removal and to recover the harmful products from liquid solutions and gas mixtures (Dąbrowski 2001; Crini and Badot 2010; Kyzas and Kostoglou 2014). Consequently, adsorption is of interest to many economic sectors and concerns areas such as chemistry, food and pharmaceutical industries, and the treatment of drinking water and industrial wastewater. Indeed, adsorption is – along with biodegradation – one of the two major treatments applied to the decontamination of water. Adsorption processes are considered the best choice compared to other methods due to their convenience, easy operation and simplicity of design, high efficiency, and also for

Table 2.1 Fundamental practical applications of adsorption-oriented processes

Fundamental practical applications of adsorption-oriented processes
Separation and purification of gas and liquid mixtures
Drying gases and liquids
Solvent recovery
Purification of air
Separation and purification of chemicals, pharmaceutical and biological substances
Removal of impurities from liquid and gas media
Decolorizing applications
Water purification (pesticides removal, arsenic elimination...)
Wastewater decontamination

their wider applicability in water pollution control (McKay 1996; Babel and Kurniawan 2003; Swami and Buddhi 2006; Crini 2006; Qu 2008; Vijayaraghavan and Yun 2008; Gadd 2009). From an industrial point of view, adsorption is both technologically simple and economically feasible while also being a process that produce high quality water, with pollutant concentrations under the legal limits for discharge waters.

In general terms, activated carbons (AC) must be thought of as being most effective adsorbents and, as such, their performance in removing contaminants such as metals, radionuclides, rare earth elements, phenolic and aromatic derivatives (including dyes and pesticides), pharmaceuticals and drugs have been examined widely (Dąbrowski et al. 2005). In addition, in the field of wastewater treatment, adsorption onto commercial AC (CAC) has proved efficient in removing colloidal substances and soluble organic substances that are non-biodegradable or chemically stable like recalcitrant synthetic molecules. Attention has also focused on adsorption onto commercial activated alumina (CAA), ion-exchange using organic polymeric resins and zeolites as other non-consumptive materials (Wang and Peng 2010). However, despite the excellence of their performance, these systems are expensive to use and, as such, cannot be thought of as a truly viable option in many parts of the world.

Because of this, attention has turned to the adsorptive properties of other non-conventional solid materials proposed as low-cost, efficient and green adsorbents for pollutant removal (Pollard et al. 1992; Ramakrishna and Viraraghavan 1997; Houghton and Quarmby 1999; Blackburn 2004; Gavrilescu 2004; Crini 2005, 2006; Li et al. 2008; Oliveira and Franca 2008; Ngah and Hanafiah 2008; Gupta and Suhas 2009; Rafatullah et al. 2010; Crini and Badot 2010). The past three decades have shown an explosion in the development of new materials including new carbons produced from wastes or natural by-products, natural or synthetic adsorbents or sorbents, and biological materials or biosorbents. Table 2.2 show the top ten most cited reviews in the ISI Web of Science database for 2000–2017 with "Adsorbents", "Wastewater and "Review" in the topic. The number of reviews appearing with these three terms in the topic in 2017, 2016, 2015, 2014, 2013 and 2012 is 66, 59, 54, 46 and 39, respectively (ISI Web of Science database).

Table 2.2 The top ten most cited reviews in the ISI Web of Science database for 2000–2017 with "Adsorbents", "Wastewater" AND "Review" in the topic (out of a total of 383 reviews appearing, December 04, 2017)

1. Crini G (**2006**) Non-conventional low-cost adsorbents for dye removal. Bioresource Technology 97:1061–1085. Times cited: 1989.
2. Babel S, Kurniawan TA (**2003**) Low-cost adsorbents for heavy metals uptake from contaminated water: A review. Journal of Hazardous Materials 97:219–243. Times cited: 1589.
3. Mohan D, Pittman CU (2007) Arsenic removal from waste/wastewater using adsorbents – A critical review. Journal of Hazardous Materials 142:1–53. Times cited: 1561.
4. Gupta VK, Suhas (**2009**) Application of low-cost adsorbents for dye removal – A review. Journal of Environmental Management 90:2313–2342. Times cited: 1285.
5. Crini (2005) Recent developments in polysaccharide-based materials used as adsorbents in wastewater treatment. Progress in Polymer Science 30:38–70. Times cited: 982.
6. Kannan N, Sundaram MM (**2001**) Kinetics and mechanism of removal of methylene blue by adsorption on various carbons. Dyes and Pigments 51:25–40. Times cited: 903.
7. Crini G, Badot PM (**2008**) Application of chitosan, a natural aminopolysaccharide, for dye removal from aqueous solutions by adsorption processes using batch studies: A review of recent literature. Progress in Polymer Science 33:399–447. Times cited: 894.
8. Rafatullah M, Sulaiman O, Hashim R, Ahmad A (**2010**) Adsorption of methylene blue on low-cost adsorbents: A review. Journal of Hazardous Materials 177:70–80. Times cited: 855.
9. Ngah WSW, Hanafiah MAKM (2008) Removal of heavy metal ions from wastewater by chemically modified plant wastes as adsorbents: A review. Bioresource Technology 99:3945–3948. Times cited: 639.
10. Wang SB, Peng YL (**2010**) Natural zeolites as effective adsorbents in water and wastewater treatment. Chemical Engineering Journal J 156:11–24. Times cited: 601.

This chapter presents adsorption processes as a decontamination method for the removal of contaminants from synthetic solutions and industrial effluents. It outlines some of the principles of contamination adsorption onto solid materials. The chapter also proposes a classification for the different types of materials used and discusses different mechanisms involved in the adsorption phenomena.

2.2 Wastewater Treatment by Adsorption

2.2.1 Definition

Adsorption is a process of separation during which the substances of a fluid, liquid or gas, bind to the exterior and interior surfaces of a solid material called the adsorbent. The separation is based on the selective adsorption (i.e. thermodynamic and/or kinetic selectivity) of the contaminants by an adsorbent owing to specific interactions between the surface of the adsorbent material and the adsorbed contaminants: simple mass transfer from the liquid phase towards the solid phase (Dubinin 1966). This surface phenomenon is a manifestation of complicated interactions among the

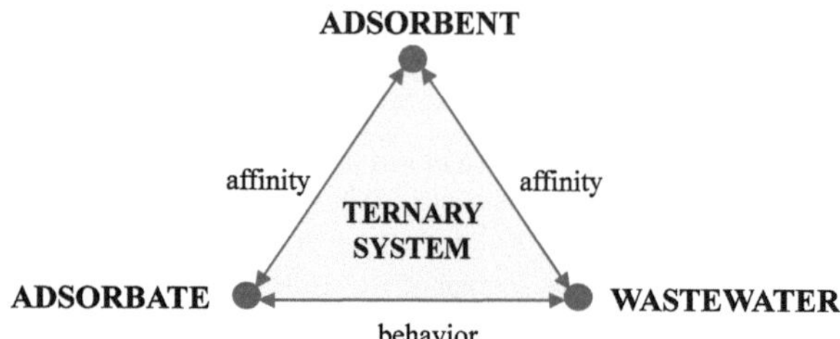

Fig. 2.1 Schema depicting the relationships between the three components of an adsorption system (Crini 2005)

three components involved, i.e. the adsorbent, the adsorbate and the wastewater (e.g. effluent, synthetic solution or water).

Figure 2.1 shows a schematic adsorption model for the three components and their interactions. Generally, in this ternary system, the affinity between the adsorbent and the adsorbate is the main interaction force controlling adsorption (Furuya et al. 1997; Crini 2005; Crini and Badot 2010). However, the affinities between the adsorbate and the solution, the adsorbent and the solution, and the contaminant molecules can also play a major role in adsorption. In aqueous solution, hydrophobic compounds have low solubility and tend to be pushed to the adsorbent surface. It is reasonable to expect that adsorption capacity will be dependent upon the interaction forces between the three adsorption components.

2.2.2 Adsorption, Sorption, Biosorption, Absorption or Bioaccumulation: What is the Most Appropriate Term?

"Adsorption", "sorption", "biosorption", "bio-adsorption", "absorption" or "bioaccumulation": What is the most appropriate term? Although this is not difficult to answer, there is a lot of confusion in the abundant literature (Dąbrowski et al. 2005; Crini 2005, 2010; Gadd 2009).

The change in the concentration of a molecule in the surface layer of a solid material in comparison with the bulk phase with respect to unit surface area is termed adsorption. Sorption is a general term used for both absorption and adsorption (Crini 2010). These terms are often confused. Absorption is the incorporation of a substance in one state into another of a different state (e.g. liquids being absorbed by a solid or gases), i.e. into a three-dimensional matrix (Gadd 2009). Adsorption is the physical adherence or bonding of molecules (or ions) onto the surface of another substance, i.e. onto a two-dimensional surface. In this case, the material accumulated at the interface is the adsorbate and the solid surface is the adsorbent.

Adsorption, strictly speaking, defines binding in terms of a physical rather than chemical surface phenomenon. In processes using carbons, adsorption is generally the preferred term (Dąbrowski 2001; Dąbrowski et al. 2005). If adsorption occurs and results in the formation of a stable molecular phase at the interface, this can be described as a surface complex. Two general kinds of surface complex exist: inner- and outer-sphere surface complexes. An interesting discussion on this subject can be found in the review by Gadd (2009). Adsorption is the most common treatment used in conventional clean-up technologies but unless it is clear which process, absorption or adsorption, is operative, sorption is a more general term, and can be used to describe any system where a sorbate (e.g. a molecular ion, a molecule, a polymer) interacts with a sorbent (i.e. a solid surface) resulting in an accumulation at the sorbate-sorbent interface.

Biosorption or bio-adsorption may be simply defined as the removal of substances from solution by biological materials (Gadd 1990; Garnham 1997; Volesky 2001; Veglio' and Beolchini 1997; Davis et al. 2003; Vijayaraghavan and Balasubramanian 2015). This is a physicochemical process and includes several mechanisms. The precise binding mechanism(s) may range from physical (i.e. electrostatic interactions, van der Waals forces, hydrogen bond) to chemical binding (i.e. ionic and covalent). Some of the reported mechanisms include absorption, (surface) adsorption, ion-exchange, binding or surface complexation, (surface) precipitation or micro-precipitation, and mineral nucleation. Biosorption is a property of both living and dead organisms, and their components. While most biosorption research concerns metals and related substances (Gadd 1990), the term is now applied to particulates and all manner of organic substances as well. Practically all biological material has an affinity for metal species (Wase and Forster 1997; Aksu 2005; Gadd 2009). However, the term "biosorption" refers to passive or physicochemical attachment of a sorbate to a biosorbent, essentially the binding of a chemical species to biopolymers. The definition, thus, specifically excludes metabolic or active uptake by living, metabolizing cells. In the literature, the term "biosorbent" includes the usage of dead biomass such as fibers, peat, rice hulls, forest by-products, chitosan, and agro-food wastes as well as living plants, fungi, algae (unicellular microalgae, cyanobacteria, multicellular macroalgae), and bacteria. Biosorbents represent cheap filter materials often with high affinity, capacity and selectivity, and they are abundant and already available in most places (Aksu 2005; Sudha and Giri Dev 2007; Vijayaraghavan and Yun 2008; Gadd 2009; Crini and Badot 2010; Michalak et al. 2013; Kyzas et al. 2013a; Lim and Aris 2014; Ong et al. 2014; Gupta et al. 2015). Some types of materials are broad range with no specific priority for metal ion bonding, while others can be specific for certain types of metal ions.

"Bioaccumulation" is also another term which induces confusion. Using biosorbents such as algae for metal ions removal, another mechanism can occur. Indeed, precipitation or crystallization of metals may occur within and around cell walls as well as the production by biomass of metal binding polysaccharides: These processes which could be considered as biosorption are better termed "bioaccumulation" (Gadd 1990; Garnham 1997). These two terms 'biosorption' and 'bioaccumulation' have been adopted for the description of the two mechanistically different types of metal sequestering by microorganisms. The first has been proposed for the sequestration by non-metabolically mediated process (inactive microorganisms) and the second for the sequestration of metal ions by metabolically mediated processes (living microorganisms). Biosorption tends to be very rapid and reversible while bioaccumulation tends to be slower and irreversible. An interesting discussion on the features of biosorption and bioaccumulation can be found in the review by Vijayaraghavan and Yun (2008). So, there are mechanistic differences between these two terms. However, the two mechanisms can co-exist in a biosorption system, and can also function independently.

2.2.3 Contacting Systems

When studying adsorption from solutions on materials it is convenient to differentiate between adsorption from dilute solution and adsorption from binary and multicomponent mixtures covering the entire range of mole fractions. To judge by the number of papers published annually on adsorption from dilute (single) solution, this subject is more important than adsorption from binary mixtures. It is also important to consider the modes of contacting the solid adsorbent and the wastewater when applying the adsorption system to both industrial large scale treatments and laboratory scale (McKay 1996; Bajpai and Rajpoot 1999; Ali 2014). There are several types of contacting systems available to obtain experimental data and for industrial applications, including batch methods, fixed-bed type processes, pulsed beds, moving mat filters and fluidized beds. However, the two most frequently used systems applied in solid/liquid adsorption processes are the batch-type contact and fixed-bed type processes (Fig. 2.2).

Adsorption processes for decontamination of wastewaters can be carried out either discontinuously in batch reactors or continuously in fixed-bed reactors or columns (Volesky and Holan 1995; Volesky 2001; Crini 2003; Ali 2014). Fixed-bed reactors or dynamic continuous-flow systems are commonly used in the industrial world while batch methods are preferred on the laboratory scale because, apart from their simplicity and ease of operation, they are limited to the treatment of small volumes of solution. Fixed-bed systems have an important advantage because adsorption depends on the concentration of the solute in the solution being treated (Ali 2014). The adsorbent is continuously in contact with fresh solution; hence the concentration in the solution in contact with a given layer of adsorbent in a column is relatively constant. Conversely, the concentration of adsorbate in contact with a given quantity of adsorbent, as in a batch system, is continuously changing due to the adsorbate being adsorbed (McKay 1996). Other advantages of employing fixed-bed columns for industrial adsorption processes are higher residence times and better heat and mass transfer characteristics than batch reactors.

Batch methods are also widely used because this technology is cheap and simple to operate and, consequently often favored for small and medium size process applications using simple and readily available mixing tank equipment. Simplicity, well-established experimental methods, and easily interpretable results are some of the main reasons frequently evoked for the extensive use of these methods. Another interesting advantage is the fact that, in batch systems, the parameters of the solution/effluent such as contact time, pH, strength ionic, temperature, etc. can be controlled and/or adjusted.

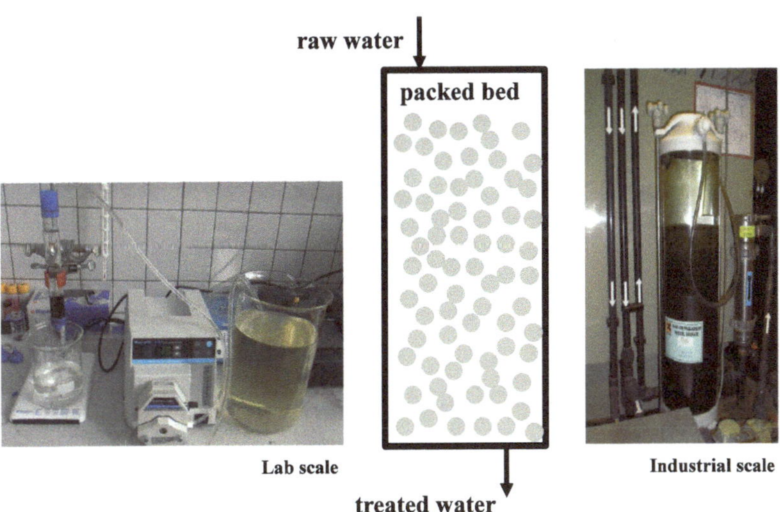

Fig. 2.2 Schematic representations of two mains schemes used for adsorption of pollutants from wastewaters: batch process and continuous process

2.2.4 *Desorption of Contaminants*

It is important to point out that adsorption using batch systems is a non-destructive technique involving only a phase change of contaminants, and hence imposes further problems in the form of sludge disposal. For fixed-bed reactors, Fig. 2.3 shows two main strategies (regeneration step or replacement) that could be used to deal with spent adsorbent after its usage. One of the important characteristics of a solid material is whether it can be regenerated if necessary. The regeneration of the adsorbent may be crucially important for keeping the process costs down and

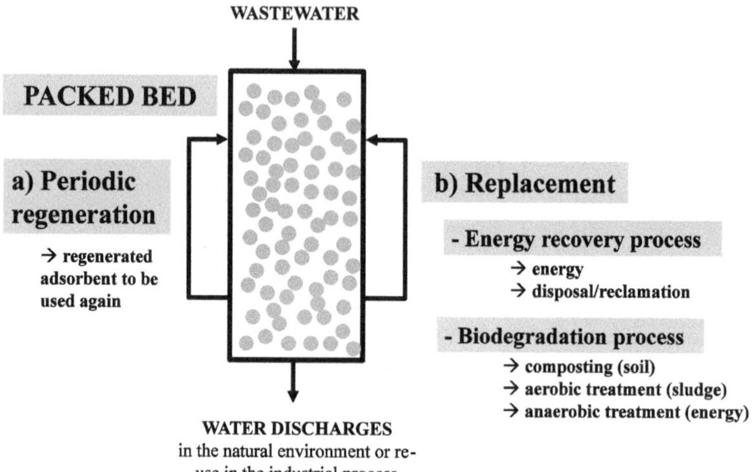

Fig. 2.3 Schematic representations of two mains strategies (regeneration step or replacement) that could be used to deal with spent adsorbent after its usage

opening the possibility of recovering the contaminant extracted from the solution. For this purpose, it is desirable to desorb the adsorbed contaminants and to regenerate the material for another cycle of application. Desorption studies also reveal the mechanism of adsorption. However, except for commercial activated carbons and organic resins, this aspect has not been adequately studied and there is little literature focusing on this topic.

2.2.5 *Control Adsorbent Performance*

In an adsorption-oriented process, separation is defined as a system that transforms a mixture of substances into two or more products that differ from each other in composition. The process is difficult to achieve because it is the opposite of mixing, a process favored by the second law of thermodynamics. For many separation processes, the separation is caused by a mass separating agent, the solid material or adsorbent (King 1980; McKay 1996; Yang 2003). Consequently, the performance of any adsorptive separation or purification process is directly determined by its quality. So, the first important step to an efficient adsorption process is the search for a solid material with high capacity, selectivity, and rate of adsorption.

In principle, as adsorption is a surface phenomenon, any porous solid having a large surface area may be an adsorbent (McKay 1996). Other requirements to be taken into account in choosing a material are based on the following criteria: low cost and readily available, suitable mechanical properties, high physical strength (not disintegrating) in solution, a long life, able to be regenerated if required, etc. The data from the literature show that the control of adsorption performances of a solid material in liquid-phase adsorption depends on the following factors: (i) the origin and nature of the solid such as its physical structure (e.g. particle size, specific surface area, porosity), chemical nature and functional groups (e.g. surface charge, pH at the point of zero charge), and mechanical properties; (ii) the activation conditions of the raw solid (e.g. physical treatment, chemical modification); (iii) the influence of process variables used in the contacting system such as contact time, initial pollutant concentration, solid dosage and stirring rate; (iv) the chemistry of the pollutant(s) (for instance, for a dye molecule, its pK_a, polarity, size and functional groups); and finally, (v) the solution conditions, referring to its pH, ionic strength, temperature, presence of multi-pollutant or impurities, and its variability (Crini 2005, 2006; Park et al. 2010; Crini and Badot 2010).

2.3 Types of Materials for Contaminant Removal

2.3.1 Adsorbents Classification

Solid materials used as adsorbents can take a broad range of chemical forms and different geometrical surface structures. This is reflected in the range of their applications in industry, or helpfulness in laboratory practice. Adsorbents can be usually classified in five categories: (1) natural materials such as sawdust, wood, fuller's earth or bauxite; (2) natural materials treated to develop their structures and properties such as activated carbons, activated alumina or silica gel; (3) manufactured materials such as polymeric resins, zeolites or aluminosilicates; (4) agricultural solid wastes and industrial by-products such as date pits, fly ash or red mud; and (5) biosorbents such as chitosan, fungi or bacterial biomass. Another classification was introduced by Dąbrowski (2001) as shown in Table 2.3. Another simplified classification, introduced by Crini (Crini 2005, 2006; Crini and Badot 2007), can be used as follows: conventional and non-conventional adsorbents. The list of conventional adsorbents includes commercial activated carbons (CAC), commercial ion-exchange resins (polymeric organic resins) and inorganic materials such as commercial activated aluminas (CAA), silica gel, zeolites and molecular sieves

Table 2.3 Basic types of industrial adsorbents

Carbon adsorbents	Mineral adsorbents	Other adsorbents
Activated carbons	Silica gels	Synthetic polymers
Activated carbon fibres	Activated alumina	Composite adsorbents (mineral-carbons)
Molecular carbon sieves	Metal oxides	Mixed adsorbents
Fullerenes	Metal hydroxides	
Carbonaceous materials	Zeolites	
	Clay minerals	
	Pillared clays	
	Inorganic nanomaterials	

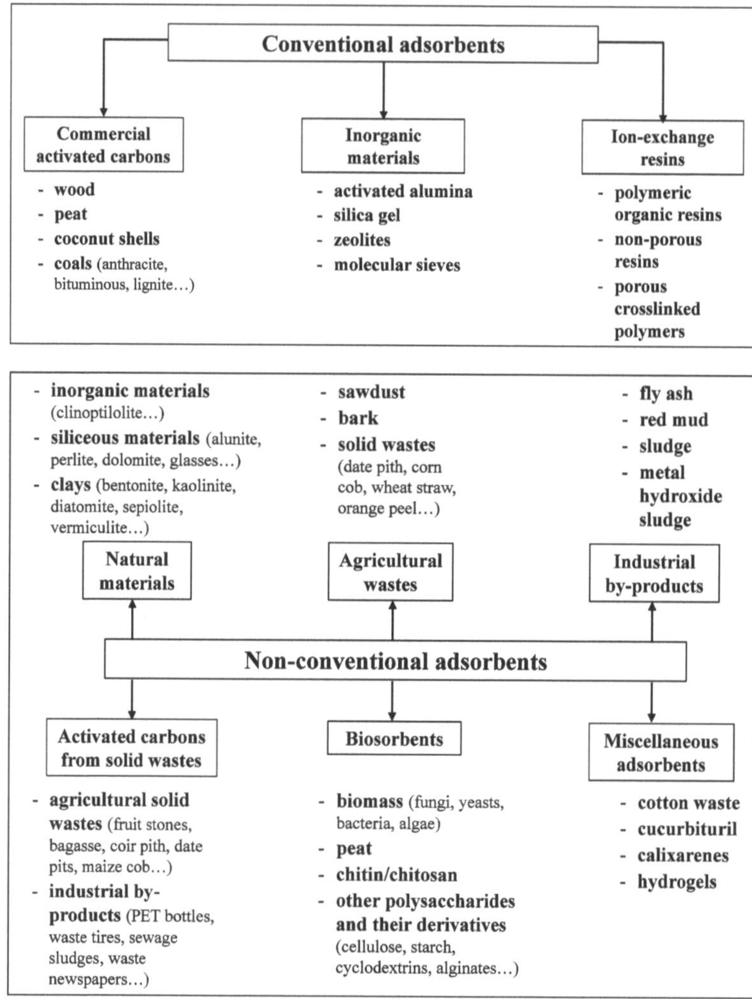

Fig. 2.4 Conventional and non-conventional adsorbents for the removal of pollutants from wastewaters according to Crini (Crini 2005, 2006; Crini and Badot 2007)

(which are formally not zeolites) (Fig. 2.4). Only four types of generic adsorbents have dominated the commercial use of adsorption: CAC >> zeolites >> silica gel > CAA (Yang 2003). The list of non-conventional adsorbents includes activated carbons (AC) obtained from agricultural solid waste and industrial by-products, natural materials such as clays, industrial by-products such as red mud, biosorbents such as chitosan, and miscellaneous adsorbents such as alginates (Fig. 2.4).

Table 2.4 Analytical techniques used in adsorption research in order to obtain information on adsorbent characterization, adsorbate characterization and adsorption mechanism

Objective/analytical technique(s)
Characterization of the adsorbent
Elemental composition and distribution: element analysis, energy dispersive X-ray spectroscopy (EDS)
Surface area, porosity (pore size, pore size distribution): nitrogen adsorption (BET measurements)
Crystallographic structure: X-ray diffraction (XRD)
(Surface, inner) morphology: surface electron microscopy (SEM), transmission electron microscopy (TEM), often coupled with EDS
Chemical structure: solid state nuclear magnetic resonance (NMR) spectroscopy, X-ray photoelectron spectroscopy (XPS), X-ray absorption spectroscopy (XAS)
Ion-exchange capacity: titration
Surface chemistry, surface acid-base characterization: titration, calorimetry, XPS
Surface properties (hydrophilicity, hydrophobicity): contact angle measurements
Determination of the active sites: titration, NMR, Fourier transform infrared spectroscopy (FT-IR), electron spin resonance spectroscopy (ESR)
Swelling capacity: (hypo)osmotic test, pure mechanical swelling
Stability of the material: thermogravimetric analysis (TGA), differential scanning calorimetry (DSC)
Characterization of the adsorbate in the aqueous solution
Determination of the contaminant concentration: atomic absorption spectroscopy (AAS), inductively coupled plasma (ICP), UV-Vis spectrophotometry, fluorescence spectroscopy, photometry (kits tests), ion selective electrode (ISE)
Separation and identification of the contaminant components existing in the solution: liquid chromatography mass spectrometry (LC-MS)
Adsorption mechanism
Chemical characterization of contaminant bound on the material: EDS, NMR, FT-IR
Chemical composition of contaminant bound on the material: XRD
Determination of the oxidation state of contaminant (metal bound): XPS, XAS

2.3.2 Analytical Techniques for the Characterization of an Adsorbent

Table 2.4 shows examples of analytical techniques used for the characterization of an adsorbent, and more generally available in adsorption research (Crini and Badot 2010; Park et al. 2010; Michalak et al. 2013; Fomina and Gadd 2014; Ramrakhiani et al. 2016). Such techniques often provide distinctive but complementary information not only on the characterization of the adsorbent used but also on adsorption of a target contaminant onto the material and its performance. For example, the characterization of the morphological structure and chemistry of a biomass-based adsorbent used for metal removal is essential for understanding the metal binding mechanism on the biomass surface (Park et al. 2010). This can be elucidated using different techniques such as potentiometric titrations, Fourier transform infrared spectroscopy (FT-IR), energy dispersive X-ray spectroscopy (EDS), X-ray diffraction (XRD), X-ray photoelectron spectroscopy (XPS), and also surface electron microscopy (SEM), transmission electron microscopy (TEM). SEM interpretation provides topography of the surface feature and elemental information/metal distribution with a virtually unlimited depth of field. TEM also provides information on the topographical, morphological, compositional and crystalline structures. In general, combination of SEM-EDS and TEM-EDS are performed in order to obtain information regarding the location of the metal. X-ray absorption spectroscopy (XAS) determines the oxidation state of a metal bound to the biosorbent and its coordination environment (Ramrakhiani et al. 2016).

It is well-known that the addition of electrolytes can increase the aggregation of the dye molecules due to reducing the electrical double layer effects to favour self-assembly. For instance, the addition of NaCl in relatively high concentrations can induce the size of the particles in the solution, reducing the space available for dye adsorption (due to occupancy of ions at active sites in competition with dye species), leading to higher aggregation, in accordance with DVLO theory (Blokzijl and Engberts 1993). Liquid chromatography mass spectrometry analysis are useful tools to separate and identify the dye components existing in the solution (Won et al. 2008). Because of their high molar absorptivity, dye molecules and their aggregates are also easily detected by spectrophotometric and photophysical techniques, even at relatively low concentrations. This approach provides information to analyze the type of adsorption mechanisms.

2.3.3 Commercial Activated Carbons

Many wastewaters contain significant levels of organic and mineral contaminants which are toxic or otherwise undesirable because they create, in particular, odor, bad taste and color (McKay 1996; Kannan and Sundaram 2001; Swami and Buddhi 2006; Crini 2006; Qu 2008). Amongst the numerous techniques of contaminant removal, liquid-solid adsorption using AC is the procedure of choice and gives the best results as it can be used to remove different types of contaminants (Dubinin 1966; Manes 1998; Dąbrowski 2001; Dąbrowski et al. 2005; Crini and Badot 2008). Indeed, carbons are one of the oldest and most widely used adsorbents in industry. Due to their great capacity to adsorb contaminants, commercial activated carbons (CAC) are the most effective adsorbents, and if the adsorption system is properly designed they give a good-quality output. This capacity is mainly due to their structural characteristics and their porous texture which gives them a large surface area, and their chemical nature which can be easily modified by chemical treatment in order to vary their properties. The processes that use these usual adsorbents are often carried out in a batch mode, by adding activated carbon to a vessel containing the contaminated solution, or by feeding the solution continuously through a packed bed of carbon.

In general, CAC are used as very good adsorbents of organic matter (OM) to reduce the organic load in secondary and/or tertiary treatment, for instance to process heavily polluted effluent (color, COD, TOC) from the textile industry. They are generally very broad spectrum adsorbents that efficiently eliminate man-made pollutants such as pesticides, aromatic and phenolic derivatives (PAH, PCB, etc.), pharmaceutics, volatile organic compounds (VOC), hydrocarbons and surfactants, minerals including metals but also the molecules that discolor water (dyes), or that cause a taste or smell. They also retain toxic organic compounds refractory to treatments upstream, or they finish off the elimination of OM before discharge in the tertiary treatment of industrial effluent. Also, macroporous CAC can be used as supports for bacteria: the bacteria then degrade part of the adsorbed OM (biological elimination) and thus participate in the *in situ* regeneration of the sorbent. This type of treatment is, in general, coupled to an ozonation stage, further improving the performance of the process. CAC biological filters are, for instance, used for the detoxification of effluent loaded with ions (removal of iron, manganese, nitrate) or for the reduction of BOD, COD and TOC (Radovic et al. 2000). An additional major advantage of treatment with CAC is the fact that no by-products are produced, unlike during treatment by chemical oxidation. To obtain cost-effective technology (especially in the field of water recycling) CAC powder is used in conjunction with an ultrafiltration membrane or with other techniques, such as oxidation. Active carbon competes favorably with nanofiltration and has completely replaced oxidation with ozone.

Although CAC have been used for a long time, development is still being pursued particularly as there is increasing demand for very clean water. Research, both fundamental and applied, is currently very active concerning (i) the possible use of new precursors for the CAC such as agricultural and industrial wastes, water bottles made of PET, scrap tyres; (ii) the development of novel classes of materials such as activated carbon cloth, nanotubes; (iii) understanding the mechanisms of activation, sorption and regeneration (e.g. microwave techniques, techniques not requiring heat) (Mui et al. 2004; Aktas and Seçen 2007; Dias et al. 2007; Li et al. 2010).

AC technology also presents several disadvantages. CAC is quite expensive (e.g. the higher the quality, the greater the cost), and non-selective. Different qualities of carbon also exist which vary not only as a function of the raw material used but also of the carbonization conditions and of the way in which activation is performed (physical or chemical). And yet, even though the high absorbing power of active carbons no longer needs to be proved, not only is there the problem of disposal of spent CAC, there is also the drawback of their rapid saturation, and thus their regeneration. This regeneration step of saturated carbon is also expensive, not straightforward, and results in loss of the adsorbent. For these reasons, their widespread use is restricted, in particular small and medium-size enterprises (SMEs) cannot employ such treatment due to high cost.

2.3.4 Other Commercial Materials

Various studies have been carried out to replace AC by other commercial adsorbent materials (Allen 1996; Allen and Koumanova 2005; Aksu 2005; Crini 2006; San Miguel et al. 2006; Sudha and Giri Dev 2007; Crini and Badot 2007; Tang et al. 2007; Vijayaraghavan and Yun 2008; Qu 2008; Wan Ngah et al. 2008; Sud et al. 2008). Those adopted on an industrial scale are zeolites, commercial activated alumina, silica gels, ion-exchange resins, and sand (Yang 2003; Crini and Badot 2007; Crini 2010). The resins reduce the discharge of polluted water into the environment and are successfully applied, for instance, in the elimination of mineral and organic contaminants including numerous types of dye molecules from rinsing water or polluting metals from pickling baths. The advantages of ion-exchange include no loss of adsorbent on regeneration, reclamation of solvent after use and the removal of soluble contaminants at trace levels. However, like commercial activated carbons, these materials are not cheap (except sand) – a factor that cannot be ignored.

2.3.5 Non-conventional Green Adsorbents

Although these commercial materials are preferred conventional adsorbents for contaminant removal, their widespread industrial use is restricted due to high cost. In addition, Streat et al. (1995) previously reported that the use of commercial carbons based on relatively expensive starting materials is unjustified for most pollution control applications and environmental purposes. As such, alternative non-conventional adsorbents, mainly products and by-products of biological (named biosorbents including biomasses), industrial and agricultural origin and from forest industries (green adsorbents), were proposed, studied and employed as inexpensive and efficient adsorbents (Volesky 1990, 2004, 2007; McKay 1996; Varma et al. 2004; Crini 2005, 2006; Gerente et al. 2007; Li et al. 2008; O'Connell et al. 2008; Oliveira and Franca 2008; Gadd 2009; Crini and Badot 2010; Elwakeel 2010; Sanghi and Verma 2013). These include algae, bacteria, fungi, and yeasts, bark, sawdust, peat, natural products (e.g. cotton, flax, hemp), polysaccharides such as starch, cellulose, chitosan, and alginates, industrial byproducts (e.g. red mud, sludge…), plants, and innovative nanomaterials.

All these non-conventional materials are interesting due to the fact that they are abundant in nature, available in large quantities, inexpensive, and may have potential as complexing materials due to their physicochemical characteristics and particular structure. However, it is important to point out that the adsorption processes using these materials are basically at the laboratory stage in spite of unquestionable progress (Gadd 2009; Crini and Badot 2010). Table 2.5 shows a selection of different reviews on non-conventional adsorbents used for pollutant removal.

Table 2.5 Selected comprehensively reviews on non-conventional adsorbents used in adsorption-oriented processes (selected references)

Adsorbent	Contaminant(s)	Reference(s)
Agricultural byproducts	Metals, dyes, PAH	Oliveira and Franca (2008), Crini and Badot (2010), Sharma et al. (2011), Nguyen et al. (2013), Rangabhashiyam et al. (2014), Lim and Aris (2014), Kharat (2015), Zhou et al. (2015), Emenike et al. (2016), and Sulyman et al. (2017)
Agri-food wastes	Metals, dyes	Demirbas (2008), Oliveira and Franca (2008), and Kumar et al. (2011)
Fruit and vegetable wastes	Metals, dyes	Swami and Buddhi (2006) and Patel (2012)
Sawdust	Metals, dyes, phenols	Shukla et al. (2002), Swami and Buddhi (2006), Larous and Meniai (2012), Kharat (2015), and Sahmoune and Yeddou (2016)
Bark	Metals, dyes, PAH, chlorinated phenols	Ahmaruzzaman (2008), Demirbas (2008), Kharat (2015), and Sen et al. (2015)
Rice husk	Metals, dyes	Chuah et al. (2005), Ahmaruzzaman (2011), Nguyen et al. (2013), Dhir (2014), and Sulyman et al. (2017)
Wheat	Metals	Ngah and Hanafiah (2008) and Farooq et al. (2010)
Sugar-beet pulp	Metals	Ngah and Hanafiah (2008), Ahmaruzzaman (2011), and Dhir (2014)
Coconut	Metals, phenols	Swami and Buddhi (2006), Bhatnagar et al. (2010), Patel (2012), and Bazrafshan et al. (2016)
Opuntia ficus-indica		Nharingo and Moyo (2016)
Coffee		Anastopoulos et al. (2017a) and Sulyman et al. (2017)
Tea factory waste	Metals	Ahmaruzzaman (2011) and Sulyman et al. (2017)
Peat	Metals, dyes, phosphorus	Brown et al. (2000), Vohla et al. (2011), and Raval et al. (2016)
Lignocellulosic wastes	Metals	Miretzky and Cirelli (2010), Abdolali et al. (2014), and De Quadros Melo et al. (2016)
Cellulose	Metals, dyes, organics	O'Connell et al. (2008), Hubbe et al. (2011), Vandenbossche et al. (2015), and Grishkewich et al. (2017)
Microbial biosorbents	Metals, dyes	Solis et al. (2012), Mudhoo et al. (2012), Ahluwalia and Goyal (2007), and Srivastava et al. (2015)
Bacterial biosorbents	Metals, dyes	Vijayaraghavan and Yun (2008) and Mudhoo et al. (2012)

(continued)

Table 2.5 (continued)

Adsorbent	Contaminant(s)	Reference(s)
Fungi	Metals, dyes	Kaushik and Malik (2009), Wang and Chen (2009), and Dhankhar and Hooda (2011)
Extracellular polymeric substances	Metals, phosphorus	More et al. (2014) and Li et al. (2015)
Algae, marine algae	Metals, dyes	Brinza et al. (2007), Wang and Chen (2009), Mudhoo et al. (2012), He and Chen (2014), and Zeraatkar et al. (2016)
Plants	Metals, dyes	Ahluwalia and Goyal (2007), Srivastava et al. (2015), and Saba et al. (2016)
Pectins	Metals, dyes	Miretzky and Cirelli (2010), Vandenbossche et al. (2015), and Zhao and Zhou (2016)
Starch	Metals, dyes	Crini (2005), Crini and Badot (2010), Panic et al. (2013), Wang et al. (2013), and Vandenbossche et al. (2015)
Chitin	Dyes	Yong et al. (2015), Barbusinski et al. (2016), and Anastopoulos et al. (2017b)
Chitosan	Metals, dyes, phenols	No and Meyers (2000), Elwakeel (2010), Li et al. (2008), Crini and Badot (2008), Kyzas et al. (2013b), Liu and Bai (2014), Vakili et al. (2014), Yong et al. (2015), Barbusinski et al. (2016), and Azarova et al. (2016)
Alginate	Metals, dyes	Vandenbossche et al. (2015) and Grishkewich et al. (2017)
Cyclodextrins	Metals, dyes, organics	Mocanu et al. (2001), Crini and Morcellet (2002), Crini (2005, 2014), Crini and Badot (2010), Landy et al. (2012a, b), and Panic et al. (2013)
Hydrogels	Metals, dyes, organics	Panic et al. (2013), Khan and Lo (2016), Mittal et al. (2016), Muya et al. (2016), and Grishkewich et al. (2017)
Clays	Metals, dyes	Bhattacharyya and Gupta (2008) and Ngulube et al. (2017)
Industrial byproducts	Metals, dyes	Crini (2005, 2006) and El-Sayed and El-Sayed (2014)
Fly ash	Metals, dyes	Swami and Buddhi (2006), Ahmaruzzaman (2010, 2011), and Raval et al. (2016)
Red mud	Phosphorus, phosphate, fluroride, nitrate, metals, metalloids, dyes, bacteria, virus	Wang et al. (2008) and Ahmaruzzaman (2011)

(continued)

Table 2.5 (continued)

Adsorbent	Contaminant(s)	Reference(s)
Municipal wastes	Metals, dyes	Bhatnagar and Sillanpää (2010)
Sludge-based adsorbents	Metals, dyes	Raval et al. (2016) and Devi and Saroha (2017)
Nanomaterials	Metals, dyes	Crini and Badot (2010), Ali (2012), Kumar et al. (2014), Zhao and Zhou (2016), and Sadegh et al. (2017)

2.3.6 Which is the Best Non-conventional Adsorbent?

Since the range of non-conventional adsorbents proposed in the literature is extremely extensive, attempting to provide a comprehensive list of potential effective materials would be unrealistic. For instance, agricultural wastes and byproducts from forest industries include tea waste, coffee, hazelnut shells, peanut hull, sawdust, barks, palm kernel husk, coconut husk, peanut skins, cellulosic and lignocellulosic wastes, hemp-based products, cotton and modified cotton, corncobs, rice hulls, apple wastes, wool fibers, olive cake, almond shells, cactus leaves, banana and orange peels, sugar beet pulp, palm fruit bunch, maize leaf, and other different by-products. Adsorption onto these biosorbents has been the focus of much attention and abundant data on their performance can be found in the literature (Sharma 2015; Crini 2015; Vandenbossche et al. 2015; Khalaf 2016; Morin-Crini and Crini 2017).

In Table 2.6 we reported the features, advantages and limitations of some selected non-conventional and conventional adsorbents (Crini 2005, 2015; Allen and Koumanova 2005; Crini and Badot 2007, 2008, 2010; Bhattacharyya and Gupta 2008; Wang and Peng 2010). Which is the best adsorbent? There is no direct answer to this question because each adsorbent has advantages and drawbacks (Crini 2006; Gadd 2009).

The comparison of adsorption performance depends on several parameters and a direct comparison of data obtained using different materials is not possible since experimental conditions are not systematically the same. Most of the information is related to a single contaminant removal individually in batch experiments and little or no data on removal of contaminants in complex form in real wastewater. Other factors such as operation difficulty, practicability, regeneration potential and environmental impact, need to be taken into consideration when selecting one adsorbent over another. Due to the scarcity of consistent cost information, cost comparisons are also difficult to estimate.

Generally, the adsorption capacity exhibited by each material relates primarily to its textural and chemical properties. It is also important to point out that a particular non-conventional adsorbent is only applicable to a particular class of contaminants. Thus, using only one type of material is difficult for the treatment of the complex mixtures of pollutant wastewaters. For instance, bentonite is an ineffective adsorbent for nonionic organic compounds in water. Chitosan without chemical modification is

Table 2.6 Principal commercial, conventional and emerging materials for contaminant removal by adsorption and/or ion-exchange processes

Adsorbent	Features/advantages/mechanisms	Limitations/comments
Activated carbons	The most effective adsorbents in industry (charcoal is the oldest material known in wastewater)	Initial cost of the carbon
Powder activated carbon (PAC)	Porous adsorbents with large surface area	The higher the quality, the greater the cost
	Versatile material	Performance is dependent on the type of carbon used
	Two main forms: powdered (PAC) forms to be used in batch experiments followed by filtration and granular (GAC) forms for use in column (more adaptable to continuous contacting)	Non-selective process
Granular activated carbon (GAC)	PAC: used in batch experiments due to low capital cost and lesser contact time requirements	Problems with hydrophilic substances
Other forms	Widely applied in the treatment of (drinking) water and wastewater	Ineffective against As(III), disperse and vat dyes
	Great capacity to adsorb a wide range of pollutants including metals and metalloids (As(V)), dyes, phenols and chlorophenols, pesticides, and pharmaceutics and drugs	Require complexing agents to improve their removal performance
	Extensively used for organic contaminant removal (COD, BOD and TOC removal)	Many problems connected with regeneration (large capital investments, expensive steps, loss of adsorbent)
	High capacity and high rate of adsorption	GAC regeneration is easier than PAC
	Fast kinetics	Identification of adsorption mechanisms (in particular for modified activated carbons)
	Produce a high-quality treated effluent	
	Interesting technology in combination with other techniques (precipitation, sand filtration, ion-exchange) or in conjunction with microorganisms	
	Physisorption mechanisms	
Activated carbons from solid wastes	Inexpensive and renewable additional sources of carbon	The performance depends on the raw material, the history of its preparation and treatment conditions
	A potential alternative to existing CAC	Reactivation results in a loss of the carbon

(continued)

Table 2.6 (continued)

Adsorbent	Features/advantages/mechanisms	Limitations/comments
Agricultural wastes	Interesting properties in terms of surface chemistry, surface charge and pore structure like CAC	Laboratory stage
Wood wastes	Efficient for a large range of pollutants	Identification of adsorption mechanisms
City wastes		
Industrial by-products		
Sand	A very common adsorbent, mainly in granular form, used in pre- or post-treatment	Construction cost depending on kinds of filters and technologies (rapid or slow processes)
	Well-known filtration technique	Requires a pre-treatment (pH adjustment, coagulation, flocculation) and also a post-treatment (disinfection)
	Large choice of filtration medium with a wide variety in size and specific gravity	Filters become clogged with flocs after a period in use
	Rapid and efficient for SS removal	Frequent cleaning required (every 24–72 h)
	No limitations regarding initial turbidity levels (if coagulant or flocculant is correctly applied)	Cost of energy (regeneration) and cost for treatment of generated sludge
	Widely applied for treating large quantities of drinking water	Not effective for viruses, fluoride, arsenic and salts
	Somewhat effective for odor, taste, bacteria and OM	
	Rapid cleaning time	
	Interesting as pretreatment in combination with CAC treatment	
	Physisorption and diffusion mechanisms	
Activated alumina	Relatively well-known and commercially available	Cost of the adsorbent
Bauxite	Highly porous materials with a high surface area and an interesting distribution of both macro and micropores	pH-dependent
	Used mainly as desiccants	Requires a pre-treatment to prevent clogging of the material bed when the water contains SS or to remove certain ions
	Efficient filter for fluoride, selenium and arsenic removal	Fluoride removal: regeneration is often required to make it cost-effective
	Also interesting for the treatment of copper, zinc, mercury, uranium and phosphates	Arsenic removal: needs replacement after four or five regeneration
	Classified by the USEPA as among one of the best available	Can accumulate bacteria

(continued)

Table 2.6 (continued)

Adsorbent	Features/advantages/mechanisms	Limitations/comments
	technologies for arsenic removal in drinking water	
	The presence of impurities (iron. . .) do not affect the performance	Use of strong acid and base solutions for regeneration step
Zeolites	Easily available and relatively cheap (the price depends on the quality of zeolite)	More than 40 natural species: adsorption properties depend on the different materials
Aluminosilicates	40 natural and over 100 synthetic materials; clinoptilolite: The most abundant and frequently studied mineral	Not suitable for reactive dyes
Clinoptilolite	Highly porous aluminosilicates with different cavity structures and unique surface chemistries (a three dimensional framework having a negatively charged lattice) and valuable physicochemical properties (cation exchange, molecular sieving, catalysis and adsorption)	Low permeability
Chabazite	Suitable adsorbents for dyes, metals, phenols and chlorophenols: high ion-exchange capacity and high selectivity	Complex adsorption mechanism
Modified materials	A high capacity to be easily regenerated (while keeping their initial properties)	Requires chemical modification
	Main applications: softening and deionization of water, waste treatment, purification of products	
	Ion-exchange mechanism	
Silica	A very common adsorbent, mainly in granular form	Cost of the adsorbent
Silica gel	A highly porous solid with mechanical stability	Hydrophilic material
Silica beads	High surface area	Low values for the pH of point of zero charge
Glasses	Numerous industrial environmental applications	High affinity for water (silica gel is a drying agent)
Silica modified	Efficient for removal of organics (toluene, xylene, dyes)	Low resistance toward alkaline solutions
Hybrid materials	Very high adsorption capacities (acid dyes)	Requires chemical modification
	Physisorption and chemisorption (ion-exchange) mechanisms	

(continued)

Table 2.6 (continued)

Adsorbent	Features/advantages/mechanisms	Limitations/comments
Siliceous materials	Abundant, available and low-cost inorganic materials	Important role of the pH of the solution
Alunite	Porous texture with high surface area	Requires physical and chemical modification
Perlite	Regeneration (alunite) is not necessary	Variable differences in composition (perlite)
Dolomite	Promising adsorbent for dyes	Results depend on the types of perlite used (expanded and unexpanded) and on its origin
Diatomite	Perlite: an amorphous siliceous mineral, inexpensive and easily available in many countries	
	Dolomite: a common double carbonate mineral consisting of alternative layers of calcite and magnesite	
	Chemisorption mechanism	
Clays	Natural well-known minerals (several classes of clays)	Not efficient for pollutants having a strong acid character
Montmorillonite	Low-cost and abundance on most continents	Requires chemical modification or activation (bentonite)
Bentonite	Layered structures with large surface area and high porosity, and high	Results are pH-dependent
Fuller's earth	chemical and mechanical stability	Identification of adsorption mechanism (for modified materials)
Sepiolite	Strong candidates for ion-exchange (high cation exchange capacity)	
Kaolinite	Considered as host materials with a strong capacities to adsorb positively charged species; they can also adsorb anionic and neutral species	
Modified materials	Efficient for basic dyes, phenols and metal ions	
	High adsorption capacities with rapid kinetics	
	Montmorillonite: clay with the largest surface area and the highest cation exchange capacity	
	Fuller' earth: a natural clay with an open porous structure	
	Bentonite: a fine powder clay with a high surface area and an efficient sorbent for acid, basic and disperse dyes (very interesting material when coupled with ultrafiltration)	
	Organobentonites: powerful adsorbents	
	Formation of dye-clay complex or organoclay	

(continued)

Table 2.6 (continued)

Adsorbent	Features/advantages/mechanisms	Limitations/comments
Commercial polymeric organic resins and synthetic organic resins	Established treatment process	Derived from petroleum-based raw materials
Porous cross-linked polymers	A large choice of commercially available materials: regular spherical beads with high surface area, a wide range of pore structure, high mechanical strength and high chemical resistance, and with chelating properties, comparable with those of AC	Commercial resins are quite expensive
Macroporous copolymers	Industrial use for adsorption and ion-exchange processes	Sensitive to particle, suspended solids, COD and oils
Hypercross-linked polymers	Effective materials – produce a high-quality treated effluent	Performance is dependent on the type of resin used
Organic resins	High adsorption capacities toward target pollutant including phenolic derivatives, metals, ionic contaminants (fluorides…) and dyes	Incapable of treating large volumes
Chelating polymers	Very interesting technology in combination with CAC treatment	pH-dependence
Hydrogels	Economically valid for precious metal recovery	Poor contact with aqueous pollution
Synthetic- or natural-based polymers	Can remove unwanted molecules to lower concentrations than CAC does	Requires a modification for enhanced the water wettability
	No loss of material on regeneration	Non-selective process (conventional resins)
	Organic resins: mechanism due to ion-exchange and/or diffusion into the porous network	Not effective for all dyes (disperse dyes)
	Hydrogels (super-swelling polymers): mechanisms involve both electrostatic interactions and diffusion into the three-dimensional polymeric structures	
Industrial by-products	Low-cost materials and local availability	Adsorption properties strongly depend on the different materials (fly ash, red mud)
Fly ash	Effective for metal and dye removal with interesting adsorption capacities	Contain hazardous substances (coal fly ash may contain harmful metal oxides)
Red mud	Fly ash: a waste material produced in great amounts in combustion processes	Low surface area
Metal hydroxide sludge	Metal hydroxide sludge: a low-cost waste material from the	Require physical and chemical modification (red mud)

(continued)

Table 2.6 (continued)

Adsorbent	Features/advantages/mechanisms	Limitations/comments
	electroplating industry containing insoluble metal hydroxides and salts (calcium, sodium)	
Blast furnace slag and sludge etc.	Red mud: efficient for metalloids and anions removal	Results are pH-dependent (metal hydroxide sludge)
	Ion-exchange mechanism and/or diffusion	Influence of salts (metal hydroxide sludge)
Agricultural solid wastes and byproducts from forest industries	Cheap and readily available resources	Adsorption properties depend on the different materials
Sawdust	Effective for many types of pollutants and pollution (metals, dyes, oils, salts)	Adsorption mechanism must be clarified (bark)
Bark		
Date pits	Bark: a polyphenol-rich material containing a high tannin content	Results are pH-dependent
Pith	Sawdust: an abundant by-product available in large quantities at zero or negligible price containing various organic compounds with polyphenolic groups	Require chemical pre-treatment to improve the adsorption capacity and enhance the efficiency (sawdust)
	Possible regeneration (sawdust)	
	Physisorption and chemisorption (ion-exchange, hydrogen bonding) mechanisms	
Biomass	Interesting competitive, cheap and effective technology	Slow process and limiting pH tolerance (algae)
Dead or living biomass	Publicly acceptable	Performance depends on the biomass species (algal species), differences in the cell wall composition of the species, cell size and morphology, and on some external factors (pH, salts, competitive adsorption, metal speciation, temperature)
Fungi	Simplicity, versatility, flexibility for a wide range of applications	Results depend on the functional groups present in the biomass
Algae	Availability of different biomasses in large quantities and at low cost	Not appropriate for column systems (an immobilization step is necessary for use in column reactors)
White-rot fungi	Effective and selective adsorbents containing a variety of functional	Technologies are still being developed

(continued)

Table 2.6 (continued)

Adsorbent	Features/advantages/mechanisms	Limitations/comments
	groups: more selective than traditional ion-exchange resins and CAC	
Yeasts	Important adsorption capacities reported for metal ions and dyes	
Agricultural wastes	Suitable for a wide range of metal concentrations from 100 ppm to 100 ppb or even less (algae)	
Food processing	Fungi can reduce pollutant concentrations to ppb levels	
Aquatic plants	Regeneration is not necessary	
	Physisorption and chemisorption mechanisms	
Peat	Plentiful, inexpensive and widely available biosorbent	Low mechanical strength
Raw material	A porous and complex soil material with a polar character	A high affinity for water
Modified materials	A low-grade carbonaceous fuel containing lignin, cellulose and humic acids	Poor chemical stability
	Excellent adsorption and ion-exchange properties (contains various functional groups); good adsorption capacities for a variety of organic and inorganic pollutants; particularly effective adsorbent for basic dyes and metal ions	A tendency to shrink and/or swell
	Chemisorption mechanisms	Requires a pre-treatment or a chemical activation step
		Influence of some factors (pH, agitation speed, initial dye concentration)
Chitosan	Abundant, renewable, biodegradable and environmentally friendly resource	Nonporous material
Chitin and derivatives	Chitin, the second most abundant natural polysaccharide next to cellulose, is fairly abundant (found in the exoskeleton of shellfish and crustaceans); considered as a byproduct of food processing	Low surface area
Chitosan-based derivatives	Low-cost biopolymer and extremely cost-effective	The performance depends on the origin of the polysaccharide, the

(continued)

Table 2.6 (continued)

Adsorbent	Features/advantages/mechanisms	Limitations/comments
		degree of N-acetylation and the treatment of the polymer
	Hydrophilic biopolymer with high reactivity and cationic properties in acidic medium	Variability in the polymer characteristics and in the materials used
	Excellent diffusion properties	Not effective for cationic dyes (except after derivatization)
	Versatile materials (powders, gels, beads, fibres. . .) with excellent chelation and complexation behavior	Requires chemical modification to improve both its performance and stability
	Outstanding metal-binding capacities (useful for the recovery of valuable metals) and extremely high affinities for many classes of dyes	Results depend on the functional groups grafted
	High efficiency and selectivity in detoxifying both very dilute or concentrated effluents; a high-quality treated effluent is obtained	Results are strongly pH-dependent
	Easy regeneration if required	Hydrogels: not appropriate for column systems (except for cross-linked beads)
	Physisorption (van der Waals attraction, hydrogen bonding, coulombic attraction) and chemisorption (chelation, complexation) mechanisms	
Miscellaneous adsorbents	Renewable resources (starches, cotton waste, cellulose), economically attractive and feasible	Cost (calixarenes, cucurbiturils, cyclodextrins)
Cellulose	A remarkably high swelling capacity in water	Low mechanical strength
Starch	Relatively low-cost materials with good adsorption capabilities for a variety of pollutants (can reduce pollutant concentrations to ppb levels)	Low surface area
Cyclodextrin	Good removal of a wide range of contaminants	Variability in the materials used
Alginates	Capable of forming host-guest complexes (cyclodextrins, calixarenes, cucurbituril)	A high affinity for water
Cotton waste	Starch: the most abundant carbohydrate (next to cellulose) with numerous biological and chemical properties – abundant biopolymer and widely available in many countries	Poor chemical stability

(continued)

Table 2.6 (continued)

Adsorbent	Features/advantages/mechanisms	Limitations/comments
Calixarenes	Cyclodextrins: natural macrocyclic oligomers having a hydrophobic cavity and a amphiphilic character; exhibit high adsorption capacities towards organic species	A tendency to shrink and/or swell
Cucurbiturils	Cucurbituril: macrocyclic ligand with interesting complexing properties; a high capacity to adsorb (textile) dyes and lanthanide cations	Not appropriate for column systems (hydrodynamic limitations, column fouling, technical constraints)
	Chemisorption mechanism (complexation, inclusion complex formation, ion-exchange)	Requires pre-treatment and/or chemical modification
		Cucurbituril: dissolution problem

also ineffective for the removal of cationic dyes. Red mud, alumina, zeolite, calcite and clay have been proposed for fluoride removal but, with fluoride concentration decreasing, these materials lose the fluoride removal capacity. Inherent limitations of raw clays as adsorbents of metals are their low loading capacity, relatively small metal ion binding constants, and low selectivity to the type of metal.

Despite the number of papers published on conventional and non-conventional adsorbents for pollutant removal from contaminated solutions, there is, as yet little literature reporting a full study of comparisons between materials. The data have not been compared systematically with commercial activated carbons or synthetic ion-exchange resins which show high removal efficiencies and rapid kinetics. Finally, despite continuing dramatic increases in published research, there has been little or no exploitation in an industrial context.

2.4 Modeling and Mechanisms of Adsorption

2.4.1 Batch Experiments

Batch experiments provide fundamental information on the behavior of the adsorbents used and are thus necessarily carried out in all adsorption studies. Indeed, these methods are widely used to describe the adsorption capacity, the adsorption kinetics and the thermochemistry of the process. The experimental protocol used is simply and easily reproducible. The solution to be treated and the adsorbent are intimately mixed in an agitated contacting reactor/tank for a set time to enable the system to approach equilibrium, following which the thin slurry is filtered to separate the solid adsorbent and adsorbate from the solution. Batch studies use the fact that the adsorption phenomenon at the solid/liquid interface leads to a change in the concentration of the solution. Adsorption isotherms are then constructed by measuring the concentration of adsorbate in the medium before and after adsorption at a fixed temperature. In this respect, in general, adsorption data including equilibrium and kinetic studies are performed using standard procedures which consist in mixing a fixed volume of contaminant solution, at a known concentration in a tightly closed flask, with a known amount of material (the adsorbent is usually applied in the form of a finely ground powder) in controlled conditions of contact time, agitation rate, temperature, and pH. The solution is stirred on a thermostatic mechanical shaker operating at a constant agitation speed. The solution is then, for example, centrifuged to remove any adsorbent particles, and the supernatant is analyzed for the final contaminant concentration: at predetermined times, the residual concentration is determined by using chemical analysis.

Contaminant concentrations in solution can be estimated quantitatively thanks to a linear regression equation obtained by plotting a calibration curve over a range of concentrations. The amount of contaminant adsorbed (i.e. adsorption capacity, contaminant uptake or abatement) is calculated by subtracting the final solution concentration from the initial value. The results are in general reproducible. Blanks containing no contaminant or adsorbent can be conducted in similar conditions as controls to evaluate possible contaminant change (for example color change in the case of dye molecules) and/or precipitation processes for both components. The amount of contaminant adsorbed at time t by the material (q_t) was obtained from the differences between the concentrations of contaminant added to that in the supernatant. q_t was calculated from the mass balance equation given by Eq. (2.1) where C_o and C_t are the initial and final adsorbate concentrations in liquid phase (mg L^{-1}), respectively, V is the volume of adsorbate solution (L) and m the mass of adsorbent used (g). When t is equal to the equilibrium time (i.e., $C_t = C_e$, $q_t = q_e$), then the amount of contaminant adsorbed at equilibrium, q_e, can be calculated by using Eq. (2.2) where C_e is the liquid phase contaminant concentration at equilibrium (mg L^{-1}). The amount of adsorbate adsorbed can be also expressed as percentage of removal (uptake) or abatement (R in %) by using Eq. (2.3). The term "abatement", used by the industrial sector, illustrates the ability of the material to reduce the pollutant load or the concentration of contaminant(s) in effluent to be treated.

$$q_t = \frac{V(C_o - C_t)}{m} \tag{2.1}$$

$$q_e = \frac{V(C_o - C_e)}{m} \tag{2.2}$$

$$R = \frac{100(C_o - C_t)}{C_o} \tag{2.3}$$

2.4.2 Modeling

Adsorption properties and equilibrium data, commonly known as adsorption isotherms, describe how pollutants interact with adsorbent materials and so are critical in optimizing the use of adsorbents. In order to optimize the design of an adsorption system to remove contaminant from solutions, it is important to establish the most appropriate correlation for the equilibrium curve. An accurate mathematical description of equilibrium adsorption capacity is indispensable for reliable prediction of adsorption parameters and quantitative comparison behavior for different materials (or for varied experimental conditions) within any given system (Crini and Badot 2007, 2008).

Adsorption equilibrium is established when the amount of contaminant being adsorbed onto the adsorbent is equal to the amount being desorbed (Giles et al. 1958). It is possible to depict the equilibrium adsorption isotherms by plotting the concentration of the contaminant in the solid phase versus that in the liquid phase. The distribution of contaminant molecule (or ion) between the liquid phase and the material is a measure of the position of equilibrium in the adsorption process and can generally be expressed by one or more of a series of isotherm models. The shape of an isotherm may be considered with a view to predicting if an adsorption process is "favorable" or "unfavorable". The isotherm shape can also provide qualitative information on the nature or the molecule-surface interaction.

There are several isotherm equations available for analyzing experimental data and for describing the equilibrium of adsorption, including the Freundlich, Langmuir, BET, Langmuir-Freundlich, Generalized, Höll-Kirch, Sips, Koble-Corrigan, Radke-Prausnitz, Dubinin-Radushkevich-Kaganer, Redlich-Peterson, Tóth, Temkin, Elovich, Kiselev, Hill-de Boer, Fowler-Guggenheim, Frumkin, Harkins-Jura, Halsey, Nitta, Myers, Henderson, Jossens, Weber-van Vliet, Fritz-Schlunder, and Baudu models (Freundlich 1906; Langmuir 1916, 1918; Radushkevich 1949; Sips 1948; Redlich and Peterson 1959; Dubinin 1960; Toth 1971; Tien 1994; Ho 2003; Allen et al. 2004; Song and Won 2005; Hamdaoui and Naffrechoux 2007a, b; Crini and Badot 2010). The different equation parameters and the underlying thermodynamic assumptions of these models often provide insight into both the adsorption mechanism and the surface properties and affinity of the adsorbent. The two most frequently used equations applied in solid/liquid systems to describe adsorption isotherms are the Langmuir and the Freundlich models and the most popular isotherm theory is the Langmuir type, although these models were initially developed for the modelling of the adsorption of gas solutes onto metallic surfaces (Langmuir 1916, 1918; Polanyi 1920; Dubinin 1966, 1972), and are based on the hypothesis of physical adsorption.

The Langmuir isotherm model is widely used in the literature due to the fact that it incorporates an easily interpretable constant which corresponds to the highest possible adsorbate uptake (i.e. the complete saturation isotherm-curve plateau). The Langmuir equation is represented by Eq. (2.4) where x is the amount of pollutant adsorbed (mg), m is the amount of adsorbent used (g), C_e (mg/L) and q_e (mg/g) are the liquid phase concentration and solid phase concentration of adsorbate at equilibrium, respectively, and K_L (L/g) and a_L (L/mg) are the Langmuir isotherm constants. The Langmuir isotherm constants, K_L and a_L are evaluated through linearisation of Eq. (2.4). By plotting C_e/q_e against C_e, it is possible to obtain the value of K_L from the intercept which is $1/K_L$ and the value of a_L from the slope which is a_L/K_L. Using these constants, it is also possible to obtain an interesting parameter widely used in the literature to promote a solid material as adsorbent, i.e., the theoretical monolayer capacity of an adsorbent (q_{max} in mg/g). Its value, numerically equal to K_L/a_L, permits to evaluate the maximum adsorption capacity of a material for the adsorption of a target pollutant (Crini 2003, 2005, 2006, 2015). At this stage, it is important to note that the uptake of a contaminant by two material adsorbents must be compared not only at the same equilibrium concentration but also in the same experimental conditions (particularly pH). Another essential feature of the Langmuir isotherm can be expressed in terms of a dimensionless constant called separation factor (R_L, also called equilibrium parameter) which is defined by the Eq. (2.5) where C_o is the initial concentration (mg/L) and a_L is the Langmuir constant related to the energy of adsorption (L/mg). The value of R_L indicates the shape of the isotherms to be either unfavorable ($R_L > 1$), linear ($R_L = 1$), favorable ($0 < R_L < 1$) or irreversible ($R_L = 0$).

$$q_e = \frac{x}{m} = \frac{K_L C_e}{1 + a_L C_e} \tag{2.4}$$

$$R_L = \frac{1}{1 + a_L C_0} \tag{2.5}$$

The Freundlich isotherm is expressed by Eq. (2.6) where C_e (mg L^{-1}) and q_e (mg g^{-1}) are the liquid phase concentration and solid phase concentration of adsorbate at equilibrium, respectively, K_F is the Freundlich constant (L^{-1} mg) and $1/n_F$ is the heterogeneity factor. The Freundlich constants are empirical constants which depend on several environmental factors. The value of $1/n_F$ ranges between 0 and 1, and indicates the degree of non-linearity between solution concentration and adsorption as follows: if the value of $1/n_F$ is equal to unity, the adsorption is linear; if the value is below unity, this implies that the adsorption process is chemical; if the value is above unity, adsorption is a favorable physical process; the more heterogeneous the surface, the closer $1/n_F$ value is to 0.

$$q_e = K_F {C_e}^{1/n_F} \tag{2.6}$$

In single-component isotherm studies, the optimization procedure also requires an error function to be defined in order to quantitatively compare the applicability of different models in fitting data. To determine isotherm constants for two-parameter isotherms such as the Langmuir and the Freundlich models, a linear method is available which is based on converting the equation into a linear form by transforming the isotherm variables (Choy et al. 1999; Ho et al. 2000, 2002, 2005; Allen et al. 2003, 2004). Indeed, the typical assessment of the quality of the isotherm fit to the experimental data is based on the magnitude of the correlation coefficient for the regression, i.e. the isotherm giving an R^2 value closest to unity is deemed to provide the best fit. Linearization using such data transformations implicitly alters the error structure however, and may also violate the error equality of variance (homoscadasticity) and normality hypotheses for standard least squares. This may help to explain earlier observations according to which isotherm parameters derived from the linearized forms of the equations are biased in that the Freundlich parameters produce isotherms which tend to fit the data better at low concentrations whereas those derived for the Langmuir isotherm tend to fit the data better at higher concentrations (Allen et al. 2003; Crini and Badot 2010).

In the literature, linear regression is the most commonly used method to estimate adsorption, and linear coefficients of determination are preferred (Ho et al. 2000, 2005; Allen et al. 2004; Crini and Badot 2010; El-Khaiary 2008; El-Khaiary and Malash 2011). However, the use of this method is limited to solving linear forms of equation which measure the difference between experimental data and theoretical data in linear plots only, but not the errors in isotherm curves. The linearization of a non-linear isotherm expression can produce different outcomes (Allen et al. 2003, 2004; Crini and Badot 2008, 2010). The values of individual isotherm constants can change with the error methodology selected.

As an alternative to the linear transformation, nonlinear optimization has also been applied to determine isotherm parameter values. It most commonly uses algorithms based on the Levenberg-Marquardt or Gauss-Newton methods. The optimization procedure requires an error function to be defined in order to enable the optimization process to determine and evaluate the fit of the isotherm to the experimental equilibrium data (Allen et al. 2003, 2004). Different error functions were proposed such as the sum of the squares of the errors (Eq. 2.7), the hybrid fractional error function (Eq. 2.8), the Marquardt's percent standard deviation (Eq. 2.9), the average relative error (Eq. 2.10) and the sum of the absolute errors (Eq. 2.11). In each case, the isotherm parameters can be determined by minimizing the respective error function across the concentration range studied when using the solver add-in with Microsoft's spreadsheet, Excel.

$$\sum_{i=1}^{p} \left(q_{e,calc} - q_{e,meas}\right)_i^2 \tag{2.7}$$

$$\frac{100}{p-n} \sum_{i=1}^{p} \left[\frac{\left(q_{e,meas} - q_{e,calc}\right)^2}{q_{e,meas}}\right]_i \tag{2.8}$$

$$100\left(\sqrt{\frac{1}{p-n}\sum_{i=1}^{p}\left[\frac{\left(q_{e,meas} - q_{e,calc}\right)}{q_{e,meas}}\right]_i^2}\right) \tag{2.9}$$

$$\frac{100}{p} \sum_{i=1}^{p} \left|\frac{\left(q_{e,calc} - q_{e,meas}\right)}{q_{e,meas}}\right|_i \tag{2.10}$$

$$\sum_{i=1}^{p} \left|q_{e,calc} - q_{e,meas}\right|_i \tag{2.11}$$

Several kinetic models can be also used to find the best fitted model for the experimental data obtained (Lagergren 1898; Boyd et al. 1947; Reichenberg 1953; Weber and Morris 1963; Weber et al. 1963; Hall et al. 1966; Blanchard et al. 1984; Treybal 1987; Al-Duri 1996; Ho and McKay 1998, 2003; McKay et al. 1999; Ho 2004, 2006; Azizian 2004; Liu and Liu 2008; Plazinski et al. 2009; Largitte and Pasquier 2016). It is well-accepted that there are several steps in a solid-liquid adsorption process: initially the adsorbate molecules/ions migrate from the bulk of the solution to the material surface (bulk diffusion); the molecules diffuse through the boundary layer to the surface of the material (film diffusion); then, the adsorbate diffuses from the surface to the interior of the particle (pore diffusion, intraparticle diffusion); and finally the molecules reacts with the active sites on the surface of the material (physical adsorption, chemical reaction).

The use of kinetic models such as diffusional models and adsorption models permit elucidation of the adsorption mechanism. The diffusional models assume that the diffusion is the rate limiting step. They are divided in two groups (Crini and Badot 2010). The first is the external mass transfer model which assume that the transfer is controlled by boundary layer diffusion (e.g. Boyd's film-diffusion equation, Spahn and Schlunder model). The adsorbate molecules must pass through the hydrodynamic layer to the surface of the material. Transportation through the boundary layer is due to molecular diffusion, and the distance the adsorbate must travel, or the thickness of the boundary layer, depends on the velocity of the bulk solution. The size of the boundary layer affects the rate of transportation (the thinner the boundary layer, the higher the rate of the transportation). The second gathers the internal diffusion models suggesting a mass transfer through the pores (e.g. Crank model, Weber and Morris model, Bangham model). The internal transport occurs after the adsorbate has passed through the boundary layer and must be transported through the pores to adsorption sites. This intraparticle transportation may occur by molecular diffusion through the solution in the pores (pore diffusion) or by diffusion along the material surface (surface diffusion) after adsorption takes places. The final step, adsorption, is the attachment of the adsorbate onto the material surface at available sites. This step is very rapid, and therefore one of the preceding diffusion steps will control the rate of mass transfer.

In the adsorption models, the adsorption is considered to be the slowest process. This is the case when the adsorbate uptake on the material is of chemical nature. These models include the pseudo-first order model (Lagergren model), the pseudo-second order model (Ho and McKay equations), the pseudo-n order model (for n different from zero), the Langmuir model, and the Elovich equation. The two most popular kinetic equations are pseudo-second-order kinetic and intraparticle diffusion equations proposed by Ho and McKay, and Weber and Morris, respectively.

The pseudo-second order model can be represented in the linear form described by Eq. (2.12) where q_t (mg/g) and q_e (mg/g) are the amount of pollutant adsorbed at time t and equilibrium, respectively, and k_2 is the equilibrium rate constant of pseudo-second order adsorption (g/mg min). The parameters k_2 and q_e can be directly obtained from the intercept and slope of the plot of t/q_t against t. This model is commonly used to describe kinetics of contaminant adsorption on solid adsorbents, although as pointed out by McKay's group (Ho and McKay 1998; McKay et al. 1999), the application of a single kinetic model to the adsorption on materials may be questionable because of the heterogeneity of the adsorbent surfaces and diversity of adsorption phenomena (i.e. transport, diffusion, reactions). Such approach has no physical significance and it is more reasonable to interpret the kinetic data in term of mass transfer.

$$\frac{t}{q_t} = \frac{1}{k_2 q_e^2} + \frac{1}{q_e} t \qquad (2.12)$$

In a batch system under rapid stirring, there is a possibility that the transport of the adsorbate from the solution into the bulk of the adsorbent is the rate controlling step. This possibility was tested in terms of a graphical relationship between the amount of pollutant adsorbed and the square root of time. According to the intraparticle diffusion model proposed by Weber and Morris, the initial rate of intraparticle diffusion is given by the Eq. (2.13) where q_t is the amount of pollutant on the surface of the sorbent at time t (mg/g), k_i is the intraparticle diffusion rate constant (mg/g min$^{1/2}$), t is the time (min) and C is the intercept (mg/g). According to Eq. (2.13), a plot of q_t versus $t^{1/2}$ should be a straight line when adsorption mechanism follows the intraparticle diffusion process. In general, the plots present a multi-linearity, which indicates that two or more steps occur in the process.

$$q_t = k_i t^{1/2} + C \qquad (2.13)$$

Finally, the adsorption characteristics of a material can be expressed in thermodynamic parameters such as ΔG, ΔH and ΔS (entropy change). These parameters can be calculated by using the thermodynamic equilibrium coefficient obtained at different temperatures and concentrations. The models of Arrhenius, Gibbs, van't Hoff and Clausius-Clapeyron can be used (Crini and Badot 2008). The values of ΔG, ΔH and ΔS provide valuable information about the thermodynamics of the adsorption process. ΔG addresses the possibility and feasibility of a certain reaction. Its negative value shows the process is feasible and spontaneous. ΔH shows the route of energy in the process and its positive value indicates an endothermic system. Some authors conclude that the nature of the contaminant adsorption is predominantly physical, involving weak interactions. ΔS can be used to describe the randomness at the solid/solution interface during the adsorption process.

There is no doubt that mathematical modeling is an invaluable tool for the analysis and design of adsorption systems and also for the theoretical evaluation and interpretation of thermodynamic parameters (Allen et al. 2004). However, two important points must be pointed out. The first is that, although these adsorption and kinetic models remain a useful and convenient tool for the comparing results from different sources due to their highly idealistic simplicity, a given plot is an empirically relationship (Liu and Liu 2008; Liu and Wang 2008; Lin and Wang 2008; Wu et al. 2009; Rudzinski and Plazinski 2009; Douven et al. 2015). An isotherm may fit experimental data accurately under one set of conditions but fail entirely under another. No single model has been found to be generally applicable. This is readily understandable in the light of the hypotheses associated with their respective deviations. In addition, the two-parameter isotherm model such as the Langmuir and the Freundlich models are based on the hypothesis of physical adsorption. In the case of dye adsorption onto a biosorbent, which is more chemical than physical, it would be more appropriate to consider pollutant adsorption with models based on chemical reactions. However, these models are complicated in nature. Simple kinetic models used in the literature are also questionable because, generally speaking, these models cannot represent the real course of adsorption and thus cannot offer useful information to gain insight in mechanism. It is more reasonable to interpret the kinetic data in term of mass transfer (homogeneous diffusion model, double exponential model, etc.) but these models are also complex and effective graphical analysis software are required to solve mathematical models. The book published by Tien (1994) can be consulted on this topic. The second point is related the abundant literature data. Despite the number of papers published, there is as yet little literature containing a full study comparing various models and this topic clearly needs further detailed research (Wase and Forster 1997; McKay 1999; Cooney 1999; Yang 2003; Hamdaoui and Naffrechoux 2007a, b; Crini and Badot 2007, 2010).

2.4.3 Mechanisms of Adsorption

In the context of adsorption, the major challenge is to select the most promising types of adsorbent, mainly in terms of low-cost, high capacity (often expressed by the q_{max} value), high adsorption rate, high selectivity, and rapid kinetics. The next real challenge is to clearly identify the adsorption mechanism(s), in particular the interactions occurring at the adsorbent/adsorbate interface (Veglio' and Beolchini 1997; Crini 2005). This is an important topic because the adsorption mechanisms involved in contaminant uptake can orientate the design of the desorption strategy (for example, the recovery of certain contaminants such as "precious" metal ions is also an important parameter for the economics of the process).

Despite the large number of papers devoted to the adsorption of contaminants onto conventional or non-conventional adsorbents, most focus on the evaluation of adsorption performances and only a few aim at gaining a better understanding of adsorption mechanisms (Veglio' and Beolchini 1997; Crini 2005). These mechanisms are not fully understood because a large number of interactions are possible. Some of the reported interactions include (Crini 2005):

- physisorption (physical adsorption)
- surface adsorption
- van der Waals interactions
- hydrogen bonding
- electrostatic interactions (attraction interactions)
- ion-exchange

– complexation (coordination)
– chelation
– acid-base interactions
– proton displacement
– precipitation (surface precipitation, microprecipitation)
– hydrophobic interactions (π-π interactions, Yoshida's interactions)
– oxidation/reduction
– inclusion complex formation
– diffusion into the network of the material
– covalent binding

An interesting question remains: Must all these interactions be taken into account to explain the adsorption mechanism? The answer to this question is not so easy. In an oriented-adsorption process using a given adsorbent, it is possible that more than one of these interactions can occur simultaneously depending on the composition of the material, the contaminant structure and its properties, and the solution conditions (pH, ionic strength, temperature).

Crini reported the simplified classification of contaminant adsorption mechanisms described in Fig. 2.5 (Crini 2005, 2006, 2010; Crini and Badot 2007). Four main mechanisms have been proposed, namely physisorption, chemisorption, ion-exchange and precipitation. Some authors consider ion-exchange process as a chemisorption mechanism. Davis et al. (2003) reported that the term ion-exchange does not explicitly identify the binding mechanism, rather it is used as an umbrella term to describe the experimental observations. The use of the "microprecipitation" term is also a source of debate. This term is used to indicate precipitation taking place

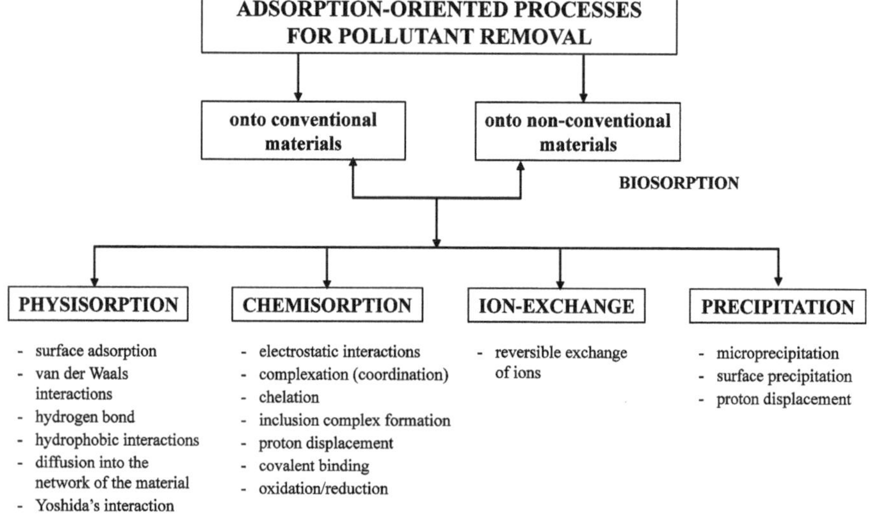

Fig. 2.5 Classification of pollutant adsorption mechanisms according to Crini (Crini 2005, 2006; Crini and Badot 2007)

locally at the surface of a biosorbent due to particular conditions. A discussion on these terms can be found in the recent review by Robalds et al. (2016). Other classifications of contaminant adsorption mechanisms can be found in the following references: Veglio' and Beolchini (1997), Srivastava and Goyal (2010), Naja and Volesky (2011), Asgher (2012), Michalak et al. (2013) and Robalds et al. (2016).

Literature based on commercial activated carbons clearly shows a greater number of studies on the adsorption of organic compounds as compared with the inorganic ones, and it is well known that carbon adsorbents are very versatile due to their high surface area, well-developed pore structure, and surface properties (Radovic et al. 1997, 2000; Ania et al. 2002). The main mechanism is physisorption. However, although extensive experimental and modeling studies on activated carbons have been reported, the subject remains highly controversial as described by Radovic et al. (2000). Much confusion exists in the literature, especially on the role of surface-oxygen functions in the adsorption of aromatic compounds (Moreno-Castilla et al. 1995; Dubinin 1966; Radovic et al. 2000; Pereira et al. 2003).

For biosorbents, the adsorption mechanism are yet not fully understood although some are now well-accepted. In the case of biosorption of metals by biomass, the mechanisms can be viewed as being extracellular or occurring discretely at the cell wall. Intracellular adsorption would normally imply bioaccumulation by a viable organism. Gadd (1990), Veglio' and Beolchini (1997), Volesky (2001), and Aksu (2005) previously discussed various mechanisms involved in biosorption using microorganisms (fungi, yeasts or bacteria, etc.). Gadd (1990) commented that "*a large variety of physical, chemical and biological mechanisms may be involved, including adsorption, precipitation, complexation and transport*". In general, two different metal-binding mechanisms have been postulated to be active in biosorption pollutant uptake: (1) chemisorption by ion-exchange, complexation (including coordination) and/or chelation (biosorption process), and (2) physical adsorption and/or (micro)precipitation (in this case, the process is termed bioaccumulation). Other interactions have been proposed, including metabolism-dependent transport, adsorption of simple ionic species, hydrogen bonding interactions, biological mechanisms, reactions involving hydrolysis products of metal ions or oxidation/reduction.

Accumulation of metals and radionuclides by algae can be described as being composed of two phases: a rapid phase of metabolism-independent binding to the cell surface (biosorption) followed by a slower phase due to simultaneous effects of growth and surface adsorption, active or intracellular uptake by passive diffusion (Garnham 1997). Biosorption- and bioaccumulation-based mechanisms have been adopted for the description of the mechanistically different types of metal sequestering by microorganisms. However, the nature of the binding processes in biosorption is yet complex and unknown, due to the complexity of most biopolymers.

Numerous authors accept that the decolourisation process using non-conventional materials generally results from two main mechanisms – adsorption and ion exchange (Allen and Koumanova 2005; Crini 2006), and is influenced by many factors including the type of adsorbents/biosorbents and dyes, and the process variables. For instance, ion-exchange and hydrogen bonding interactions are the main mechanisms for the removal of metals by sawdust (Shukla et al. 2002). The cell walls of sawdust mainly consist of cellulose and lignin, with many hydroxyl groups such as those on tannins or other phenolic compounds. All these components are active ion-exchange compounds.

Polysaccharide-based materials are mainly used as a chelating or complexing ion-exchange media. These biopolymers (i.e. starch, cellulose, chitin and alginates) and their derivatives (e.g. chitosan, cyclodextrin) contain a variety of functional groups, which can chelate ionic species of a specific size and charge. Such materials are thus often much more selective than traditional ion-exchange resins and can reduce heavy metal ion concentrations to ppb levels (Kentish and Stevens 2001; Crini and Badot 2008). Crini (2015), reviewing dye removal by starch-based materials reported that the adsorption mechanisms were not fully understood because numerous interactions were possible, including ion-exchange, acid-base interactions, precipitation, hydrogen bonding, hydrophobic interactions and physisorption.

For cyclodextrin-based adsorbents, inclusion complex formation and diffusion into the polymer network are the preferred concepts in biosorption because they account for many of the observations made during contaminant uptake studies (Crini 2014). In a comprehensive review, Morin-Crini et al. (2017) recently reported that, in spite of the abundance of literature and conclusive results obtained at the laboratory scale, interpreting the mechanism of pollutant elimination remains an interesting source of debate and sometimes of contradiction.

Plant fibers used as biosorbents consist mainly of cellulose, hemicelluloses, lignin and some pectin and extractives (fats, waxes, etc.). Metal ions mainly adsorb to carboxylic (primarily present in hemicelluloses, pectin and lignin), phenolic (lignin and extractives) and to some extent hydroxyl (cellulose, hemicelluloses, lignin, extractives and pectin) and carbonyl groups (lignin). Strong bonding of metal ions by the hydroxyl, phenolic and carboxylic groups often involves complexation and ion-exchange (Crini and Badot 2010).

2.5 Concluding Remarks

Preserving the environment, and in particular the problem of water contamination, has become a major preoccupation for everyone – the public, industry, scientists and researchers as well as decision-makers on a national, European, or international level. The public demand for pollutant-free waste discharge to receiving waters has made decontamination of industrial wastewaters a top priority. This is a difficult and challenging task. It is also difficult to define a universal method that could be used for the elimination of all contaminants from wastewaters. Among the numerous and various treatment processes currently cited for wastewater treatment, only a few are commonly used by the industrial sector. However, it is now accepted that adsorption is the procedure of choice and gives the best results as it can be used to remove many different types of contaminants.

In this chapter, a general overview on adsorption processes for contaminant removal has been presented, including a classification for the different types of conventional and non-conventional adsorbents used for this purpose and the interactions proposed to explain adsorption mechanisms. Commercial activated carbon is extensively used not only for adsorbing contaminants from drinking water sources but also for removing pollutants from wastewater streams. Other conventional materials include organic resins, activated alumina, zeolites, and sand. Non-conventional adsorbents can be also obtained and employed as low-cost and efficient solid adsorbents. These materials have not yet been significantly commercialized even though they possess numerous advantages over currently available carbon and organic resins for pollutant removal. Perhaps one reason why non-conventional adsorbents have not been widely used in industry is the lack of knowledge about the engineering of such materials. We think that, for novel adsorbents to be accepted by industry, it will be necessary to adopt a multidisciplinary approach in which chemists, engineers, material scientists, biologists, microbiologists, and computer scientists work together. The opportunity now exists to consider other solid materials such as biomass, cellulose and chitosan for emerging applications. They will undoubtedly be at the centre of some extremely profitable commercial activities in the future although their development requires further investigation in the direction of mechanisms (modeling), of regeneration of the adsorbent material (if necessary), and of testing materials with real industrial effluents.

References

Abdolali A, Guo WS, Ngo HH, Chen SS, Nguyen NC, Tung KL (2014) Typical lignocellulosic wastes and by-products for biosorption process in water and wastewater treatment: a critical review. Bioresour Technol 160:57–66. https://doi.org/10.1016/j.biortech.2013.12.037

Ahluwalia SS, Goyal D (2007) Microbial and plant derived biomass for removal of heavy metals from wastewater. Bioresour Technol 98:2243–2257. https://doi.org/10.1016/j.biortech.2005.12.006

Ahmaruzzaman M (2008) Adsorption of phenolic compounds on low-cost adsorbents: a review. Adv Colloid Int Sci 143:48–67. https://doi.org/10.1016/j.cis.2008.07.002

Ahmaruzzaman M (2010) A review on the utilization of fly ash. Prog Energy Combust Sci 36:327–363. https://doi.org/10.1016/j.pecs.2009.11.003

Ahmaruzzaman M (2011) Industrial wastes as low-cost potential adsorbents for the treatment of wastewater laden with heavy metals. Adv Colloid Int Sci 166:36–59. https://doi.org/10.1016/j.cis.2011.04.005

Aksu Z (2005) Application of biosorption for the removal of organic pollutants: a review. Process Biochem 40:997–1026. https://doi.org/10.1016/j.procbio.2004.04.008

Aktas Ö, Çeçen F (2007) Bioregeneration of activated carbon: a review. Int Biodeterior Biodegrad 59:257–272. https://doi.org/10.1016/j.ibiod.2007.01.003

Al-Duri B (1996) Chapter 7: Adsorption modeling and mass transfer. In: McKay G (ed) Use of adsorbents for the removal of pollutants from wastewaters. CRC Press, Boca Raton, pp 133–173

Ali I (2012) New generation adsorbents for water treatment. Chem Rev 112:5073–5091. https://doi.org/10.1021/cr300133d

Ali I (2014) Water treatment by adsorption columns: evaluation at ground level. Sep Purif Rev 43:175–205. https://doi.org/10.1080/15422119.2012.748671

Allen SJ (1996) Chapter 5: Types of adsorbent materials. In: McKay G (ed) Use of adsorbents for the removal of pollutants from wastewaters. CRC Press, Boca Raton, pp 59–97

Allen SJ, Koumanova B (2005) Decolourisation of water/wastewater using adsorption (review). J Univ Chem Technol Metall 40:175–192

Allen SJ, Gan Q, Matthews R, Johnson PA (2003) Comparison of optimised isotherm models for basic dye adsorption by kudzu. Bioresour Technol 88:143–152

Allen SJ, McKay G, Porter JF (2004) Adsorption isotherm models for basic dye adsorption by peat in single and binary component systems. J Colloid Int Sci 280:322–333

Anastopoulos I, Karamesouti M, Mitropoulos AC, Kyzas GZ (2017a) A review for coffee adsorbents. J Mol Liq 229:555–565. https://doi.org/10.1016/j.molliq.2016.12.096

Anastopoulos I, Bhatnagar A, Bikiaris DN, Kyzas GZ (2017b) Chitin adsorbents for toxic metals: a review. Int J Mol Sci 18:1–11. https://doi.org/10.3390/ijms18010114

Ania CO, Parra JB, Pis JJ (2002) Influence of oxygen-containing functional groups on active carbon adsorption of selected organic compounds. Fuel Process Technol 79:265–271. https://doi.org/10.1016/S0378-3820(02)00184-4

Asgher M (2012) Biosorption of reactive dyes: a review. Water Air Soil Pollut 223:2417–2435. https://doi.org/10.1007/s11270-011-1034-z

Azarova YA, Pestov AV, Bratskaya SZ (2016) Application of chitosan and its derivatives for solid-phase extraction of metal and metalloid ions: a mini-review. Cellulose 23:2273–2289. https://doi.org/10.1007/s10570-016-0962-6

Azizian S (2004) Kinetic models of sorption: a theoretical analysis. J Colloid Int Sci 276:47–52. https://doi.org/10.1016/j.jcis.2004.03.048

Babel S, Kurniawan TA (2003) Low-cost adsorbents for heavy metals uptake from contaminated water: a review. J Hazard Mater 97:219–243. https://doi.org/10.1016/S0304-3894(02)00263-7

Bajpai AK, Rajpoot M (1999) Adsorption techniques – a review. J Sci Ind Res 58:844–860

Barbusinski K, Salwiczek S, Paszewska A (2016) The use of chitosan for removing selected pollutants from water and wastewater - short review. Archit Civil Eng Environ 9:107–115

Bazrafshan E, Amirian P, Mahvi AH, Ansari-Moghaddam A (2016) Application of adsorption processes for phenolic compounds removal from aqueous environments: a systematic review. Global NEST J 18:146–163

Berefield LD, Judkins JF, Weand BL (1982) Process chemistry for water and wastewater treatment. Prentice-Hall, Englewood Cliffs, 510 p

Bhatnagar A, Sillanpää M (2010) Utilization of agro-industrial and municipal waste materials as potential adsorbents for water treatment – a review. Chem Eng J 157:277–296. https://doi.org/10.1016/j.cej.2010.01.007

Bhatnagar A, Vilar VJP, Bothelo CMS, Rui B, Boaventura RAR (2010) Coconut-based biosorbents for water treatment – a review. Adv Colloid Int Sci 160:1–15. https://doi.org/10.1016/j.cis.2010.06.011

Bhattacharyya KG, Gupta SS (2008) Adsorption of a few heavy metals on natural and modified kaolinite and montmorillonite: a review. Adv Colloid Int Sci 140:114–131. https://doi.org/10.1016/j.cis.2007.12.008

Blackburn RS (2004) Natural polysaccharides and their interactions with dye molecules: applications in effluent treatment. Environ Sci Technol 38:4905–4909. https://doi.org/10.1021/es049972n

Blanchard G, Maunaye M, Martin G (1984) Removal of heavy metals from waters by means of natural zeolites. Water Res 18:1501–1507

Blokzijl W, Engberts JBFN (1993) Hydrophobic effects. Opinions and facts. Angew Chem 32:1545–1579. https://doi.org/10.1002/anie.199315451

Boyd GE, Adamson AW, Myers LS (1947) The exchange adsorption of ions from aqueous solutions by organic zeolites. II. Kinetics. J Am Chem Soc 69:2836–2848

Brinza L, Dring MJ, Gavrilescu M (2007) Marine micro and macro algal species as biosorbents for heavy metals. Environ Eng Manag J 6:237–251
Brown PA, Gill SA, Allen SJ (2000) Metal removal from wastewater using peat. Water Res 34:3907–3916
Choy KKH, McKay G, Porter JF (1999) Sorption of acid dyes from effluents using activated carbon. Resour Conserv Recycl 27:57–71. https://doi.org/10.1016/S0921-3449(98)00085-8
Chuah TG, Jumasiah A, Azni I, Katayon S, Choong SYT (2005) Rice husk as a potentially low-cost biosorbent for heavy metal and dye removal: an overview. Desalination 175:305–316. https://doi.org/10.1016/j.desal.2004.10.014
Cooney DO (1999) Adsorption design for wastewater treatment. Lewis Publishers, Boca Raton, 208 p
Cox M, Négré P, Yurramendi L (2007) Industrial liquid effluents. INASMET Tecnalia, San Sebastian, p 283
Crini G (2003) Studies on adsorption of dyes on beta-cyclodextrin polymer. Bioresour Technol 90:193–198. https://doi.org/10.1016/S0960-8524(03)00111-1
Crini G (2005) Recent developments in polysaccharide-based materials used as adsorbents in wastewater treatment. Prog Polym Sci 30:38–70. https://doi.org/10.1016/j.progpolymsci.2004.11.002
Crini G (2006) Non-conventional low-cost adsorbents for dye removal. Bioresour Technol 97:1061–1085. https://doi.org/10.1016/j.biortech.2005.05.001
Crini G (2010) Chapter 2: Wastewater treatment by sorption. In: Sorption processes and pollution. PUFC, Besançon, pp 39–78
Crini G (2014) Review: a history of cyclodextrins. Chem Rev 114:10940–10975. https://doi.org/10.1021/cr500081p
Crini G (2015) Non-conventional adsorbents for dye removal. In: Sharma SK (ed) Green chemistry for dyes removal from wastewater. Scrivener Publishing LLC, Hoboken, pp 359–407
Crini G, Badot PM (2007) Traitement et épuration des eaux industrielles polluées. PUFC, Besançon, 353 p (in French)
Crini G, Badot PM (2008) Application of chitosan, a natural aminopolysaccharide, for dye removal from aqueous solutions by adsorption processes using batch studies: a review of recent literature. Prog Polym Sci 33:399–447. https://doi.org/10.1016/j.progpolymsci.2007.11.001
Crini G, Badot PM (eds) (2010) Sorption processes and pollution. PUFC, Besançon, 489 p
Crini G, Morcellet M (2002) Synthesis and applications of adsorbents containing cyclodextrins. J Sep Sci 25:789–813
Dąbrowski A (2001) Adsorption – from theory to practice. Adv Colloid Int Sci 93:135–224
Dąbrowski A, Podkościelny P, Hubicki Z, Barczak M (2005) Adsorption of phenolic compounds by activated carbon - a critical review. Chemosphere 58:1049–1070. https://doi.org/10.1016/j.chemosphere.2004.09.067
Davis TA, Volesky B, Mucci A (2003) A review of the biochemistry of heavy metal biosorption by brown algae. Water Res 37:4311–4330. https://doi.org/10.1016/S0043-1354(03)00293-8
De Quadros Melo D, De Oliveira Sousa Neto V, De Freitas Barros FC, Raulino GSC, Vidal CB, Do Nascimento RF (2016) Chemical modifications of lignocellulosic materials and their application for removal of cations and anions from aqueous solutions. J Appl Polym Sci 133:1–22. https://doi.org/10.1002/APP.43286
Demirbas A (2008) Heavy metal adsorption onto agro-based waste materials: a review. J Hazard Mater 157:220–229. https://doi.org/10.1016/j.jhazmat.2008.01.024
Devi P, Saroha AK (2017) Utilization of sludge based adsorbents for the removal of various pollutants: a review. Sci Total Environ 578:13–33. https://doi.org/10.1016/j.scitoten.2016.10.220
Dhankhar R, Hooda A (2011) Fungal biosorption – an alternative to meet the challenges of heavy metal pollution in aqueous solutions. Environ Technol 32:467–491. https://doi.org/10.1080/09593330.2011.572922

Dhir B (2014) Potential of biological materials for removing heavy metals from wastewater. Environ Sci Pollut Res 21:1614–1627. https://doi.org/10.1007/s11356-013-2230-8
Dias JM, Alvim-Ferraz MCM, Almeida MF, Rivera-Utrilla J, Sánchez-Polo M (2007) Waste materials for activated carbon preparation and its use in aqueous-phase treatment: a review. J Environ Manag 85:833–846. https://doi.org/10.1016/j.jenvman.2007.07.031
Douven S, Paez CA, Gommes CJ (2015) The range of validity of sorption kinetic models. J Colloid Int Sci 448:437–450. https://doi.org/10.1016/j.jcis.2015.02.053
Dubinin MM (1960) The potential theory of adsorption of gases and vapors for adsorbents with energetically non-uniform surfaces. Chem Rev 60:235–241
Dubinin MM (1966) Porous structure and adsorption properties of activated carbons. In: Walker PL (ed) Chemistry and physics of carbon, vol 2. Marcel Dekker, New-York, pp 51–120
Dubinin MM (1972) Physical adsorption of gases and vapors in micropores. In: Cadenhead DA (ed) Progress in Surface and Membrane Science, vol 9. Academic, New York, pp 1–70
El-Khaiary M (2008) Least-squares regression of adsorption equilibrium data: comparing the options. J Hazard Mater 158:73–87. https://doi.org/10.1016/j.hazmat.2008.01.052
El-Khaiary M, Malash GF (2011) Common data analysis errors in batch adsorption studies. Hydrometallurgy 105:314–320. https://doi.org/10.1016/j.hydromet.2010.11.005
El-Sayed HEM, El-Sayed MMH (2014) Assessment of food processing and pharmaceutical industrial wastes as potential biosorbents: a review. Biomed Res Int 146769:1–25. https://doi.org/10.1155/2014/146769
Elwakeel KZ (2010) Environmental application of chitosan resins for the treatment of water and wastewater: a review. J Dispers Sci Technol 31:273–288. https://doi.org/10.1080/01932690903167178
Emenike PC, Omole DO, Ngene BU, Tenebe IT (2016) Potentiality of agricultural adsorbent for the sequestering of metal ions from wastewater. Global J Environ Sci Manag 2:411–442. https://doi.org/10.22034/gjesm.2016.02.04.010
Farooq U, Kozinski JA, Khan MA, Athar M (2010) Biosorption of heavy metal ions using wheat biosorbents – a review of the recent literature. Bioresour Technol 101:5043–5053. https://doi.org/10.1016/biortech.2010.02.030
Fomina M, Gadd GM (2014) Biosorption: current perspectives on concept, definition and application. Bioresour Technol 160:3–14. https://doi.org/10.1016/j.biortech.2013.12.102
Freundlich HMF (1906) Über die adsorption in lösungen. Z Phys Chem 57:385–471
Furuya EG, Chang HT, Miura Y, Noll KE (1997) A fundamental analysis of the isotherm for the adsorption of phenolic compounds on activated carbon. Sep Purif Technol 11:69–78
Gadd GM (1990) Biosorption. Chem Ind 13:421–426
Gadd GM (2009) Biosorption: critical review of scientific rationale, environmental importance and significance for pollution treatment. J Chem Technol Biotechnol 84:13–28. https://doi.org/10.1002/jctb.1999
Garnham GW (1997) Chapter 2: The use of algae as metal biosorbents. In: Wase J, Forster C (eds) Biosorbents for metal ions. Taylor & Francis Ltd, London, pp 11–37
Gavrilescu M (2004) Removal of heavy metals from the environment by biosorption. Eng Life Sci 4:219–232. https://doi.org/10.1002/elsc.200420026
Gerente C, Lee VKC, Le Cloirec P, McKay G (2007) Application of chitosan for the removal of metals from wastewaters by adsorption – mechanisms and models review. Crit Rev Environ Sci Technol 37:41–127. https://doi.org/10.1080/10643380600729089
Giles CH, Hassan ASA, Subramanian RVR (1958) Adsorption at organic surfaces IV – adsorption of sulphonated azo dyes by chitin from aqueous solution. J Soc Dyers Colour 74:682–688. https://doi.org/10.1111/j.1478-4408.1958.tb02221.x
Grishkewich N, Mohammed N, Tang JT, Tam KC (2017) Recent advances in the application of cellulose nanocrystals. Curr Opin Colloid Int Sci 29:32–45. https://doi.org/10.1016/j.cocis.2017.01.005
Gupta VK, Suhas (2009) Application of low-cost adsorbents for dye removal – a review. J Environ Manag 90:2313–2342. https://doi.org/10.1016/j.jenvman.2008.11.017

Gupta VK, Nayak A, Agarwal S (2015) Bioadsorbents for remediation of heavy metals: current status and their future prospects. Environ Eng Res 20:1–18. https://doi.org/10.4491/eer.2014.018

Hall KR, Eagleton LC, Acrivos A, Vermeulen T (1966) Pore- and solid-diffusion kinetics in fixed-bed adsorption under constant-pattern conditions. Ind Eng Chem Fundam 5:212–223

Hamdaoui O, Naffrechoux E (2007a) Modeling of adsorption isotherms of phenol and chlorophenols onto granular activated carbon. Part I: Two-parameters models and equations allowing determination of thermodynamic parameters. J Hazard Mater 147:381–394. https://doi.org/10.1016/j.hazmat.2007.01.021

Hamdaoui O, Naffrechoux E (2007b) Modeling of adsorption isotherms of phenol and chlorophenols onto granular activated carbon. Part II: Models with more than two parameters. J Hazard Mater 147:401–411. https://doi.org/10.1016/j.hazmat.2007.01.023

Harvey PJ, Campanella BF, Castro PM, Harms H, Lichtfouse E, Schäffner AR, Smrcek S, Werck-Reichhart D (2002) Phytoremediation of polyaromatic hydrocarbons, anilines and phenols. Environ Sci Pollut Res Int 9:29–47

He J, Chen P (2014) A comprehensive review on biosorption of heavy metals by algal biomass: materials, performances, chemistry, and modeling simulation tools. Bioresour Technol 160:67–78. https://doi.org/10.1016/j.biortech.2014.01.068

Ho YS (2003) Removal of copper ions from aqueous solution by tree fern. Water Res 37:2323–2330

Ho YS (2004) Citation review of Lagergren kinetic rate equation on adsorption reactions. Scientometrics 59:171–177. https://doi.org/10.1023/B:SCIE.0000013305.99473.cf

Ho YS (2006) Review of second-order models for adsorption systems. J Hazard Mater 136:681–689. https://doi.org/10.1016/j.jhazmat.2005.12.043

Ho YS, McKay G (1998) A comparison of chemisorption kinetic models applied to pollutant removal on various sorbents. Process Safe Environ Prot 76:332–340

Ho YS, McKay G (2003) Sorption of dyes and copper ions onto biosorbents. Process Biochem 38:1047 1061

Ho YS, Ng JCT, McKay G (2000) Kinetics of pollutant sorption by biosorbents: review. Sep Purif Method 29:189–232

Ho YS, Porter JF, McKay G (2002) Equilibrium isotherm studies for the sorption of divalent metal ions onto peat: copper, nickel and lead single component systems. Water Air Soil Pollut 141:1–33. https://doi.org/10.1023/A:1021304828010

Ho YS, Chiu WT, Wang CC (2005) Regression analysis for the sorption isotherms of basic dyes on sugarcane dust. Bioresour Technol 96:1285–1291. https://doi.org/10.1016/j.biortech.2004.10.021

Houghton JI, Quarmby J (1999) Biopolymers in wastewater treatment. Curr Opin Biotechnol 10:259–262. https://doi.org/10.1016/S0958-1669(99)80045-7

Hubbe MA, Hasan SH, Ducoste JJ (2011) Cellulosic substrates for removal of pollutants from aqueous systems: a review. 1. Metals. Bioresources 6:2161–U2914

Kannan N, Sundaram MM (2001) Kinetics and mechanism of removal of methylene blue by adsorption on various carbons. Dyes Pigments 51:25–40. https://doi.org/10.1016/S0143-7208(01)00056-0

Kaushik P, Malik A (2009) Fungal dye decolourization: recent advances and future potential. Environ Int 35:127–141. https://doi.org/10.1016/j.envint.2008.05.010

Kentish SE, Stevens GW (2001) Innovations in separations technology for the recycling and re-use of liquid waste streams. Chem Eng J 84:149–159

Khalaf MN (2016) Green polymers and environmental pollution control. CRC Press; Apple Academic Press, Inc, Oakville, 436 p

Khan M, Lo IMC (2016) A holistic review of hydrogel applications in the adsorptive removal of aqueous pollutants: recent progress, challenges, and perspectives. Water Res 106:259–271. https://doi.org/10.1016/j.watres.2016.10.008

Kharat DS (2015) Preparing agricultural residue based adsorbents for removal of dyes from effluents – a review. Braz J Chem Eng 32:1–12. https://doi.org/10.1590/0104-6632.20150321s00003020
King CJ (1980) Separation processes, 2nd edn. McGraw-Hill, New York
Kumar J, Balomajumder C, Mondal P (2011) Application of agro-based biomasses for zinc removal from wastewater – a review. Clean Soil Air Water 39:641–652. https://doi.org/10.1002/clen.201000100
Kumar S, Ahlawat W, Bhanjana G, Heydarifard S, Nazhad MM, Dilbaghi N (2014) Nanotechnology-based water treatment strategies. J Nanosci Nanotechnol 14:1838–1858. https://doi.org/10.1166/jnn.2014.9050
Kyzas GZ, Kostoglou M (2014) Green adsorbents for wastewaters: a critical review. Materials 7:333–364. https://doi.org/10.3390/ma7010333
Kyzas GZ, Fu J, Matis KA (2013a) The change from past to future for adsorbent materials in treatment of dyeing wastewaters. Materials 6:5131–5158. https://doi.org/10.3390/ma6115131
Kyzas GZ, Kostoglou M, Lazaridis NK, Bikiaris N (2013b) Chapter 7: Decolorization of dyeing wastewater using polymeric adsorbents – an overview. In: Günay M (ed) Eco-friendly textile dyeing and finishing. Intech, p 177205. https://doi.org/10.5772/52817
Lagergren S (1898) Zur theorie der sogenannten adsorption gelöester stoffe (About the theory of so-called adsorption of soluble substances). Kungliga Svenska Vetenskapsakademiens Handlingar 4:1–39
Landy D, Mallard I, Ponchel A, Monflier E, Fourmentin S (2012a) Cyclodextrins for remediation technologies. In: Lichtfouse E, Schwarzbauer J, Robert D (eds) Environmental chemistry for a sustainable world: nanotechnology and health risk, vol 1. Springer, Berlin, pp 47–81
Landy D, Mallard I, Ponchel A, Monflier E, Fourmentin S (2012b) Remediation technologies using cyclodextrins: an overview. Environ Chem Lett 10:225–237. https://doi.org/10.1007/s10311-011-0351-1
Langmuir I (1916) The constitution and fundamental properties of solids and liquids. Part I. Solids. J Am Chem Soc 38:2221–2295
Langmuir I (1918) The adsorption of gases on plane surfaces of glass, mica and platinum. J Am Chem Soc 40:1361–1403
Largitte L, Pasquier R (2016) A review of the kinetics adsorption models and their application to the adsorption of lead by an activated carbon. Chem Eng Res Des 109:495–504. https://doi.org/10.1016/j.cherd.2016.02.006
Larous S, Meniai AH (2012) The use of sawdust as by product adsorbent of organic pollutant from wastewater: adsorption of phenol. Terragreen 2012: clean energy solutions for sustainable environment. Book series. Energy Procedia 18:905–914. https://doi.org/10.1016/j.egypro.2012.05.105
Li CB, Hein S, Wang K (2008) Biosorption of chitin and chitosan. Mater Sci Technol 24:1088–1099. https://doi.org/10.1179/17438408X341771
Li L, Liu S, Zhu T (2010) Application of activated carbon derived from scrap tires for adsorption of Rhodamine B. J Environ Sci 22:1273–1280. https://doi.org/10.1016/S1001-0742(09)60250-3
Li WW, Zhang HL, Sheng GP, Yu HQ (2015) Roles of extracellular polymeric substances in enhanced biological phosphorus removal process. Water Res 86:85–95. https://doi.org/10.1016/j.watres.2015.06.034
Lim AP, Aris AZ (2014) A review on economically adsorbents on heavy metals removal in water and wastewater. Rev Environ Sci Biotechnol 13:163–181. https://doi.org/10.1007/s11157-013-9330-2
Lin CI, Wang LH (2008) Rate equations and isotherms for two adsorption models. J Chin Insit Chem Eng 39:579–585. https://doi.org/10.1016/j.jcice.2008.04.003
Liu C, Bai R (2014) Recent advances in chitosan and its derivatives as adsorbents for removal of pollutants from water and wastewater. Curr Opin Chem Eng 4:62–70. https://doi.org/10.1016/j.coche.2014.01.004
Liu DHF, Liptak BG (eds) (2000) Wastewater treatment. CRC Press, Boca Raton

Liu Y, Liu YJ (2008) Biosorption isotherms, kinetics and thermodynamics. Sep Purif Technol 61:229–242. https://doi.org/10.1016/j.seppur.2007.10.002

Liu Y, Wang ZW (2008) Uncertainty of preset-order kinetic equations in description of biosorption data. Bioresour Technol 99:3309–3312. https://doi.org/10.1016/j.biortech.2007.06.026

Manes M (1998) Activated carbon adsorption fundamentals. In: Meyers RA (ed) Encyclopedia of environmental analysis and remediation, vol 1. Wiley, New York, pp 26–68

McKay G (1996) Use of adsorbents for the removal of pollutants from wastewaters. CRC Press, Boca Raton, 208 p

McKay G, Ho YS, Ng JCY (1999) Biosorption of copper from waste waters: a review. Sep Purif Methods 28:87–125

Michalak I, Chojnacka K, Witek-Krowiak A (2013) State of the art for the biosorption process – a review. Appl Biochem Biotechnol 170:1389–1416. https://doi.org/10.1007/s12010-013-0269-0

Miretzky P, Cirelli AF (2010) Cr(VI) and Cr(III) removal from aqueous solution by raw and modified lignocellulosic materials: a review. J Hazard Mater 180:1–19. https://doi.org/10.1016/j.jhazmat.2010.04.060

Mittal H, Ray SS, Okamoto M (2016) Recent progress on the design and applications of polysaccharide-based graft copolymer hydrogels as adsorbents for wastewater purification. Macromol Mater Eng 301:496–522. https://doi.org/10.1002/mame.201500399

Mocanu G, Vizitiu D, Carpov A (2001) Cyclodextrin polymers. J Bioact Compat Polym 16:315–342. https://doi.org/10.1106/JJUV-8F2K-JGYF-HNGF

Mohan D, Pittman CU (2007) Arsenic removal from waste/wastewater using adsorbents – a critical review. J Hazard Mater 142:1–53. https://doi.org/10.1016/j.jhazmat.2007.01.006

More TT, Yadav JSS, Yan S, Tyagi RD, Surampalli RY (2014) Extracellular polymeric substances of bacteria and their potential environmental applications. J Environ Manag 144:1–25. https://doi.org/10.1016/j.jenvman.2014.05.010

Moreno-Castilla C, Ferro-García MA, Joly JP, Bautista-Toledo I, Carrasco-Marín F, Rivera-Utrilla J (1995) Activated carbon surface modifications by nitric acid, hydrogen peroxide, and ammonium peroxydisulfate treatments. Langmuir 11:4386–4392

Morin-Crini N, Crini G (eds) (2017) Eaux industrielles contaminées. PUFC, Besançon, 513 p (in French)

Morin-Crini N, Winterton P, Fourmentin S, Wilson LD, Fenyvesi E, Crini G (2017) Water-insoluble β-cyclodextrin-epichlorohydrin polymers for removal of pollutants from aqueous solutions by sorption processes using batch studies: a review of inclusion mechanism. Prog Polym Sci 78:1. https://doi.org/10.1016/j.progpolymsci.2017.07.004

Mudhoo A, Garg VK, Wang SB (2012) Removal of heavy metals by biosorption. Environ Chem Lett 10:109–117. https://doi.org/10.1007/s10311-011-0342-2

Mui ELK, Ko DCK, McKay G (2004) Production of active carbons from waste tyres – a review. Carbon 42:2789–2805. https://doi.org/10.1016/j.carbon.2004.06.023

Muya FN, Sunday CE, Baker P, Iwuoha E (2016) Environmental remediation of heavy metal ions from aqueous solution through hydrogel adsorption: a critical review. Water Sci Technol 73:983–992. https://doi.org/10.2166/wst.2015.567

Naja G, Volesky B (2011) The mechanism of metal cation and anion biosorption. In: Kotrba P, Mackova M, Macek T (eds) Microbial biosorption of metals. Springer, Dordrecht, pp 19–58. https://doi.org/10.1007/978-94-007-0443-5_3

Ngah WSW, Hanafiah MAKM (2008) Removal of heavy metal ions from wastewater by chemically modified plant wastes as adsorbents: a review. Bioresour Technol 99:3945–3948. https://doi.org/10.1016/j.biortech.2007.06.011

Ngulube T, Gumbo JR, Masindi V, Maity A (2017) An update on synthetic dyes adsorption onto clay based minerals: a state-of-art review. J Environ Manag 191:35–57. https://doi.org/10.1016/j.jenvman.2017.12.031

Nguyen TAH, Ngo HH, Guo WS, Zhang J, Liang S, Yue QY, Li Q, Nguyen TV (2013) Applicability of agricultural waste and by-products for adsorptive removal of heavy metals from wastewater. Bioresour Technol 148:574–585. https://doi.org/10.1016/j.biortech.2013.08.124

Nharingo T, Moyo M (2016) Application of Opuntia ficus-indica in bioremediation of wastewaters. A critical review. J Environ Manag 166:55–72. https://doi.org/10.1016/j.jenvman.2015.10.005

No HK, Meyers SP (2000) Application of chitosan for treatment of wastewaters. Rev Environ Contam Toxicol 63:1–28. https://doi.org/10.1007/978-1-4757-6429-1_1

O'Connell DW, Birkinshaw C, O'Dwyer TF (2008) Heavy metal adsorbents prepared from the modification of cellulose: a review. Bioresour Technol 99:6709–6724. https://doi.org/10.1016/j.biortech.2008.01.036

Oliveira LS, Franca AS (2008) Low cost adsorbents from agro-food wastes. In: Greco LV, Bruno MN (eds) Food science and technology: new research. Nova Publishers, New York, pp 1–39

Ong ST, Keng PS, Lee SL, Hung YT (2014) Low cost adsorbents for sustainable dye containing-wastewater treatment. Asian J Chem 26:1873–1881. https://doi.org/10.14233/ajchem.2014.15653

Panic VV, Seslija SI, Nesic AR, Velickovic SJ (2013) Adsorption of azo dyes on polymer materials. Hemijska Industija 67:881–900. https://doi.org/10.2298/HEMIND121203020P

Park D, Yun YS, Park JM (2010) The past, present, and future trends of biosorption. Biotechnol Bioprocess Eng 15:86–102. https://doi.org/10.1007/s12257-009-0199-4

Patel S (2012) Potential of fruit and vegetable wastes as novel biosorbents: summarizing the recent studies. Rev Environ Sci Biotechnol 11:365–380. https://doi.org/10.1007/s11157-012-9297-4

Pereira MFR, Soares SF, Órfão JJM, Figueiredo JL (2003) Adsorption of dyes on activated carbons: influence of surface chemical groups. Carbon 41:811–821. https://doi.org/10.1016/S0008-6223(02)00406-2

Plazinski W, Rudzinski W, Plazinska A (2009) Theoretical models of sorption kinetics including a surface reaction mechanism: a review. Adv Colloid Int Sci 152:2–13. https://doi.org/10.1016/j.cis.2009.07.009

Polanyi M (1920) Neueres über adsorption und ursache der adsorptionkräfte. Z Elektrochem 26:370–374

Pollard SJT, Fowler GD, Sollars CJ, Perry R (1992) Low-cost adsorbents for waste and wastewater treatment: a review. Sci Total Environ 116:31–52. https://doi.org/10.1016/0048-9697(92)90363-W

Qu J (2008) Research progress of novel adsorption processes in water purification: a review. J Environ Sci 20:1–13. https://doi.org/10.1016/S1001-0742(08)60001-7

Radovic LR, Silva IF, Ume JI, Menéndez JA, Leon Y Leon CA, Scaroni AW (1997) An experimental and theoretical study of the adsorption of aromatics possessing electron-withdrawing and electron-donating functional groups by chemically modified activated carbons. Carbon 35:1339–1348

Radovic LR, Moreno-Castilla C, Rivera-Utrilla J (2000) Carbon materials as adsorbents in aqueous solutions. Chem Phys Carbon 27:227–405

Radushkevich LV (1949) Potential theory of sorption and structures of carbons. Zhurnal Fizicheskoi Khimii 23:1410–1420

Rafatullah M, Sulaiman O, Hashim R, Ahmad A (2010) Adsorption of methylene blue on low-cost adsorbents: a review. J Hazard Mater 177:70–80. https://doi.org/10.1016/j.jhazmat.2009.12.047

Ramakrishna KR, Viraraghavan T (1997) Dye removal using low cost adsorbents. Water Sci Technol 36:189–196. https://doi.org/10.1016/S0273-1223(97)00387-9

Ramrakhiani L, Ghosh S, Majumdar S (2016) Surface modification of naturally available biomass for enhancement of heavy metal removal efficiency, upscaling prospects, and management aspect of spent biosorbents: a review. Appl Biochem Biotechnol 180:41–78. https://doi.org/10.1007/s12010-016-2083-y

Rangabhashiyam S, Suganya E, Selvaraju N, Varghese LA (2014) Significance of exploiting non-living biomaterials for the biosorption of wastewater pollutants. World J Microbiol Biotechnol 30:1669–1689. https://doi.org/10.1007/s11274-014-1599-y

Raval NP, Shah PU, Shah NK (2016) Adsorptive removal of nickel(II) ions from aqueous environment: a review. J Environ Manag 179:1–20. https://doi.org/10.1016/j.jenvman.2016.04.045

Redlich O, Peterson DL (1959) A useful adsorption isotherm. J Phys Chem 63:1024

Reichenberg D (1953) Properties of ion-exchange resins in relation to their structure. III. Kinetics. J Am Chem Soc 75:589–594

Robalds A, Naja GM, Klavins M (2016) Highlighting inconsistencies regarding metal biosorption. J Hazard Mater 304:553–556. https://doi.org/10.1016/j.hazmat.2015.10.042

Rudzinski W, Plazinski W (2009) On the applicability of the pseudo-second order equation to represent the kinetics of adsorption at solid/solution interfaces: a theoretical analysis based on the statistical rate theory. Adsorption 15:181–192. https://doi.org/10.1007/s10450-009-9167-8

Saba B, Christy AD, Jabeen M (2016) Kinetic and enzymatic decolorization of industrial dyes utilizing plant-based biosorbents: a review. Environ Eng Sci 33:601–614. https://doi.org/10.1089/ees.2016.0038

Sadegh H, Ali GAM, Gupta VK, Makhlouf ASH, Shahryari-Ghoshekandi R, Nadagouda MN, Sillanpää M, Megiel E (2017) The role of nanomaterials as effective adsorbents and their applications in wastewater treatment. J Nanostruct Chem 7:1–14. https://doi.org/10.1007/s40097-017-0219-4

Sahmoune MN, Yeddou AR (2016) Potential of sawdust materials for the removal of dyes and heavy metals: examination of isotherms and kinetics. Desalin Water Treat 57:24019–24034. https://doi.org/10.1080/19443994.2015.1135824

San Miguel G, Lambert SD, Graham NJD (2006) A practical review of the performance of organic and inorganic sorbents for the treatment of contaminated waters. J Chem Technol Biotechnol 81:1685–1696

Sanghi R, Verma P (2013) Decolorisation of aqueous dye solutions by low-cost adsorbents: a review. Coloration Technol 129:85–108. https://doi.org/10.1111/cote.12019

Sen A, Pereira H, Olivella MA, Villaescusa I (2015) Heavy metals removal in aqueous environments using bark as a biosorbent. Int J Environ Sci Technol 12:391–404. https://doi.org/10.1007/s13762-014-0525-z

Sharma SK (2015) Green chemistry for dyes removal from wastewater. Scrivener Publishing LLC Wiley, Beverley, 496 p

Sharma P, Kaur H, Sharma M, Sahore V (2011) A review on applicability of naturally available adsorbents for the removal of hazardous dyes from aqueous waste. Environ Monit Assess 183:151–195. https://doi.org/10.1007/s10661-011-1914-0

Shukla A, Zhang YH, Dubey P, Margrave JL, Shukla SS (2002) The role of sawdust in the removal of unwanted materials from water. J Hazard Mat B95:137–152

Sips R (1948) On the structure of a catalyst surface. J Chem Phys 16:490–495

Solis M, Solis A, Perez HI, Manjarrez N, Flores M (2012) Microbial decolouration of azo dyes: a review. Process Biochem 47:1723–1748. https://doi.org/10.1016/j.procbio.2012.08.014

Song DI, Won SS (2005) Three-parameter empirical isotherm model: its application to sorption onto organoclays. Environ Sci Technol 39:1138–1143. https://doi.org/10.1021/es048800n

Srivastava S, Goyal P (2010) Novel biomaterials. Decontamination of toxic metals from wastewater. Springer, New York, 140 p. https://doi.org/10.1007/978-3-642-11329-1

Srivastava S, Agrawal SB, Mondal MK (2015) A review on progress of heavy metal removal using adsorbents of microbial and plant origin. Environ Sci Pollut Res 22:15386–15415. https://doi.org/10.1007/s11356-015-5278-9

Streat M, Patrick JW, Perez MJC (1995) Sorption of phenol and para-chlorophenol from water using conventional and novel activated carbons. Water Res 29:467–472. https://doi.org/10.1016/0043-1354(94)00187-C

Sud D, Mahajan G, Kaur MP (2008) Agricultural waste material as potential sorbent for sequestering heavy metal ions from aqueous solutions – a review. Bioresour Technol 99:6017–6027. https://doi.org/10.1016/j.biortech.2007.11.064

Sudha S, Giri Dev VR (2007) Low cost-sorbents – an overview. Synth Fibre 36:5–9

Sulyman M, Namiesnik J, Gierak A (2017) Low-cost adsorbents derived from agricultural by-products/wastes for enhancing contaminant uptakes from wastewater: a review. Pol J Environ Stud 26:479–510. https://doi.org/10.15244/pjoes/66769

Swami D, Buddhi D (2006) Removal of contaminants from industrial wastewater through various non-conventional technologies: a review. Int J Environ Poll 27:324–346. https://doi.org/10.1504/IJEP.2006.010576

Tang X, Zhang X, Zhou A (2007) Research progresses on adsorbing heavy metal ions with crosslinked chitosan. Ion Exch Adsorpt 23:378–384

Tien C (1994) Adsorption calculations and modeling. Butterworth-Heinemann College, Newton, 288 p

Toth J (1971) State equations of the solid gas interface layer. Acta Chemical Academia Science Hungaria 69:311–317

Treybal RE (1987) Mass transfer operations. McGraw-Hill, New York, 800 p

Vakili M, Rafatullah M, Salamatinia B, Abdullah AZ, Ibrahim MH, Tan KB, Gholami Z, Amouzgar P (2014) Application of chitosan and its derivatives as adsorbents for dye removal from water and wastewater: a review. Carbohydr Polym 113:115–130. https://doi.org/10.1016/j.carbpol.2014.07.007

Vandenbossche M, Jimenez M, Casetta M, Traisnel M (2015) Remediation of heavy metals by biomolecules: a review. Crit Rev Environ Sci Technol 45:1644–1704. https://doi.org/10.1080/10643389.2014.966425

Varma AJ, Deshpande SV, Kennedy JF (2004) Metal complexation by chitosan and its derivatives: a review. Carbohydr Polym 55:77–93. https://doi.org/10.1016/j.carbpol.2003.08.005

Veglio' F, Beolchini F (1997) Removal of metals by biosorption: a review. Hydrometallurgy 44:301–316. https://doi.org/10.1016/S0304-386X(96)00059-X

Vijayaraghavan K, Balasubramanian R (2015) Is biosorption suitable for decontamination of metal-bearing wastewaters? A critical review on the state-of-the-art of biosorption processes and future directions. J Environ Manag 160:283–296. https://doi.org/10.1016/j.jenvman.2015.06.030

Vijayaraghavan K, Yun YS (2008) Bacterial biosorbents and biosorption. Biotechnol Adv 26:266–291. https://doi.org/10.1016/j.biotechadv.2008.02.002

Vohla C, Koiv M, Bavor HJ, Chazarenc F, Mander U (2011) Filter materials for phosphorus removal from wastewater in treatment wetlands – a review. Ecol Eng 37:70–89. https://doi.org/10.1016/j.ecoleng.2009.08.003

Volesky B (1990) Biosorption of metals. CRC Press, Boca Raton, 408 p

Volesky B (2001) Detoxification of metal-bearing effluents: biosorption for the next century. Hydrometallurgy 59:203–216. https://doi.org/10.1016/S0304-386X(00)00160-2

Volesky B (2004) Sorption and biosorption. BV-Sorbex, Inc, Montreal, 316 p

Volesky B (2007) Biosorption and me. Water Res 41:4017–4029. https://doi.org/10.1016/j.watres.2007.05.062

Volesky B, Holan ZR (1995) Biosorption of heavy metals. Biotechnol Prog 11:235–250. https://doi.org/10.1021/bp00033a001

Wan Ngah WS, Hanafiah MAKM (2008) Removal of heavy metal ions from wastewater by chemically modified plant wastes as sorbents: a review. Bioresour Technol 99:3935–3948. https://doi.org/10.1016/j.biortech.2007.06.011

Wang J, Chen C (2009) Biosorbents for heavy metals and their future. Biotechnol Adv 27:195–226. https://doi.org/10.1016/j.biotechadv.2008.11.002

Wang SB, Peng YL (2010) Natural zeolites as effective adsorbents in water and wastewater treatment. Chem Eng J 156:11–24. https://doi.org/10.1016/j.cej.2009.10.029

Wang S, Ang HM, Tadé MO (2008) Novel applications of red mud as coagulant, adsorbent and catalyst for environmentally benign processes. Chemosphere 72:1621–1635. https://doi.org/10.1016/j.chemosphere.2008.05.013

Wang X, Zhu CS, Chen CZ, Zhang H, Wang D, Ma YY (2013) Recent developments in treatment of chromium-contaminated wastewater by starch-based absorbents. Adv Chem Res 781-784:2120–2123. https://doi.org/10.4028/www.scientific.net/AMR.781-784.2120

Wase J, Forster C (1997) Biosorbents for metal ions. Taylor & Francis, Bristol, 249 p

Weber WJ, Morris JC (1963) Kinetics of adsorption on carbon from solution. J Sanit Eng Div 89:31–60

Weber WJ, Asce AM, Morris JC (1963) Kinetics of adsorption on carbon from solution. J Sanit Eng Div Proc Am Soc Civil Eng SA2:31–59

Won SW, Han MH, Yun YS (2008) Different binding mechanism in biosorption of reactive dyes according to their reactivity. Water Res 42:4847–4855. https://doi.org/10.1016/j.watres.2008.09.003

Wu FC, Tseng RL, Huang SC, Juang RS (2009) Characteristics of pseudo-order kinetic model liquid-phase adsorption: a mini-review. Chem Eng J 151:1–9. https://doi.org/10.1016/j.cej.2009.02.024

Yang TR (2003) Adsorbents: fundamentals and applications. Wiley-Interscience, Hoboken, 424 p

Yong SK, Shrivastava M, Srivastava P, Kunhikrishnan A, Bolan N (2015) In: Whitacre DM (ed) Environmental applications of chitosan and its derivatives. Book series: Reviews of environmental contamination and toxicology, vol 233. Springer, Cham, pp 1–43. https://doi.org/10.1007/978-3-319-10479-9_1

Zeraatkar AK, Ahmadzadeh H, Talebi AF, Moheimani NR, McHenry MP (2016) Potential use of algae for heavy metal bioremediation, a critical review. J Environ Manag 181:817–831. https://doi.org/10.1016/j.jenvman.2016.06.059

Zhao XJ, Zhou ZQ (2016) Synthesis and applications of pectin-based nanomaterials. Curr Nanosci 12:103–109. https://doi.org/10.2174/1573413711666150818224020

Zhou Y, Zhang L, Cheng ZJ (2015) Removal of organic pollutants from aqueous solution using agricultural wastes: a review. J Mol Liq 212:739–762. https://doi.org/10.1016/j.molliq.2015.10.023

Chapter 3
A Systematic Analysis and Review of the Fundamental Acid-Base Properties of Biosorbents

Pablo Lodeiro, María Martínez-Cabanas, Roberto Herrero, José L. Barriada, Teresa Vilariño, Pilar Rodríguez-Barro, and Manuel E. Sastre de Vicente

Contents

P. Lodeiro
GEOMAR Helmholtz Centre for Ocean Research Kiel, Kiel, Germany

M. Martínez-Cabanas · R. Herrero · J. L. Barriada · T. Vilariño · P. Rodríguez-Barro
M. E. Sastre de Vicente (✉)
Departamento de Química, Universidade da Coruña, A Coruña, Spain
e-mail: eman@udc.es

G. Crini, E. Lichtfouse (eds.), *Green Adsorbents for Pollutant Removal*, Environmental Chemistry for a Sustainable World 18,
https://doi.org/10.1007/978-3-319-92111-2_3

Abstract A broad variety of materials of biological origin have been successfully used in recent decades for the removal of pollutants from solution. These biosorbents present a range of natural polymers that play a key role on their adsorption capacity. It is therefore critical to understand the physicochemical properties of the chemical groups that form these polymers. According to bibliography, less than 3% of biosorption papers include studies on proton binding. The acid-base properties of biomass are affected by pH, ionic strength and medium composition. Nevertheless, these crucial parameters are not always considered during biosorption studies. This review outlines the major advances on proton binding data interpretation and modelling on biosorbents. In addition, we propose some experimental considerations that cover all issues raised in this review concerning the acid-base properties of biosorbents. Only 30% of the reviewed papers that study algae, agricultural wastes or lignocellulosic materials use Donnan or double-layer surface models to account for electrostatic interactions on proton binding. Expressions for activity coefficients, such as Debye-Hückel or Pitzer equations, are shown only in c.a. 15% of these papers. Moreover, studies investigating a range of ionic strengths represent a 40%, while this variable is not even considered in 20% of the papers. We could not find any biosorption study related to specific salt or Hofmeister effects. Moreover, in 6 out of 10 papers there is important experimental information missing such as the calibration of the electrodes. We consider therefore that there is an important need for reviewing the role of proton binding on biosorption studies.

Symbols

A Specific surface area

CA Condensation approximation

C_a Concentration of acid solution

C_A Total concentration of acidic sites in a biosorbent

$c_{\pm}$ Ionic distribution at an interface

E^* Formal potential

emf, E Electromotive force

EXAFS Extended X-ray absorption fine structure

F Faraday constant, 96485.34 C mol^{-1}

f Weighted sum of local isotherms

FTIR Fourier transform infrared spectroscopy

$(H^+)_0$ Local proton ion activity at the binding site, e.g. surface proton ion activity

$[H^+]_0$ Proton ion concentration at the local binding site

(H^+) Experimental accessible bulk activity of the proton ion

I Ionic strength

K^* Stoichiometric proton dissociation constant

K_{app}	Apparent conditional dissociation constant
K_{int}	Intrinsic dissociation constant
K^T	Thermodynamic proton dissociation constant
LFER	Linear free energy relationships
MSA	Mean spherical approximation
NICA	Non-ideal competitive and thermodynamically consistent adsorption
NMR	Nuclear magnetic resonance
NOM	Natural organic matter
pK_m	Empirical, ionic strength dependent pK
$p(K)$	Affinity spectrum
pzc	Potential of zero charge
Q	Charge of a species net charge
$Q(\gamma_i)$	Ratio of activity coefficients of the species in the equilibrium
SCM	Surface complexation model
SIT	Specific interaction theory
$U_\pm$	Dispersion-dependent energy term
v	Volume added in a titration
V_D	Active Donnan volume
V_0	Initial volume
WHAM	Windermere humic aqueous model
XANES	X-ray absorption near edge structure
X_i	Co-ions and counterions
z_i	Charge of a species
α	Degree of dissociation
$\gamma_\pm$	Mean ionic activity coefficient
γ_{eff}	Effective activity coefficient
γ_i	Activity coefficient of species i
ΔG	Gibbs free energy
ΔG^{AB}	Lewis acid-base contribution to Gibbs energy
ΔG_{ads}	Gibbs free energy of adsorption
ΔG_{diss}	Gibbs free energy in a dissociation equilibrium
ΔG_{elec}	Electrostatic Gibbs free energy
ΔG_{int}	Intrinsic Gibbs free energy
ΔG^{LW}	Lifshitz-van der Waals contribution to Gibbs energy
$\Delta G_{non-elec}$	Non-electrostatic Gibbs free energy
θ	Coverage fraction of binding sites
ρ_0	Charge in the region occupied by the biosorbent in the absence of mobile ions
σ	Charge density
Ψ	Electrostatic potential
Ψ_0	Electrostatic potential at the binding site

3.1 Introduction

The removal of pollutants (e.g. heavy metals, phenols, dyes, endocrine disruptors, etc.) from contaminated waters is an issue of current concern. In 2015 the percentage of untreated wastewaters in high- and low-income countries was c.a. 30 and 92%, respectively (Koncagül et al. 2017). The Convention for the Protection of the Marine Environment of the North-East Atlantic showed time trends for metals and organic contaminants in Europe during the period 2006–2011. For example, Cu, Zn and Pb show increasing trends in the majority of areas sampled; Hg and As trends are mainly descendent, while Cd does not show a clear trend (Larsen and Fryer 2012). The contribution of EU industry to pollutant emissions to water is substantial, with e.g. 36,875 kg for Cd, 498.2 tons for Pb and 1028.5 tons for phenols in 2010 (Lawton et al. 2014).

The search for an efficient, affordable and easy-to-handle technology has produced a potential alternative to traditional wastewater treatments such as ionic exchange or precipitation: that is the use of green adsorbents. These materials of biological origin are usually referred to as biosorbents, and the technique involving their use for pollutant removal from waters, biosorption.

Different representative definitions of biosorption with variable implications have been proposed in the last decade. Bohumil Volesky (2007) reported a broad definition in 2007 when biosorption was already a well-established technique: "*Biosorption has been defined as the property of certain biomolecules (or types of biomass) to bind and concentrate selected ions or other molecules from aqueous solutions. As opposed to a much more complex phenomenon of bioaccumulation based on active metabolic transport, biosorption by dead biomass (or by some molecules and/or their active groups) is passive and based mainly on the "affinity" between the (bio-) sorbent and sorbate*". B. Volesky included in his definition not only biomass but also biomolecules. Therefore, according to B. Volesky the accumulation of contaminants by polymers of biological origin, such as alginate or chitosan, should be considered as biosorption.

Later on Geoffrey M. Gadd (2009) pointed towards a wider biosorption definition including living and dead biomass: "*Thus, the term biosorption can describe any system where a sorbate (e.g. an atom, molecule, a molecular ion) interacts with a biosorbent (i.e. a solid surface of a biological matrix) resulting in an accumulation at the sorbate – biosorbent interface, and therefore a reduction in the solution sorbate concentration*". In contrast to the definitions generally suggested in literature (Volesky 2007; Michalak et al. 2013) both, passive and active accumulation of sorbates are implicitly included in Gadd's description.

More recently, I. Michalak et al. (2013) have proposed another representative definition of biosorption: "*Biosorption is a subcategory of adsorption, where the sorbent is a biological matrix. Biosorption is a process of rapid and reversible binding of ions from aqueous solutions onto functional groups that are present on the surface of biomass. This process is independent of cellular metabolism*". The distinction between biosorption and adsorption proposed by Michalak et al.

constitutes a novel aspect. Nevertheless, it is worth mentioning that biosorption is not only driven by adsorption but many other mechanisms such as ionic exchange, redox reactions, etc.

We suggest here a simple definition of biosorption as the removal of pollutants from solution using dead, non-metabolically active, biomass of biological origin (namely biosorbents), which preserve its pristine active chemical structure or active groups. We therefore consider that if the biomass is significantly processed, for example when preparing biochars by pyrolysis, the accumulation of sorbates involving those materials should not be defined as biosorption. Therefore, only papers that fit in the proposed definition of biosorption are reviewed here.

A broad variety of materials have been used in biosorption studies in recent decades (De Gisi et al. 2016): e.g. algae (Davis et al. 2003), bacteria (Vijayaraghavan and Yun 2008a; Gupta and Diwan 2017), fungi (Kapoor and Viraraghavan 1995), agricultural by-products (Bhatnagar and Sillanpää 2010) or chitin and wood derivatives (Gerente et al. 2007; Abdolali et al. 2014). The biosorbents present a heterogeneous matrix that constitutes the biomass structure, and is formed of polysaccharides (Crini 2005). The excellent pollutant adsorption capacity reported for many biosorbents results from the presence of specific chemical groups in their biopolymers chains. It is therefore critical to understand the physicochemical properties of these natural polymers present in the biomass used in biosorption studies.

The knowledge about the biosorption mechanism is the base for proposing models, process understanding and experimental optimization. The biosorption mechanism is complex due to the heterogeneity and structure of the biosorbents (Aksu 2005; Robalds et al. 2016). Different chemical active compounds, such as carboxyl, hydroxyl, sulfonate, acetamide or amino groups are present in the polysaccharides that form the structure of biosorbents (Volesky 2003). Those chemical groups are responsible for pollutant removal in solution. To understand the chemical nature of the biosorption process it is necessary to determine which chemical groups on the biosorbent are involved in the binding of different contaminants. The removal of contaminants from solution depends on the affinity between the binding sites and the pollutants (namely, the specific equilibrium constants), the availability of the binding sites, i.e. chemical state of the groups, their quantity and accessibility (Schiewer and Volesky 2000). Therefore, the biosorption mechanism is also influenced by factors such as the pH, temperature and solution composition, or the concentration and type of contaminant. The dominant biosorption mechanisms usually proposed are: ionic exchange (Schiewer 1999), chemical binding (ionic and covalent) (Schiewer 1999) and redox reactions (Lopez-Garcia et al. 2013). Other phenomena, such as physical binding (electrostatic and van der Waals forces) or microprecipitation, may also have some contribution (Fiol and Villaescusa 2009). Those mechanisms can act in combination, with different contribution degrees, and even one binding site can participate in different binding mechanisms. For example, using algae as biosorbents, the proton binding has been considered to have a dominant covalent bound character at ionic strength > 0.1 mol kg^{-1} and above pH 3, while some electrostatic effects are noticed at lower ionic strength and pH < 3 (Rey-Castro et al. 2003). Protons and covalently bound contaminants (e.g. heavy metals) compete for the same binding sites. This competition, together with pollutant speciation, makes the solution pH a key parameter in biosorption studies. In addition, the binding of contaminants to biosorbents can be largely influenced by charge behaviour, also regulated by the solution pH (Dewit et al. 1993). When biosorbents are fully protonated many chemical groups (carboxyl, hydroxyl, sulfonate, etc.) present no charge, while those groups are negatively charged when deprotonated. At pH values higher than the pK of the binding groups, they can attract positively charged species in solution. On the contrary, groups such as amine, amide or imidazole are positively charge when protonated and neutral when deprotonated, so at pH values lower than the pK of those chemical groups the attraction of negatively charged species is favoured.

Since protons are always present in solution, the study of proton binding to biosorbents and its dependence on pH, ionic strength and medium composition should constitute the first step in any biosorption study (Fig. 3.1).

Besides, competitive adsorption effects and biomass adsorption capacity are also related to the acid-base properties of the biosorbent. Despite its critical importance, few authors have included a systematic investigation of the acid-base properties of biomass in their biosorption studies (Schiewer and Volesky 1997a; Ravat et al. 2000; Bouanda et al. 2002; Rey-Castro et al. 2003; Li and Englezos 2005; Martín-Lara et al. 2008; Vilar et al. 2009; Liu et al. 2013). A simple bibliographic search using Scopus and Web of Science databases shows that < 3% of the total peer-reviewed literature contain any of the key words related to acid-base studies when searching for "biosorption" or "biosorbent" (Fig. 3.2).

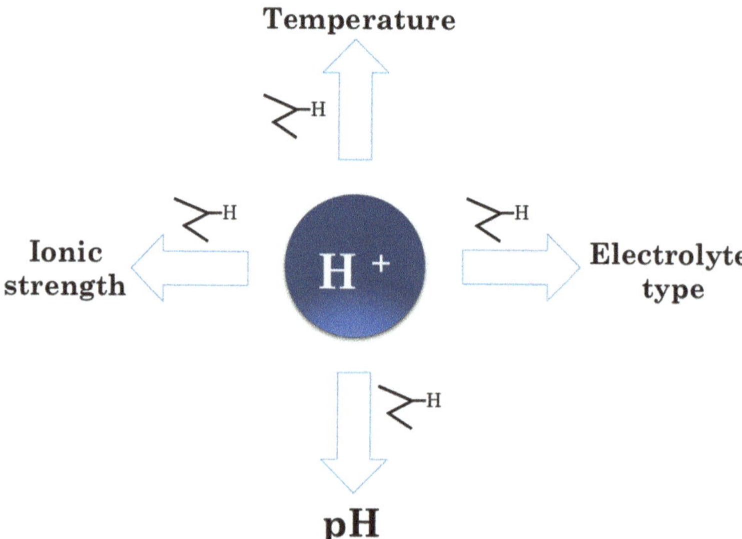

Fig. 3.1 Schematic representation of the main variables that affect proton binding to bioadsorbents. That is, ionic strength, pH, medium composition or electrolyte type, and temperature. In order to properly understand dissociation/binding reactions at constant temperature, different experiments at several pHs, ionic strengths and/or electrolyte types should be performed

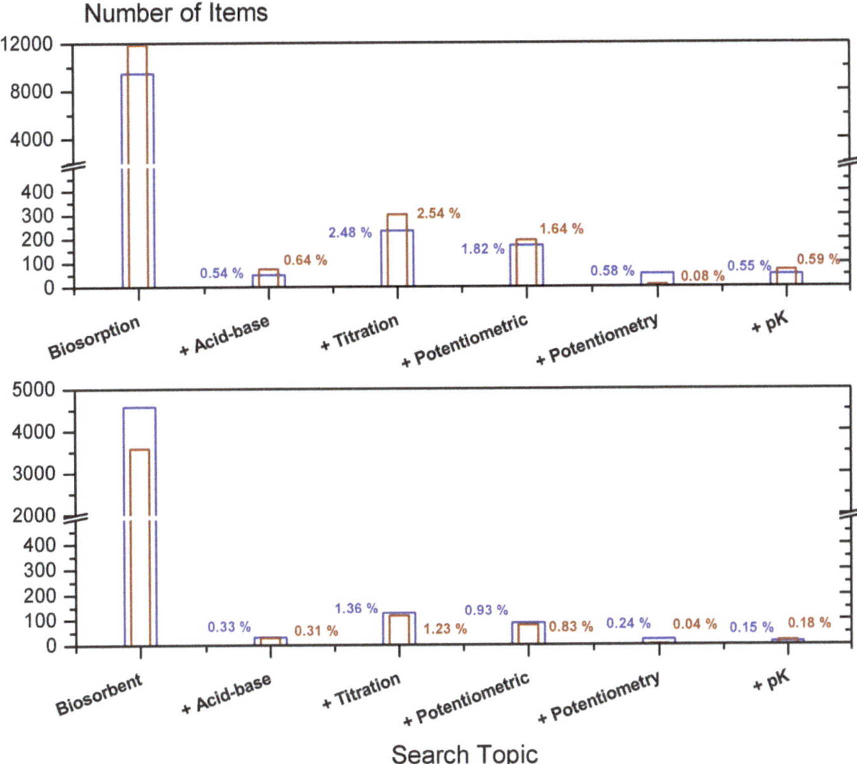

Fig. 3.2 Total bibliography search scores (January 2018) using Scopus (blue bars) and Web of Science (red bars) databases for "biosorption " and "biosorbent", containing any of the key words related to acid-base studies: acid-base, titration, potentiometric, potentiomery or pK. The figure shows that less than 3% of biosorption papers include studies on proton binding

The experimental approach needed for studying proton binding to biosorbents is simple and readily accessible. A potentiometric titration, where the biomass is titrated using a base or acid in a neutral electrolyte solution, constitutes the basics to determine the acidic properties of functional groups in biosorbents. Two crucial parameters can be obtained from potentiometric titrations of biosorbents: the maximum proton-exchange capacity, and the equilibrium dissociation constants of the chemical groups. The former measures the concentration of ionizable functional groups, usually referred to as active binding sites. It is therefore possible to calculate the abundance of potential binding chemical groups involved on pollutant adsorption. As proton binding is predominantly covalent, only other compounds with more covalent character (e.g. heavy metals) will displace the protons from the chemical groups. Therefore, the number of acidic groups obtained from a potentiometric titration is usually considered a potential maximum of binding sites for contaminants, such as metals, endocrine disruptors or dyes. The other key parameter that can be easily obtained from potentiometric titrations, the proton equilibrium dissociation constant, provides the nature of the active chemical sites in the biosorbent. The biosorbents present a high heterogeneity with a range of dissociation constants of their chemical groups (Volesky 2003). This chemical heterogeneity can be also calculated modelling the potentiometric titrations. Nevertheless, the heterogeneity together with the similarity in the equilibrium constant of the chemical groups, make the accurate determination of specific contributions from each binding site challenging (Dewit et al. 1993).

An alternative to the experimental determination of proton binding constants consists in their semi-empirical estimation using Linear Free Energy Relationships (LFER) (Matynia et al. 2010). This methodology has been used for natural organic matter (e.g. fulvic and humic compounds), including the determination of metal-ligand constants (Carbonaro et al. 2011).

In order to ensure that the chemical groups are occupied by protons and not by light metals (Na, K, Mg, Ca), the biosorbents should be fully protonated before performing a potentiometric titration. Light metals are present in natural waters, industrial runoffs and sewage treatment plant effluents. The presence of these elements in solution then influences the binding of other species (e.g. pollutants) competing for the same chemical sites. Typical "hard" counterions (Na, K, Mg, Ca) form electrostatic bounds with negatively charged chemical groups, and reduce the local concentration of other ions (e.g. protons and metals) until convergence with their concentration in the bulk solution. The electrostatically bound counterions cannot displace covalently bound ions, but can reduce their local concentration, and then also decrease the covalent binding. The ionic strength (I), a function of the concentration and charge of ions in solution, is therefore another key parameter, together with pH, to consider during biosorption studies. A medium of constant ionic strength is required to perform the potentiometric titration of biosorbents. The ionic strength does not influence the number of acidic groups obtained from an acid-base titration, but it strongly affects the apparent proton binding constant values of those chemical groups. Besides, the medium composition should also be considered during the calibration of the pH electrodes used on titration studies.

The main aim of this review is to characterize the proton binding equilibria, as an extremely important and preliminary step, for a correct interpretation of biosorption results. We first analyze the basics of the acid-base properties of simple substances in saline solutions as starting point for a proper interpretation of the more complex physicochemical behaviour of polyelectrolytes and biosorbents. We evaluate the role of key parameters such as the ionic strength, pH or medium composition. Following, we show and discuss what has been done so far regarding acid-base characterisation of biosorbents and which models have been commonly used to describe the proton binding equilibria. Moreover, the role that acid-base properties of biosorbents play on pollutant removal is also discussed. Finally, we propose some experimental considerations for future works that cover all issues raised in this review concerning the acid-base studies of biosorbents.

It is worth mentioning that for the sake of simplicity most of the discussions and analysis shown in this review are focused on the acid-base properties of biosorbents and their implications for metal biosorption. Nevertheless, proton binding also influences the biosorption of other pollutants such as organic compounds. The basic interaction principles are similar for either metals or organic substances. Moreover, most of the theoretical and practical considerations described in this review allows for the description of the biosorption of any pollutant. However, the extrapolation of the results and models considered here to pollutants others than metals should be considered cautiously.

3.2 Acid-Base Properties in Solution: pH, Ionic Strength and Medium Composition as Relevant Variables

Organic functional groups are part of the polysaccharides that form the structure of the biosorbents. The study of the acid-base properties of these simple organic compounds is therefore of great importance.

Considering the dissociation of an acid, AH, in aqueous solution (Eq. 3.1) different operational equilibrium constants can be defined:

$$AH \leftrightarrows A^- + H^+ \tag{3.1}$$

$$K^T = \frac{(A^-)(H^+)}{(AH)} = \frac{[A^-][H^+]}{[AH]} \frac{\gamma_{A^-}\gamma_{H^+}}{\gamma_{AH}} = K^* \frac{\gamma_{A^-}\gamma_{H^+}}{\gamma_{AH}} \tag{3.2}$$

where K^T and K^* are the thermodynamic and stoichiometric proton dissociation constants, respectively; the brackets represent activities, the square brackets represent concentrations, and γ_i are the activity coefficients.

The thermodynamic proton dissociation constant of simple ligands depends on the medium composition, namely on the ionic strength and electrolyte type. As described in Eq. 3.2, this dependency is a function of the activity coefficients of the species involved, which are ions and neutral molecules:

$$K^T(P, T, solvent) = K^* \cdot Q(\gamma_i) \tag{3.3}$$

where $Q(\gamma_i)$ represents the ratio of activity coefficients of the species involved in the equilibrium. The activity coefficients are considered unity at infinite dilution or zero ionic strength.

Alternatively, taking logarithms and multiplying by RT, Eq. 3.3 can be expressed in terms of the Gibbs free energy (ΔG):

$$\Delta G^* = \Delta G^0 + (\Delta G_{non-elec} + \Delta G_{elec}) = \Delta G_{int} + \Delta G_{elec} \tag{3.4}$$

where the energetic terms associated to the activity coefficients have been split into two parts: electrostatic and non-electrostatic (Sastre De Vicente 1997; Lodeiro et al. 2007) (see Table 3.1 and Sect. 3.2.1); and the term ΔG_{int} includes all the contributions to Gibbs free energy different from the electrostatic ones.

Table 3.1 Main models proposed in bibliography to calculate activity coefficients (γ)

Model	Function	Comments	Type	References
Brönsted	$\log\gamma_{\pm} = -3\alpha\sqrt{c} - 2\beta \cdot c$	Applicable at moderate concentration, c	Empirical	Bronsted (1922a, b)
Limiting Debye-Hückel	$\log\gamma_{\pm} = -A \mid z_{\pm}z_{-} \mid \sqrt{I}$	Applicable at very diluted ionic strength	Theoretical	Debye and Huckel (1923a, b)
Extended Debye-Hückel	$\log\gamma_{\pm} = \frac{-A\lvert z_{+}z_{-}\rvert\sqrt{I}}{1+Ba\sqrt{I}}$	Applicable at diluted ionic strength, size effects	Debye-Hückel based	Debye and Huckel (1923a, b)
Guggenheim	$\ln\gamma_{MX} = -\frac{A'\lvert z_M z_X\rvert\sqrt{I}}{1+\sqrt{I}} + \frac{2v_M}{v_M+v_X}\sum_a \beta_{Ma}m_a + \frac{2v_X}{v_M+v_X}\sum_c \beta_{cX}m_c$	Applicable at moderate ionic strength/ concentration	Debye-Hückel based	Guggenheim (1935)
Hückel	$\log\gamma_{\pm} = \frac{-A\lvert z_{+}z_{-}\rvert\sqrt{I}}{1+Ba\sqrt{I}} + \beta c$	Applicable at moderate concentration/ ionic strength	Debye-Hückel based	Huckel (1925)
MSA	$\log\gamma_i = -\frac{\alpha^2 z_i^2}{4\pi}\frac{\Gamma}{1+\Gamma\sigma_i} - \frac{\alpha^2 z_i\sigma_i}{4(1+\Gamma\sigma_i)}\frac{P_n}{\Delta} + \frac{\pi\alpha^2\sigma_i^3}{16}\left(\frac{1}{1+\Gamma\sigma_i} - \frac{1}{3}\right)\frac{P_n^2}{\Delta^2} - \ln\Delta + \frac{\sigma_i^3 X_0 + 3\sigma_i^2 X_1 + 3\sigma_i X_2}{\Delta} + \frac{3\sigma_i^3 X_1 X_2 + 9/2\sigma_i^2 X_2^2}{\Delta^2} + \frac{3\sigma_i^3 X_2^3}{\Delta^3}$	Inclusion of size and charge effects. Applicable at moderate concentrations/ionic strengths	Statistical mechanics approach based in hard spheres.	Blum (1975)

Pitzer	$\ln \gamma_{M^+} = z_M^2 f^\gamma + 2\sum_a m_a \left[B_{Ma} + \left(\sum mz\right) C_{Ma}\right] + 2\sum_c m_c \theta_{Mc} + -\sum_c \sum_a m_c m_a \left(z_M^2 B'_{ca} + z_M C_{ca} + \psi_{Mca}\right) + \frac{1}{2}\sum_a \sum_{a'} m_a m_{a'} \left(z_M^2 \theta'_{aa'} + \psi_{Maa'}\right) + + \frac{z_M^2}{2}\sum_c \sum_{c'} m_c m_{c'} \theta'_{cc'}$	Applicable at very high concentration/ ionic strength	Debye-Hückel based	Pitzer (1973, 1991) and Pitzer and Mayorga (1973)
Sammartano	$\log\gamma_z = \frac{-Az^2\sqrt{I}}{1+B\sqrt{I}} + C_z I + D_z I^{3/2}$	Empirical modification of the extended Debye-Hückel limiting law. Applicable at moderate to high ionic strengths	Debye-Hückel based	Daniele et al. (1997)

The table includes the name and type of model, the equation, information about the range of applicability and relevant bibliography
*For symbol interpretation see references mentioned in the table and references therein

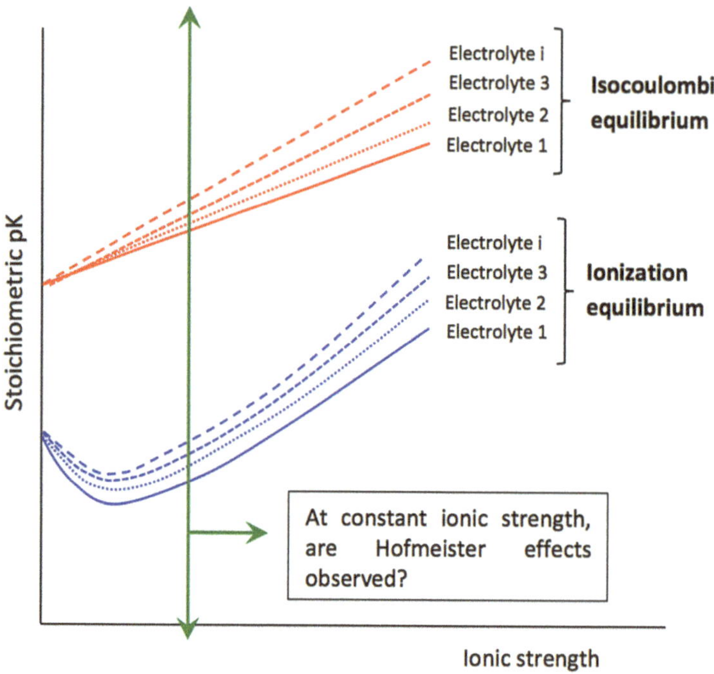

Fig. 3.3 pK* versus ionic strength plot for isocoulombic and ionization acid-base equilibrium. For isocoulombic equilibria, pK* is a linear function of the ionic strength (red lines). On the contrary, the plot shows that for ionization equilibrium curves passes through a minimum (blue lines). Hofmeister effects could be identified at constant ionic strength

Taking into account that activity coefficients in saline solutions can be expressed as a function of the ionic strength and the specific parameters of the system (Table 3.1 and references therein), the following equation is obtained at constant P, T and background electrolyte:

$$pK^{*} = pK^{T} + f(I, system\ parameters) \tag{3.5}$$

The representation of pK^{*} *versus* ionic strength depends on the electrostatics involved in the acid-base equilibrium. Therefore, in a typical ionization (charge separation) equilibrium (Eq. 3.1), a plot of the pK* dependence on ionic strength commonly passes through a minimum. Nevertheless, the pK^{*} is usually a linear function of the ionic strength when isocoulombic equilibria are involved; this is commonly observed, for example, for amine protonation reactions: $BH^{+} \leftrightarrows B + H^{+}$. An example of these two behaviours can be seen in Fig. 3.3.

The relative proportion of anionic/neutral species, and therefore the balance of the existing interactions for the general proton dissociation equilibrium represented in Eq. 3.1, changes with the pH and ionic strength of a particular solution as follows:

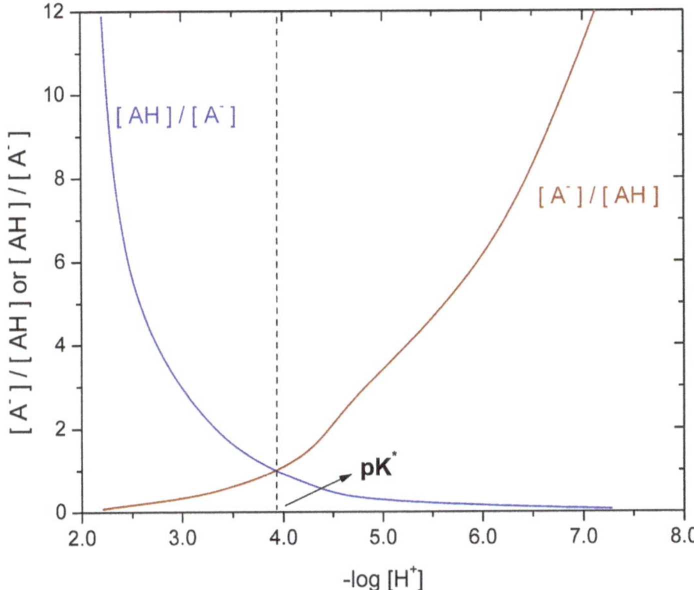

Fig. 3.4 Evolution of the proportion of neutral/anionic (blue line) and anionic/neutral (red line) species with pH. The data represents a potentiometric titration of the brown seaweed *Sargassum muticum* in $NaNO_3$ 0.05 mol·L^{-1} at 25 °C. The intersection of both plots corresponds to the stoichiometric protonation constant, pK*, of the acidic groups present in the seaweed

$$\frac{[A^-]}{[AH]} = \frac{K^*(I)}{[H^+]} = f(I, pH) \tag{3.6}$$

Hence, mainly the pH but also the ionic strength are relevant variables tuning the interactions between solutes present in a specific medium composition (see Fig. 3.4).

3.2.1 *Models for the Activity Coefficients of Species in Solution*

Different equations have been proposed to obtain expressions for the activity coefficients (log γ_i) according to the theory of electrolytes (Table 3.1). As mentioned in Sect. 3.2 (Eq. 3.4), the activity coefficient of an electrolyte can be split into two contributions: long-range Coulomb's interactions and short-range specific interactions. The former are a function of the ionic strength, and are independent of the electrolyte nature. Although, the short range interactions represent pairwise or three particle interactions in solution, hence they are electrolyte dependent. Most of the models used to calculate activity coefficients in electrolyte systems are based on the Debye-Hückel limiting law, which is only valid for very dilute concentrations i.e. < 0.001 mol kg^{-1}. This model considers that the interactions between ions are exclusively electrostatic, that is dependent on ionic strength. The ionic strength is therefore considered as a key variable in the study of electrolyte systems, even nearly hundred years after its introduction in chemistry by G. N. Lewis and M. Randall (Sastre De Vicente 1997, 2004). Equations that extend the validity range of activity coefficient calculations to moderate or high ionic concentrations, should take into account not only the ionic strength, but also electrolyte specific effects. Different approaches have been proposed to account for non-electrostatic interactions between ions. The simplest models are based on the Specific Interaction Theory (SIT) of BrØnsted-Guggenheim, e.g. the Pitzer's equations are a representative example (Pitzer 1991). The Pitzer formulation has been extensively used in the literature for different ligands in simple electrolytes and complex mixtures, such as seawater (Daniele et al. 1997; De Stefano et al. 2002; Grenthe 2002; Millero and Pierrot 2002; Turner et al. 2016). A more elaborated theory, the Mean Spherical Approximation (MSA), allows the calculation of the activity coefficients term, $Q(\gamma_i)$, including explicitly the ion charge, size and concentration, and the temperature, as parameters in its formulation (Sastre de Vicente and Vilariño 2002). Therefore, the MSA theory allows, for example, studying size effects on chemical equilibria, which is not possible using Specific Interaction Theory expressions.

3.3 Gibbs Free Energy of Proton Binding: Electrostatic and Non-electrostatic Contributions

About 40 years ago, in an already classical work on ionizable surfaces, Healy and White (1978) presented a reaction of dissociation as:

$$AH^{(Q-1)^-} \rightleftarrows A^{Q^-} + H^+ \tag{3.7}$$

For this dissociation process, or its thermodynamically equivalent proton adsorption/binding reaction, the interaction free energy can be split into two: on the one hand an electrostatic term associated with double layer interactions, on the other, contributions including dispersion and other non-electrostatic forces. The Gibbs free energy in adsorption processes usually involves a wide range of reaction energies, which can be grouped into non-electrostatic and electrostatic terms in analogy to Eq. 3.4 (Moreno-Castilla 2004):

$$\Delta G_{diss} = \Delta G_{non-elec} + \Delta G_{elec} \tag{3.8}$$

In addition, according to Van Oss (2006; Van Oss and Giese 2011), the ΔG_{diss} (dissociation) of interactions between two different entities e.g. molecules, particles, surfaces, etc. in aqueous solution can be expressed as:

$$\Delta G_{diss} = \Delta G^{LW} + \Delta G^{AB} + \Delta G_{elec} \tag{3.9}$$

where ΔG^{LW} and ΔG^{AB} represent Lifshitz-van der Waals and Lewis Acid-Base (including hydrogen bonding) energies, respectively (Goss and Schwarzenbach 2001). Both free energy terms can be either attractive or repulsive. The comparison of Eqs. 3.8 and 3.9 allows the identification of the relevant contributions of non-covalent interactions in the $\Delta G_{non\text{-}elec}$ term, also called intrinsic free energy (ΔG_{int}).

The protons are a master variable that control any acid-base system and influence practically all processes in aqueous chemistry (Stumm and Morgan 1996). The pH, as discussed in Sect. 3.2, becomes therefore an extremely important parameter affecting the proportion of neutral/charged sites in an adsorbent. This effect appears irrespective to the presence of metals or other substances in solution. Therefore, as previously stated for single acid-base equilibria (Eqs. 3.4 and 3.5), the pH also affects the interactions involved in the different Gibbs free energy contributions (Eq. 3.9). In addition, other parameters such as the ionic strength, and the specific electrolyte nature, influence, as discussed for Eq. 3.4, the energetic terms in Eq. 3.9 to a variable degree.

Therefore, for a given couple sorbent/sorbate in aqueous solution, the Gibbs free energy of adsorption (ΔG_{ads}) can be expressed as:

$$\Delta G_{ads} = \Delta G_{ads}(pH, I, electrolyte\ nature) \tag{3.10}$$

The pH and ionic strength are generic variables independent of the characteristic of the electrolytes present in solution, and both contribute to ΔG_{elec}. These variables modulate the electrical properties of the interface sorbate-solution/sorbent, acting as a sort of charge regulators (Trefalt et al. 2016). Moreover, the influence of the nature of dissolved salts in solution is associated to Hofmeister lyotropic or salting in *versus* salting out effects. Those terms are widely used to describe specific electrolyte effects on many physicochemical properties (Salis and Ninham 2014).

Equation 3.10 is a general expression that can be applied to adsorption processes involving different sorbates, e.g. protons, metals or organic substances, with distinct speciation characteristics and variable structural complexity, i.e. different polarity or charge, degree of hydrophobicity, etc.

Equations 3.9 and 3.10 also indicate that changes in the pH, ionic strength and/or the nature of salts in solution, will affect the value of ΔG_{ads}. Therefore, in order to properly understand dissociation/binding reactions, different experiments at several pHs, ionic strengths and/or electrolyte types should be performed. These three key variables are not usually changed simultaneously: for example, a potentiometric titration is usually performed maintaining the nature of the electrolyte and ionic strength constant while changing pH. Afterwards, a series of experiments can be performed at different ionic strengths using the same electrolyte at identical solution temperature. An adequate interpretation of the obtained results leads to important physicochemical information of the process such as intrinsic equilibrium constants. However, in most cases data interpretation involves the use of different models. Besides, when modelling data specific physical properties of the adsorbent/biomass e.g. volume, size, texture, etc. are required or assumed.

3.4 Acid-Base Properties in Macromolecular Systems: A Complex Problem

The study of the equilibrium binding shows significant differences between fully characterized simple inorganic ions or organic compounds (e.g. acetic acid), and complex, not well-defined, macromolecules (e.g. humic/fulvic acids, polysaccharides or polyelectrolytes). The study of the acid-base properties of simple organic ligands is straightforward. Nevertheless, the study of ion binding to biosorbents is challenging and requires considering the following singularities (Fig. 3.5):

1. High quantity of binding sites: the number of chemical species involved can be in the order of hundreds to thousands or more. It is convenient then to describe the binding to biosorbents in terms of distribution of species.

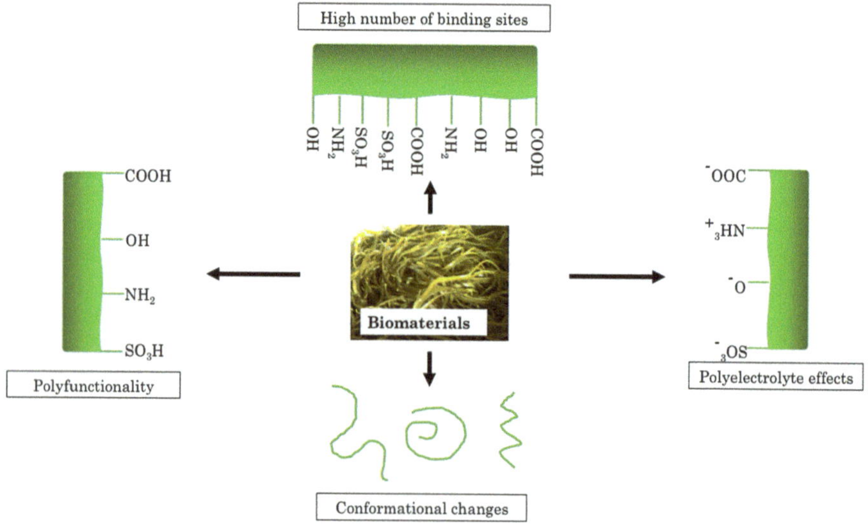

Fig. 3.5 Schematic representation of the main singularities that affect the acid-base and binding properties of biosorbents when compare to simple compounds. That is, presence of high number of binding sites, presence of many different chemical groups (polyfunctionality), conformational changes and polyelectrolyte effects associated with the presence of charged functional groups

2. High polyfunctionality: presence of many different chemical groups such as carboxyl, hydroxyl or amino that can also have different environments, for example linked through aliphatic or aromatic groups. This is also known as chemical heterogeneity.
3. Conformational changes: possible polymer chain reconfigurations depending on solution variables such as pH or ionic strength. The changes in the steric disposition of the binding sites can make them either more or less accessible, and lead to titration hysteresis.
4. Polyelectrolyte effects: the presence of charged functional groups in the biosorbents may constrain further dissociation of binding sites with increasing or decreasing pH values due to a progressive change of electrostatic attractions; as a consequence, the apparent dissociation constant also change.

All those singularities can occur simultaneously and affect to a different extent the acid-base and binding properties of biosorbents, and should then be considered in biosorption studies.

3.5 Classification of Biosorbents Based on Their Functional Groups

Based on the natural sources of biomass in nature, biosorbents are classified broadly into two major categories (Bailey et al. 1999): microscopic biomass that mainly involves microalgae, fungi and bacteria (Sag 2001; Aksu 2005; Vijayaraghavan and Yun 2008a; Wang and Chen 2009); and macroscopic biomass, which comes from a wide variety of sources, among them forestry-based biomass and agricultural residues (Brown et al. 2000; Gardea-Torresdey et al. 2004; Hubbe et al. 2011; Kyzas et al. 2013), marine biomass (e.g. green, red and brown algae) (Plazinski 2013a) or biomass from the seafood processing industry, including crustacean shells and arthropods (Fig. 3.6) (Guibal 2004; Tudor et al. 2006; Morris and Sneddon 2011; Muzzarelli 2011).

If they are classified by majority chemical composition, most of these materials owe their properties to the most abundant biopolymers found in nature, i.e. cellulose, chitin and lignin. Apart from them, alginates, naturally occurring polymers in the cell-wall of brown algae, are showing themselves as one of the most powerful adsorbents of metal ions and organic molecules (Davis et al. 2003). The alginates are chain-forming heteropolysaccharides made up of blocks of mannuronic and guluronic acid. Besides, carrageenans from red and green algae, pectins from fruit, tannins from forestry biomass, fucoidans, xanthates, starch, lipids or proteins are also noteworthy as adsorbents.

A further chemical characterization of the major molecules shows that their interaction with pollutants occurs through certain functional groups that are repeated many times in their polymeric structure. This interaction is responsible for both the adsorption process and the acid-base properties of the biosorbents. It is worth

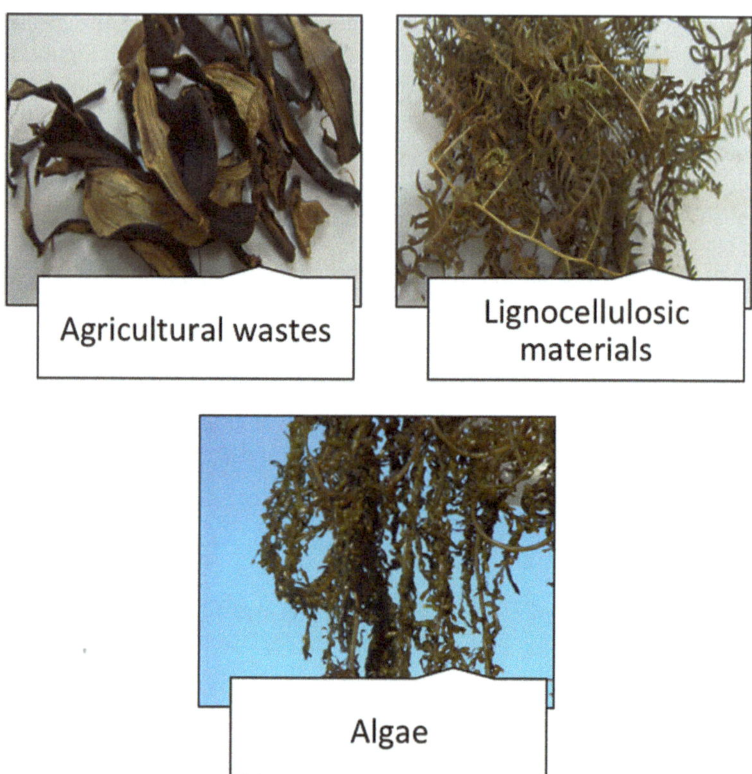

Fig. 3.6 Examples of representative biosorbents obtained from agricultural wastes (banana skin), lignocellulosic materials (fern) and marine biomass (macroalga)

mentioning the carboxyl, hydroxyl, phenolic, amine, amide, phosphonate, acetamide and sulfonate groups, both for their abundance, and for their key role in the acid-base and adsorbent properties of natural biomass.

Large numbers of biosorption studies have been carried out using all sorts of biomass to eliminate different pollutants. However, studies dealing with the determination of the acid-base properties of these materials are scarce, even though they are the basis for interpretation of the main biosorption mechanisms.

The characterization of the acid-base properties is mainly carried out by a potentiometric study of the biomass to determine the type and number of binding sites. These studies allow also equilibrium constants to be obtained and they can be supplemented by other techniques, such as Fourier transform infraRed spectroscopy (FTIR), X-ray absorption near edge structure (XANES), fluorescence spectroscopy, extended X-ray absorption fine structure (EXAFS) or nuclear magnetic resonance (NMR), to identify all the functional groups in the biomass and their role in the adsorption process (Sun and Berg 2003; Joud and Barthés-Labrousse 2015).

3.6 Modelling the Proton Binding Equilibria in Biosorbents

The following discussion is based on the split of the adsorption energy into electrostatic and other non-electrostatic contributions proposed in Eq. 3.8. This division allows simplifying and correctly interpret the proton binding equilibria in biosorption processes under different experimental conditions.

3.6.1 *Electrostatic Effects: Influence of pH and Ionic Strength*

In most cases, the biosorbents present a negative charge associated to the dissociation of their acidic groups. As mentioned in Sect. 3.4, one of the main differences in studying the acid-base properties of simple organic ligands and biosorbents is due to the higher charge associated with the latter (polyelectrolyte effect). Moreover, the interactions of biosorbents with other species such as metals or organics in solution depend on the acid-base properties of the biomass and the chemical speciation. Therefore, the study of variables such as the pH, ionic strength or sorbate/binding-sites ratio is of great importance. These variables regulate the relative significance of the observed effects, mainly those associated with electrostatic interactions.

As discussed in Sect. 3.3, a simple way to model coulombic effects of proton binding to biosorbents consists of splitting the intrinsic (non-electrostatic) and electrostatic energy contributions to the binding according to Eq. 3.8. The biosorbents contain natural biopolymers; considering therefore the biosorbent as a polyelectrolyte, the electrostatic work involved in bringing a proton from the bulk solution to the biding site can be written as (Morel et al. 1993):

$$\Delta G_{elec} = nF\psi_0 = F\psi_0 \tag{3.11}$$

where F is the Faraday constant and ψ_0 is the electrostatic potential at the location of the binding site. In terms of equilibrium constants, the equation reads:

$$K_{elec} = e^{-\frac{\Delta G_{elec}}{RT}} = e^{-\frac{F\psi_0}{RT}} \tag{3.12}$$

Considering the proton dissociation reaction presented in Eq. 3.2, a biosorbent of charge Q will present an apparent dissociation constant (K_{app}) given by:

$$K_{app} = \frac{\left[A^{Q^-}\right]\left[H^+\right]}{\left[AH^{(Q-1)^-}\right]} = \frac{\left(A^{Q^-}\right)\left(H^+\right)}{\left(AH^{(Q-1)^-}\right)}\frac{\gamma_{AH^{(Q-1)^-}}}{\gamma_{A^{Q^-}}\gamma_{H^+}} = K_{int}\frac{\gamma_{AH^{(Q-1)^-}}}{\gamma_{A^{Q^-}}\gamma_{H^+}} \tag{3.13}$$

The proton and ratio of the biosorbent activity coefficients correspond then to the corrections to the intrinsic dissociation constant, K_{int}. If Q>> > 1, the following effective activity coefficient can be defined as (Morel et al. 1993):

$$\ln \gamma_{eff} = \ln \frac{\gamma_{A^{Q^-}}}{\gamma_{AH^{(Q-1)^-}}} = -\frac{F\psi_0}{RT} \quad (3.14)$$

The intrinsic dissociation constant can be calculated from:

$$K_{int} = K_{app}\gamma_{H^+}\gamma_{eff} = \frac{[A^{Q^-}]}{[AH^{(Q-1)^-}]}(H^+)_0 \quad (3.15)$$

The local ion activity of the proton at the binding site, $(H^+)_0$, is then given by its experimentally accessible bulk activity, (H^+), multiplied by a Boltzmann factor:

$$(H^+)_0 = [H^+]\gamma_{H^+}e^{-\frac{F\psi_0}{RT}} = (H^+)e^{-\frac{F\psi_0}{RT}} \quad (3.16)$$

Therefore, the surface proton activity, $(H^+)_0$, or concentration can be obtained from the electrostatic potential at the active binding site. The value of the electrostatic potential can be estimated using different models, as shown below. When considering the proton activity, but not the concentration, a correction for the activity coefficient of the proton in solution is required. Therefore, a suitable model for the activity coefficient (Table 3.1) should be chosen depending on experimental conditions, especially at low or high ionic strengths.

In addition to geometrical constraints the potential around a charged species in an electrolyte solution is a function of the ionic strength. Equation 3.16 constitutes the basis for carrying out electrostatic corrections, which present different dependencies on ionic strength.

The presence of an electrolyte in solution can affect the binding in a direct and indirect way. The former reduces the occurrence of other ions near binding sites, while the indirect way is due to the fact that intraparticle activities are higher than bulk activities. Nevertheless, according to the interpretation and definition of the Debye length, it is worth mentioning that in most cases electrostatic effects should be suppressed at ionic strengths c.a. 0.5–1 mol kg^{-1} (Israelachvili 2011). However, a minimum in the Debye length has been observed around these values for some systems (Smith et al. 2016).

By analogy with models initially developed for humic and fulvic acids (Saito et al. 2005), two different approaches have been mainly used to account for electrostatic effects in biosorbents, namely one and two phase models (Fig. 3.7). Two-phase models consider the active polyelectrolyte (biosorbent) sites as a three-dimensional permeable structure or Donnan volume; while in one-phase models an active rigid surface (two-dimensional double-layer) is assumed. Calculating the contribution of electrostatic effects to free energies usually involves solving the appropriate Poisson-Boltzmann equation (Eq. 3.17), which relates the Laplacian of the electrostatic potential (ψ) to the charge density in the medium (Bartschat et al. 1992).

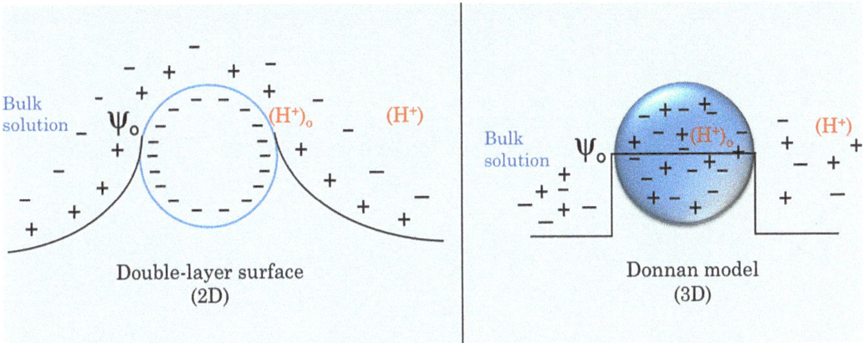

Fig. 3.7 Schematic representation of the double layer surface model (left panel) and Donnan model (right panel). The former considers the active binding sites as a two-dimensional rigid structure. On the contrary, on the later the active sites are represented as a three-dimensional permeable volume

$$\nabla^2\psi = -\frac{1000F}{\varepsilon}\left(\sum_i z_i[X_i]e^{-\frac{F\psi}{RT}} + \rho_0\right) \tag{3.17}$$

where ρ_0 (mol·L^{-1}) is the charge in the region occupied by the biosorbent in the absence of mobile ions, and the summation term is the charge density produced by the distribution of co- and counterions (X_i) in the potential field. Equation 3.17 is valid either for surface double-layer ($\rho_0 = 0$) or Donnan models, where $\rho_0 \neq 0$.

The Surface Complexation Model (SCM) is one of the most well-known and used surface models. The works of Borrok et al. (2005) and Goldberg and Criscenti (2008) constitute a good review for this matter. Table 3.2 shows several examples where the SCM model and the alternative Donnan approach were used in biosorption studies. It is worth mentioning that most of the references shown in Table 3.2 do not include any ionic strength (electrostatic) correction term when analysing the proton binding. Moreover, significant differences can be found regarding important experimental conditions, such as the electrode calibration or the ionic strength adjustment during the proton titration performed by different authors.

The application of the finite difference method allows the more general non-linear Poisson-Boltzmann equation to be solved, with important implications on diverse biological and chemical phenomena (Honig and Nicholls 1995).

The macroscopic Donnan model can be considered as a special case of the microscopic Poisson-Boltzmann theory (Dahnert and Huster 1999). In fact, both models represent essentially the same electrostatic and osmotic phenomena. Nevertheless, the Donnan model is simpler and does not require a solution to the Poisson–Boltzmann equation; the Donnan model relates the electrostatic effects to the difference between the bulk and local concentrations multiplied by the volume of the binding sites.

Table 3.2 Compilation of potentiometric experiments for different biosorbents including experimental and modelling titration conditions

Biosorbent	Electrostatic proton binding model	Specific Interaction Model	Master curve approach	Chemical proton binding model	Electrode calibration	pH range	Ionic strength mol/L	Electrolyte	References
Brown algae	Donnan/diffuse layer	Pitzer/no	Yes	Continuous (L-F)	Proton concentration	2–10/11.5	0.05–2	NaCl and KNO_3	Rey-Castro et al. (2003, 2004a)
Brown algae	No	No	No	Discrete (one site)/continuous (L-F)/Katchalsky	Proton concentration	2.2–6.5/11	0.05	$NaNO_3$	Lodeiro et al. (2004, 2005a, b, 2006a)
Brown algae	No	No	No	No	No info	3–10	0.001	NaCl	Fourest and Volesky (1997)
Brown algae	No	No	No	Discrete (one site)	No info	2.5–4.5	0.1	$NaNO_3$	Seki and Suzuki (1998)
S. Polycystum (brown alga)	No	Davies equation	No	Discrete (three sites)	No info	2–10.5	0.1 and 1	$LiNO_3$	Yun and Volesky (2003) and Yun (2004)
Sargassum sp. (brown alga)	Donnan	No	No/pH-pNa	Discrete (one site)	No info	2–6	0–1	$NaNO_3$	Schiewer and Volesky (1997b)
S. fluitans (brown alga)	Donnan/CHEM model	No	No	Discrete (one site)	No info	2.2–4	0–5	$NaNO_3$	Schiewer (1999)
S. fluitans (brown alga)	No	No	No	No	No info	3–10	0.001	NaCl	Fourest and Volesky (1996)
Brown/green algae	Donnan	No	No	Discrete (one site)	No info	2–8.5	0–0.1	Deionised water/$NaNO_3$	Schiewer and Wong (2000)

P. caniculata (brown alga)	No	No	No	Discrete (two sites)-continuous sips (L-F)	Buffers pH 4, 7 and 10	2.5–9.5	0.001–0.1	$CaCl_2$	Costa et al. (2010)
Alginate	Donnan	No	Yes	NICA (L-F) (one site)	Buffers pH 4, 7 and 9	3–8	0.005–0.3	KNO_3	Lamelas et al. (2005)
U. pinnatifida (green alga)	Gouy-chapman	No	No	Discrete (two sites)	No info	2.5–11	0.0001–1	NaCl	Kim et al. (1998)
P. subcapitata (green alga)	No	No	No	Discrete (four sites)	No info	3–10	0.1	$NaClO_4$	Kaulbach et al. (2005)
C. vulgaris (green alga)	No	No	No	Discrete (four sites)-continuous sips (L-F)	No info	2–12	0.1	$NaNO_3$	Pagnanelli et al. (2013)
Vaucheria sp. (green alga)	No	No	No	No	No info	3–11	Not fixed	Deionised water	Crist et al. (1981)
Ulva lactuca (green alga)	No	Extended Debye-Hückel	No	Discrete (three sites)	Proton concentration	2–10	0.01–5	NaCl	Schijf and Ebling (2010)
G. sesquipedale (red alga)	No	No	No	Discrete (two sites)-continuous sips (L-F)	Proton concentration	3.5–10.5	0.005–0.1	$NaNO_3$	Vilar et al. (2009)
Phyllophora crispa (red alga)	No	No	No	Discrete (three sites)/Katchalsky	No info	3–10	0.01	NaCl	Meichik et al. (2011)
Chlorella miniata (microalga)	No	No	No	Discrete (three sites)	No info	3.5–10.5	Not fixed	Deionised water	Han et al. (2006)
Olive Pomace	No	Davies equation	No	Discrete (two sites) -continuous L-sips	No info	2–10	Not fixed 0.07–0.09	Deionised water	Pagnanelli et al. (2005b)
Olive Pomace	No	Davies equation	No	Discrete (one-two sites) -continuous L-sips- NICA (L-F)	Buffers pH 4 and 7	2–	Not fixed	Deionised water	Pagnanelli et al. (2005a)

(continued)

Table 3.2 (continued)

Biosorbent	Electrostatic proton binding model	Specific Interaction Model	Master curve approach	Chemical proton binding model	Electrode calibration	pH range	Ionic strength mol/L	Electrolyte	References
Olive Pomace	No	No	No	Continuous L-sips	No info	2.5–9.5/ 11.5	Not fixed	Deionised water	Pagnanelli et al. (2008)
Olive Pomace	No	No	No	Discrete (two sites) -continuous L-sips	No info	2–12	Not fixed	Deionised water	Pagnanelli et al. (2003)
Olive Pomace	No	No	No	Discrete (two sites) -continuous sips	No info	2–9.5/ 10	Not fixed	Deionised water	Martín-Lara et al. (2008)
Olive oil production waste	No	No	No	Discrete (three sites)	No info	3–11	Not fixed	Deionised water	Calero et al. (2010)
Sugar beet pulp	Diffuse layer	No info	No	Discrete (three sites)	No info	3–11	0.001–0.1	$NaNO_3$	Reddad et al. (2002)
Wheat bran substrate	Donnan	No	Yes	NICA (L-F) (two sites)	Proton concentration	3.5–8	0.04–1	$NaNO_3$	Bouanda et al. (2002)
Wheat bran substrate	Diffuse layer	Davies equation	No	Discrete (two sites)	Buffers pH 4, 7 and 10	2.5–11	0.01–0.1	$NaNO_3$	Ravat et al. (2000)
Citrus peels	No	No	No	Continuous pK spectrum (FOCUS), (4 sites)	No info	2.5–11	Not fixed	Nano-pure water	Schiewer and Patil (2008a)
Orange peel/ bracken fern	No	No	No	Discrete (two – Three sites)	Proton concentration	3–11.5	0.05	$NaNO_3$	Lodeiro et al. (2008)
Banana skin	No	No	No	Discrete (two sites)	Proton concentration	3.5–11	0.1	KNO_3	Lopez-Garcia et al. (2013)
Bracken fern	No	No	No	Katchalsky	Proton concentration	No info	0.05	KNO_3	Barriada et al. (2009)
Festuca rubra (grass)	No	No	No	Discrete (three sites)	Buffers pH 2, 4, 7 and 10	2.2–9.7	0.1	$NaClO_4$	Ginn et al. (2008)

Cumin	No	No	No	Discrete (two sites)	No info	2–10	0.1	$NaNO_3$	Komy (2004)
Bamboo sawdust	Donnan	No	Yes	NICA (L-F) (two sites)	No info	3.5–10.5	0.01, 0.1, 1	KNO_3	Zhao et al. (2015)
Peat	Donnan	No	No	NICA (L-F) (two sites)	Proton concentration	3.5–10	0.1–0.3	KNO_3	Lopez et al. (2011)
Modified peat	No	No	No	Discrete (four sites)	No info	4–11	0.01	$NaNO_3$	Randelovic et al. (2016)
C. glutamicum (bacteria)	No	No	No	Discrete (three sites)	No info	3–11	Not fixed	Water	Vijayaraghavan and Yun (2008b)
Bacillus subtilis (bacteria)	No	Davies equation	No	Discrete (five sites)	Buffers pH 4, 7 and 10	4–10	0.025 and 0.1	$NaNO_3$	Cox et al. (1999)
Bacillus subtilis (bacteria)	Constant capacitance	No	No	Discrete (three sites)	Proton concentration	2.1–9.7	0.1	NaCl	Leone et al. (2007)
Bacillus subtilis (bacteria)	Constant capacitance	No	No	Discrete (three sites)	No info	3.5–10.5	0.1 and 0.3	$NaNO_3$	Fein et al. (1997)
Bacillus subtilis (bacteria)	Constant capacitance	No	No	Discrete (two-four sites)-continuous L-F	Proton concentration/buffers	2–10	0.01–0.3	$NaClO_4$	Fein et al. (2005)
Bacillus subtilis (bacteria)	Donnan	Davies equation	No	No	No info	2.4–9	0.001, 0.01 and 0.1	KNO_3	Yee et al. (2004)
A. flavithermus (bacteria)	Donnan	Davies equation	No	Discrete (three sites)	No info	3.5/4–9/10	0.001, 0.01 and 0.1	$NaNO_3$	Burnett et al. (2006)
A. flawithermus (bacteria)	Donnan	No	No	Discrete (two sites)	No info	3–10	0.01/0.5	$NaNO_3$/NaCl	Heinrich et al. (2007)
Pseudomonas sp. (bacteria)	Diffuse layer/triple layer	Davies equation	No	Discrete (four sites)	No info	2.5–9.7	0.01, 0.1 and 0.5	$NaClO_4$	Borrok and Fein (2005)

(continued)

Table 3.2 (continued)

Biosorbent	Electrostatic proton binding model	Specific Interaction Model	Master curve approach	Chemical proton binding model	Electrode calibration	pH range	Ionic strength mol/L	Electrolyte	References
P. aeruginosa (bacteria)	No	No	No	Discrete (three sites)	No info	3–10	0.1	$NaNO_3$	Komy et al. (2006)
P. pseudo-alcaligenes (bacteria)	Constant capacitance	No	No	Discrete (one-three sites)	No info	2.5–10.5	0.01, 0.1 and 0.5	$NaNO_3$	Liu et al. (2013)
A. manzaensis (bacteria)	Donnan	No	No	Discrete (three sites)	No info	2–10	0.001, 0.01 and 0.1	$NaNO_3$	He et al. (2013)
S. putrefaciens (bacteria)	Constant capacitance	No	No	Discrete (three sites)	No info	4.2–9.5	0.1	NaCl	Haas (2004)
S. natans (bacteria)	Donnan	No	No	Discrete (two sites) -continuous L-sips	No info	3–10	0.1	$NaNO_3$	Pagnanelli et al. (2004)
Bacteria sp.	Constant capacitance	No	No	Discrete (five sites)	No info	3–11	0.01	$NaNO_3$	Ngwenya et al. (2009)
Bacteria sp.	No	No	No	Discrete (four sites)	No info	3–10	0.1	$NaClO_4$	Kenney and Fein (2011)
Bacteria sp.	Donnan	Yes (no info)	Yes	Continuous pK spectrum (FOCUS), (4 sites)	Buffers pH 4, 7 and 10	3–10	0.01–0.5	KNO_3	Martinez et al. (2002)
Bacteria sp.	No	No	No	Discrete (four sites)	No info	2.7–9.5	0.1	$NaClO_4$	Mishra et al. (2010)
Bacteria sp.	Donnan	No	Yes	Continuous L-F	Proton concentration	3–10.5	0.01, 0.1 and 1	$NaNO_3$	Plette et al. (1995)
Bacteria sp.	No	No	No	Discrete (three sites)	No info	4–10	0.001	NaCl	Guine et al. (2006)

Gloeocapsa sp. cyanobacteria	No	No	No	Discrete (three sites)	No info	3–11	0.01–1	$NaNO_3$/ NaCl	Pokrovsky et al. (2008)
Cyanobacteria	No	No	No	Discrete (three sites)	No info	0.5–11.5	0.1	$NaNO_3$	Gagrai et al. (2013)
Pico-cyanobacteria strains	No	No	No	Discrete (two-three sites)	Buffers pH 4, 7 and 10	2.9–10	0.01	$NaNO_3$	Dittrich and Sibler (2005)
Aspergillus niger (fungi)	No info	No	No	Discrete (two sites)	No info	3–11	Not fixed	Distil water	Mukhopadhyay (2008)
Saccharomyces Cerevisiae (fungi)	Constant capacitance	No	No	Discrete (four sites)	No info	3–9.7	0.001, 0.1 and 0.3	$NaClO_4$	Naeem et al. (2006)
S. cerevisiae (yeast)	No	No	No	Discrete (three sites)	No info	2–10	0.1 and 1	$NaNO_3$	Di Caprio et al. (2014)
Archeon	Yes (no info)	No	No	Discrete (three sites)	No info	3.5–9.5	0.01	$NaNO_3$	Daughney et al. (2010)
Diatoms	Constant capacitance	No	No	Discrete (three sites)	No info	2.3–11.5	0.001–1	NaCl/ $NaNO_3$	Gelabert et al. (2004)
Phytoplankton sp.	No	No	No	Discrete (three sites)	No info	2–12	0.7	NaCl	Gonzalez-Davila et al. (2000)

On this model the electrostatic potential at the active binding site is given by Wonders et al. (1997):

$$\psi_0 = arcsinh\left(\frac{-Q/V_D}{2I}\right) \tag{3.18}$$

where V_D is the active Donnan volume. Equation 3.18 is obtained considering a homogeneously distributed charge within a finite volume, assuming overall electroneutrality and that the active volume dimension is much larger than the Debye length. A more general solution of the Poisson-Boltzmann equation has been proposed by Ohshima and Kondo (1991).

On the contrary, theoretical models based on electrostatics, such as double-layer surface models, require the resolution of the Poisson–Boltzmann equation. This equation can be more or less easy to solve depending on the particular geometry and charge distribution considered for the system. For example, the SCM accounts for the electrostatic effects by a term including the potential at the plane of the adsorption, which is considered as an infinitive flat surface. In this case, the electrostatic potential at the active binding site is given by (Lyklema 1995):

$$\psi_0 = 2arcsinh\left(\frac{-\overline{\sigma}}{\sqrt{I}}\right) \tag{3.19}$$

$$\overline{\sigma} = \frac{\sigma}{\sqrt{8 \times 10^3 \varepsilon \varepsilon_0 RT}} \tag{3.20}$$

where $\sigma = -QF/A$, is the charge density and A the specific surface area.

A suitable model describing the electrostatic contribution to the binding has to be based on the biosorbent properties. The Donnan-type models are applied when the biosorbent has a permeable structure that shrinks and swells, its size is larger than the double-layer or Debye thickness and presents a large surface charge uniformly distributed within the biomass. Those characteristics are typical, for example, of many marine algae (Schiewer and Volesky 1997a, b; Schiewer 1999; Schiewer and Wong 2000; Rey-Castro et al. 2003; Pagnanelli et al. 2004; Rey-Castro et al. 2004b), and lignocellulosic agriculture derivatives (Bouanda et al. 2002; Lopez et al. 2011; Zhao et al. 2015). On the contrary, the SCM is used when the biosorbent is considered to have an impenetrable, rigid surface; for example, surface complexation models have been used to describe metal and proton binding on bacterial surfaces (Fein et al. 1997; Grenthe 2002; Haas 2004; Rey-Castro et al. 2004a; Borrok and Fein 2005; Fein et al. 2005; Leone et al. 2007; Ngwenya et al. 2009; Liu et al. 2013). Despite the different structural considerations on which those models are based, fitting similarities between Donnan and double-layer surface models have been reported when describing experimental binding data (Rey-Castro et al. 2004a, b; De Stefano et al. 2005). In fact, Donnan models have been successfully applied to bacterial biomass (Plette et al. 1995; Martinez et al. 2002; Pagnanelli et al. 2004; Yee et al. 2004; Burnett et al. 2006; Heinrich et al. 2007; He et al. 2013), and double layer models to agriculture derivatives (Ravat et al. 2000; Reddad et al. 2002) or algae biomass (Kim et al. 1998).

3.6.2 *Validation of the Accuracy of Electrostatic Models*

The electroneutrality condition for solutions states that the addition of positive and negative electric charges, considering the solution as a whole, must be zero. The electroneutrality condition is not a fundamental law of nature, but despite its ambiguous definition is very useful when studying chemical processes in solution. Moreover, it constitutes an excellent approximation to reality on the basis of the Poisson's equation of electrostatics (Sastre and Santaballa 1989).

The master curve approach, for example, is built on the electroneutrality condition. This approach consists on performing potentiometric titrations against a simple ion, usually the proton, in an electrolyte solution at different ionic strengths (see Fig. 3.8).

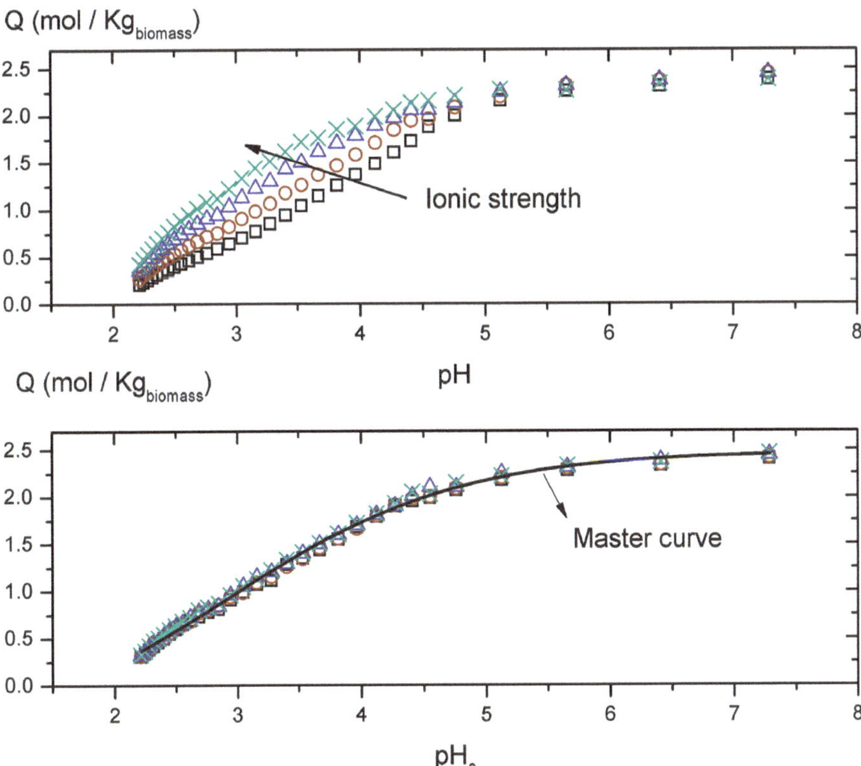

Fig. 3.8 Simulated data of charge versus pH curves (top graph) and calculated master curve (bottom graph) in a specific electrolyte solution. The master curve approach is built on the electroneutrality condition, so the proton binding data can be transformed into data of net charge. If the electrostatic model is correct, the dependence of the binding on ionic strength vanish; therefore, the corrected binding curve merge into the master curve that is independent of the ionic strength as can be observed in the bottom graph

The main goal is to obtain a validation test of the electrostatic model used to describe the charged biosorbent. Therefore, the proton binding data, obtained from experimental titrations, can be transformed using the electroneutrality condition into data of net charge (Q), proton coverage or dissociation degree of the biosorbent *versus* pH (see Sect. 3.9 for details).

As stated in the previous section, the pH and ionic strength modulate the net charge of the biosorbent, then:

$$Net\ charge = function(pH, I) = function_b(pH_0) \tag{3.21}$$

where pH_0 ($-\log [H^+]_0\gamma_{H+}$), the pH at the local binding site, can be obtained from Eq. 3.16.

Therefore, the charge-pH curves obtained over a range of ionic strength are used to optimise the parameters of the selected electrostatic model. If the electrostatic model used to calculate ψ is correct, the dependence of the binding on ionic strength should vanish and the corrected binding curve will merge into the so-called master curve that is independent of the ionic strength (Eq. 3.21).

More advanced theoretical treatments based on Monte Carlo simulations allow the study, not only on the influence of the ionic strength but also on the effect of ion size and surface charge models, of potentiometric titration of ionizable polyelectrolytes (Madurga et al. 2009).

3.6.3 Non-electrostatic (Intrinsic) Effects: Hofmeister Series

Electrostatic interactions alone cannot provide with an explanation of ion-specific interactions and their associated outcomes. This is because pure electrostatic treatments predict that ions of the same valence provide the same results, irrespective of their chemical nature. The so-called non-electrostatic or intrinsic effects are associated to specific ion or salt effects in solutions or interfaces of different electrolytes. The intrinsic effects are involved in many phenomena including colloid, polymer and interface science in the fields of chemistry or biology (Cacace et al. 1997; Lo Nostro and Ninham 2012).

Franz Hofmeister was a pioneer of specific salts effects with his work on the precipitation of proteins (Kunz et al. 2004a). At the end of the nineteenth-century, Hofmeister investigated the concentration requirements of distinct salts in precipitating egg white lecithin and found the following standard sequence for a fixed cation: $SO_4^{2-} > OH^- > F^- > Cl^- > Br^- > NO_3^- > I^- > SCN^- > ClO_4^-$, and for a fixed anion: $NH_4^+ > K^+ > Na^+ > Cs^+ > Li^+ > Rb^+ > Mg^{2+} > Ca^{2+} > Ba^{2+}$.

These non-electrostatic interactions or dispersion forces are associated to the specific nature of ions, their size and their polarizability. These effects, represented by the terms ΔG^{LW} and ΔG^{AB} in Eq. 3.9 are interpreted as differences in the properties of salts in solution, usually at concentrations higher than 0.1 M (Ninham and Yaminsky 1997).

Dispersion forces as a whole have a relevant role. Nevertheless, their analysis through theoretical developments is challenging, and unquantifiable terms such as hydration, hydrophilic, hydrophobic, π-cation interactions, hydrogen bonding, soft-hard ions, chaotropic, cosmotropic, etc. have been used in the derivation of these dispersion forces. However, the use of some simple empirical rules, such as the "law of matching water affinities" (Collins 2004; Collins et al. 2007; Vlachy et al. 2009), allows for the description of an important number of the above mentioned properties.

From a theoretical point of view, the electrostatic and dispersion forces must be equally treated. One of the current approaches consists of including an additional term of dispersion, which is added to the conventional electrostatic potential, in the Poisson-Boltzmann equation. The ionic distribution at an interface is given then by (Kunz et al. 2004b; Parsons et al. 2011; Salis and Ninham 2014):

$$c_{\pm} = ce^{\pm\frac{\left(z_i e\psi + U_{\pm}\right)}{kT}} \tag{3.22}$$

Therefore, the Eq. 3.16 that describes the proton activity at the local binding site would be also modified including an energetic dispersion-dependent term ($U_{\pm}$), thus providing a more realistic picture of the forces involved.

Quantitative studies on Hofmeister effects are scarce (Parsons 2016), and most works rely on qualitative or semi-quantitative analysis of results. Most of these papers are particularly focused on the adsorption of organic substances (Para and Warszynski 2007; Nelson and Schwartz 2013) where the complexity of electrostatic and non-electrostatic interactions is always present (Bauerlein et al. 2012). Some recent simulation studies on ion binding to carboxylic groups considering Hofmeister effects have also been performed (Schwierz et al. 2015; Stevens and Rempe 2016). These studies are of interest for biosorption due to the relevance of the carboxylic groups, which usually form part of the polysaccharide structure in many types of biomasses.

Despite pH or ionic strength effects that are studied in biosorption, to the best of our knowledge, there are no studies related to specific salt or Hofmeister effects.

3.6.4 Empirical Models to Describe the Proton Binding in Biosorbents

The complex and heterogeneous nature of biosorbents makes the investigation and interpretation of their acid-base properties challenging. The local binding interface is supposed to be constituted of a charged polyelectrolyte. Therefore, the proton dissociation of an acid group in a biosorbent can be described using a reaction formally identical to Eq. 3.1 but considering AH as a whole (not only a specific acid site). The apparent conditional dissociation constant (K_{app}) can be written then as:

$$pK_{app} = pH - \log\frac{\alpha}{1-\alpha} \tag{3.23}$$

where the degree of dissociation, α, is given by:

$$\alpha = \frac{[A^-]}{[A^-] + [AH]} \tag{3.24}$$

Ideally, as previously mentioned, the relative contribution from each of the effects concerning equilibrium binding in biosorbents should be accounted for by means of an appropriate model.

A first approach, suggested in several biosorption studies, is to fit the potentiometric titration data with a set of previously defined discrete ligand constants. A further approach involves the use of a Gaussian distribution of ion binding constants, considering the well-known heterogeneity of the biosorbents. Despite the simplicity of those models, they can provide useful information regarding the acid-base and complexation equilibria under specific conditions of pH, ionic strength, temperature and medium composition. Nevertheless, these empirical models fail when relating the specific properties of biosorbents, such as size or charge distribution, to the model fit parameters.

In the majority of cases, the equilibrium constants determined in biosorption studies are conditional stoichiometric constants, K^* (see Eq. 3.2), valid only for the specific ionic strength at which they have been determined. These conditional constants indirectly include all unspecific interactions among ions, namely, activity coefficients.

The modified Henderson-Hasselbach equation (Eq. 3.25) is one of the empirical models widely used to describe the dependence of the protonation constants of polyelectrolytes or biosorbents on the degree of dissociation:

$$pH = pK_m + n \log\frac{\alpha}{1-\alpha} \tag{3.25}$$

where pK_m and n are empirical constants that change with ionic strength. Therefore, the relationships between pK_{app} and α or pH can be easily obtained:

$$pK_{app} = pK_m - (n-1) \log\frac{\alpha}{1-\alpha} \tag{3.26}$$

$$pK_{app} = pK_m - \left(\frac{n-1}{n}\right) pH \tag{3.27}$$

Note that $pK_m = pK_{app}$ for $\alpha=0.5$ and $n > 1$. Potentiometric titrations of biosorbents do not provide a simple set of discrete dissociation constants, as when using simple ligands, but a continuous distribution of binding sites. This fact, together with the polyelectrolyte and associated effects, results in flatter titration curves with not well-defined end points (Fig. 3.9).

The best description of the binding properties of biosorbents has been provided therefore, when the model explicitly includes both, heterogeneity and

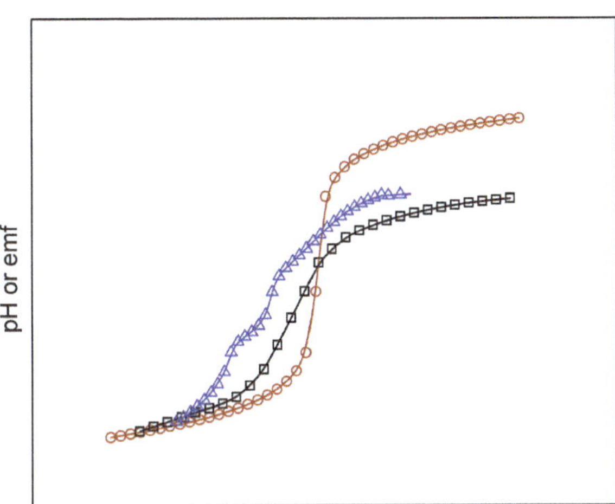

Fig. 3.9 Experimental titration curves for the simple organic compound tris(hydroxymethyl) aminomethane, Tris (red), the macroalga *Sargassum muticum* (black) and the lignocellolosic material *Pteridium aquilinum*, fern (blue). The graph evidences the stepper character and simplicity of the Tris titration curve compared to the obtained for a biosorbent with multifunctional binding sites

polyelectrolytic effects (despite conformational changes, not explicitly reflected). Therefore, reorganising Eq. 3.16 and substituting in Eq. 3.23 provides with an expression for the solution pH ($-\log [H^+]\gamma_{H+}$) or pK_{app}:

$$pH = pK_{int} + log\left(\frac{\alpha}{1-\alpha}\right) - \frac{1}{\ln 10 \frac{e\psi_0}{kT}} \tag{3.28}$$

$$pK_{app} = pK_{int} - \frac{1}{\ln 10 \frac{e\psi_0}{kT}} \tag{3.29}$$

It is worth mentioning that on the Donnan and surface charge models pK_{int} refers to the limit of high ionic strength, where ψ tends to zero, as the reference state. Nevertheless, for simple ligands the ion activity coefficients, so the pK_{int}, are refer to zero ionic strength or infinite dilution, where no interactions between ions are assumed. It is therefore important to consider this difference when comparing intrinsic binding constants of polyelectrolytes or biosorbents with the ones obtained for simple ligands. Moreover, for simple ligands, the Debye-Hückel law imposes a linear increase of pK with $\sqrt{I}$ at low ionic strengths, whereas for a biosorbent or polyelectrolyte an approximately linear increase of pK with log I is expected as the ionic strength decreases.

3.6.5 *Description of the Chemical Heterogeneity*

Once the electrostatic and/or non-electrostatic effects have been explicitly accounted for, a set of intrinsic binding constants that only depend on the chemical heterogeneity can be obtained. The biosorbents present many different chemical groups that can also have different steric and chemical environments. Therefore, a model for the description of the chemical heterogeneity is required. For the particular case of proton binding reactions, the coverage fraction of binding sites (θ) is given by:

$$\theta = 1 - \alpha = \frac{[AH]}{[A_{tot}]} = \frac{[AH]}{[A^-] + [AH]} \tag{3.30}$$

The plot of θ *vs* [H$^+$] is called the biding curve. The models used to account for chemical heterogeneity usually describe the coverage fraction of binding sites as a weighted sum of local isotherms (f), which describe the binding in each site. If it is assumed that the binding sites do not interact each other, and all have the same local isotherm:

$$\theta([H^+]) = \int_0^\infty p(K) f([H^+], K) dK \tag{3.31}$$

where *p(K)* is a probability density function known as the affinity spectrum, which represents the fraction of binding sites with a value of the microscopic affinity constant between *K* and *K* + *dK*. The simple Langmuir or Langmuir-Freundlich equations are commonly used as local isotherms. Therefore, the proton affinity distribution can be calculated using a simplified approximation of the local isotherm by using, for example, the condensation approximation (CA) method. In this method, the local isotherm is replaced by a step function, which is the first derivative of the binding curve:

$$f_{CA}(\log K = -\log[H^+]) = \frac{d\theta}{d\log[H^+]} \tag{3.32}$$

Alternatively, the experimental binding curve data can be described by means of an arbitrary empirical isotherm, using a conventional fitting procedure. The NICA (Non-Ideal Competitive and thermodynamically consistent Adsorption) isotherm model has been extensively used to describe heterogeneity and competition on ion binding to humic/fulvic substances (Kinniburgh et al. 1999) and many different biosorbents (Lodeiro et al. 2006b; Herrero et al. 2011; Lopez et al. 2011; Zhao et al. 2015). If only proton binding is considered (absence of competing ions) the NICA equation leads to the well know Langmuir-Freundlich isotherm.

On account of the importance of the ion exchange mechanism in biosorption, it is worth mentioning that analogies and differences between competitive adsorption models and ion exchange models has been discussed and analysed for metal biosorption systems, concluding that both descriptions are equivalent if only equilibrium properties are compared (Rudzinski and Plazinski 2010; Plazinski 2013b). Moreover, heterogeneity effects considering a continuous function of binding site, the stoichiometry of the ion exchange reaction responsible for the “apparent” heterogeneity and a site discrete model, have been also studied (Plazinski and Rudzinski 2009, 2011).

It should finally be mentioned that in addition to the NICA-Donnan model, there are other models commonly used for studying the metal-natural organic matter (NOM) interactions; those are the Windermere humic aqueous model (WHAM), including three different versions (V, VI and VII) (Tipping 2002; Groenenberg and Lofts 2014), and the Stockholm humic model (Gustafsson 2001). The WHAM model is a discrete site model used to describe the binding of protons and metal cations to humic substances. This model has been mainly applied to speciation studies in freshwaters; although both WHAM and NICA-Donnan models, have been applied for the first time at ionic strengths greater than 1 mol kg^{-1} (Marsac et al. 2017). The WHAM models can provide, in general, similar results to the ones obtained using the NICA-Donnan continuous distribution site approach. Nevertheless, only the later model is applied to adsorption/biosorption studies. However, both models have been used for proton binding determinations, and several comparisons between them can be found in the literature (Christensen et al. 1998; Tipping 2002; Dudal and Gerard 2004; Merdy et al. 2006).

3.7 Nature, Abundance and Strength of Functional Sites in Biosorbents

Despite the wide variety of biopolymers and, therefore, of functional groups, almost all potentiometric studies practically agree that the acid-base properties of different types of natural biomass are mainly due to the contribution of the surface carboxyl and phenolic groups.

Several authors have studied the acid-base properties of a lignocellulosic substrate extracted from wheat bran. Ravat et al. (2000) characterized the solid by IR and ^{13}C-NMR and studied the proton binding by potentiometric titrations. These authors found that the substrate can be represented by two acid groups, a carboxylic and a phenolic one, at a concentration of 0.08 $mmol{\cdot}g^{-1}$ and 0.28 $mmol{\cdot}g^{-1}$, respectively. By using a discrete two-pK model combined with an electrostatic double-layer model, the surface acidity constants were found to be 3.37 and 8.34, which was attributed to the pK of carboxylic and phenolic groups, respectively. Bouanda et al. (2002) also studied a lignocellulosic substrate obtained from wheat bran. The potentiometric and conductimetric techniques were both used to quantify the number of acid functional groups and to determine the proton-binding constants with the NICA-Donnan model. The number of acid groups found was 0.54 $mmol{\cdot}g^{-1}$ for the carboxylic groups and 0.31 $mmol{\cdot}g^{-1}$, for the phenolic groups, whereas the acidity constants were 5.51 and 7.22, respectively. Bouanda's results are somewhat different from those of Ravat et al. That might be due to the different proportion of fatty acids on the substrates studied, because of the different physicochemical treatment used in the extraction, and the different models used to analyse the data. The NICA-Donnan model takes into account both the swelling behaviour and the heterogeneity of the lignocellulosic substrate. The NICA-Donnan model was also employed by Li and Englezos to describe the proton binding to lignocellulosic substrate extracted from wheat bran (Li and Englezos 2005). The authors reported 0.68 $mmol{\cdot}g^{-1}$ of carboxylic groups and 0.23 $mmol{\cdot}g^{-1}$ of phenolic groups, with protonation constants values of 6.30 and 7.61, respectively. Fern is another lignocellulosic substrate that has been often employed in biosorption studies, given its abundance in nature. Barriada et al. (2009) studied the acid-base properties and heavy metal adsorption capacity of bracken fern (*Pteridium aquilinum*). The analysis of acid-base titration data by using the Katchalsky model allowed the authors to obtain a total number of acid groups of 0.432 $mmol{\cdot}g^{-1}$and a pK value of 4.37. Lodeiro et al. (2008) also investigated fern biomass; a total number of acid groups of 0.67 $mmol{\cdot}g^{-1}$ and a pK value of 4.24 were obtained. The results from both studies agree with the contribution of the carboxylate groups for the lignocellulosic substrates previously described. However, Lodeiro et al. proposed a model with two and three functional groups to take into account the heterogeneity of the material. Both models provided similar results. When two functional groups were considered, the results agreed with those found for any lignocellulosic material.

Many authors have also investigated the acid-base properties of different agricultural wastes. Reddad et al. (2002) studied the sugar beet pulp, that consists essentially of polysaccharides (72% of the dry matter); glucose, arabinose and galacturonic acid are the main components, all of them in similar proportions and close to 20%. A discrete model with three surface functional groups combined with a diffuse double layer model was proposed. The contribution of the carboxylic groups was split into two, strong and weak contributions, and the third functional group corresponds to the phenols. The authors reported 0.246 $mmol{\cdot}g^{-1}$ of strong carboxyl groups, 0.220 $mmol{\cdot}g^{-1}$ of weak carboxyl groups and 0.109 $mmol{\cdot}g^{-1}$ of phenolic groups, and the corresponding protonation constants were 3.43, 6.05 and 7.89, respectively.

Fruit wastes have also been studied as promising biosorbents. Peels derived from several fruits consist primarily of cellulosic materials rich in pectin, a polysaccharide based on poly-galacturonic acid. Schiewer and Patil (2008b) studied the behaviour of citrus peels. The use of a continuous model revealed four acidic groups with pK values of 3.8, 6.4, 8.4 and 10.7 and a total site quantity of 1.14 $mmol{\cdot}g^{-1}$. Lodeiro et al. (2008) studied acid-base properties of orange peels. The use of a discrete model with two types of binding sites positions resulted in the pK values of 4.00 and 10.35 and the concentrations of 0.49 $mmol{\cdot}g^{-1}$ and 1.43 $mmol{\cdot}g^{-1}$ for each functional group. López-García et al. (2013) obtained similar results when banana skin was studied.

Pagnanelli et al. (2003, 2005a, 2008) have carried out an extensive work to study the acid-base and the complexation properties of a very abundant waste from olive oil production plants, olive pomace. This solid residue consists of fibre (as cellulose), lignin and uronic acids along with oily wastes and polyphenolic compounds. The analysis of the native material, as well as its different fractions and chemical modifications, allows the carboxyl and phenolic groups to be identified as the main active sites responsible for the acid-base behaviour and the metal complexation. From the native material, the authors reported concentrations of 0.17 $mmol{\cdot}g^{-1}$ of carboxyl groups and 0.49 $mmol{\cdot}g^{-1}$ of phenolic groups with pK values of 4.0 and 8.9, respectively (Pagnanelli et al. 2008).

Many different types of biomass have been investigated, but marine biomass, particularly brown algae, is probably the most widely studied adsorbent substrate. Brown algae have been found to be very effective for metal binding due to their high content of alginic acid in the cell wall, which may comprise between 14 and 40% of the dry weight (Percival and McDowell 1967; Davis et al. 2003). Alginic acid is a natural polysaccharide containing β-D_mannuronic and α-L guluronic acid residues arranged in a non-regular linear chain. Its acid-base properties have been studied by a number of authors. Haug (Haug 1961) studied alginates from different sources, mainly Laminaria algae, and he found that the proportion of the two uronic residues in the alginate determine the acid-base/physical properties and reactivity of the polysaccharide. The pK values of 3.38 and 3.65 were found for mannuronic and guluronic acid, respectively. De Stefano et al. (2005) investigated the acid-base behaviour of sodium alginate by potentiometric and calorimetric titration measurements. The dependence on the ionic strength of the protonation constants were analysed by a modified specific interaction theory (SIT) model. Lin and Marinsky (1993) studied the saline effect on the basis of a Gibbs-Donnan based approach. Rey-Castro et al. (2004b) also investigated this effect based on the Gibbs-Donnan and the specific-ion interaction theories.

The interaction of protons and metals with algae was first investigated by Crist et al. in the 1980s (Crist et al. 1981, 1988). Their studies were focussed on different types of green algae, macroscopic freshwater ones, such as *Vaucheria*, or filamentous ones, such as *Spyrogyra* or *Oedogonium* (Crist et al. 1988). These authors found that adsorption occurs due to the electrostatic interaction of the protons and the metal ions with the carboxylic groups from the cell wall pectin of the green algae.

Volesky and co-workers have been the largest contributors to the study of adsorption processes by use of marine biomass. Their research covers the biosorption of pollutant metals, precious metals, radionuclides, anions, the adsorption mechanisms, continuous processes, modelling tools and also the study of the acid-base properties of algae.

Fourest and Volesky (1996) investigated the two potential ligands present in brown algae: carboxyl and sulfonate groups. The brown seaweed of *Sargassum fluitans* was firstly studied. Simultaneous potentiometric and conductimetric titrations, together with chemical analysis, gave information concerning the amount of strong and weak acidic functional groups in the biomass, 0.25 $mmol{\cdot}g^{-1}$ for the sulfonate groups from fucoidans and 2.00 $mmol{\cdot}g^{-1}$ for the carboxylic groups from alginates, respectively; a third contribution due to polyphenols was also found. Later on, the study was extended to four different brown algae *Sargassum fluitans, Ascophyllum nodosum, Fucus vesiculosus*, and *Laminaria japonica,* that were characterized by using potentiometric titrations, ^{13}C-NMR, chemical analysis, and viscosity measurements (Fourest and Volesky 1997).

Schiewer and Volesky (1997b) studied the brown alga *Sargassum*. The Donnan model, that had been previously applied to humic and fulvic acids, was used for interpretation of potentiometric data at different ionic strengths. Furthermore, these authors also studied the swelling of *Sargassum* particles, which was explained by a simple linear relationship between swelling and pH.

Schiewer and Wong (2000) studied the green alga *Ulva Fascia* and the brown algae *Sargassum hemiphyllum*, *Petalonia fascia*, and *Colpomenia sinuosa*. From potentiometric measurements, the total amount of binding sites was determined. The number of carboxylic groups were found to decrease in the order *Petalonia* (2.9 $mmol{\cdot}g^{-1}$) > *Sargassum* (2.6 $mmol{\cdot}g^{-1}$) > *Colpomenia* (1.5 $mmol{\cdot}g^{-1}$) > *Ulva* (1.1 $mmol{\cdot}g^{-1}$). The titration curves for all algae showed a marked effect of ionic strength and the Donnan model was successfully used to account for this effect. The same pK_{app} value 3.0 can be used for all algae.

Sastre de Vicente and co-workers have also contributed significantly to studying the adsorption properties of marine algae, and their acid-base behaviour in particular.

Rey-Castro et al. (2003) carried out an extensive study on the acid-base properties of three brown algae: *Sargassum muticum, Cystoseira baccata* and *Saccorhiza polyschides*. The proton binding equilibria of the three seeweeds was studied potentiometrically and the effect of pH, ionic strength and composition of the medium was investigated. The Donnan model combined with the master curve approach was used to interpret the influence of the ionic strength. Different empirical expressions that describe the swelling behaviour of the sorbents were tested. The results showed very little influence of the type of electrolyte. The dependence of proton binding capacity on the ionic strength was very similar for the three algae. The maximum proton binding capacities obtained ranged between 2.4 and 2.9 $mmol{\cdot}g^{-1}$ and the average intrinsic proton affinity constants ranged between 3.1 and 3.3. These data were then re-analysed to compare two opposite and ideal electrostatic models, the Donnan model and the so-called surface charge model (Rey-Castro et al. 2004a). The Donnan model assumes the interphase to behave as a permeable three-dimensional gel that shows an effective Donnan volume, whereas the surface charge model describes the system as a non-permeable two-dimensional surface with a specific surface area as main feature. Both models seemed to be almost equivalent, although the Donnan model provided slightly better results and a simpler way to account for the effect of activity coefficients and the non-specific binding.

Further studies on *Sargassum muticum* have been considered of great interest, as this alga is an invasive species in European waters. Different approximations, e.g. NICA model or Katchalsky model (Katchalsky et al. 1954) were applied to analyse the acid-base properties (Lodeiro et al. 2004, 2005b). Katchalsky model was also applied to study the brown algae *Cystoseira baccata* (Lodeiro et al. 2006a), *Bifurcaria bifurcata*, *Saccorhiza polyschides*, *Ascophyllum nodosum*, *Laminaria ochroleuca* and *Pelvetia caniculata* (Lodeiro et al. 2005a). Similar results were found for all the brown algae. The total number of weak acid groups was large and in the range from 2.43 to 3.33 $mmol{\cdot}g^{-1}$. All the pK_{app} values were found to be around 3.5, which can be identified with the alginic acid, specifically with the mannuronic and guluronic acids.

3.8 The Role of the Acid-Base Properties of Biosorbents on Metals Removal

This chapter is focused on some selected examples where the interactions of biomass with metals are described. The main reasons for the specific selection of metal binding biosorption studies, and not others dealing with the interactions of organic pollutants (e.g. dyes, phenols or endocrine disruptors) are explained below.

As previously mentioned, the main goal of this review is to study the interactions of protons with biosorbents, i.e. their acid-base properties, both from a theoretical and experimental point of view. These acid-base properties are the main factor responsible for the interactions of either metals or organic substances with the surface of the biosorbents. The great majority of c.a. 16,000 papers (SciFinder database) related to biosorption studies investigated metal-biomass interactions, with a minority of studies dedicated to the investigation of organic pollutants. Moreover, the interpretation of the interactions of organic substances with biosorbents is more challenging than the interpretation of the respective metal interactions. As in the case of metals, the presence of ionized organic substances in solution results in electrostatic effects. These charge-related effects are influenced by pH, ionic strength, and electrolyte type. Moreover, the organic substances, e.g. pesticides, dyes, phenols, nitro compounds or endocrine disrupting chemicals, present a great structural variety. This chemical diversity influences the presence of van der Waals forces, hydrophobic effects and hydrogen bonding, in addition to the electrostatic forces already mentioned. Therefore, pH not only determines the ionic/neutral species ratio that influences the interaction with the surface of biosorbents (Xiao and Pignatello 2014), but also balances all the interactions specifically associated to the organic compounds. Therefore, quantitative analysis and interpretation of the interactions between biosorbents and organic pollutants are not straightforward (Healy and White 1978; Nelson and Schwartz 2013) and would require a dedicated review (Luthy et al. 1997; Aksu 2005; Shon et al. 2006; Higgins and Luty 2007; Richter et al. 2009; Kushwaha et al. 2013; Webster 2014; Van Son et al. 2015; Zhou et al. 2015; Christl et al. 2016).

The relationship between the acid-base properties of biosorbents and their adsorption capacity is probably the key question in many biosorption studies. This issue is not a simple one because of the nature of biomass, which consists of a varied and complex mixture of polymeric species. However, the detailed investigation of acid-base properties reveals that the polymer in largest proportion determines the fundamental behaviour.

The analysis of the adsorption capacity is an even more complex issue, mostly because of the failure to elucidate the adsorption mechanism. Biosorption consist of several mechanisms (Chen and Jianlong 2009; Javanbakht et al. 2014) mainly physical adsorption, ionic exchange, complexation, chelation, reduction or microprecipitation (Veglio and Beolchini 1997; Schiewer and Volesky 2000; Crini 2005).

The interaction, and thus the adsorption, is strongly dependent on the solution conditions, which are decisive for the biomass surface, the metal speciation or the competition of other ions or organic molecules. The direct consequence is that it is difficult to explain the adsorption by one single mechanism. Therefore, it is quite possible that some of these mechanisms are acting to varying degrees simultaneously, most commonly, the ionic exchange, complexation, reduction/oxidation reactions and metal precipitation.

The occurrence of the functional groups involved in the ion exchange and complexation mechanisms are usually the same as those that account for acid-base properties. When these two mechanisms govern the adsorption, there is a direct correlation with the acid-base properties of the biomass (Schiewer and Volesky 1995).

When reduction of the metal ions and resulting metal precipitation play the key role, the adsorption is attributed to those functional groups that are easily oxidisable, without need for being related to the acid-base properties.

In any case, as indicated above, it is most likely to find a complex mechanism in which some of the above-mentioned processes participate simultaneously.

Despite the problems mentioned, a large number of studies report the correlation of the adsorption capacity with the number of protonated groups in the biomass. Lodeiro et al. (2008) studied the Cr(III)-binding capacity of three different types of biomass, the brown *Sargassum muticum* macroalgae, orange peel and bracken fern. On the one hand, the authors found that the maximum Cr(III) uptake capacity is approximately equal to the number of carboxyl functional groups determined potentiometrically. On the other hand, the complexation constants were determined and a very close value is obtained for the three materials ($\log K_{Cr} = 2.9–3.1$), that reinforces the hypothesis of the implication of the same functional group, e.g. carboxyl groups, in metal uptake. Barriada et al. (2009) studied the adsorption of Cd (II) and Pb (II) on bracken fern. Maximum uptake was found to be the same for both metals (0.410 $mmol \cdot g^{-1}$), which is very similar to the number of acidic groups determined for this material that was found to be 0.432 $mmol \cdot g^{-1}$. Once again, the results indicate that acidic groups were responsible of the sequestration of both metal ions.

The analysis of the effect of pH on adsorption capacity, further supportive evidences for the implication of acidic groups on metal binding. For example, an S-Shaped curve centred at pH 3–4 is usually found for metal adsorption (Figure 3.10) (Schiewer and Volesky 1995; Ravat et al. 2000; Reddad et al. 2002; Lodeiro et al. 2004, 2005a). At pH values below 2.0, the metal uptake is very low, but not negligible, which is related to the presence of a relatively low amount of very strong acid groups such as sulfonic groups, which are present in the fucoidans of brown algae. The change in the ionic state of the carboxyl functional groups, which are associated with the polymers of the cell wall, explains the dramatic increase in adsorption of metals from pH 2 to 4. Above pH 4 the metal sorption capacity levels off at a maximum value (Haug and Smidsrod 1970; Rey-Castro et al. 2004b).

The fact that the same functional groups, i.e. the same sites, are used for the proton and metal binding is evident when acid-base and metal adsorption properties

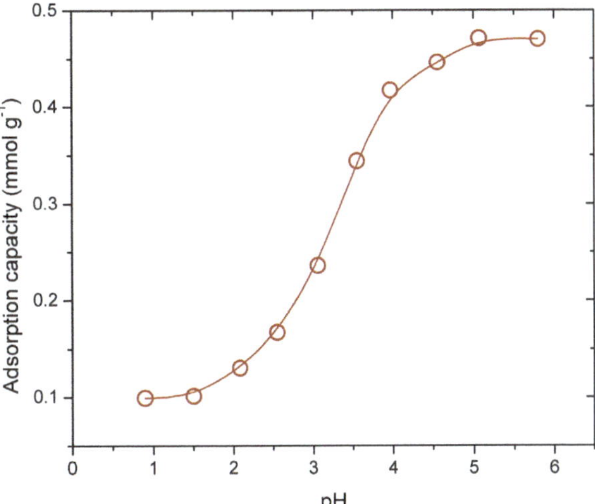

Fig. 3.10 Dependence of the Cd adsorption capacity of the macroalga *Cystoseira baccata* on pH. The circles represent experimental data points and the line guides the eye. The graph shows the typical S-shaped curve that supports evidence for the implication of acidic groups on metal binding

are modelled simultaneously, by use of competitive proton-metal models. The proposed equations are able to describe both proton and metal experimental data satisfactorily. Models of varying degrees of complexity have been proposed; some of them take into account different isothermal models, others several binding sites, heterogeneity or different stoichiometric proton/metal ratios.

Langmuir competitive model with a single binding site is one of the simplest models that has been successfully applied by Schiewer and Wong (1999) to Ni and Cu adsorption by several types of algae, assuming 1:2 binding stoichiometry. Lodeiro et al. (2005b) investigated the Cd adsorption by biomass of the brown marine algae *Sargassum muticum*. The authors compared Langmuir competitive models, assuming 1: 1 and 1: 2 stoichiometries. The NICA model can adequately explain all the experimental data, both concentration and pH dependence of cadmium uptake, employing the same constants attained through proton binding studies. Pagnanelli et al. (2005a) obtained similar results using the NICA model to reproduce the Cu and Cd biosorption experiments on olive pomace. Li and Englezos (2005) employed the NICA-Donnan model to describe the interaction of protons and metal ions, Cu (II), Pb (II), Fe (III) and Mn (II), and the lignin extracted from wheat bran and kraft pulp. They were able to reproduce with great accuracy the experimental data, assuming only two types of sites for the binding of protons or metal ions to lignin, considered to be due to carboxylic-type and phenolic-type groups.

Herrero et al. (2011) studied the Cu(II) uptake by the macroalga *Sargassum muticum*. A simple Langmuir or Langmuir-Freundlich isotherm can be used to accurately describe equilibrium experiments. However, only the NICA model allows a good description of all equilibrium experiments tested, i.e. isotherm, pH influence and competition between Cu and Cd, employing the same constants attained through proton binding studies.

All studies above make clear that protons and metal ions compete for the same adsorption positions on different types of biomass.

3.9 Potentiometric Determination of the Acid-Base Properties of Biosorbents

The determination of acid-base properties of the biosorbents provides very useful information about the physicochemical behaviour of these substances, and consequently their performance in adsorption processes. Not only the total number of acidic sites can be quantified, but also their proton binding affinities (Pagnanelli et al. 2000, 2004; Schiewer and Patil 2008a; Li et al. 2014). In conjunction with other analytical measurements such as calorimetric data or NMR, the acid-base characterization of biosorbents can provide significant information about the functional groups present in these substances, and the contribution of these groups to overall adsorption. However acid-base characterization is not limited to these two aspects, and it can also be used to determine the potential of zero charge (pzc) of the biosorbent (Fiol and Villaescusa 2009; Lodeiro et al. 2012; Pagnanelli et al. 2013; Li et al. 2014), which is another interesting property of these materials and with important effects on the adsorption.

Broadly speaking, the determination of acid-base properties of biosorbents does not differ from the determination of the acid-base behaviour of any other simple substances. That is, during a titration a typical s-shape curve will be obtained, with one or several inflection points, depending on the nature of the biosorbent (Naja et al. 2005; Schiewer and Patil 2008a). The analysis of these curves will provide the corresponding acid-base information of the substance under study. But it has to be kept in mind that any biosorbent can be considered as a heterogeneous mixture of multifunctional polymers. Consequently, the analysis of the titration curves is not as trivial as that obtained for a single, pure substance (Lenoir and Manceau 2010). Besides, being multifunctional biopolymers, the corresponding titration curves are not as sharp as the ones obtained for simple substances and the s-shape curves have inflection points not always well defined (Fig. 3.9).

Different techniques can be used to obtain acid-base information of substances, among others NMR measurements, UV-Vis spectrophotometric readings, conductivity measurements or calorimetric data (Gans et al. 1996, 2008; Bouanda et al. 2002; De Stefano et al. 2005). Nevertheless, potentiometry with glass electrode is the most important one, widely used especially in biosorbent acid-base analysis; most likely due to its simplicity and accuracy. Moreover, potentiometry requires very simple, commonly employed equipment present in any laboratory, and is also reasonably easy to automate (Barriada et al. 2009; Lodeiro et al. 2012; Lopez-Garcia et al. 2013), decreasing the manipulation of the sample and the preparation of different mixtures.

Potentiometric titrations are based in the measurement of the electromotive force (emf) appearing between a glass electrode and a reference electrode in a solution. In the case of a biosorbent determination, instead of working with a solution, a suspension of the biomaterial is present; that is the major change compared to a traditional titration. As in any titration, the pH of the suspension is modified by addition of an acid or base, and the evolution of the pH of the mixture (in fact, the evolution of the emf readings) is followed. The pH or emf and volume of titrant added constitute the starting data employed during analysis.

3.9.1 Experimental Set-Up

The typical potentiometric experiment set-up can be described as a vessel where the biosorbent suspension is allowed to attain equilibrium after each titrant addition (Barriada et al. 2009; Lodeiro et al. 2012; Lopez-Garcia et al. 2013). Temperature control using a thermostated vessel is particularly useful, especially if the effect of the temperature is going to be analysed. The titration vessel is typically closed with a lid with several ports where the glass electrode, an inert gas bubbler and a tip for adding the titrant are introduced. A temperature probe is optional. A reference electrode is also required, but if just proton activity is going to be followed, a combination glass electrode is typically used. If other species activities are going to be measured (e.g. metal cations) a common reference electrode has to be shared between the glass electrode and the ion selective electrode in order to avoid reading interferences between the two electrodes. Nowadays the most common reference electrode is the Ag|AgCl electrode, either in a saturated chloride solution or at 3 M concentration chloride solution. An inert gas bubbler must be used during the titration; it helps to remove the interference of gases present in the suspension, mainly CO_2 and O_2. A vigorous stirring procedure is often employed, typically magnetic stirring, to achieve a correct mixture of the suspension. However, it has to be kept in mind that many of the biosorbents studied are rather soft materials, so caution with stirring has to be taken into consideration, since too vigorous stirring could disaggregate the biomaterial under study. This is especially important if high temperatures are going to be used during the titration. Figure 3.11 shows a typical potentiometric titration set-up.

Finally, another common procedure in potentiometric acid-base titrations, also for biomaterials, is the usage of an inert electrolyte (Bouanda et al. 2002; De Stefano et al. 2005; Barriada et al. 2009). A glass electrode is sensitive to proton activity, therefore in order to minimize the change in proton activity coefficient, a suspension of the biomaterial in a solution containing an inert electrolyte is employed rather than a suspension in pure water. Both the suspension of the biosorbent and the titrant added during the determination are prepared in a solution with the same ionic strength. The inert electrolyte should be present at 0.1 M or more in order to buffer changes in the proton activity coefficient. This allows the response of the glass electrode to be related to the concentration of free protons in solution rather than to their activity. This facilitates the subsequent derivation of the apparent equilibrium constants of the acidic groups involved in the acid-base equilibria.

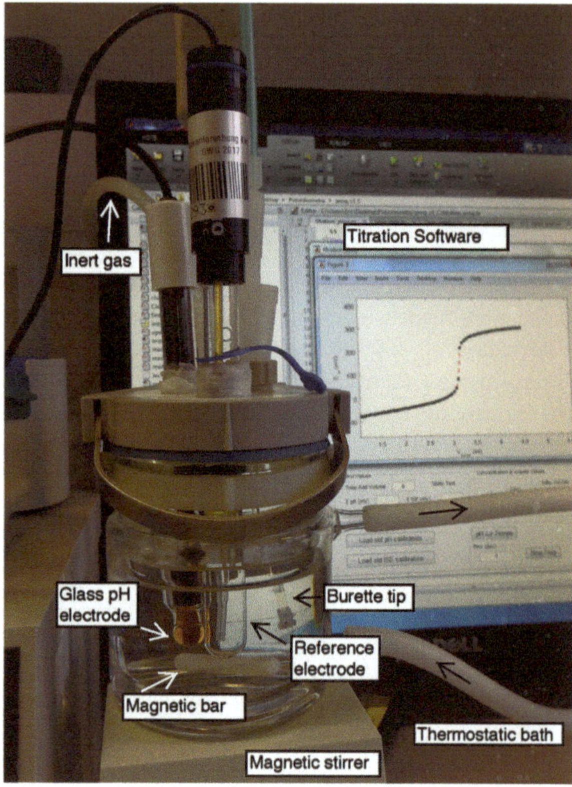

Fig. 3.11 Typical titration set-up for biomass acid-base properties determination. The experimental set-up is formed by a thermostated vessel, a pH glass electrode, a reference electrode, an inert gas bubbler and a burette for adding the titrant. Magnetic stirring is used to achieve a correct mixture of the biomass suspension

3.9.2 Calibrations

As it has been mentioned above, the most common procedure in potentiometric acid-base titrations consist of using an inert electrolyte. This electrolyte does not interact with the species being titrated, and buffers the ionic strength of the solution. In this situation, the response of the electrode depends proportionally on proton concentration, while the activity coefficient of this species remains almost constant during the whole titration. In order to obtain a relationship between the proton concentration and the emf readings, a calibration procedure is followed. The simplest calibration can be performed using a solution of the inert electrolyte, at the same ionic strength that is going to be used in the titration of the biosorbent, where aliquots of a strong acid e.g. HCl with electrolyte at the same ionic strength are added. If the concentration of the strong acid is accurately known, it is possible to calculate the concentration of the protons after each addition. A direct relationship between the emf readings and the proton concentration can be easily obtained using the following equation:

$$E = E^* + p\ log[H^+] \tag{3.33}$$

where E is the electromotive force, E* is the so-called formal potential which encompasses the standard potential of the electrode system, the potential of asymmetry and the liquid junction potential; moreover, it also depends on the activity coefficient of the proton, a constant when fixed ionic strength solutions are used. The slope in the representation, *p*, theoretically should be the Nernstian slope, 59.16 mV per decade in aqueous solutions at 25 °C. Nevertheless, in practice, it is considered a fitting parameter together with the formal potential. When a glass electrode is providing an accurate response, a plot of log $[H^+]$ vs. E renders a straight line (May et al. 1982) with a slope that should not be very different from the theoretical value (59.16 at 25 °C). For every addition the concentration of protons is calculated as:

$$[H^+] = C_a \cdot \frac{v}{V_0 + v} \tag{3.34}$$

where C_a is the concentration of the acid in the titrant solution, v is the total volume of the aliquots of acid added for each emf reading and V_0 is the initial volume of the electrolyte solution.

Calculation of proton concentration for every addition using Eq. 3.34 together with the emf readings allow fitting of the experimental data to Eq. 3.33, obtaining both the slope *p* and the formal potential.

Calibration of glass electrode following this procedure must be done periodically, since both formal potential and slope change over time, especially if the glass electrode has been stored dry for a considerable time. Frequent calibrations of a glass electrode will evolve to stable values of formal potential and slope, which in turn can be used to transform the readings of a biomass titration into values of free proton concentration, which are used in the calculation of the proton affinity constants of the functional groups present in the biomass.

Working with the free proton concentration scale ($p[H^+] = -\log[H^+]$) (May et al. 1982; Fiol et al. 1992; Brandariz et al. 2001, 2004) might look cumbersome compared to working with pH ($pH = -\log a_{H^+}$, where a_{H^+} stands for activity of protons). However, it has to be kept in mind that pH calibration of glass electrodes using the commercially available pH buffers is done in an unknown ionic media that is going to be totally different from the media used in the biomass titration. Hence an unknown uncertainty regarding proton activity coefficient will be introduced using the pH scale. Thus, use of free proton concentration scale is recommended.

3.9.3 Measurements

Although the potentiometric titrations of biosorbents are very similar to acid-base titrations of simple substances, some important differences have to be discussed. As has been indicated above, a biosorbent titration is done with a suspension of the biosorbent. Therefore, a heterogeneous mixture is used rather than working with a homogenous solution, as is the case of a simple substance titration. This fact makes stirring an important factor, since the biosorbent will tend to separate from solution by gravity. Furthermore, if the biosorbent is decanted at the bottom of the titration vessel it might not react with the titrant added, or it will take longer to react. The time required to achieve a good hydration of the sorbent material is also important. In some cases, e.g. biomass from fern, the biomaterial is stored dry; when this material is used in a titration, it is quite water-repellent and it takes a considerable time to become fully hydrated.. Other materials, such as alga, do not show this problem. They are hydrated within minutes, but they tend to swell and become very soft, so caution should be taken when stirring to avoid destruction of the sorbent.

Another point to take into consideration is the protonation state of the biosorbent. Depending on the nature of the sorbent and the environment where it is found, the functional groups that constitute the active sites of the material can be in different protonation states. This might make it difficult to interpret the potentiometric titration data, since it can significantly influence in the number of total acid sites determined. A common procedure consists in doing an acid wash of the material in order to start the titration with the biosorbent on its fully protonated state before proceeding with the titration (Rey-Castro et al. 2003; Schiewer and Patil 2008a; Barriada et al. 2009). Other researchers used a different approach. Starting with the native, untreated material, an accurate amount of acid is added to the suspension and the mixture is allowed to reach equilibrium (Pagnanelli et al. 2013). At that moment, the titration is started by adding base of known concentration. The amount of acid initially added is taken into account in the subsequent calculations. Of course, this description corresponds to a "forward" titration, where the pH of the mixture is progressively increased by addition of a base of known concentration. It is also possible to do "backward" titrations, starting with the biomass in a fully deprotonated state, and perform the titration adding as titrant an acid of accurately known concentration.

It is also important to consider that during biomass titrations the glass electrode readings are much more unstable than in a common acid-base titration. Electromotive force tends to drift significantly, not only because a heterogeneous medium is involved, but also a complex biopolymer mixture is being titrated. Therefore, conformational modifications of the biomolecules can be expected during the titration besides complex electrostatic effects. As a consequence of this drift in the glass electrode readings, the titrations take a considerable time, and a criterion for "stable" readings must be adopted. Usually researchers considered that the glass electrode is "stable" if the change in the emf is small (below 1 mV) during a moderate time interval (2–5 min) (Bouanda et al. 2002; Naja et al. 2005). This drift in the readings is more significant when the titration is reaching an inflection point and a typical titration can take hours depending on the system under study. Dealing with such long times, and having to wait for stable readings, biomass titrations are time-consuming, and therefore automation of the titration is desirable. A computer controlled titration system can easily cope with the recording of experimental data that will be used in data analysis (Barriada et al. 2009; Lodeiro et al. 2012).

3.9.4 Data Analysis

The analysis of the data obtained during a titration is usually based on a charge balance equation; that is, the electroneutrality condition: the charges of all the positive species has to be compensated by the charge of the negative species. As an example, let us suppose that a titration of a biosorbent is done with sodium hydroxide and a known amount of hydrochloric acid has been added to start the titration in acidic conditions. In this situation, at any point of the titration the electroneutrality condition can be written as it follows:

$$[Na^+] + [H^+] - [Cl^-] - [OH^-] + Q = 0 \tag{3.35}$$

where Q is the charge concentration present in the biosorbent at each point of the titration. If the Q value is positive, it implies that the biosorbent has a neat positive charge. On the contrary, the biosorbent is negatively charged. The Eq. 3.35 can be easily transformed into charge-pH or master curves (see Sect. 3.6).

The concentrations of the cations and anions from the background electrolyte are not considered here since they mutually cancel each other out. If the glass electrode has been calibrated in the free proton scale, the emf readings can be easily transformed into proton concentration $[H^+]$ (Eq. 3.33). Also, hydroxide concentration can be obtained from the equilibrium constant of water once the hydrogen concentration is known. The concentrations of sodium and chloride ions can be easily calculated from knowledge of the concentration of the titrant, the initial volume of the mixture and the concentration of acid added. Therefore, the charge concentration of the biosorbent can be computed for each titrant addition (Fig. 3.11).

In fact, this is a procedure to determine the potential of zero charge of the biosorbent. If the titration is made in the same conditions but for different masses of biosorbent, a plot of pH *versus* charge should cross at a single point corresponding to the potential of zero charge. Since at this potential the biomass shows a neat zero charge at its surface, this point will be independent of the mass of biomaterial used in the titration and the intersection of the curves will provide the pzc value. Figure 3.12 shows a typical plot that can be used to determine the pzc value of a substance. In this case instead of plotting the charge *versus* the pH, a plot of pH *versus* titrant added is represented. In any case, the intersection of the curves indicates the pH at which the material used has a zero net charge in its surface.

If the charge so calculated corresponds to the charge of an acidic group once it gets deprotonated, according to the following equilibrium:

$$R - AH \rightleftharpoons R - A^- + H^+ \tag{3.36}$$

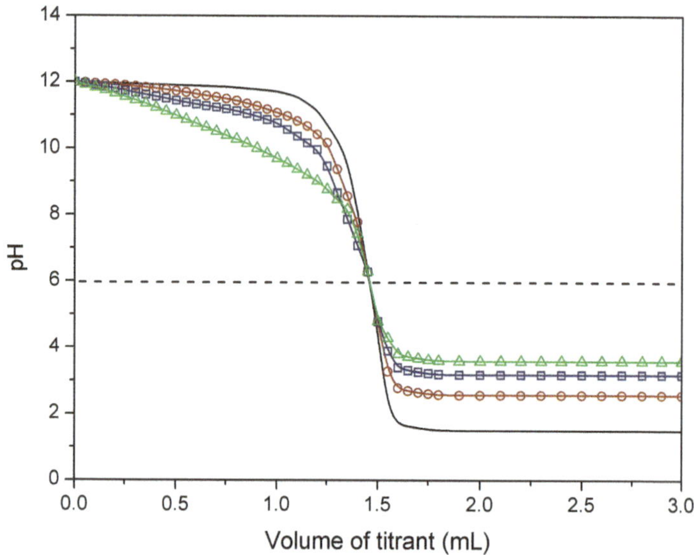

Fig. 3.12 Simulated acid-base titration of a biomaterial using different masses: e.g. 10 (red), 25 (blue) and 50 (green) g·L^{-1}. The black line represents a control titration without biomass. The intersection point, in this case ca. pH 6, corresponds to the potential of zero charge (pzc) of the biomaterial under study. At this pH, the biomass should show a neat zero charge at its surface, so this point is independent of the mass of biomaterial

then, the dissociation degree, α, can be calculated as:

$$\alpha = \frac{[R - A^{-}]}{[R - AH]_{tot}} = \frac{Q}{C_A} \tag{3.37}$$

where $[R - AH]_{tot} = C_A$ represents the total number of acidic sites present in the biosorbent, which can be calculated from the inflection point of the titration curve, or estimated by other means such as the barium hydroxide method (Fukushima et al. 1999).

The apparent protonation constant (K_{app}) of the acidic groups involved in the equilibrium can then be computed with Eq. 3.25, which is based on a modification of Henderson-Hasselbach equation (Katchalsky and Spitnik 1947; De Stefano et al. 2005). The plot of $p[H^{+}]$ vs. $log\left(\frac{1-\alpha}{\alpha}\right)$ should be a straight line that intersects (i.e. when α=0.5) the x axes at $p[H^{+}] = \log K_m = \log K_{app}$.

The procedure described above corresponds to a situation where only one major functional group (e.g. carboxylic acid) seems to be present in the biomass, i.e. only one inflection point is observed in the titration curve. If more than one inflection point are observed in the curves, different approaches are required. Sometimes a continuous affinity spectra model is used; in that case, the affinity constants of the functional groups present in the biomass are considered to vary in a continuous range of values. A discrete values model can be also used, where between 2 and 4 types of functional monoprotic acidic sites are modelled in order to described the experimental curves. Generally speaking, differences regarding to experimental curve fitting between the continuous model and a 3 sites model are minimal. Titrations of biomass with several inflection points are usually well described considering strong acidic sites with apparent pK values below 4, weak acidic sites with pK values between 4 and 7 and very weak acidic sites with pK values above 7. Correct identification of these functional sites is more complicated, considering the very heterogeneous nature of the biomaterials, but quite often the strong acidic sites are associated with sulfonate or phosphoric functional groups, weak sites are associated with carboxylic moieties and very weak with phenolic or amine groups present in the biopolymers.

3.10 Conclusions and Future Research Needs

1. A deep knowledge of the acid-base properties of biosorbents is the basis to understand the mechanisms of biosorption. Moreover, a full comprehension of basic acid-base processes helps to perform better the experimental set-up on biosorption studies. For example, the acid-base chemistry knowledge supports the design of biosorbents for a specific use, or the establishment of adequate experimental conditions for the optimum behaviour of a biosorbent.
2. Different external parameters have to be modified to obtain relevant information on the properties of the biosorbent. The most important variables at fixed temperature are the pH, the ionic strength and the nature of the electrolyte.
3. Ionization/protonation of biosorbents can be interpreted, as a first approach, in terms of Gibbs free energies including electrostatic and non-electrostatic interactions. This conceptual division allows associating the interaction terms with molecular parameters, which have relevant effects on the biosorption process. This approach is underdeveloped and future research is needed.
4. The analysis of possible Hofmeister series could help to quantify non-electrostatic effects associated with the interaction between biosorbent and adsorbate. The consideration of Hofmeister effects is not only valid for proton binding but also for interactions between biosorbents and other species such as metals or organic substances.
5. The proton is a master variable in aqueous chemistry that is involved in ion exchange, precipitation, complexation and other chemical processes. Proton experimental control is relatively easy by potentiometry. However, potentiometric measurements should be supplemented with other techniques that confirm the nature of the functional groups involved in the biosorbent acid-base behaviour.
6. The potentiometric technique also allows the determination of zero charge point values of the biosorbents. This aspect is well developed and discussed for minerals. Nevertheless, the interpretation and coherency of the values obtained for biomass materials is far from being clear. A systematic determination and investigation of zero charge values for biosorbents is therefore needed.

7. Basic interaction laws must be equal at different scales, although with unexpected effects. Therefore, the simple pH-metric technique should be also useful for the study and interpretation of the acid-base behaviour of nanosorbents. This use of potentiometry could provide with relevant information regarding acid-base properties of specific nanomaterials that should be considered for future studies.

References

Abdolali A, Guo WS, Ngo HH, Chen SS, Nguyen NC, Tung KL (2014) Typical lignocellulosic wastes and by-products for biosorption process in water and wastewater treatment: a critical review. Bioresour Technol 160:57–66. https://doi.org/10.1016/j.biortech.2013.12.037

Aksu Z (2005) Application of biosorption for the removal of organic pollutants: a review. Process Biochem 40:997–1026. https://doi.org/10.1016/j.procbio.2004.04.008

Bailey SE, Olin TJ, Bricka RM, Adrian DD (1999) A review of potentially low-cost sorbents for heavy metals. Water Res 33:2469–2479. https://doi.org/10.1016/S0043-1354(98)00475-8

Barriada JL, Caridad S, Lodeiro P, Herrero R, Sastre de Vicente ME (2009) Physicochemical characterisation of the ubiquitous bracken fern as useful biomaterial for preconcentration of heavy metals. Bioresour Technol 100:1561–1567. https://doi.org/10.1016/j.biortech.2008.09.027

Bartschat BM, Cabaniss SE, Morel FMM (1992) Oligoelectrolyte model for cation binding by humic substances. Environ Sci Technol 26:284–294. https://doi.org/10.1021/es00026a007

Bauerlein PS, Mansell JE, ter Laak TL, de Voogt P (2012) Sorption behavior of charged and neutral polar organic compounds on solid phase extraction materials: which functional group governs sorption? Environ Sci Technol 46:954–961. https://doi.org/10.1021/es203404x

Bhatnagar A, Sillanpää M (2010) Utilization of agro-industrial and municipal waste materials as potential adsorbents for water treatment—a review. Chem Eng J 157:277–296. https://doi.org/10.1016/j.cej.2010.01.007

Blum L (1975) Mean spherical model for asymmetric electrolytes .1. Method of solution. Mol Phys 30:1529–1535. https://doi.org/10.1080/00268977500103051

Borrok DM, Fein JB (2005) The impact of ionic strength on the adsorption of protons, Pb, Cd, and Sr onto the surfaces of gram negative bacteria: testing non-electro static, diffuse, and triple-layer models. J Colloid Interface Sci 286:110–126. https://doi.org/10.1016/j.jcis.2005.01.015

Borrok D, Turner BF, Fein AB (2005) A universal surface complexation framework for modeling proton binding onto bacterial surfaces in geologic settings. Am J Sci 305:826–853. https://doi.org/10.2475/ajs.305.6-8.826

Bouanda J, Dupont L, Dumonceau J, Aplincourt M (2002) Use of a NICA–Donnan approach for analysis of proton binding to a lignocellulosic substrate extracted from wheat bran. Anal Bional Chem 373:174–182. https://doi.org/10.1007/s00216-002-1305-z

Brandariz I, Barriada JL, Taboada-Pan C, Sastre de Vicente ME (2001) Estimating the change in liquid junction potential on glass electrodes. Electroanalysis 13:1110–1114. https://doi.org/10.1007/s00706-004-0239-x

Brandariz I, Barriada JL, Vilarinio T, Sastre de Vicente ME (2004) Comparison of several calibration procedures for glass electrodes in proton concentration. Monatsh Chem 135:1475–1488. https://doi.org/10.1007/s00706-004-0239-x

Bronsted JN (1922a) Calculation of the osmotic and activity functions in solutions of uni-univalent salts. JACS 44:938–948. https://doi.org/10.1021/ja01426a003

Bronsted JN (1922b) Studies on solubility. IV. The principle of the specific interaction of ions. J Am Chem Soc 44:877–898. https://doi.org/10.1021/ja01426a001

Brown PA, Gill SA, Allen SJ (2000) Metal removal from wastewater using peat. Water Res 34:3907–3916. https://doi.org/10.1016/s0043-1354(00)00152-4

Burnett PG, Heinrich H, Peak D, Bremer PJ, McQuillan AJ, Daughney CJ (2006) The effect of pH and ionic strength on proton adsorption by the thermophilic bacterium *Anoxybacillus flavithermus*. Geochim Cosmochim Acta 70:1914–1927. https://doi.org/10.1016/j.gca.2006.01.009

Cacace MG, Landau EM, Ramsden JJ (1997) The Hofmeister series: salt and solvent effects on interfacial phenomena. Q Rev Biophys 30:241–277. https://doi.org/10.1017/s0033583597003363

Calero M, Hernainz F, Blazquez G, Angeles Martin-Lara M (2010) Potentiometric titrations for the characterization of functional groups on solid wastes of the olive oil production. Environ Prog Sustain Energy 29:249–258. https://doi.org/10.1002/ep.10376

Carbonaro RF, Atalay YB, Di Toro DM (2011) Linear free energy relationships for metal-ligand complexation: Bidentate binding to negatively-charged oxygen donor atoms. Geochim Cosmochim Acta 75:2499–2511. https://doi.org/10.1016/j.gca.2011.02.027

Chen C, Jianlong W (2009) General mechanisms of biosorption. Liu Y, Wang JL (eds) Fundamentals and applications of biosorption isotherms, kinetics and thermodynamics. Nova Science Publishers, New York, pp 155–212

Christensen JB, Tipping E, Kinniburgh DG, Gron C, Christensen TH (1998) Proton binding by groundwater fulvic acids of different age, origins, and structure modeled with the Model V and NICA-Donnan model. Environ Sci Technol 32:3346–3355. https://doi.org/10.1021/es971134o

Christl I, Ruiz M, Schinidt JR, Pedersen JA (2016) Clarithromycin and tetracycline binding to soil humic acid in the absence and presence of calcium. Environ Sci Technol 50:9933–9942. https://doi.org/10.1021/acs.est.5b04693

Collins KD (2004) Ions from the Hofmeister series and osmolytes: effects on proteins in solution and in the crystallization process. Methods 34:300–311. https://doi.org/10.1016/j.ymeth.2004.03.021

Collins KD, Neilson GW, Enderby JE (2007) Ions in water: characterizing the forces that control chemical processes and biological structure. Biophys Chem 128:95–104. https://doi.org/10.1016/j.bpc.2007.03.009

Costa JF, Vilar VJ, Botelho CM, da Silva EA, Boaventura RA (2010) Application of the Nernst-Planck approach to lead ion exchange in Ca-loaded *Pelvetia canaliculata*. Water Res 44:3946–3958. https://doi.org/10.1016/j.watres.2010.04.033

Cox JS, Smith DS, Warren LA, Ferris FG (1999) Characterizing heterogeneous bacterial surface functional groups using discrete affinity spectra for proton binding. Environ Sci Technol 33:4514–4521. https://doi.org/10.1021/es990627l

Crini G (2005) Recent developments in polysaccharide-based materials used as adsorbents in wastewater treatment. Prog Polym Sci 30:38–70. https://doi.org/10.1016/j.progpolymsci.2004.11.002

Crist RH, Oberholser K, Shank N, Ming N (1981) Nature of bonding between metallic ions and algal cell walls. Environ Sci Technol 15:1212–1217. https://doi.org/10.1021/es00092a010

Crist RH, Oberholser K, Schwartz D, Marzoff J, Ryder D, Crist DR (1988) Interactions of metals and protons with algae. Environ Sci Technol 22:755–760. https://doi.org/10.1021/es00172a002

Dahnert K, Huster D (1999) Comparison of the Poisson-Boltzmann model and the Donnan equilibrium of a polyelectrolyte in salt solution. J Colloid Interface Sci 215:131–139. https://doi.org/10.1006/jcis.1999.6238

Daniele PG, De Stefano C, Foti C, Sammartano S (1997) The effect of ionic strength and ionic medium on the thermodynamic parameters of protonation and complex formation. Curr TopSolut Chem 2:253–274

Daughney CJ, Hetzer A, Heinrich HTM, Burnett P-GG, Weerts M, Morgan H, Bremer PJ, McQuillan AJ (2010) Proton and cadmium adsorption by the archaeon *Thermococcus zilligii*: generalising the contrast between thermophiles and mesophiles as sorbents. Chem Geol 273:82–90. https://doi.org/10.1016/j.chemgeo.2010.02.014

Davis TA, Volesky B, Mucci A (2003) A review of the biochemistry of heavy metal biosorption by brown algae. Water Res 37:4311–4330. https://doi.org/10.1016/S0043-1354(03)00293-8

De Gisi S, Lofrano G, Grassi M, Notarnicola M (2016) Characteristics and adsorption capacities of low-cost sorbents for wastewater treatment: a review. Sustain Mater Technol 9:10–40. https://doi.org/10.1016/j.susmat.2016.06.002

De Stefano C, Foti C, Gianguzza A, Piazzese D, Sammartano S (2002) Binding ability of inorganic major components of sea water towards some classes of ligands, metal and organometallic cations. In: Gianguzza A, Pelizzetti E, Sammartano S (eds) Chemistry of marine water and sediments (chapter 9). Springer, Berlin, pp 221–262

De Stefano C, Gianguzza A, Piazzese D, Sammartano S (2005) Modelling of proton and metal exchange in the alginate biopolymer. Anal Bional Chem 383:587–596. https://doi.org/10.1007/s00216-005-0025-6

Debye P, Hückel E (1923a) The theory of electrolytes I. the lowering of the freezing point and related occurrences. Phys Z 24:185–206

Debye P, Hückel E (1923b) The theory of the electrolyte II – the border law for electrical conductivity. Phys Z 24:305–325

Dewit JCM, Vanriemsdijk WH, Koopal LK (1993) Proton binding to humic substances .1. Electrostatic effects. Environ Sci Technol 27:2005–2014. https://doi.org/10.1021/es00047a004

Di Caprio F, Altimari P, Uccelletti D, Pagnanelli F (2014) Mechanistic modelling of copper biosorption by wild type and engineered *Saccharomyces cerevisiae biomasses*. Chem Eng J 244:561–568. https://doi.org/10.1016/j.cej.2014.01.098

Dittrich M, Sibler S (2005) Cell surface groups of two picocyanobacteria strains studied by zeta potential investigations, potentiometric titration, and infrared spectroscopy. J Colloid Interface Sci 286:487–495. https://doi.org/10.1016/j.jcis.2005.01.029

Dudal Y, Gerard F (2004) Accounting for natural organic matter in aqueous chemical equilibrium models: a review of the theories and applications. Earth-Sci Rev 66:199–216. https://doi.org/10.1016/j.earscirev.2004.01.002

Fein JB, Daughney CJ, Yee N, Davis TA (1997) A chemical equilibrium model for metal adsorption onto bacterial surfaces. Geochim Cosmochim Acta 61:3319–3328. https://doi.org/10.1016/s0016-7037(97)00166-x

Fein JB, Boily JF, Yee N, Gorman-Lewis D, Turner BF (2005) Potentiometric titrations of *Bacillus subtilis* cells to low pH and a comparison of modeling approaches. Geochim Cosmochim Acta 69:1123–1132. https://doi.org/10.1016/j.gca.2004.07.033

Fiol N, Villaescusa I (2009) Determination of sorbent point zero charge: usefulness in sorption studies. Environ Chem Lett 7:79–84. https://doi.org/10.1007/s10311-008-0139-0

Fiol S, Arce F, Armesto XL, Penedo F, Sastre de Vicente ME (1992) Analysis of systematic-errors in calibrating glass electrodes with H as a concentration probe. Fresenius J Anal Chem 343:469–472. https://doi.org/10.1007/bf00322151

Fourest E, Volesky B (1996) Contribution of sulfonate groups and alginate to heavy metal biosorption by the dry biomass of *Sargassum fluitans*. Environ Sci Technol 30:277–282. https://doi.org/10.1021/es950315s

Fourest E, Volesky B (1997) Alginate properties and heavy metal biosorption by marine algae. Appl Biochem Biotechnol 67:215–226. https://doi.org/10.1007/bf02788799

Fukushima M, Tatsumi K, Wada S (1999) Evaluation of the intrinsic acid-dissociation constant of alginic acid by considering the electrostatic effect. Anal Sci 15:1153–1155. https://doi.org/10.2116/analsci.15.1153

Gadd GM (2009) Biosorption: critical review of scientific rationale, environmental importance and significance for pollution treatment. J Chem Technol Biotechnol 84:13–28. https://doi.org/10.1002/jctb.1999

Gagrai MK, Das C, Golder AK (2013) Non-ideal metal binding model for Cr(III) sorption using *Spirulina platensis* biomass: experimental and theoretical approach. Can J Chem Eng 91:1904–1912. https://doi.org/10.1002/cjce.21772

Gans P, Sabatini A, Vacca A (1996) Investigation of equilibria in solution. Determination of equilibrium constants with the HYPERQUAD suite of programs. Talanta 43:1739–1753. https://doi.org/10.1016/0039-9140(96)01958-3

Gans P, Sabatini A, Vacca A (2008) Simultaneous calculation of equilibrium constants and standard formation enthalpies from calorimetric data for systems with multiple equilibria in solution. J Solut Chem 37:467–476. https://doi.org/10.1007/s10953-008-9246-6
Gardea-Torresdey JL, de la Rosa G, Peralta-Videa JR (2004) Use of phytofiltration technologies in the removal of heavy metals: a review. Pure Appl Chem 76:801–813. https://doi.org/10.1351/pac200476040801
Gelabert A, Pokrovsky OS, Schott J, Boudou A, Feurtet-Mazel A, Mielczarski J, Mielczarski E, Mesmer-Dudons N, Spalla O (2004) Study of diatoms/aqueous solution interface. I. Acid-base equilibria and spectroscopic observation of freshwater and marine species. Geochim Cosmochim Acta 68:4039–4058. https://doi.org/10.1016/j.gca.2004.01.011
Gerente C, Lee VKC, Le Cloirec P, McKay G (2007) Application of chitosan for the removal of metals from wastewaters by adsorption – mechanisms and models review. Crit Rev Environ Sci Technol 37:41–127. https://doi.org/10.1080/10643380600729089
Ginn BR, Szymanowski JS, Fein JB (2008) Metal and proton binding onto the roots of *Fescue rubra*. Chem Geol 253:130–135. https://doi.org/10.1016/j.chemgeo.2008.05.001
Goldberg S, Criscenti LJ (2008) Modeling adsorption of metals and metalloids by soil components. Wiley, Hoboken/New Jersey
Gonzalez-Davila M, Santana-Casiano JM, Laglera LM (2000) Copper adsorption in diatom cultures. Mar Chem 70:161–170. https://doi.org/10.1016/s0304-4203(00)00020-7
Goss KU, Schwarzenbach RP (2001) Linear free energy relationships used to evaluate equilibrium partitioning of organic compounds. Environ Sci Technol 35:1–9. https://doi.org/10.1021/es000996d
Grenthe I (2002) Equilibrium analysis, the ionic medium method and activity factors. Springer-Verlag, Berlin
Groenenberg JE, Lofts S (2014) The use of assemblage models to describe trace element partitioning, speciation, and fate: a review. Environ Toxicol Chem 33:2181–2196. https://doi.org/10.1002/etc.2642
Guggenheim EA (1935) The specific thermodynamic properties of aqueous solutions of strong electrolytes. Philos Mag 19:588–643
Guibal E (2004) Interactions of metal ions with chitosan-based sorbents: a review. Sep Purif Technol 38:43–74. https://doi.org/10.1016/j.seppur.2003.10.004
Guine V, Spadini L, Sarret G, Muris M, Delolme C, Gaudet JP, Martins JMF (2006) Zinc sorption to three gram-negative bacteria: combined titration, modeling, and exafs study. Environ Sci Technol 40:1806–1813
Gupta P, Diwan B (2017) Bacterial exopolysaccharide mediated heavy metal removal: a review on biosynthesis, mechanism and remediation strategies. Biotechnol Rep 13:58–71. https://doi.org/10.1016/j.btre.2016.12.006
Gustafsson JP (2001) Modeling the acid-base properties and metal complexation of humic substances with the Stockholm humic model. J Colloid Interface Sci 244:102–112. https://doi.org/10.1006/jcis.2001.7871
Haas JR (2004) Effects of cultivation conditions on acid-base titration properties of *Shewanella putrefaciens*. Chem Geol 209:67–81. https://doi.org/10.1016/j.chemgeo.2004.04.022
Han X, Wong YS, Tam NFY (2006) Surface complexation mechanism and modeling in Cr(III) biosorption by a microalgal isolate, *Chlorella miniata*. J Colloid Interface Sci 303:365–371. https://doi.org/10.1016/jjcis.2006.08.028
Haug A (1961) Dissociation of alginic acid. Acta Chem Scand 15:950. https://doi.org/10.3891/acta.chem.scand.15-0950
Haug A, Smidsrod O (1970) Selectivity of some anionic polymers for divalent metal ions. Acta Chem Scand 24:843–854. https://doi.org/10.3891/acta.chem.scand.24-0843

He Z, Yang Y, Zhou S, Zhong H, Sun W (2013) The effect of culture condition and ionic strength on proton adsorption at the surface of the extreme thermophile *Acidianus manzaensis*. Colloids Surf B: Biointerfaces 102:667–673. https://doi.org/10.1016/j.colsurfb.2012.09.028

Healy TW, White LR (1978) Ionizable surface group models of aqueous interfaces. Adv Colloid Interf Sci 9:303–345. https://doi.org/10.1016/0001-8686(78)85002-7

Heinrich HTM, Bremer PJ, Daughney CJ, McQuillan AJ (2007) Acid-base titrations of functional groups on the surface of the thermophilic bacterium *Anoxybacillus flavithermus*: comparing a chemical equilibrium model with ATR-IR spectroscopic data. Langmuir 23:2731–2740. https://doi.org/10.1021/la062401j

Herrero R, Lodeiro P, Garcia-Casal LJ, Vilarino T, Rey-Castro C, David C, Rodriguez P (2011) Full description of copper uptake by algal biomass combining an equilibrium NICA model with a kinetic intraparticle diffusion driving force approach. Bioresour Technol 102:2990–2997. https://doi.org/10.1016/j.biortech.2010.10.007

Higgins CP, Luty RG (2007) Modeling sorption of anionic surfactants onto sediment materials: an a priori approach for perfluoroalkyl surfactants and linear alkylbenzene sulfonates. Environ Sci Technol 41:3254–3261. https://doi.org/10.1021/es062449j

Honig B, Nicholls A (1995) Classical electrostatics in biology and chemistry. Science 268:1144–1149. https://doi.org/10.1126/science.7761829

Hubbe MA, Hasan SH, Ducoste JJ (2011) Cellulosic substrates for removal of pollutants from aqueous systems: a review. 1. Metals. Bioresources 6:2161–2914

Hückel E (1925) The theory of concentrated, aqueous solutions of strong electrolytes. Phys Z 26:93–147

Israelachvili JN (2011) Intermolecular and surface forces, 3rd edn. Academic, Burlington

Javanbakht V, Alavi SA, Zilouei H (2014) Mechanisms of heavy metal removal using microorganisms as biosorbent. Water Sci Technol 69:1775–1787. https://doi.org/10.2166/wst.2013.718

Joud J-C, Barthés-Labrousse M-G v (2015) Physical chemistry and acid-base properties of surfaces. Wiley, London

Kapoor A, Viraraghavan T (1995) Fungal biosorption – an alternative treatment option for heavy metal bearing wastewaters: a review. Bioresour Technol 53:195–206. https://doi.org/10.1016/0960-8524(95)00072-1

Katchalsky A, Spitnik P (1947) Potentiometric titrations of polymethacrylic acid. J Polym Sci 2:432–446. https://doi.org/10.1002/pol.1947.120020409

Katchalsky A, Shavit N, Eisenberg H (1954) Dissociation of weak polymeric acids and bases. J Polym Sci 13:69–84. https://doi.org/10.1002/pol.1954.120136806

Kaulbach ES, Szymanowski JES, Fein JB (2005) Surface complexation modeling of proton and Cd adsorption onto an algal cell wall. Environ Sci Technol 39:4060–4065. https://doi.org/10.1021/es0481833

Kenney JPL, Fein JB (2011) Importance of extracellular polysaccharides on proton and Cd binding to bacterial biomass: a comparative study. Chem Geol 286:109–117. https://doi.org/10.1016/j.chemgeo.2011.04.011

Kim YH, Park JY, Yoo YJ (1998) Modeling of biosorption by marine brown *Undaria pinnatifida* based on surface complexation mechanism. Korean J Chem Eng 15:157–163. https://doi.org/10.1007/bf02707068

Kinniburgh DG, van Riemsdijk WH, Koopal LK, Borkovec M, Benedetti MF, Avena MJ (1999) Ion binding to natural organic matter: competition, heterogeneity, stoichiometry and thermodynamic consistency. Colloid Surf A Physicochem Eng Asp 151:147–166. https://doi.org/10.1016/s0927-7757(98)00637-2

Komy ZR (2004) Determination of acidic sites and binding toxic metal ions on cumin surface using nonideal competitive adsorption model. J Colloid Interface Sci 270:281–287. https://doi.org/10.1016/j.jcis.2003.08.046

Komy ZR, Gabar RM, Shoriet AA, Mohammed RM (2006) Characterisation of acidic sites of pseudomonas biomass capable of binding protons and cadmium and removal of cadmium via

biosorption. World J Microbiol Biotechnol 22:975–982. https://doi.org/10.1007/s11274-006-9143-3
Koncagül E, Tran M, Connor R, Uhlenbrook S, Cordeiro Ortigara A R. 2017. The United Nations world water development report. Facts and figures. UNESCO
Kunz W, Henle J, Ninham BW (2004a) 'Zur lehre von der wirkung der salze' (about the science of the effect of salts): Franz Hofmeister's historical papers. Curr Opin Colloid Interface Sci 9:19–37. https://doi.org/10.1016/j.cocis.2004.05.005
Kunz W, Lo Nostro P, Ninham BW (2004b) The present state of affairs with Hofmeister effects. Curr Opin Colloid Interface Sci 9:1–18. https://doi.org/10.1016/j.cocis.2004.05.004
Kushwaha S, Soni H, Ageetha V, Padmaja P (2013) An insight into the production, characterization, and mechanisms of action of low-cost adsorbents for removal of organics from aqueous solution. Crit Rev Environ Sci Technol 43:443–549. https://doi.org/10.1080/10643389.2011.604263
Kyzas GZ, Fu J, Matis KA (2013) The change from past to future for adsorbent materials in treatment of dyeing wastewaters. Materials 6:5131–5158. https://doi.org/10.3390/ma6115131
Lamelas C, Avaltroni F, Benedetti M, Wilkinson KJ, Slaveykova VI (2005) Quantifying Pb and Cd complexation by alginates and the role of metal binding on macromolecular aggregation. Biomacromolecules 6:2756–2764. https://doi.org/10.1021/bm050252y
Larsen M M, Fryer R (2012) Levels and trends in marine contaminants and their biological effects – cemp assessment report. OSPAR commission. ISBN 978-1-909159-29-7
Lawton K, Cherrier V, Grebot B, Zglobisz N, Esparrago J, Ganzleben C, Kallay T, Farmer A (2014) Study on: contribution of industry to pollutant emissions to air and water. Amec environment & infrastructure uk limited in partnership with bio intelligence service, milieu, ieep and rec. KH-04-14-737-EN-N
Lenoir T, Manceau A (2010) Number of independent parameters in the potentiometric titration of humic substances. Langmuir 26:3998–4003. https://doi.org/10.1021/la9034084
Leone L, Ferri D, Manfredi C, Persson P, Shchukarev A, Sjoberg S, Loring J (2007) Modeling the acid-base properties of bacterial surfaces: a combined spectroscopic and potentiometric study of the gram-positive bacterium *Bacillus subtilis*. Environ Sci Technol 41:6465–6471. https://doi.org/10.1021/es070996e
Li X-S, Englezos P (2005) Application of the NICA–Donnan approach to calculate equilibrium between proton and metal ions with lignocellulosic materials. J Colloid Interface Sci 281:267–274. https://doi.org/10.1016/j.jcis.2004.08.141
Li M, Liu Q, Lou Z, Wang Y, Zhang Y, Qian G (2014) Method to characterize acid–base behavior of biochar: site modeling and theoretical simulation. ACS Sustain Chem Eng 2:2501–2509. https://doi.org/10.1021/sc500432d
Lin FG, Marinsky JA (1993) A Gibbs-Donnan-based interpretation of the effect of medium counterion concentration levels on the acid dissociation properties of alginic acid and chondroitin sulfate. React Polym 19:27–45. https://doi.org/10.1016/0923-1137(93)90008-4
Liu R, Song Y, Tang H (2013) Application of the surface complexation model to the biosorption of Cu(II) and Pb(II) ions onto *Pseudomonas pseudoalcaligenes* biomass. Adsorpt Sci Technol 31:1–16. https://doi.org/10.1260/0263-6174.31.1.1
Lo Nostro P, Ninham BW (2012) Hofmeister phenomena: an update on ion specificity in biology. Chem Rev 112:2286–2322. https://doi.org/10.1021/cr200271j
Lodeiro P, Cordero B, Grille Z, Herrero R, Sastre de Vicente ME (2004) Physicochemical studies of cadmium(II) biosorption by the invasive alga in europe, *Sargassum muticum*. Biotechnol Bioeng 88:237–247. https://doi.org/10.1002/bit.20229
Lodeiro P, Cordero B, Barriada JL, Herrero R, Sastre de Vicente ME (2005a) Biosorption of cadmium by biomass of brown marine macroalgae. Bioresour Technol 96:1796–1803. https://doi.org/10.1016/j.biortech.2005.01.002
Lodeiro P, Rey-Castro C, Barriada JL, Sastre de Vicente ME, Herrero R (2005) Biosorption of cadmium by the protonated macroalga *Sargassum muticum*: binding analysis with a nonideal,

competitive, and thermodynamically consistent adsorption (NICA) model. J Colloid Interface Sci 289:352–358. https://doi.org/10.1016/j.jcis.2005.04.002
Lodeiro P, Barriada JL, Herrero R, Sastre de Vicente ME (2006a) The marine macroalga *Cystoseira baccata* as biosorbent for cadmium(II) and lead(II) removal: kinetic and equilibrium studies. Environ Pollut 142:264–273. https://doi.org/10.1016/j.envpol.2005.10.001
Lodeiro P, Herrero R, Sastre de Vicente ME (2006b) Thermodynamic and kinetic aspects on the biosorption of cadmium by low cost materials: a review. Environ Chem 3:400–418. https://doi.org/10.1071/en06043
Lodeiro P, Barriada JL, Herrero R, Sastre de Vicente ME (2007) Electrostatic effects in biosorption: the role of the electrochemistry. Port Electrochim Acta 25:43–54
Lodeiro P, Fuentes A, Herrero R, Sastre de Vicente ME (2008) Cr-III binding by surface polymers in natural biomass: the role of carboxylic groups. Environ Chem 5:355–365. https://doi.org/10.1071/en08035
Lodeiro P, Lopez-Garcia M, Herrero L, Barriada JL, Herrero R, Cremades J, Barbara I, Sastre de Vicente ME (2012) A physicochemical study of Al(+3) interactions with edible seaweed biomass in acidic waters. J Food Sci 77:C987–C993. https://doi.org/10.1111/j.1750-3841.2012.02855.x
Lopez R, Gondar D, Antelo J, Fiol S, Arce F (2011) Proton binding on untreated peat and acid-washed peat. Geoderma 164:249–253. https://doi.org/10.1016/j.geoderma.2011.06.018
Lopez-Garcia M, Lodeiro P, Herrero R, Barriada JL, Rey-Castro C, David C, Sastre de Vicente ME (2013) Experimental evidences for a new model in the description of the adsorption-coupled reduction of Cr(VI) by protonated banana skin. Bioresour Technol 139:181–189. https://doi.org/10.1016/j.biortech.2013.04.044
Luthy RG, Aiken GR, Brusseau ML, Cunningham SD, Gschwend PM, Pignatello JJ, Reinhard M, Traina SJ, Weber WJ, Westall JC (1997) Sequestration of hydrophobic organic contaminants by geosorbents. Environ Sci Technol 31:3341–3347. https://doi.org/10.1021/es970512m
Lyklema J (1995) Fundamentals of interface and colloid science. Academic, London
Madurga S, Lluis Garces J, Companys E, Rey-Castro C, Salvador J, Galceran J, Vilaseca E, Puy J, Mas F (2009) Ion binding to polyelectrolytes: Monte Carlo simulations versus classical mean field theories. Theor Chem Accounts 123:127–135. https://doi.org/10.1007/s00214-009-0550-z
Marsac R, Banik NL, Luetzenkirchen J, Catrouillet C, Marquardt CM, Johannesson KH (2017) Modeling metal ion-humic substances complexation in highly saline conditions. Appl Geochem 79:52–64. https://doi.org/10.1016/j.apgeochem.2017.02.004
Martinez RE, Smith DS, Kulczycki E, Ferris FG (2002) Determination of intrinsic bacterial surface acidity constants using a Donnan shell model and a continuous pK_a distribution method. J Colloid Interface Sci 253:130–139. https://doi.org/10.1006/jcis.2002.8541
Martín-Lara MA, Pagnanelli F, Mainelli S, Calero M, Toro L (2008) Chemical treatment of olive pomace: effect on acid-basic properties and metal biosorption capacity. J Hazard Mater 156:448–457. https://doi.org/10.1016/j.jhazmat.2007.12.035
Matynia A, Lenoir T, Causse B, Spadini L, Jacquet T, Manceau A (2010) Semi-empirical proton binding constants for natural organic matter. Geochim Cosmochim Acta 74:1836–1851. https://doi.org/10.1016/j.gca.2009.12.022
May PM, Williams DR, Linder PW, Torrington RG (1982) The use of glass electrodes for the determination of formation-constants .1. A definitive method for calibration. Talanta 29:249–256. https://doi.org/10.1016/0039-9140(82)80108-2
Meichik NR, Popova NI, Nikolaeva II, Ermakov IP, Kamnev AN (2011) Ion-exchange properties of cell walls of red seaweed *Phyllophora crispa*. Appl Biochem Microbiol 47:176–181. https://doi.org/10.1134/S000368381102013X
Merdy P, Huclier S, Koopal LK (2006) Modeling metal-particle interactions with an emphasis on natural organic matter. Environ Sci Technol 40:7459–7466. https://doi.org/10.1021/es0628203
Michalak I, Chojnacka K, Witek-Krowiak A (2013) State of the art for the biosorption process-a review. Appl Biochem Biotechnol 170:1389–1416. https://doi.org/10.1007/s12010-013-0269-0
Millero FJ, Pierrot D (2002) Speciation of metals in natural waters. Springer-Verlag, Berlin

Mishra B, Boyanov M, Bunker BA, Kelly SD, Kemner KM, Fein JB (2010) High- and low-affinity binding sites for Cd on the bacterial cell walls of *Bacillus subtilis* and *Shewanella oneidensis*. Geochim Cosmochim Acta 74:4219–4233. https://doi.org/10.1016/j.gca.2010.02.019

Morel FO, Hering JG, Morel FO (eds) (1993) Principles and applications of aquatic chemistry. Wiley, New York

Moreno-Castilla C (2004) Adsorption of organic molecules from aqueous solutions on carbon materials. Carbon 42:83–94. https://doi.org/10.1016/j.carbon.2003.09.022

Morris A, Sneddon J (2011) Use of crustacean shells for uptake and removal of metal ions in solution. Appl Spectrosc Rev 46:242–250. https://doi.org/10.1080/05704928.2011.557458

Mukhopadhyay M (2008) Role of surface properties during biosorption of copper by pretreated *Aspergillus niger* biomass. Colloid Surf A Physicochem Eng Asp 329:95–99. https://doi.org/10.1016/j.colsurfa.2008.06.052

Muzzarelli RAA (2011) Potential of chitin/chitosan-bearing materials for uranium recovery: an interdisciplinary review. Carbohydr Polym 84:54–63. https://doi.org/10.1016/j.carbpol.2010.12.025

Naeem A, Woertz JR, Fein JB (2006) Experimental measurement of proton, Cd, Pb, Sr, and Zn adsorption onto the fungal species *Saccharomyces cerevisiae*. Environ Sci Technol 40:5724–5729. https://doi.org/10.1021/es0606935

Naja G, Mustin C, Volesky B, Berthelin J (2005) A high-resolution titrator: a new approach to studying binding sites of microbial biosorbents. Water Res 39:579–588. https://doi.org/10.1016/j.watres.2004.11.008

Nelson N, Schwartz DK (2013) Specific ion (Hofmeister) effects on adsorption, desorption, and diffusion at the solid-aqueous interface. J Phys Chem Lett 4:4064–4068. https://doi.org/10.1021/jz402265y

Ngwenya BT, Tourney J, Magennis M, Kapetas L, Olive V (2009) A surface complexation framework for predicting water purification through metal biosorption. Desalination 248:344–351. https://doi.org/10.1016/j.desal.2008.05.074

Ninham BW, Yaminsky V (1997) Ion binding and ion specificity: the Hofmeister effect and Onsager and Lifshitz theories. Langmuir 13:2097–2108. https://doi.org/10.1021/la960974y

Ohshima H, Kondo T (1991) On the electrophoretic mobility of biological cells. Biophys Chem 39:191–198. https://doi.org/10.1016/0301-4622(91)85021-h

Pagnanelli F, Petrangeli Papini M, Trifoni M, Vegliò F (2000) Biosorption of metal ions on *Arthrobacter sp.*: biomass characterization and biosorption modeling. Environ Sci Technol 34:2773–2778. https://doi.org/10.1021/es991271g

Pagnanelli F, Mainelli S, Vegliò F, Toro L (2003) Heavy metal removal by olive pomace: biosorbent characterisation and equilibrium modelling. Chem Eng Sci 58:4709–4717. https://doi.org/10.1016/j.ces.2003.08.001

Pagnanelli F, Vegliò F, Toro L (2004) Modelling of the acid–base properties of natural and synthetic adsorbent materials used for heavy metal removal from aqueous solutions. Chemosphere 54:905–915. https://doi.org/10.1016/j.chemosphere.2003.09.003

Pagnanelli F, Mainelli S, De Angelis S, Toro L (2005a) Biosorption of protons and heavy metals onto olive pomace: Modelling of competition effects. Water Res 39:1639–1651. https://doi.org/10.1016/j.watres.2005.01.019

Pagnanelli F, Mainelli S, Toro L (2005b) Optimisation and validation of mechanistic models for heavy metal bio-sorption onto a natural biomass. Hydrometallurgy 80:107–125. https://doi.org/10.1016/j.hydromet.2005.07.008

Pagnanelli F, Mainelli S, Toro L (2008) New biosorbent materials for heavy metal removal: product development guided by active site characterization. Water Res 42:2953–2962. https://doi.org/10.1016/j.watres.2008.03.012

Pagnanelli F, Jbari N, Trabucco F, Martínez ME, Sánchez S, Toro L (2013) Biosorption-mediated reduction of Cr(VI) using heterotrophicallygrown *Chlorella vulgaris*: active sites and ionic strength effect. Chem Eng J 231:94–102. https://doi.org/10.1016/j.cej.2013.07.013

Para G, Warszynski P (2007) Cationic surfactant adsorption in the presence of divalent ions. Colloid Surf A Physicochem Eng Asp 300:346–352. https://doi.org/10.1016/j.colsurfa.2007.01.052
Parsons DF (2016) The impact of nonelectrostatic physisorption of ions on free energies and forces between redox electrodes: ion-specific repulsive peaks. Electrochim Acta 189:137–146. https://doi.org/10.1016/j.electacta.2015.12.090
Parsons DF, Bostrom M, Nostro PL, Ninham BW (2011) Hofmeister effects: interplay of hydration, nonelectrostatic potentials, and ion size. Phys Chem Chem Phys 13:12352–12367. https://doi.org/10.1039/C1CP20538B
Percival E, McDowell RH (eds) (1967) Chemistry and enzymology of marine algal polysaccharides. Academic, London/New York
Pitzer KS (1973) Thermodynamics of electrolytes. 1. Theoretical basis and general equations. J Phys Chem 77:268–277. https://doi.org/10.1021/j100621a026
Pitzer KS (1991) Activity coefficients in electrolyte solutions, 2nd edn. CRC Press, Boca Raton
Pitzer KS, Mayorga G (1973) Thermodynamics of electrolytes. 2. Activity and osmotic coefficients for strong electrolytes with one or both ions univalent. J Phys Chem 77:2300–2308. https://doi.org/10.1021/j100638a009
Plazinski W (2013a) Binding of heavy metals by algal biosorbents. Theoretical models of kinetics, equilibria and thermodynamics. Adv Colloid Interf Sci 197:58–67. https://doi.org/10.1016/j.cis.2013.04.002
Plazinski W (2013b) Equilibrium and kinetic modeling of metal ion biosorption: on the ways of model generalization for the case of multicomponent systems. Adsorpt J Int Adsorpt Soc 19:659–666. https://doi.org/10.1007/s10450-013-9489-4
Plazinski W, Rudzinski W (2009) Modeling the effect of surface heterogeneity in equilibrium of heavy metal ion biosorption by using the ion exchange model. Environ Sci Technol 43:7465–7471. https://doi.org/10.1021/es900949e
Plazinski W, Rudzinski W (2011) Biosorption of heavy metal ions: ion-exchange versus adsorption and the heterogeneity of binding sites. Adsorpt Sci Technol 29:479–486. https://doi.org/10.1260/0263-6174.29.5.479
Plette ACC, Vanriemsdijk WH, Benedetti MF, Vanderwal A (1995) pH dependent charging behavior of isolated cell-walls of a gram-positive soil bacterium. J Colloid Interface Sci 173:354–363. https://doi.org/10.1006/jcis.1995.1335
Pokrovsky OS, Martinez RE, Golubev SV, Kompantseva EI, Shirokova LS (2008) Adsorption of metals and protons on *Gloeocapsa sp.*cyanobacteria: a surface speciation approach. Appl Geochem 23:2574–2588. https://doi.org/10.1016/j.apgeochem.2008.05.007
Randelovic M, Momcilovic M, Purenovic M, Zarubica A, Bojic A (2016) The acid-base, morphological and structural properties of new biosorbent obtained by oxidative hydrothermal treatment of peat. Environ Earth Sci 75. https://doi.org/10.1007/s12665-016-5242-0
Ravat C, Dumonceau J, Monteil-Rivera F (2000) Acid/base and Cu(II) binding properties of natural organic matter extracted from wheat bran:modeling by the surface complexation model. Water Res 34:1327–1339. https://doi.org/10.1016/s0043-1354(99)00255-9
Reddad Z, Gerente C, Andres Y, Le Cloirec P (2002) Modeling of single and competitive metal adsorption onto a natural polysaccharide. Environ Sci Technol 36:2242–2248. https://doi.org/10.1021/es010237a
Rey-Castro C, Lodeiro P, Herrero R, Sastre de Vicente ME (2003) Acid−base properties of brown seaweed biomass considered as a Donnan gel. A model reflecting electrostatic effects and chemical heterogeneity. Environ Sci Technol 37:5159–5167. https://doi.org/10.1021/es0343353
Rey-Castro C, Herrero R, de Vicente MES (2004a) Surface charge and permeable gel descriptions of the ionic strength influence on proton binding to seaweed biomass. Chem Speciat Bioavailab 16:61–69. https://doi.org/10.3184/095422904782775117
Rey-Castro C, Herrero R, Sastre de Vicente ME (2004b) Gibbs–Donnan and specific-ion interaction theory descriptions of the effect of ionic strength on proton dissociation of alginic acid. J Electroanal Chem 564:223–230. https://doi.org/10.1016/j.jelechem.2003.10.023

Richter MK, Sander M, Krauss M, Christl I, Dahinden MG, Schneider MK, Schwarzenbach RP (2009) Cation binding of antimicrobial sulfathiazole to leonardite humic acid. Environ Sci Technol 43:6632–6638. https://doi.org/10.1021/es900946u

Robalds A, Naja GM, Klavins M (2016) Highlighting inconsistencies regarding metal biosorption. J Hazard Mater 304:553–556. https://doi.org/10.1016/j.jhazmat.2015.10.042

Rudzinski W, Plazinski W (2010) How does mechanism of biosorption determine the differences between the initial and equilibrium adsorption states? Adsorpt J Int Adsorpt Soc 16:351–357. https://doi.org/10.1007/s10450-010-9244-z

Sag Y (2001) Biosorption of heavy metals by fungal biomass and modeling of fungal biosorption: a review. Sep Purif Methods 30:1–48. https://doi.org/10.1081/spm-100102984

Saito T, Nagasaki S, Tanaka S, Koopal LK (2005) Electrostatic interaction models for ion binding to humic substances. Colloid Surf A Physicochem Eng Asp 265:104–113. https://doi.org/10.1016/j.colsurfa.2004.10.139

Salis A, Ninham BW (2014) Models and mechanisms of Hofmeister effects in electrolyte solutions, and colloid and protein systems revisited. Chem Soc Rev 43:7358–7377. https://doi.org/10.1039/c4cs00144c

Sastre De Vicente ME (1997) Ionic strength effects on acid-base equilibria. A review. Curr Top Solution Chem 2:157–181

Sastre de Vicente ME (2004) The concept of ionic strength eighty years after its introduction in chemistry. J Chem Educ 81:750. https://doi.org/10.1021/ed081p750

Sastre de Vicente ME, Vilariño T (2002) Chemistry of marine water and sediments. Springer-Verlag, Berlin

Sastre M, Santaballa JA (1989) A note on the meaning of the electroneutrality condition for solutions. J Chem Educ 66:403. https://doi.org/10.1021/ed066p403

Schiewer S (1999) Modelling complexation and electrostatic attraction in heavy metal biosorption by *Sargassum* biomass. J Appl Phycol 11:79–87. https://doi.org/10.1023/a:1008025411634

Schiewer S, Patil SB (2008a) Modeling the effect of pH on biosorption of heavy metals by citrus peels. J Hazard Mater 157:8–17. https://doi.org/10.1016/j.jhazmat.2007.12.076

Schiewer S, Patil SB (2008b) Pectin-rich fruit wastes as biosorbents for heavy metal removal: equilibrium and kinetics. Bioresour Technol 99:1896–1903. https://doi.org/10.1016/j.biortech.2007.03.060

Schiewer S, Volesky B (1995) Modeling of the proton-metal ion exchange in biosorption. Environ Sci Technol 29:3049–3058. https://doi.org/10.1021/es00012a024

Schiewer S, Volesky B (1997a) Ionic strength and electrostatic effects in biosorption of divalent metal ions and protons. Environ Sci Technol 31:2478–2485. https://doi.org/10.1021/es960751u

Schiewer S, Volesky B (1997b) Ionic strength and electrostatic effects in biosorption of protons. Environ Sci Technol 31:1863–1871. https://doi.org/10.1021/es960434n

Schiewer S, Volesky B (2000) Biosorption processes for heavy metal removal. In: Environmental microbe metal interactions. ASM Press, Washington, DC, pp 329–362

Schiewer S, Wong MH (1999) Metal binding stoichiometry and isotherm choice in biosorption. Environ Sci Technol 33:3821–3828. https://doi.org/10.1021/es981288j

Schiewer S, Wong MH (2000) Ionic strength effects in biosorption of metals by marine algae. Chemosphere 41:271–282. https://doi.org/10.1016/s0045-6535(99)00421-x

Schijf J, Ebling AM (2010) Investigation of the ionic strength dependence of Ulva lactuca acid functional group pK s by manual alkalimetrictitrations. Environ Sci Technol 44:1644–1649 https://doi.org/10.1021/es9029667

Schwierz N, Horinek D, Netz RR (2015) Specific ion binding to carboxylic surface groups and the pH dependence of the Hofmeister series. Langmuir 31:215–225. https://doi.org/10.1021/la503813d

Seki H, Suzuki A (1998) Biosorption of heavy metal ions to brown algae, *Macrocystis pyrifera*, *Kjellmaniella crassiforia*, and *Undaria pinnatifida*. J Colloid Interface Sci 206:297–301. https://doi.org/10.1006/jcis.1998.5731
Shon HK, Vigneswaran S, Snyder SA (2006) Effluent organic matter (EfOM) in wastewater: constituents, effects, and treatment. Crit Rev Environ Sci Technol 36:327–374. https://doi.org/10.1080/10643380600580011
Smith AM, Lee AA, Perkin S (2016) The electrostatic screening length in concentrated electrolytes increases with concentration. J Phys Chem Lett 7:2157–2163. https://doi.org/10.1021/acs.jpclett.6b00867
Stevens MJ, Rempe SLB (2016) Ion-specific effects in carboxylate binding sites. J Phys Chem B 120:12519–12530. https://doi.org/10.1021/acs.jpcb.6b10641
Stumm W, Morgan JJ (1996) Aquatic chemistry: chemical equilibria and rates in natural waters, 3rd edn. Wiley, New York
Sun CH, Berg JC (2003) A review of the different techniques for solid surface acid-base characterization. Adv Colloid Interf Sci 105:151–175. https://doi.org/10.1016/s0001-8686(03)00066-6
Tipping E (2002) Cation binding by humic substances, 1st edn. Cambridge University Press, New York
Trefalt G, Behrens SH, Borkovec M (2016) Charge regulation in the electrical double layer: ion adsorption and surface interactions. Langmuir 32:380–400. https://doi.org/10.1021/acs.langmuir.5b03611
Tudor HEA, Gryte CC, Harris CC (2006) Seashells: detoxifying agents for metal-contaminated waters. Water Air Soil Pollut 173:209–242. https://doi.org/10.1007/s11270-005-9060-3
Turner DR, Achterberg EP, Chen C-TA, Clegg SL, Hatje V, Maldonado MT, Sander SG, van den Berg CMG, Wells M (2016) Toward a qualitycontrolled and accessible Pitzer model for seawater and related systems. Front Mar Sci 3:132. https://doi.org/10.3389/fmars.2016.00139
Van Oss CJ (2006) Interfacial forces in aqueous media, 2nd edn. Taylor & Francis, Boca Raton
Van Oss CJ, Giese RF (2011) Role of the polar properties of water in separation methods. Sep Purif Rev 40:163–208. https://doi.org/10.1080/15422119.2011.555215
Van Son T, Huu Hao N, Guo W, Zhang J, Liang S, Cuong T-T, Zhang X (2015) Typical low cost biosorbents for adsorptive removal of specific organic pollutants from water. Bioresour Technol 182:353–363. https://doi.org/10.1016/j.biortech.2015.02.003
Veglio F, Beolchini F (1997) Removal of metals by biosorption: a review. Hydrometallurgy 44:301–316. https://doi.org/10.1016/s0304-386x(96)00059-x
Vijayaraghavan K, Yun Y-S (2008a) Bacterial biosorbents and biosorption. Biotechnol Adv 26:266–291. https://doi.org/10.1016/j.biotechadv.2008.02.002
Vijayaraghavan K, Yun Y-S (2008b) Competition of Reactive red 4, Reactive orange 16 and Basic blue 3 during biosorption of Reactive blue 4 by polysulfone-immobilized *Corynebacterium glutamicum*. J Hazard Mater 153:478–486. https://doi.org/10.1016/j.jhazmat.2007.08.079
Vilar VJP, Botelho CMS, Pinheiro JPS, Domingos RF, Boaventura RAR (2009) Copper removal by algal biomass: biosorbents characterization and equilibrium modelling. J Hazard Mater 163:1113–1122. https://doi.org/10.1016/j.jhazmat.2008.07.083
Vlachy N, Jagoda-Cwiklik B, Vacha R, Touraud D, Jungwirth P, Kunz W (2009) Hofmeister series and specific interactions of charged headgroups with aqueous ions. Adv Colloid Interf Sci 146:42–47. https://doi.org/10.1016/j.cis.2008.09.010
Volesky B (2003) Sorption and biosorption. BV Sorbex, Montreal
Volesky B (2007) Biosorption and me. Water Res 41:4017–4029. https://doi.org/10.1016/j.watres.2007.05.062
Wang J, Chen C (2009) Biosorbents for heavy metals removal and their future. Biotechnol Adv 27:195–226. https://doi.org/10.1016/j.biotechadv.2008.11.002
Webster EM (2014) Models of the equilibrium distribution of organic chemicals between water and solid phases of environmental media. Environ Rev 22:430–444. https://doi.org/10.1139/er-2013-0079

Wonders J, VanLeeuwen HP, Lyklema J (1997) Metal- and proton-binding properties of a core-shell latex: interpretation in terms of colloid surface models. Colloid Surf A-Physicochem Eng Asp 120:221–233. https://doi.org/10.1016/s0927-7757(96)03680-1

Xiao F, Pignatello JJ (2014) Effect of adsorption nonlinearity on the pH–adsorption profile of ionizable organic compounds. Langmuir 30:1994–2001. https://doi.org/10.1021/la403859u

Yee N, Fowle DA, Ferris FG (2004) A Donnan potential model for metal sorption onto *Bacillus subtilis*. Geochim Cosmochim Acta 68:3657–3664. https://doi.org/10.1016/j.gca.2004.03.018

Yun YS (2004) Characterization of functional groups of protonated *Sargassum polycystum* biomass capable of binding protons and metal ions. J Microbiol Biotechnol 14:29–34

Yun YS, Volesky B (2003) Modeling of lithium interference in cadmium biosorption. Environ Sci Technol 37:3601–3608. https://doi.org/10.1021/es011454e

Zhao X-T, Zeng T, Li X-Y, Gao H-W (2015) Modeling and mechanism of the adsorption of proton and copper to natural bamboo sawdust using the NICA-Donnan model. J Dispers Sci Technol 36:703–713. https://doi.org/10.1080/01932691.2014.917358

Zhou Y, Zhang L, Cheng Z (2015) Removal of organic pollutants from aqueous solution using agricultural wastes: a review. J Mol Liq 212:739–762. https://doi.org/10.1016/j.molliq.2015.10.023

Chapter 4
Fractal-Like Kinetic Models for Fluid–Solid Adsorption

Marco Balsamo and Fabio Montagnaro

Contents

Abstract The journey of an adsorbate molecule from the bulk of the fluid phase to the active centre of the adsorbent particle is characterised by complex transport and capture mechanisms. These need to be described by accurate mathematical expressions for a proper design of a fluid–solid adsorption unit. In turn, this is a fundamental prerequisite to comply with strict emission limits of pollutants in fluid streams.

Consider a chemical process evolving in a system with homogeneous spatial distribution of reactants. Classical rate equations reported in literature include time-invariant kinetic constants and transport parameters. This is not appropriate for diffusion-limited heterogeneous processes. Actually, they take place into spaces where the reactants are constrained by phase boundaries or force fields such as it

M. Balsamo
Dipartimento di Ingegneria Chimica, dei Materiali e della Produzione Industriale, Università degli Studi di Napoli Federico II, Naples, Italy
e-mail: marco.balsamo@unina.it

F. Montagnaro (✉)
Dipartimento di Scienze Chimiche, Università degli Studi di Napoli Federico II, Complesso Universitario di Monte Sant'Angelo, Naples, Italy
e-mail: fabio.montagnaro@unina.it

G. Crini, E. Lichtfouse (eds.), *Green Adsorbents for Pollutant Removal*, Environmental Chemistry for a Sustainable World 18,
https://doi.org/10.1007/978-3-319-92111-2_4

happens in porous solids. Rate coefficients can show temporal memories with a decrease in time that leads to the definition of fractal-like kinetic expressions. This scenario is likely to be applied when adsorption of a pollutant from fluid phase on a porous sorbent is considered.

In this article, we investigate and discuss adsorption from liquid phase. This with an original focus on interrelationships among pollutants and adsorbents characteristics, dynamic adsorption results and fractal-like modelling of a selection of experimental data present in literature.

Among all models taken into consideration, fractal pseudo-first-order was the most statistically accurate. The value of the hybrid fractional error function ranged between 3×10^{-4} and 1×10^{-2}. It resulted in general one order of magnitude lower than what obtained for a canonical pseudo-first-order model, that appeared to overestimate the values of degree of surface coverage by 10–20% under definite case-studies. Correspondingly this model predicted a faster attainment of equilibrium conditions, up to three-times earlier than experimentally recorded.

The fractal model involves the presence of an instantaneous rate coefficient. It ranged between orders 10^{-3} and 1 min^{-1}. Adsorbents showing different features have been analysed, for example with variations in particle size (order 100 μm), specific surface area (order 10^2 m^2 g^{-1}), total pore volume (order 10^{-1} cm^3 g^{-1}). For short adsorption times, the instantaneous rate coefficient was positively affected by (i) a decrease in particle size; (ii) an increase in specific surface area and total pore volume; (iii) an increase in specific mesopore volume; (iv) an increase in average pore size.

The heterogeneity parameter of the fractal model influences the time-decay rate of the former coefficient. This parameter ranged between 0.138 and 0.478, and it was higher when: (i) the pore space is more crowded by already-adsorbed molecules; (ii) a larger degree of surface chemical heterogeneity is determined; (iii) the mean micropore size is smaller; (iv) the specific volume of ultramicropores is larger; (v) the pore size distribution is more polydispersed. Higher values for this parameter means that the exploration of the intraparticle environment is more hindered for the adsorbate.

4.1 Introduction

Adsorption is one of the most developed and widespread technologies in the chemical industry for the separation of target chemical species from fluid phases. This is due to the operating flexibility of the process, the potentially high selectivity towards a specific pollutant and the limited – if any – generation of by-products (Ruthven 1984; Suzuki 1990; Do 1998; Crini and Badot 2010). The presence of a solid adsorbent makes the process heterogeneous. In particular, the most commonly employed one is activated carbon. The process relies on intrinsic kinetic and thermodynamic aspects related to the interaction between the compound to be removed, namely the adsorbate, and the solid adsorbent. These aspects mostly depend on operating temperature and pressure together with the nature of the adsorbate/adsorbent pair. In addition, transport phenomena associated with the journey of the adsorbate molecule from the bulk of the fluid phase to the active centre of the adsorbent particle should always be taken into account. This in order to accomplish the proper design of the process on both the microscopic, i.e. at particle level, and the macroscopic, i.e. at adsorber level, standpoint.

Industrial-scale adsorption processes are usually carried out either in cyclic batch systems, where an adsorbent fixed bed is alternately operated in adsorption and regeneration mode, or in continuous flow units based on a countercurrent contact between the polluted stream and the adsorbent. The accurate description of the dynamic response of a fluid–solid adsorption unit (and, consequently, its sizing) requires adequate knowledge of the complex transport and capture mechanisms occurring into the porous structure of the adsorbent particle. Accurate mathematical expressions for both adsorption equilibrium and kinetics have to be developed to establish the nature of pollutant–adsorbent interactions and the capture mechanism. The prediction of the adsorbate removal rate is considered a more complicated problem than the theoretical analysis of the adsorption equilibrium. Actually the latter only characterises the final distribution of the contaminant between the solid and fluid phases, regardless of the complex time-dependent transport and capture steps involved in the process (Plazinski et al. 2009).

In previous works of this research group (Andini et al. 2006, 2008; Montagnaro and Santoro 2009, 2010; Balsamo et al. 2010, 2011, 2012) the topic has been faced, e.g., by investigating the case-study of the removal of metal compounds from wastewaters using adsorbents of different nature. These materials derived from coal combustion ash properly pre-treated, for instance by mechanical sieving, demineralisation, gasification, to improve its physico-chemical properties, red mud, clay, bentonite, montmorillonite, tyre waste and canonical biomasses. Biomasses were based, for example, on almond, hazelnut, egg and oyster shells, bone char, orange waste, rice husk. The analysis considered works carried out by the same or other Authors. Characteristic times for adsorbate (i) external fluid-to-particle mass transfer, (ii) adsorbent macropore and (iii) adsorbent micropore intraparticle diffusion have been defined and evaluated (see also Balsamo et al. 2011). These characteristic times depend on the mean diameter of adsorbent particle, Sherwood number, bulk, intraparticle macropore and micropore diffusivity, particle porosity, pore mean radius and tortuosity. The evaluation of intraparticle micropore diffusion, here generally indicating an activated transport mechanism, passes through the analysis of kinetic experimental data using f.i. the Reichenberg and Vermeulen approaches. By working out data reported in the pertinent literature (see Refs. listed in Montagnaro and Santoro 2010 and Balsamo et al. 2012), it has been observed that the external fluid-to-particle mass transfer hardly affects adsorption, in particular for liquid–solid systems. Moreover, in many cases also the intraparticle macropore diffusion has a shorter characteristic time than those experimentally observed for the process to fully take place. Thus, the fluid–solid adsorption rate is likely controlled by intraparticle micropore diffusion, see Fig. 4.7 in Montagnaro and

Santoro (2010) and Fig. 4.6 in Balsamo et al. (2012). If one assumes an "intraparticle micropore diffusion regime", the characteristic times of this transport process are well comparable with those experimentally seen until equilibrium is attained at least in the wide spectrum of cases taken into account in the previous publications (Montagnaro and Santoro 2010; Balsamo et al. 2012).

The above observations pave the way to the present communication. Firstly, a survey dealing with fractal concepts in process dynamics will be presented in general terms and then with specific reference to adsorption. Then, we will investigate and discuss adsorption from liquid phase. This has been chosen as the intraparticle transport mechanisms are in general slower than what may happen for gas–solid adsorption. The whole analysis will consider pollutants and adsorbents characteristics, dynamic adsorption results and original fractal-like modelling of a selection of experimental data present in literature. A discussion aiming at inter-relating parameters of fractal-like kinetic models to physico-chemical and microstructural properties of the adsorbent particle represents one of the original focuses of this critical review. Examples of adsorbent properties are porosimetric structure- and particle size-related features, and surface functional groups.

4.2 Fractal Concepts in Process Dynamics

Classical rate equations adopted to describe the dynamic evolution of chemical processes include time-invariant kinetic constants and transport properties. The implicit assumption of canonical kinetic models is that the process evolves in a dilute system with homogenous spatial distribution of reactants. The latter condition is guaranteed, for example, in perfectly-mixed reactors (Savageau 1995; Berry 2002; Dokoumetzidis and Macheras 2011). Kopelman was one of the pioneers in highlighting that classical reaction kinetics – with constant parameters – is not applicable for diffusion-limited heterogeneous processes taking place: (i) onto surfaces exhibiting fractal geometries, or (ii) in Euclidean spaces where the reactants are constrained by phase boundaries or force fields such as in porous solids (Kopelman 1986, 1988). Observations retrieved from batch experimental tests, scaling theory and Monte Carlo simulations performed for different processes occurring in fractal media or onto energetically heterogeneous surfaces, demonstrated rate coefficients with temporal "memories" (Kopelman 1986, 1988):

$$k(t) = k' t^{-h} \text{ with } h \in [0; 1] \text{ and } t \geq 1 \tag{4.1}$$

In Eq. (4.1), $k(t)$ is the instantaneous rate coefficient and the fractal exponent h is a heterogeneity parameter related to the spectral dimension of fractal systems (Kopelman 1988; Pippa et al. 2013; Wang et al. 2016). In the limiting case $h = 0$ the canonical kinetic formulation is obtained ($k = k'$, i.e. time-invariant rate coefficient), valid for diffusion in homogeneous environments (Pippa et al. 2013). The time decrease of the rate coefficient for diffusion-controlled batch processes which exhibit fractal-like kinetics according to Eq. (4.1) arises from the non-re-randomisation of the reactants location. This is due to dimensional or topological constraints (Kopelman 1988; Dokoumetzidis and Macheras 2011). The random walkers mostly fluctuate around their original positions, and h depends on the probability for a random walker to come back to its original position after a certain time. This ends up into segregation and self-ordering phenomena with an associated decrease of the reaction probability with time (Kopelman 1988). Actually, the local geometry leaves its hallmark on the rate coefficient determining its time-dependence. In the case of steady-state processes kinetically governed by reactants diffusion, the fractal-like kinetics predicts a reaction order which is not equivalent to the molecularity of the process for an elementary reactive event. Rather, this order depends on the spectral dimension of the reaction medium (Kopelman 1986, 1988; Savageau 1995).

In a more recent work (Brouers and Sotolongo-Costa 2006), while recognising the outcomes of the former articles by Kopelman, a more generalised theoretical approach based on statistical concepts has been developed to describe fractal-like kinetics. Those Authors derived an expression for the instantaneous rate coefficient that depends on the characteristic time of the process, its reaction order and a parameter called α. The latter is a fractional time index that expresses memory effects due to the energetic heterogeneity of the system. This somewhat conceptually resembles the role assumed by h in the Kopelman's approach. The interested Reader is referred to that article for extensions. In the present communication we adopt the phenomenological Eq. (4.1) proposed by Kopelman. It shows a more explicit decreasing dependence of the instantaneous rate coefficient with the process time. This represents a scenario characteristic of diffusion-limited heterogeneous processes such as liquid–solid adsorption.

The inclusion of fractal-like rate expressions in mathematical modelling of heterogeneous reactive systems dynamics has been only recently recognised as fundamental. This approach would allow a more realistic understanding of the mechanisms ruling the process evolution at microscopic scale, which in turn is a prerequisite for the reliable design of large-scale chemical reactors (Savageau 1995; Xu and Ding 2007; Dokoumetzidis and Macheras 2011; Montagnaro et al. 2016). Most scientific studies deal with the application of fractal-like kinetics in pharmaceutical sciences and biological conversion and catalysis, with particular reference to drug metabolism and absorption, enzyme kinetics, anaerobic digestion, pharmacokinetics and pharmacodynamics (Savageau 1995; Xu and Ding 2007; Dokoumetzidis and Macheras 2011; Pippa et al. 2013; Wang et al. 2016). A fractal-like Michaelis–Menten equation was proposed for describing heterogeneous enzymatic reactions occurring in disordered media and under spatially constrained conditions (Xu and Ding 2007). In the framework of gas–solid reactive systems, Montagnaro et al. (2016) have recently validated a fractal-like random pore model. It was applied to desulphurisation processes with the rate being controlled by both diffusion in the layer of solid product – formulated in a fractal-like fashion – and kinetics of reaction.

4.3 Fractal-Like Kinetic Models Applied to Adsorption

An important contribution for the application to adsorption dynamics of fractal concepts as proposed by Kopelman has been provided in the works by Harifar and Azizian (2012, 2014). In the case of homogeneous surfaces, these Authors ascribed the fractal dependence of the rate coefficient to the progressive occupation of adsorption sites via slower pathways available for the adsorbate. Instead, for heterogeneous surfaces, the decrease of the kinetic parameter with time was related to the progressive adsorption onto sites characterised by greater activation energies.

Be $q(t)$ the specific capacity of adsorption in time, expressed as quantity of pollutant captured per unit quantity of adsorbent, and q_{eq} its equilibrium value. The degree of surface coverage is here defined as:

$$\theta(t) = \frac{q(t)}{q_{eq}} \tag{4.2}$$

being $\theta(0) = 0$ (i.e., virgin adsorbent at the beginning of the process).

According to the pseudo-first-order (PFO) model (often termed Lagergren model, Lagergren 1898), it is:

$$\frac{d\theta(t)}{dt} = k_1[1 - \theta(t)] \tag{4.3}$$

where k_1 is the kinetic constant of the PFO model, with dimensions [t^{-1}]. The solution of Eq. (4.3) is:

$$\theta(t) = 1 - exp(-k_1 t) \quad \text{PFO} \tag{4.4}$$

By now substituting, into Eq. (4.3), k_1 with its fractal formulation (see Eq. 4.1):

$$k_{1\text{F}}(t) = k'_{1\text{F}} t^{-h} \tag{4.5}$$

where $k'_{1\text{F}}$ is expressed in [$t^{-(1-h)}$], Eq. (4.3) becomes:

$$\frac{d\theta(t)}{dt} = k'_{1\text{F}} t^{-h}[1 - \theta(t)] \tag{4.6}$$

with solution:

$$\theta(t) = 1 - exp\left(-\frac{k'_{1\text{F}}}{1-h} t^{1-h}\right) \quad \text{FPFO} \tag{4.7}$$

We will refer in the following to Eq. (4.7) also as "fractal pseudo-first-order" (FPFO) model.

Similarly, according to the pseudo-second-order (PSO) model as proposed in Ho and McKay (1998), the adsorption rate reads:

$$\frac{d\theta(t)}{dt} = k_2 q_{eq}[1 - \theta(t)]^2 \tag{4.8}$$

with k_2 being the kinetic constant ruling the PSO model, having dimensions [t^{-1} q^{-1}] if with [q] we indicate the dimensions of q. The integral form of Eq. (4.8) is:

$$\theta(t) = \frac{k_2 q_{eq} t}{1 + k_2 q_{eq} t} \quad \text{PSO} \tag{4.9}$$

Equation (4.8) can be expressed in fractal formulation by substituting k_2 with:

$$k_{2F}(t) = k'_{2F} t^{-h} \tag{4.10}$$

where k'_{2F} is expressed in [$t^{-(1-h)}$ q^{-1}]. It is therefore:

$$\frac{d\theta(t)}{dt} = k'_{2F} t^{-h} q_{eq}[1 - \theta(t)]^2 \tag{4.11}$$

whose solution is:

$$\theta(t) = \frac{k'_{2F} q_{eq} t^{1-h}}{(1-h) + k'_{2F} q_{eq} t^{1-h}} \quad \text{FPSO} \tag{4.12}$$

namely the expression valid for the "fractal pseudo-second-order" (FPSO) model.

In addition to the PFO and PSO models widely applied in literature, the homogeneous solid diffusion model is used as well for the description of adsorption dynamics in case of diffusion-controlled regimes. The pollutant material balance in a differential element of an adsorbent particle, considered as spherical, can be expressed as (Qiu et al. 2009):

$$\frac{\partial\theta(r,t)}{\partial t} = \frac{D}{r^2}\frac{\partial}{\partial r}\left[r^2 \frac{\partial\theta(r,t)}{\partial r}\right] \tag{4.13}$$

where D [L^2 t^{-1}] is the intraparticle diffusivity and r [L] the radial coordinate of the particle. Different approximate solutions of Eq. (4.13) have been proposed, and most of them generally provide an accurate description of kinetic data only in specific time ranges. In this context, the Vermeulen equation for adsorption kinetics is a diffusion-based model derived as approximate solution of Eq. (4.13). It can be used to predict experimental data in the whole time range. Its integral form, here named VER, is (Vermeulen 1953):

$$\theta(t) = \sqrt{1 - exp\left(-\frac{4\pi^2 D}{d_S^2} t\right)} \quad \text{VER} \tag{4.14}$$

where d_S [L] is the mean Sauter diameter of the adsorbent particle size population. The path drawn in Balsamo and Montagnaro (2015) was devoted to account for a non-uniform distribution of the pollutant in the adsorbent pores system during adsorption due to diffusion limitations. Mimicking Eq. (4.1), the following fractal-like time-dependence for intraparticle diffusivity is proposed:

$$D_F(t) = D'_F t^{-h} \tag{4.15}$$

where D'_F [$L^2\, t^{-(1-h)}$] represents the fractal diffusion kinetic constant. By substituting D in Eq. (4.14) with Eq. (4.15), one obtains the following fractal-like form of the Vermeulen model (named FVER):

$$\theta(t) = \sqrt{1 - exp\left(-\frac{4\pi^2 D'_F}{d_S^2} t^{1-h}\right)} \quad \text{FVER} \tag{4.16}$$

Let us compare the time-dependence of the intraparticle diffusivity according to the proposed Eq. (4.15) with respect to other literature mathematical analyses derived in case of pure diffusion into porous systems, i.e. in absence of adsorption phenomena. According to Sen (2004), the intraparticle diffusivity approaches a non-zero value for $t \rightarrow \infty$ for diffusion in well-connected systems. This since the walker displacement is hindered by collisions with pore walls only. On the contrary, the proposed fractal-like dependence produces the limiting case $D_F \rightarrow 0$ for long adsorption times. From a physical point of view this could be justified as follows. When approaching saturation conditions, the force fields – determined by both pore walls and environment crowded by already adsorbed species – strongly confine the adsorbate thus hindering its exploration of the pore space.

The fractal-like models proposed above were chosen as they derive from canonical expressions widely utilised in literature. Other relevant fractal-like equations have been proposed. For example, the generalised fractal kinetics obtained by Brouers and Sotolongo-Costa (2006) or Kopelman's inspired formulations based on statistical rate theory and integrated kinetic Langmuir equation (Rudzinski and Plazinski 2006; Marczewski 2010; Harifar and Azizian 2012).

4.4 Liquid–Solid Adsorption

Aim of the present communication is to provide deeper insights into relationships between parameters of fractal-like kinetic models and physico-chemical and micro-structural properties of adsorbent particles. Examples in literature focusing on specific cases are here quoted. Gaspard et al. (2006) found a relationship between the fractional time index α and the fractal dimension in the mesopore range for different activated carbons. The same research group (Figaro et al. 2009) observed a dependence of α on the pH of the solution treated by adsorption. Harifar and Azizian (2012, 2014) highlighted the influence of stirring rate and initial concentration of pollutant on *h*, the fractal exponent. Finally, the present Authors (Montagnaro and Balsamo 2014; Balsamo and Montagnaro 2015) have focused their attention on the effect that characteristics as the pore size distribution of different fly ash-based adsorbents can have on *h*.

4.4.1 Experimental Conditions Under Analysis

Table 4.1 lists information concerning pollutants and adsorbents chosen in the present investigation, together with some relevant experimental condition. All these data were retrieved from literature, with references reported in table as well. From inspection of Table 4.1 it can be observed that references were selected so as to have a range of pollutants (heavy metals, anions, emerging contaminants) and adsorbents (activated carbons of various nature, silica-based materials, zinc/aluminium hydroxides), different among each other in nature and properties.

- Mohan et al. (2001) selected three different granulometric ranges (100–150, 150–200 and 200–250 mesh, i.e. 0.104–0.152, 0.076–0.104 and 0.066–0.076 mm, respectively) of an activated carbon derived from fertiliser waste to remove Hg^{2+}.
- Kim et al. (2008) focused on three materials termed MB40, MB22 and MB04 consisting into mesoporous molecular sieves functionalised with one (MB40, MB04) or two (MB22) quaternary ammonium ions. Their target pollutant was ClO_4^-.
- In the paper by Mestre et al. (2009), ibuprofen was adsorbed on an adsorbent termed LSN (namely a physically activated wood then wet oxidised in boiling nitric acid) or on another one termed Q (namely a physically activated coal).
- The case of the article by Li et al. (2011) is singular in this review as the same adsorbent, thiol-functionalised magnetic mesoporous silica, was employed for the removal of two different heavy metals, namely Pb^{2+} and Hg^{2+}.
- Finally Zhou et al. (2011) worked with Zn/Al layered double hydroxides (LDH) synthesised by hydrothermal method using urea as precipitating agent under two different concentration values, to give the adsorbents termed LDH(0.1) and LDH (0.4). These adsorbents were investigated for the removal of PO_4^{3-} from liquid phase.

It is in general observed that in some of these works other adsorbents were analysed as well. For the sake of brevity and paper readability, not all the materials studied in the quoted articles have been reported here. Rather, the selection of materials shown in Table 4.1 should be regarded as representative of adsorbents characterised by different properties. As specified in the following, these features determine the fractal nature of the adsorption dynamics.

Table 4.1 Liquid–solid adsorption: pollutants and adsorbents, together with some information on experimental conditions, chosen as case study for investigation

Article by	Pollutant	Adsorbent(s) nature	Name of adsorbent	Pollutant initial concentration	pH	Temperature
Mohan et al. (2001)	Hg^{2+}	Activated carbon derived from fertiliser waste	100–150 mesh	10^{-2} M	2.0	27 °C
			150–200 mesh			
			200–250 mesh			
Kim et al. (2008)	ClO_4^-	Ammonium-functionalised mesoporous molecular sieve (SBA-15)	MB40	10^{-4} M	n.a.	Room
			MB22			
			MB04			
Mestre et al. (2009)	Ibuprofen	(1) Physico-chemically activated wood	(1) LSN	4.4×10^{-4} M	Circa 4	30 °C
		(2) Physically activated coal	(2) Q			
Li et al. (2011)	(1) Pb^{2+}	Thiol-functionalised magnetic mesoporous silica	SH-mSi@Fe_3O_4	(1) 4.8×10^{-6} M (Pb^{2+})	6.5	25 °C
	(2) Hg^{2+}			(2) 5×10^{6} M (Hg^{2+})		
Zhou et al. (2011)	PO_4^{3-}	Zn/Al layered double hydroxides	LDH(0.1)	10^{-4} M	7.0	30 °C
			LDH(0.4)			

4.4.2 Fractal-Like Modelling Results

Adsorption experimental data, selected from quoted articles as will be detailed for each specific case, have been statistically analysed via non-linear regression. With reference to the three canonical (PFO, PSO, VER) and fractal-like (FPFO, FPSO, FVER) kinetic models presented above, best-fitting theoretical profiles have been evaluated through the hybrid fractional error function:

$$hybrid = \frac{1}{N-P}\sum\nolimits_{t_i}\left\{\frac{\left[\theta^{th}(t_i)-\theta^{exp}(t_i)\right]^2}{\theta^{exp}(t_i)}\right\} \tag{4.17}$$

This function is considered a more reliable statistical tool with respect to the classic determination coefficient for non-linear modelling (Ncibi 2008). In Eq. (4.17), N is the number of experimental data point while P the number of model parameters. P is equal to 1 for canonical and 2 for fractal-like models. The apexes "th" and "exp" stand for theoretical – i.e., model-predicted – and experimental values, respectively, for the degree of surface coverage at time t_i. A preliminary analysis focused on the three fractal-like kinetic models presented in this work and referring to the articles here selected was carried out. It emerged that FPFO (Eq. (4.7)) was, in any case, the model with the lowest value for hybrid and therefore that best-performing over FPSO and FVER. Thus, in the following, fractal-like modelling results will be presented using FPFO model only. Moreover, for the sake of a critical comparison, PFO (Eq. (4.4)) among the three canonical kinetic models will be used as example of not fractal-like fitting of experimental data. In some cases the Authors signing the quoted articles (Table 4.1) have demonstrated in their works – through linear regression analysis – that other canonical models performed better than PFO. Nonetheless, it has been here verified the major accuracy of FPFO with respect not only to the other two fractal-like models but also to all the canonical models (PFO, PSO, VER) subject of investigation. This holds for all the case-studies but Li et al. (2011), for which (vide infra) the FPFO model reduced to PFO one.

Figure 4.1 shows experimental data of Hg^{2+} adsorption on the three-granulometric-size activated carbons studied by Mohan et al. (2001) as reported in their Fig. 4.7. They observed that the finer the particle size, the larger the specific capacity of mercury adsorption. Data fitting was here extended up to the longest experimental time chosen by those Authors, i.e. 8 h. Please the Reader observe that in the present work the adsorption time has been always expressed in minutes. Data fitting is reported with reference to PFO (Fig. 4.1a) and FPFO (Fig. 4.1b) models. The companion Table 4.2 lists the values for the single PFO (k_1) and for the two FPFO (k'_{1F} and h) fitting parameters, along with the corresponding values for hybrid function. It is observed the better accuracy of the fractal-like formulation of the kinetic model, with hybrid values lower than for PFO of about one order of magnitude. The fractal exponent h ranges from nearly 0.26 to 0.29 as a function of the particle size range. The better FPFO performance is directly visible when comparing Fig. 4.1a, b. PFO overestimates θ in particular at long adsorption times

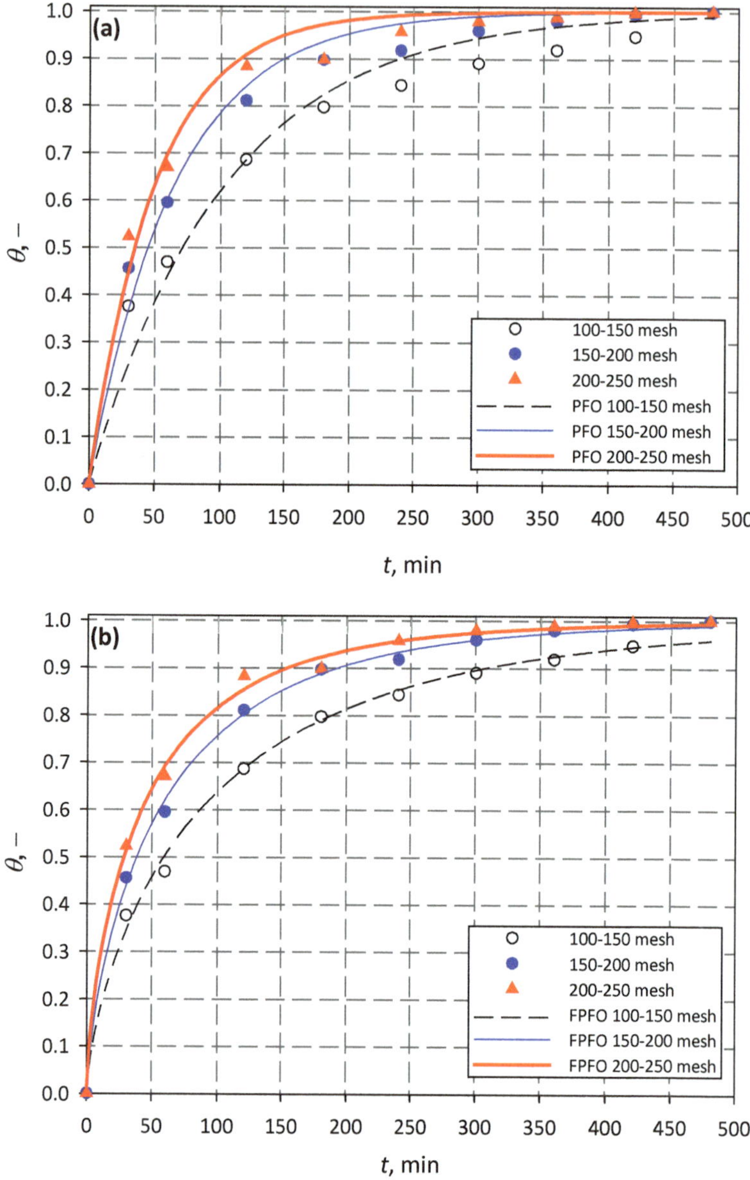

Fig. 4.1 Degree of surface coverage (θ) as a function of adsorption time (t): experimental data re-adapted from Mohan et al. (2001) together with pseudo-first-order, PFO (Fig. 4.1a) and fractal pseudo-first-order, FPFO (Fig. 4.1b) fitting

thus predicting a faster attainment of equilibrium conditions with respect to experimental data. This feature is not present in Fig. 4.1b. As a matter of fact the fractal-like model (see Eq. 4.1) embodies the slowing down of the process also due to the

Table 4.2 Best-fitting values for kinetic parameters (k_1, k'_{1F} and h) and related values for hybrid error function for pseudo-first-order (PFO) and fractal pseudo-first-order (FPFO) models applied to experimental data published by Mohan et al. (2001)

	PFO		FPFO		
Name of adsorbent	k_1 [min^{-1}]	Hybrid	k'_{1F} [min$^{-(1-h)}$]	h	Hybrid
100–150 mesh	9.6×10^{-3}	7×10^{-3}	0.026	0.269	1×10^{-3}
150–200 mesh	1.5×10^{-2}	3×10^{-3}	0.034	0.257	3×10^{-4}
200–250 mesh	2.0×10^{-2}	2×10^{-3}	0.046	0.292	4×10^{-4}

decrease of the kinetic coefficient as long as adsorption proceeds. On the contrary, the canonical model takes into account the decrease of the driving force only (see Eq. 4.3). To support the above analysis, consider for example the adsorption time of 300 min and the 100–150 mesh adsorbent: through Eq. (4.5) and data in Table 4.2 it is $k_{1F} = 5.6 \times 10^{-3}$ min^{-1} (FPFO), while from Table 4.2 the higher $k_1 = 9.6 \times 10^{-3}$ min^{-1} (PFO) value is read.

Kim et al. (2008) analysed, in their Fig. 4.4, perchlorate adsorption on ammonium-functionalised mesoporous molecular sieves. The investigation on their results is here restricted to times not longer than 20 min. Figure 4.2a, b report, in addition to the experimental data points, PFO and FPFO fitting, respectively. ClO_4^- removal by adsorption was more efficient for the bifunctionalised adsorbent (MB22) than for the two porous media functionalised with only one quaternary ammonium ion (MB04, MB40). The Authors suggested that the simultaneous existence of two functional groups in MB22 could enhance the pollutant removal performance. From the inspection of Table 4.3, the better accuracy of the FPFO model is again highlighted in terms of hybrid function. The direct comparison of adsorption trends predicted from canonical (Fig. 4.2a) and fractal-like (Fig. 4.2b) models allows one to observe the following. The differences between the two models performance are not very marked in general terms, as the characteristic equilibrium time of the adsorption process is quite short for all the adsorbents under the operating conditions chosen by Kim and co-workers. Furthermore, while the initial adsorption rate is equivalently predicted by both models, for intermediate adsorption times the canonical model slightly overestimates the degree of surface coverage. Finally (Table 4.3), it is noticed that for FPFO model the fractal exponent ranged from nearly 0.14 (MB40) to 0.30 (MB22).

The paper published in 2009 by Mestre et al. dealt with adsorption of ibuprofen onto LSN and Q materials. Experimental data retrieved from their Fig. 4.1 and for times up to 1 h are reported here in Fig. 4.3a, b. The Authors observed a larger adsorption capacity for the activated coal (material termed Q). The Q material results characterised by a more pronounced basic nature and by larger mean dimension of micropores than LSN. Moreover, in the present Fig. 4.3 we highlight a better behaviour of their activated wood (LSN) in terms of adsorption rate, see for example the PFO values for k_1 in Table 4.4. For LSN wood, Table 4.4 lists the same value for the hybrid function when comparing PFO and FPFO models (for the last, it is $h = 0.40$). From a closer inspection of Fig. 4.3 profiles, it is observed that when

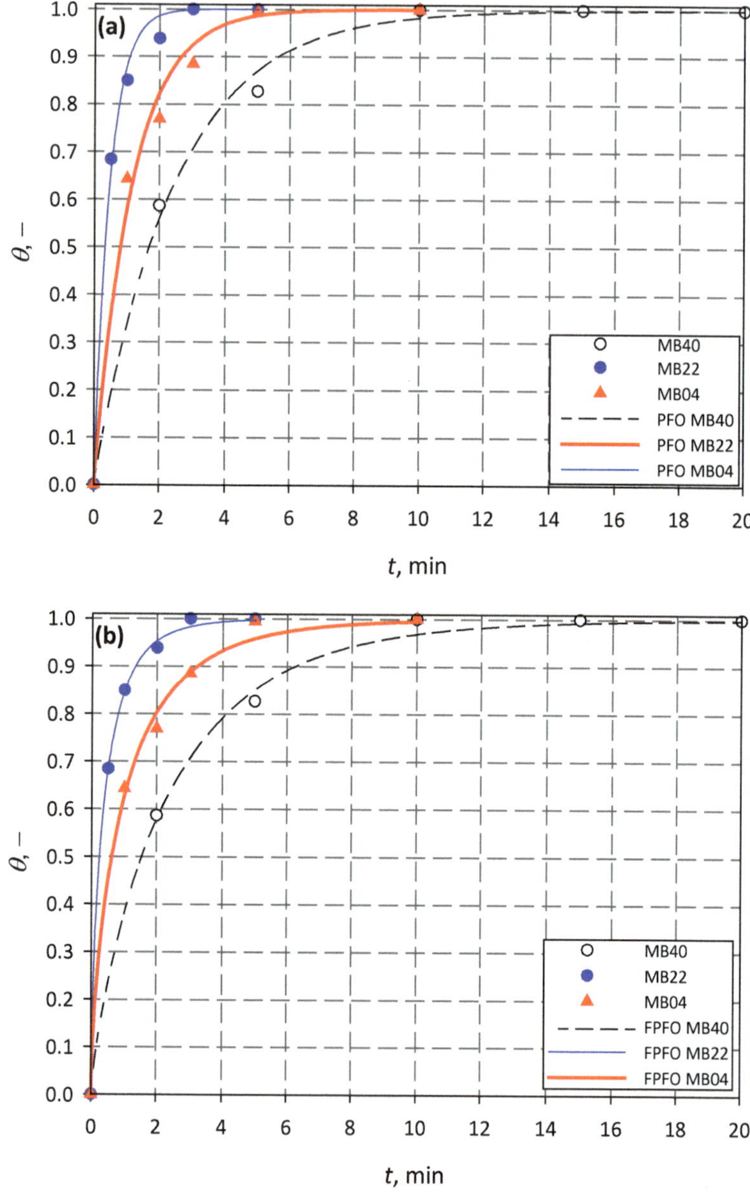

Fig. 4.2 Degree of surface coverage (θ) as a function of adsorption time (t): experimental data re-adapted from Kim et al. (2008) together with pseudo-first-order, PFO (Fig. 4.2a) and fractal pseudo-first-order, FPFO (Fig. 4.2b) fitting

Table 4.3 Best-fitting values for kinetic parameters (k_1, k'_{1F} and h) and related values for hybrid error function for pseudo-first-order (PFO) and fractal pseudo-first-order (FPFO) models applied to experimental data published by Kim et al. (2008)

	PFO		FPFO		
Name of adsorbent	k_1 [min^{-1}]	Hybrid	k'_{1F} [min$^{-(1-h)}$]	h	Hybrid
MB40	0.408	1×10^{-3}	0.409	0.138	6×10^{-4}
MB22	2.126	1×10^{-3}	1.316	0.298	2×10^{-4}
MB04	0.863	3×10^{-3}	0.714	0.273	1×10^{-3}

$t = 10$ min the fractal-like model underpredicts LSN experimental data. On the other hand PFO foresees a faster attainment of LSN equilibrium conditions, i.e. about 20 min vs. 60 min experimentally recorded. These two effects likely counterbalance each other in determining an analogous statistical performance for the two models under consideration. Switching to the activated coal Q, FPFO is characterised by a smaller value for hybrid (in addition, here $h = 0.29$) with respect to PFO. The same general considerations developed for LSN and referring to both $t = 10$ min and long adsorption times here apply. In addition, for $t = 20$ min major differences arise in favour of FPFO as the canonical model noticeably overpredicts the experimental datum ($\theta^{th} = 0.97$ vs. $\theta^{exp} = 0.90$). This last observation definitely contributes in determining the better overall accuracy of the fractal-like kinetic model.

The post-processing of data present in Fig. 4.4 of the work by Li et al. (2011) and dealing with Pb^{2+} and Hg^{2+} adsorption onto thiol-functionalised mesoporous silica is reported in Fig. 4.4 and Table 4.5. At odds with previous findings, the analysis highlighted now the same modellistic outcome when applying either PFO or FPFO. As a matter of fact, the best-fitting scenario for FPFO was attained for $h = 0$ that is for fractal-like formulation reducing into the canonical one. Therefore, Fig. 4.4 and Table 4.5 only report data referred to the PFO case. Li and colleagues have observed a greater adsorption capacity in the case of mercury ions capture than lead ions. In our Fig. 4.4, it can be deduced the faster rate of adsorption for Hg^{2+} rather than Pb^{2+}. This could be ascribed, being the adsorbent material the same in the two cases, to the smaller ionic radius for the former (Shannon 1976). Correspondingly, it is $k_1 = 2.3 \times 10^{-2}$ min^{-1} for Hg^{2+} adsorption versus $k_1 = 1.9 \times 10^{-2}$ min^{-1} for Pb^{2+} capture.

Results of PO_4^{3-} adsorption on two Zn/Al layered double hydroxides – termed LDH(0.1) and LDH(0.4) – reported by Zhou et al. (2011) in their Fig. 4.9 have been analysed, up to 30 h-adsorption time, in the present Fig. 4.5. Modelling results are also listed in Table 4.6 for both PFO and FPFO cases. The Authors observed an undoubtable greater specific capacity of adsorption for the sample treated with the highest value of urea concentration, i.e. LDH(0.4). This material also showed better porosimetric features in terms of specific surface area and total pore volume. In our Fig. 4.5, a turning point-feature is highlighted. LDH(0.4) is characterised by a faster capture dynamics at shorter process times than LDH(0.1). The opposite holds for times longer than about 120 min. The fractal-like model showed lower values for hybrid than PFO did. For FPFO it was $h = 0.26$ and 0.48 for LDH(0.1) and LDH (0.4), respectively. Comparing fitting curves in Fig. 4.5a, b it is clearly seen that the

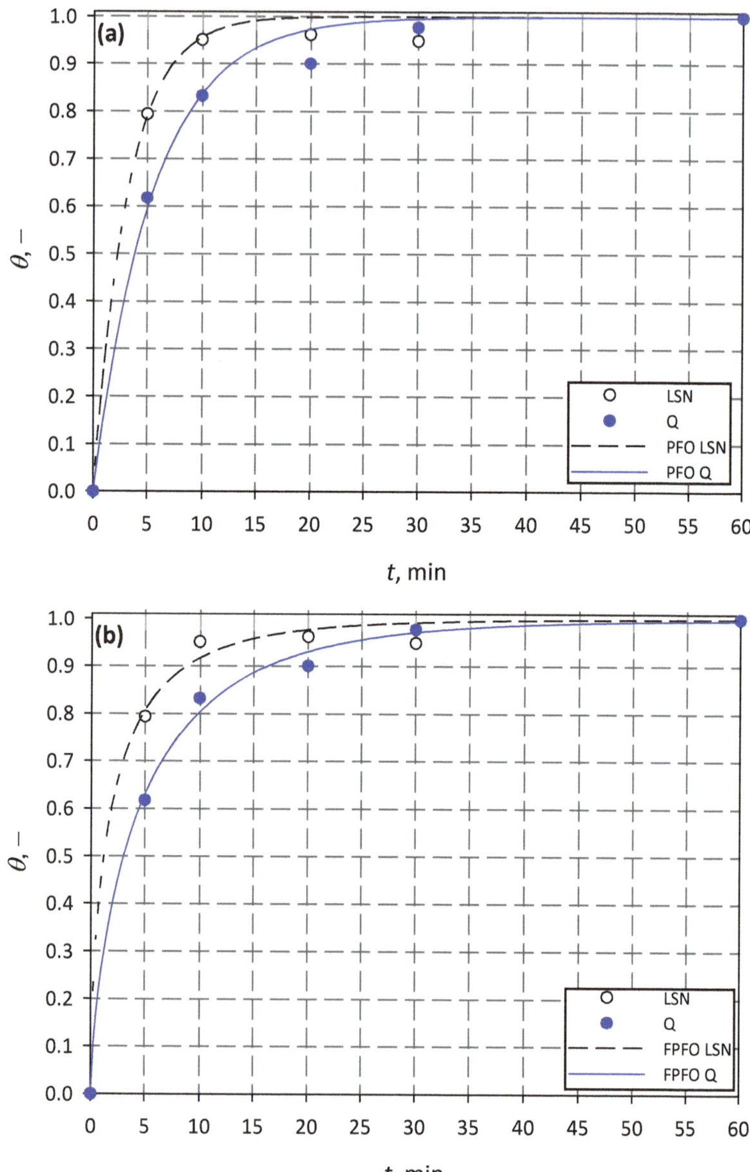

Fig. 4.3 Degree of surface coverage (θ) as a function of adsorption time (t): experimental data re-adapted from Mestre et al. (2009) together with pseudo-first-order, PFO (Fig. 4.3a) and fractal pseudo-first-order, FPFO (Fig. 4.3b) fitting

Table 4.4 Best-fitting values for kinetic parameters (k_1, k'_{1F} and h) and related values for hybrid error function for pseudo-first-order (PFO) and fractal pseudo-first-order (FPFO) models applied to experimental data published by Mestre et al. (2009)

	PFO		FPFO		
Name of adsorbent	k_1 [min^{-1}]	Hybrid	k'_{1F} [min$^{-(1-h)}$]	h	Hybrid
LSN	0.313	1×10^{-3}	0.376	0.404	1×10^{-3}
Q	0.181	2×10^{-3}	0.226	0.288	8×10^{-4}

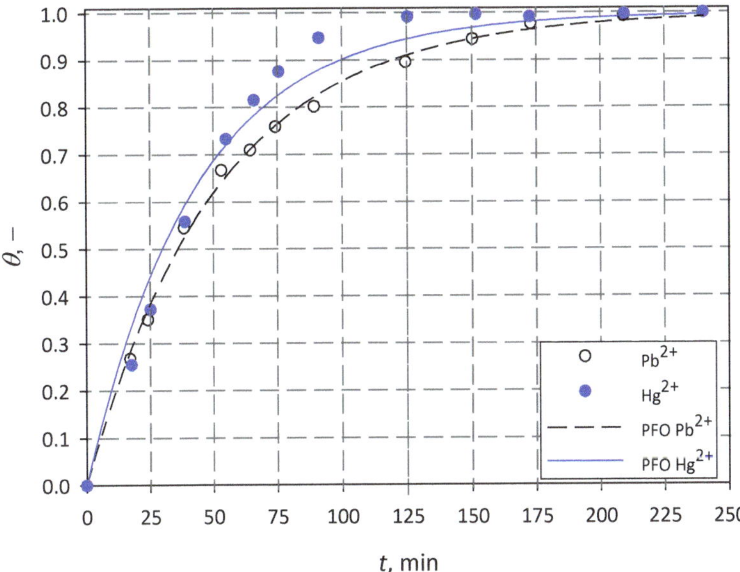

Fig. 4.4 Degree of surface coverage (θ) as a function of adsorption time (t): experimental data re-adapted from Li et al. (2011) together with pseudo-first-order (PFO) fitting

Table 4.5 Best-fitting values for kinetic parameter (k_1) and related values for hybrid error function for pseudo-first-order (PFO) model applied to experimental data published by Li et al. (2011)

	PFO	
Pollutant	k_1 [min^{-1}]	Hybrid
Pb^{2+}	0.019	4×10^{-4}
Hg^{2+}	0.023	5×10^{-3}

Fractal pseudo-first-order model, best-fitting case, reduced to PFO model as it resulted $h = 0$

turning point-feature is predicted by FPFO but not by PFO. Actually, the latter model foresees a higher adsorption rate for LDH(0.4) over the whole time range, see also the greater value for the kinetic constant in the LDH(0.4) case, Table 4.6. Moreover, the canonical model indicates an earlier saturation time than what

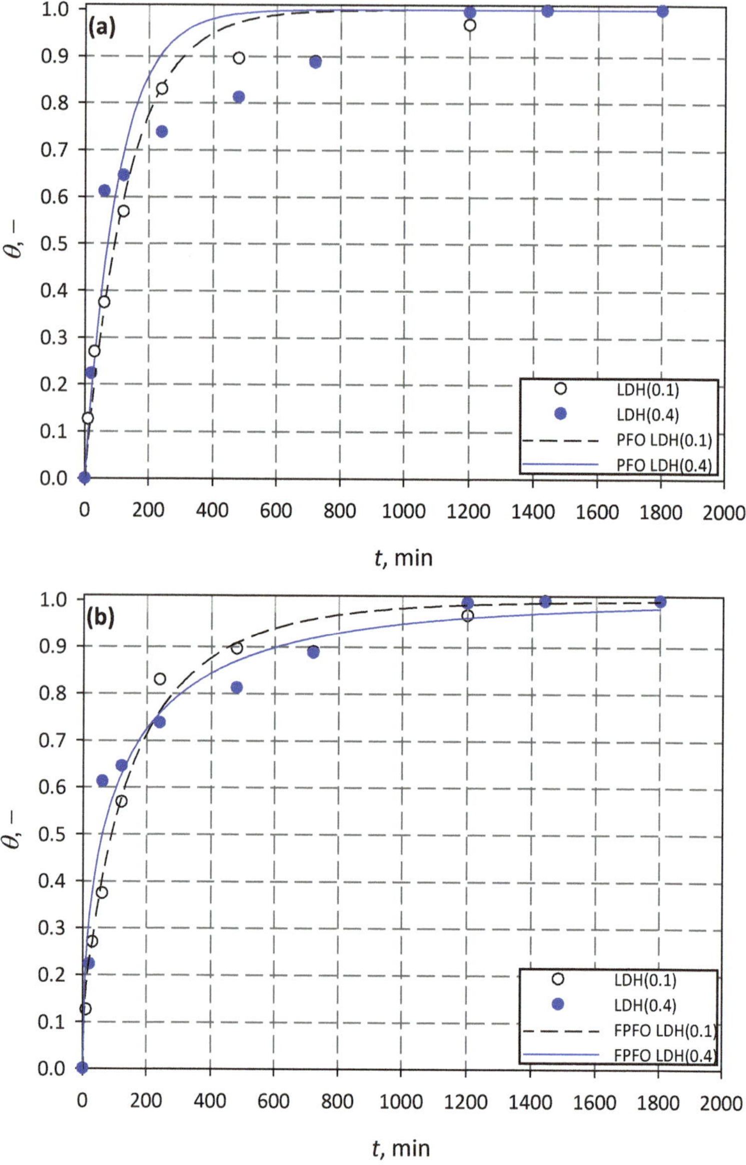

Fig. 4.5 Degree of surface coverage (θ) as a function of adsorption time (t): experimental data re-adapted from Zhou et al. (2011) together with pseudo-first-order, PFO (Fig. 4.5a) and fractal pseudo-first-order, FPFO (Fig. 4.5b) fitting

Table 4.6 Best-fitting values for kinetic parameters (k_1, k'_{1F} and h) and related values for hybrid error function for pseudo-first-order (PFO) and fractal pseudo-first-order (FPFO) models applied to experimental data published by Zhou et al. (2011)

	PFO		FPFO		
Name of adsorbent	k_1 [min^{-1}]	Hybrid	k'_{1F} [min$^{-(1-h)}$]	h	Hybrid
LDH(0.1)	7.5×10^{-3}	7×10^{-3}	0.019	0.262	2×10^{-3}
LDH(0.4)	9.8×10^{-3}	2×10^{-2}	0.042	0.478	1×10^{-2}

experimentally observed. Finally, at intermediate adsorption times the discrepancy between PFO theoretical values and adsorption data points is noticeable. Suffice it to say that, f.i., at 480 min for LDH(0.4) the value for θ^{th} is more than 20% higher than θ^{exp}.

4.4.3 *Discussion on Relationships Between Properties of Adsorbents and Modelling Outcomes*

In the previous section we have shown that the fractal-like modelling, represented here by FPFO case, generally appears more reliable than the canonical PFO formulation in predicting experimental data. Therefore, in this paragraph the discussion will rely on the results concerning the best-fitting parameters named k'_{1F} and h as listed in Tables from 4.2 to 4.6. The combination of these terms (see Eq. 4.5) gives the time-dependence of k_{1F}, i.e. the instantaneous rate coefficient drawn in Figs. 4.6, 4.7, 4.8, and 4.9 for the different cases under scrutiny and expressed in min^{-1}. Please observe that, in Figs. 4.6, 4.7, 4.8, and 4.9 (semi-log plots), a vertical "dash-dot-dot" line is added and indicates the earliest value of t that we considered to process experimental data reported by the Authors. Moreover, trends of k_{1F} have been extrapolated down to $t = 1$ min for reference purposes (cf. Eq. 4.1) and up to the longest experimental time retrieved from the parent works.

Let us start the discussion with reference to the article by Mohan et al. (2001), Fig. 4.6 and Table 4.2. Values for k_{1F}, though obviously decreasing in time, fall in order of magnitude ranges comparable to the one which k_1 (canonical PFO model) belongs to. This is a feature pertaining to all the investigated case-studies here reported. Moreover, it is observed that the larger the mean particle size the smaller the instantaneous rate coefficient. As underlined in the past (Montagnaro and Santoro 2010; Balsamo et al. 2012), (i) external fluid-to-particle mass transfer and (ii) intraparticle diffusion are likely to rule the adsorption process for (i) short and (ii) long times, respectively. In addition, the characteristic time for both phenomena (i) and (ii) depends on the square of the mean particle size: coarser particles are bound to show longer transport characteristic times so being more penalised in terms of adsorption rate. Coherently, the k_{1F} trend for the 200–250 mesh material lays above the values for the other two cases throughout the entire time-range investigated. Furthermore, Mohan and co-workers have observed that – under the

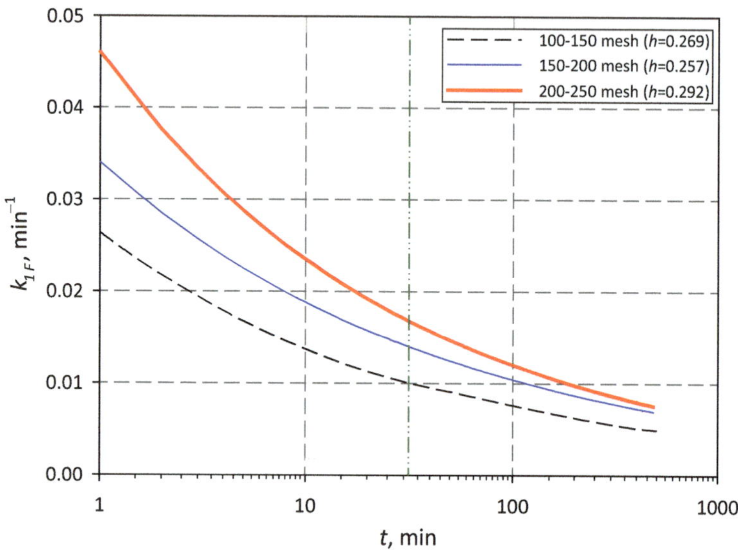

Fig. 4.6 Instantaneous rate coefficient (k_{1F}) as a function of adsorption time (t), fractal pseudo-first-order model, for materials investigated by Mohan et al. (2001). The corresponding values for the fractal exponent (h) are reported as well

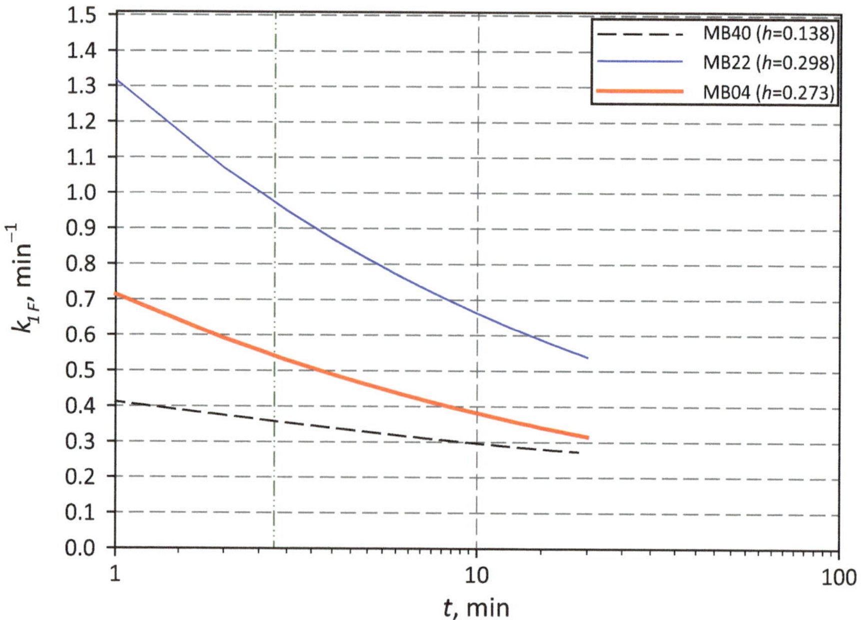

Fig. 4.7 Instantaneous rate coefficient (k_{1F}) as a function of adsorption time (t), fractal pseudo-first-order model, for materials investigated by Kim et al. (2008). The corresponding values for the fractal exponent (h) are reported as well

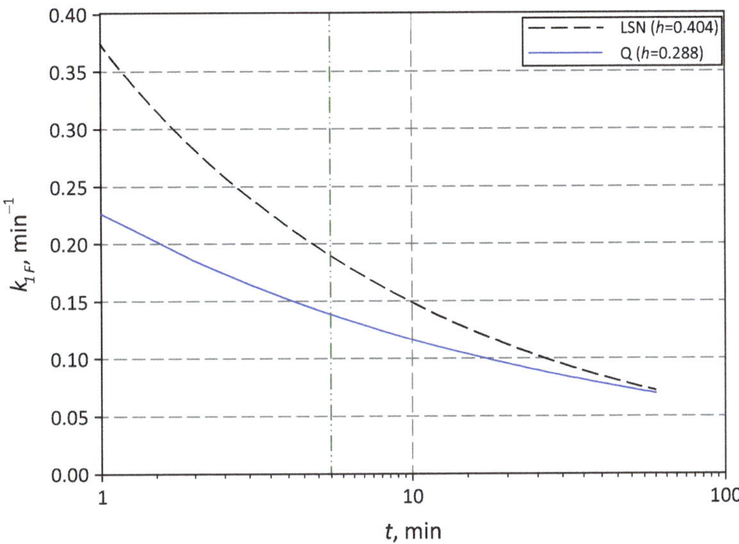

Fig. 4.8 Instantaneous rate coefficient (k_{1F}) as a function of adsorption time (t), fractal pseudo-first-order model, for materials investigated by Mestre et al. (2009). The corresponding values for the fractal exponent (h) are reported as well

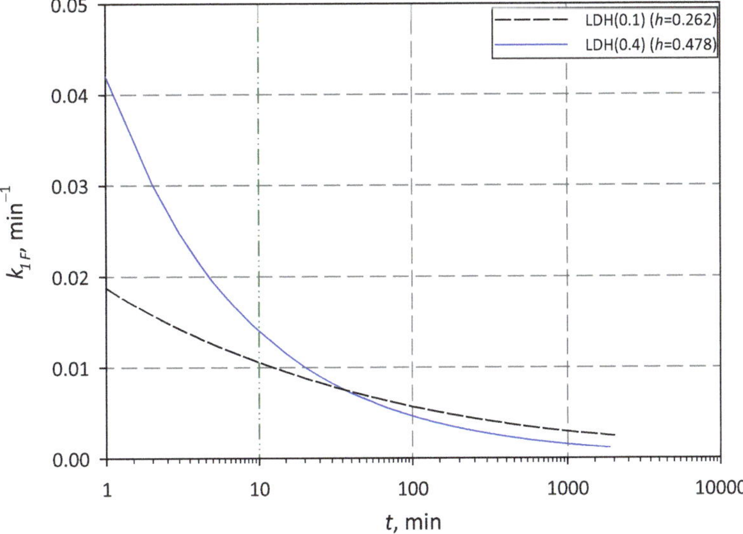

Fig. 4.9 Instantaneous rate coefficient (k_{1F}) as a function of adsorption time (t), fractal pseudo-first-order model, for materials investigated by Zhou et al. (2011). The corresponding values for the fractal exponent (h) are reported as well

experimental conditions selected in this survey – the intraparticle diffusion can be considered as limiting regime. In the framework of this scenario, the fractal-like approach foresees the decrease in time of the rate coefficient ruled by h. This to account for the probability for a random walker to come back to its original position after a certain time. The higher h, the higher this probability so hindering the even exploration of the intraparticle environment by the molecule to be adsorbed. This last aspect negatively affects the adsorption rate coefficient k_{1F}. The finest 200–250 mesh adsorbent, being the one showing the largest Hg^{2+} capture capacity, would be that with the most crowded pore space among the three materials as t increases. This limits the exploration of the pore space by the pollutant to reach free adsorption sites. In fact, h for 200–250 mesh adsorbent is the highest among the three calculated in Table 4.2. Therefore, the time-decrease of k_{1F} for 200–250 mesh material is more marked than for the other two adsorbents: after 480 min, the values of k_{1F} for 200–250 and 150–200 mesh materials practically coincide.

Figure 4.7 and Table 4.3 refer to the paper by Kim et al. (2008). The bifunctionalised MB22 adsorbent is characterised by the greatest values for k_{1F} throughout the time-range analysed. Intermediate is the case for MB04, while the smallest values are observed for MB40. For MB04 and MB40, for example at $t = 2$ min, the value of k_{1F} results about 45% and 65% lower than for MB22, respectively. The ranking of k_{1F} vs. t for the three functionalised molecular sieves mirrors that reported by the parent Authors in terms of specific Brunauer–Emmett–Teller (BET) surface area and total pore volume. For example, it is 336.6 and 171.2 $m^2\ g^{-1}$ in terms of surface area, 0.50 and 0.27 $cm^3\ g^{-1}$ in terms of pore volume for MB22 and MB40, respectively. Actually, these characteristics can promote a faster capture rate of the pollutant. In terms of heterogeneity parameter, the Reader can observe a higher value for the bifunctionalised adsorbent. The presence of two ammonium-based functional groups in MB22 (rather than only one) would ascribe a higher degree of surface chemical heterogeneity to this material with respect to monofunctionalised MB04 and MB40. This determines for MB22 a more relevant fractal-like nature of the adsorption dynamics. Actually, while for both monofunctionalised MB04 and MB40 – less heterogeneous – and bifunctionalised MB22 – more heterogeneous – one can observe a clear decay of k_{1F} in time for the reasons discussed in Harifar and Azizian (2014) as reported in Sect. 4.3, the effect of time on k_{1F} is apparently more noticeable for MB22.

To the activated wood (LSN) investigated by Mestre et al. (2009), a larger value for k'_{1F} was ascribed with respect to the activated coal (Q) studied in the same communication, Fig. 4.8 and Table 4.4 in the present work. The former Authors have reported a specific mesopore volume of 0.158 and 0.143 $cm^3\ g^{-1}$ for LSN and Q, respectively. This textural parameter is likely to rule the initial rate of adsorption, in line with findings here reported in terms of k'_{1F}. On the one hand, the larger value of k'_{1F} for activated wood makes the $k_{1F}(t)$ trend for LSN laying above the one for Q. On the other, from Fig. 4.8 it is evident the more extensive decay rate of k_{1F} for LSN itself as (cf. Table 4.4) the value of h for LSN is far larger than that for Q. More in detail, after 5 min (i.e. the earliest experimental time here analysed) it is $k_{1F} = 0.196$

and 0.142 min^{-1} for LSN and Q, respectively, namely a k_{1F} value for LSN about 40% larger than for Q. This difference is nearly halved for $t = 20$ min. The larger value of the heterogeneity parameter for activated wood is here related to: (i) the smaller mean micropore size for LSN, see Fig. 4.4 in the original paper; (ii) the major specific volume of ultramicropores that the Authors did observe for LSN in their Table 4.1, 0.241 vs. 0.182 $cm^3 g^{-1}$ for Q. Both aspects should be regarded in the light of the following consideration. Very small pores can act as bottlenecks for the diffusion of the pollutant molecule. This would hinder the establishment of uniform composition profiles along the particle radial coordinate as already observed in Balsamo and Montagnaro (2015). In addition, Mestre et al. (2009) make reference to phenomena of electrostatic repulsive interactions given by the acidic surface of LSN adsorbent. In the present scenario, this contributes in slowing down the ibuprofen capture on LSN and eventually enhances the heterogeneity degree of the process dynamics in conformity with the higher h value for LSN.

The case investigated after Li et al. (2011) shows us again the direct correlation between degree of system heterogeneity and fractal-like nature of the process dynamics. The Authors, in functionalising by thiol their mesoporous silica, have discussed about cross-linked organic groups homogeneously distributed within the silica matrix. This likely gives rise to chemically homogeneous porous environments in which diffusion phenomena take place. Furthermore it is consistent with our modelling outcomes which ended up into $h = 0$, i.e. memory-less rate coefficient.

Zhou et al. (2011), in describing in their Table 4.1 the porosimetric features of the Zn/Al layered double hydroxides LDH(0.1) and LDH(0.4), observed for the latter greater values for specific BET surface area and pore volume, and for average pore size. This is again related to the more significant initial rate of adsorption for the sample characterised by larger values of these textural parameters (Table 4.6 and Fig. 4.9). In addition, when inspecting Fig. 4.3 in the original paper, a more heterogeneous pore size distribution for LDH(0.4) can be noticed. This aspect can be directly linked to the value of h which results larger for LDH(0.4). This translates into a faster time-decline for k_{1F}, in line with findings discussed by Balsamo and Montagnaro (2015) when referring to cadmium adsorption onto coal fly ash beneficiated by different methods. In the present case, this ends up into k_{1F} for LDH(0.4) larger than for LDH(0.1) at shorter process times only (and up to 40 min) while the trend is reversed further on.

4.5 Conclusion

Case-studies of liquid–solid adsorption investigated in this manuscript have concerned pollutants (heavy metals, anions, emerging contaminants) and adsorbents (activated carbons of various nature, silica-based materials, zinc/aluminium hydroxides) of different nature.

Non-linear regression of experimental data has been performed with reference to different canonical and fractal-like kinetic models. Fractal pseudo-first-order (FPFO) resulted the most statistically accurate model among all those taken into consideration, except for one case for which FPFO reduced to the canonical pseudo-first-order (PFO) model. In general PFO, representative of canonical models and here chosen for the sake of comparison with FPFO, overestimates the values of degree of surface coverage. This leads to the prediction of a faster attainment of equilibrium conditions with respect to experimental data.

A first remark stemming from these results is the following. If the complex mechanisms ruling the adsorption phenomenon are not adequately described by mathematical expressions such as in the case where a canonical rather than fractal-like kinetic model is used, one would risk to under-design the adsorption unit. Pollutant concentrations in the downstream effluent possibly larger than the theoretically predicted value could be therefore expected.

In general, we have verified that a fractal-like nature arises for adsorption processes in which the adsorbents exhibit chemical heterogeneities and/or transport phenomena play a significant role. To support this analysis, the case-study for which the adsorbate diffused into chemically homogeneous porous environments was the only one where fractal-like features were not recognised.

More in detail, with reference to the fractal-like formulation of the rate coefficient (Eq. 4.1):

- the instantaneous rate coefficient for short adsorption times appears positively affected by: (i) a decrease in particle size; (ii) an increase in specific BET surface area and total pore volume; (iii) an increase in specific mesopore volume; (iv) an increase in average pore size;
- the heterogeneity parameter, or fractal exponent, is higher, thus mirroring stricter limits to the even exploration of the intraparticle environment by the adsorbate, when for the adsorbent: (i) the pore space is more crowded by already-adsorbed molecules; (ii) a larger degree of surface chemical heterogeneity is determined; (iii) the mean micropore size is smaller; (iv) the specific volume of ultramicropores is larger; (v) the pore size distribution is more polydispersed. It is here recalled that the value of the fractal exponent strongly influences the time-decay rate of the instantaneous rate coefficient.

The above observations should be read with care as a limited number of case-studies have been analysed in this paper. To define which parameters are able to mostly influence the fractal exponent and related functional dependencies, the Scientist should perform a complete experimental and modelling characterisation of the adsorption process. This should be done under both kinetic and thermodynamic viewpoint for different operating conditions, f.i. pH and temperature, coupled with a wide textural and chemical analysis of the adsorbent properties. With reference to the chemical characterisation, the time-dependent role of active sites should be investigated through real-time diagnostic techniques, e.g. FTIR, to assess the effective adsorbent chemical heterogeneity during the adsorption process. Once again, this advanced knowledge is devoted to the proper design of a continuous fluid–solid adsorption unit.

References

Andini S, Cioffi R, Montagnaro F, Pisciotta F, Santoro L (2006) Simultaneous adsorption of chlorophenol and heavy metal ions on organophilic bentonite. Appl Clay Sci 31:126–133. https://doi.org/10.1016/j.clay.2005.09.004

Andini S, Cioffi R, Colangelo F, Montagnaro F, Santoro L (2008) Adsorption of chlorophenol, chloroaniline and methylene blue on fuel oil fly ash. J Hazard Mater 157:599–604. https://doi.org/10.1016/j.jhazmat.2008.01.025

Balsamo M, Montagnaro F (2015) Fractal-like Vermeulen kinetic equation for the description of diffusion-controlled adsorption dynamics. J Phys Chem C 119:8781–8785. https://doi.org/10.1021/acs.jpcc.5b01783

Balsamo M, Di Natale F, Erto A, Lancia A, Montagnaro F, Santoro L (2010) Arsenate removal from synthetic wastewaters by adsorption onto fly ash. Desalination 263:58–63. https://doi.org/10.1016/j.desal.2010.06.035

Balsamo M, Di Natale F, Erto A, Lancia A, Montagnaro F, Santoro L (2011) Cadmium adsorption by coal combustion ashes-based sorbents – relationship between sorbent properties and adsorption capacity. J Hazard Mater 187:371–378. https://doi.org/10.1016/j.jhazmat.2011.01.029

Balsamo M, Di Natale F, Erto A, Lancia A, Montagnaro F, Santoro L (2012) Steam- and carbon dioxide-gasification of coal combustion ash for liquid phase cadmium removal by adsorption. Chem Eng J 207–208:66–71. https://doi.org/10.1016/j.cej.2012.07.003

Berry H (2002) Monte Carlo simulations of enzyme reactions in two dimensions: fractal kinetics and spatial segregation. Biophys J 83:1891–1901. https://doi.org/10.1016/S0006-3495(02)73953-2

Brouers F, Sotolongo-Costa O (2006) Generalized fractal kinetics in complex systems (application to biophysics and biotechnology). Physica A 368:165–175. https://doi.org/10.1016/j.physa.2005.12.062

Crini G, Badot PM (eds) (2010) Sorption processes and pollution: conventional and non-conventional sorbents for pollutant removal from wastewaters. Presses Universitaires de Franche-Comté, Besançon ISBN:978-2-84867-304-2

Do DD (1998) Adsorption analysis: equilibria and kinetics. Imperial College Press, London ISBN:978-1-86094-382-9

Dokoumetzidis A, Macheras P (2011) The changing face of the rate concept in biopharmaceutical sciences: from classical to fractal and finally to fractional. Pharm Res 28:1229–1232. https://doi.org/10.1007/s11095-011-0370-4

Figaro S, Avril JP, Brouers F, Ouensanga A, Gaspard S (2009) Adsorption studies of molasse's wastewaters on activated carbon: modelling with a new fractal kinetic equation and evaluation of kinetic models. J Hazard Mater 161:649–656. https://doi.org/10.1016/j.jhazmat.2008.04.006

Gaspard S, Altenor S, Passe-Coutrin N, Ouensanga A, Brouers F (2006) Parameters from a new kinetic equation to evaluate activated carbons efficiency for water treatment. Water Res 40:3467–3477. https://doi.org/10.1016/j.watres.2006.07.018

Harifar M, Azizian S (2012) Fractal-like adsorption kinetics at the solid/solution interface. J Phys Chem C 116:13111–13119. https://doi.org/10.1021/jp301261h

Harifar M, Azizian S (2014) Fractal-like kinetics for adsorption on heterogeneous solid surfaces. J Phys Chem C 118:1129–1134. https://doi.org/10.1021/jp4110882

Ho YS, McKay G (1998) Sorption of dye from aqueous solution by peat. Chem Eng J 70:115–124. https://doi.org/10.1016/S0923-0467(98)00076-1

Kim TH, Jang M, Park JK (2008) Bifunctionalized mesoporous molecular sieve for perchlorate removal. Micropor Mesopor Mat 108:22–28. https://doi.org/10.1016/j.micromeso.2007.03.023

Kopelman R (1986) Rate processes on fractals: theory, simulations, and experiments. J Stat Phys 42:185–200. https://doi.org/10.1007/BF01010846

Kopelman R (1988) Fractal reaction kinetics. Science 241:1620–1626. https://doi.org/10.1126/science.241.4873.1620

Lagergren S (1898) Zur theorie der sogenannten adsorption gelöster stoffe. Kungliga Svenska Vetenskapsakad Handl 24:1–39

Li G, Zhao Z, Liu J, Jiang G (2011) Effective heavy metal removal from aqueous systems by thiol functionalized magnetic mesoporous silica. J Hazard Mater 192:277–283. https://doi.org/10.1016/j.jhazmat.2011.05.015

Marczewski AW (2010) Analysis of kinetic Langmuir model. Part I: Integrated kinetic Langmuir equation (IKL): a new complete analytical solution of the Langmuir rate equation. Langmuir 26:15229–15238. https://doi.org/10.1021/la1010049

Mestre AS, Pires J, Nogueira JMF, Parra JB, Carvalho AP, Ania CO (2009) Waste-derived activated carbons for removal of ibuprofen from solution: role of surface chemistry and pore structure. Bioresour Technol 100:1720–1726. https://doi.org/10.1016/j.biortech.2008.09.039

Mohan D, Gupta VK, Srivastava SK, Chander S (2001) Kinetics of mercury adsorption from wastewater using activated carbon derived from fertilizer waste. Colloid Surf A 177:169–181. https://doi.org/10.1016/S0927-7757(00)00669-5

Montagnaro F, Balsamo M (2014) Deeper insights into fractal concepts applied to liquid-phase adsorption dynamics. Fuel Process Technol 128:412–416. https://doi.org/10.1016/j.fuproc.2014.07.021

Montagnaro F, Santoro L (2009) Reuse of coal combustion ashes as dyes and heavy metal adsorbents: effect of sieving and demineralization on waste properties and adsorption capacity. Chem Eng J 150:174–180. https://doi.org/10.1016/j.cej.2008.12.022

Montagnaro F, Santoro L (2010) Chapter 11: Non-conventional adsorbents for the removal of metal compounds from wastewaters. In: Crini G, Badot PM (eds) Sorption processes and pollution: conventional and non-conventional sorbents for pollutant removal from wastewaters. Presses Universitaires de Franche-Comté, Besançon, pp 297–312 ISBN:978-2-84867-304-2

Montagnaro F, Balsamo M, Salatino P (2016) A single particle model of lime sulphation with a fractal formulation of product layer diffusion. Chem Eng Sci 156:115–120. https://doi.org/10.1016/j.ces.2016.09.021

Ncibi MC (2008) Applicability of some statistical tools to predict optimum adsorption isotherm after linear and non-linear regression analysis. J Hazard Mater 153:207–212. https://doi.org/10.1016/j.jhazmat.2007.08.038

Pippa N, Dokoumetzidis A, Demetzos C, Macheras P (2013) On the ubiquitous presence of fractals and fractal concepts in pharmaceutical sciences: a review. Int J Pharm 456:340–352. https://doi.org/10.1016/j.ijpharm.2013.08.087

Plazinski W, Rudzinski W, Plazinska A (2009) Theoretical models of sorption kinetics including a surface reaction mechanism: a review. Adv Colloid Interf Sci 152:2–13. https://doi.org/10.1016/j.cis.2009.07.009

Qiu H, Lv L, Pan BC, Zhang QJ, Zhang WM, Zhang QM (2009) Critical review in adsorption kinetic models. J Zhejiang Univ Sci A 10:716–724. https://doi.org/10.1631/jzus.A0820524

Rudzinski W, Plazinski W (2006) Kinetics of solute adsorption at solid/solution interfaces: a theoretical development of the empirical pseudo-first and pseudo-second order kinetic rate equations, based on applying the statistical rate theory of interfacial transport. J Phys Chem B 110:16514–16525. https://doi.org/10.1021/jp061779n

Ruthven DM (1984) Principles of adsorption and adsorption processes. Wiley, New York ISBN:978-0-471-86606-0

Savageau MA (1995) Michaelis-Menten mechanism reconsidered: implications of fractal kinetics. J Theor Biol 176:115–124. https://doi.org/10.1006/jtbi.1995.0181

Sen PN (2004) Time-dependent diffusion coefficient as a probe of geometry. Concept Magn Reson Part A 23A:1–21. https://doi.org/10.1002/cmr.a.20017

Shannon RD (1976) Revised effective ionic radii and systematic studies of interatomic distances in halides and chalcogenides. Acta Crystallogr A32:751–767. https://doi.org/10.1107/S0567739476001551

Suzuki M (1990) Adsorption engineering. Elsevier Science, Amsterdam ISBN:978-0444988027

Vermeulen T (1953) Theory for irreversible and constant pattern solid diffusion. Ind Eng Chem 45:1664–1670. https://doi.org/10.1021/ie50524a025

Wang ZW, Xu F, Manchala KR, Sun Y, Li Y (2016) Fractal-like kinetics of the solid-state anaerobic digestion. Waste Manag 53:55–61. https://doi.org/10.1016/j.wasman.2016.04.019

Xu F, Ding H (2007) A new kinetic model for heterogeneous (or spatially confined) enzymatic catalysis: contributions from the fractal and jamming (overcrowding) effects. Appl Catal A-Gen 317:70–81. https://doi.org/10.1016/j.apcata.2006.10.014

Zhou J, Yang S, Yu J, Shu Z (2011) Novel hollow microspheres of hierarchical zinc–aluminium layered double hydroxides and their enhanced adsorption capacity for phosphate in water. J Hazard Mater 192:1114–1121. https://doi.org/10.1016/j.jhazmat.2011.06.013

Chapter 5
Carbonaceous Porous Materials for the Adsorption of Heavy Metals: Chemical Characterization of Oxidized Activated Carbons

Paola Rodríguez-Estupiñán, Liliana Giraldo, and Juan Carlos Moreno-Piraján

Contents

Abstract The contamination of water sources by heavy metals is one of the main environmental concerns today, due to the intrinsic toxicity, long lifetime and bioaccumulation properties of these kinds of pollutants. The use of carbonaceous porous materials as adsorbents for preconcentration of pollutants is an application that emerges as an alternative that draws attention in the treatment of industrial effluents. A thorough knowledge of surface chemistry enables the preparation of adsorbents with appropriate characteristics resulting in new useful properties for

P. Rodríguez-Estupiñán · J. C. Moreno-Piraján (✉)
Facultad de Ciencias, Departamento de Química, Grupo de Investigación en Sólidos Porosos y Calorimetría, Universidad de los Andes, Bogotá, Colombia
e-mail: jp.rodrigueze@uniandes.edu.co; jumoreno@uniandes.edu.co

L. Giraldo
Departamento de Química, Facultad de Ciencias, Universidad Nacional de Colombia, Bogotá, Colombia

G. Crini, E. Lichtfouse (eds.), *Green Adsorbents for Pollutant Removal*, Environmental Chemistry for a Sustainable World 18,
https://doi.org/10.1007/978-3-319-92111-2_5

specific applications, it is possible to modify the chemistry surface of carbonaceous porous materials by introducing oxygenated groups while essentially maintaining their original porous texture. Different processes are used for this purpose: oxidation, functionalization and thermal treatments.

The importance of surface chemistry of the porous solids on the adsorption of metal ions from aqueous solution is mainly due to specific interactions between surface groups and dissolved species in solution through different mechanisms including: the formation of metal complexes, reactions of acceptor-donor electrons, and ionic exchange. The present work deals with the effect of chemical modification on the surface of a granular activated carbon, by oxidation with nitric acid, sodium hypochlorite, and hydrogen peroxide, in ions metal adsorption from aqueous solution. The surface chemistry was evaluated by Boehm and potentiometric titrations, groups total evaluated by Boehm titration varies between 247.7 and 529.7 $\mu mol.g^{-1}$, thermogravimetric analysis coupled to mass spectroscopy and point of zero charge, showing the effect of the oxidizing agent in textural parameters such as BET surface area and pore volumes which were assessed by gas adsorption. The BET surface area values of the solids are between 687 and 889 $m^2\ g^{-1}$.

5.1 Introduction

Textural properties, like surface area, pore volume and pore size distributions, have been constantly related to the adsorption capacity of porous solids. However, the surface chemical composition plays an important role in the adsorption when specific interaction/reactions take place. According to this, an ideal adsorbent should have certain characteristics, like: a high surface area, a uniform distribution of the adsorption sites with open pore structure in order to favor the kinetics of the process, easy access to the adsorption sites, simple preparation and stability, and regeneration capacity (Saini et al. 2010; Silvestre-Albero et al. 2012). In addition to these requirements, an ideal adsorbent for environmental applications should provide specific binding or interactions sites in the material, as well as affinity and selectivity to the species to be removed.

Activated carbon is one of the most used porous solids, due to its adsorption capability and the relative easiness of using pre and post treatments to modify and/or design its chemical properties. Additionally, the research on materials development has focused the interest of the scientific community because of the possibility of introducing a wide variety of functional groups (Figueiredo et al. 1999; Smith et al. 2005; Onida et al. 2008; Goertzen et al. 2012; Ashish et al. 2012; Treviño-Cordero et al. 2013).

Nowadays, the adsorption of contaminants dissolved in aqueous solution on porous solids is a technique considered as a good alternative in the processes of effluents decontamination. There are many physical and chemical factors responsible for the adsorption process, their influence on the process is not totally clear, which is a significant disadvantage in the establishment of the conditions that allow to control and optimize this process.

The variety of mechanisms that can occur in the metal ions adsorption process on a solid requires to control several factors that may affect the process, e.g.: concentration and surface groups type of the solid; pH in the point of zero charge; nature of the metal ion considering its speciation diagram, solubility and size (Figueiredo et al. 1999; Onida et al. 2008; Ashish et al. 2012; Treviño-Cordero et al. 2013).

The surface chemistry of porous solids can generally determine the moisture content, the acidic-basic character of the surface, the adsorption capacity of dissolved species in solution and its polarity. These features are related to the presence of functional groups on the solid surface. In the case of the carbonaceous materials, oxygenate surface groups have been extensively studied because the modification of surface can determine the affinity for the molecules that are to be removed. According to the above the control of surface chemistry is an important aspect in the preparation of specific porous solids for a target adsorbate. The surface groups are attached at the edges of the graphenic layers and are analogous to organic functional groups (carboxylic acids, lactones, phenols, carbonyls, aldehydes, ethers).

According to the strength of the group these can be classified as acidic, neutral or basic. The type and concentration of surface functional groups have been studied using analytical methods such as Boehm titration, potentiometric titration and mass titration to determinate the point of zero charge, infrared spectroscopy, X-ray photoelectron spectroscopy and thermal analysis coupled to mass spectroscopy (TA -MS) (Radovic et al. 2004; Bandosz and Ania 2006; Kim et al. 2012).

This chapter is designed: Firstly, to review some techniques to characterize the chemical surface of carbonaceous materials, and to assess the effects of the surface groups in the improvement of the interaction between the adsorbent and the target adsorbate. Secondly, to study the behavior of a microporous activated carbon obtained from coconut shell by physical activation, as well as its behavior when oxidized with three different oxidants, nitric acid (HNO_3), hydrogen peroxide (H_2O_2) and sodium hypochlorite (NaClO). Textural and chemical characteristics induced by the oxidation were analyzed by: titration methods, pH_{PZC} and N_2 adsorption isotherm. Finally, to present possible mechanisms that may generate the chemical modification of the surface of activated carbons based on the oxidation reactions of small organic molecules. To exemplify some surface complex formations in adsorption processes. Some topics and principles of the techniques that have been used to characterize the surface chemistry of carbonaceous solids are detailed next.

5.1.1 Chemical Characterization: Carbonaceous Materials

The control of surface chemistry is an important aspect in the preparation of porous solids with specific capacities to a target adsorbate. The proper use of this chemical character of the surface opens the possibilities for various applications, such as ionic species. It is important to correlate the concentration and nature of the functional groups, and the resulting properties of the carbonaceous solids.

Titration methods, Boehm and potentiometric, determination of the point zero charge, and thermal analysis coupled to mass spectroscopy, are the most reported techniques among the characterization methods. A brief introduction to titration methods, pH_{PZC} and TGA-MS is discussed for the characterization of functional groups of carbonaceous solids subjected to a modification process. The typical methodology for these techniques and the interpretation of the results are discussed in the following sections.

5.1.1.1 Boehm Titrations

The Boehm is a chemical method used to identify surface functional groups. It is based on the neutralization of the acid groups present on the surface by using basic solutions of different strength such as: sodium ethoxide, sodium hydroxide, sodium carbonate and bicarbonate of sodium, to distinguish between surface acid groups of different strength (Boehm 1966).

The method is limited to phenol, lactone and carboxyl groups. Solutions of different strength are used due to the diversity of surface acid groups. Therefore, their different values of pKa, which are dependent on the location of a group regarding other groups, can exert an inductive effect (Figueiredo et al. 1999). Although this technique provides useful information, it is convenient to clarify that it does not take into account other oxygenated surface groups such as esters and ethers.

It is generally accepted that sodium hydroxide ($pK_{NaOH} = 15.74$) is the strongest base and it is assumed that it neutralizes all Bronsted acids (including phenols, lactones and carboxylic acids). On the other hand, sodium carbonate ($pK_{Na_2CO_3} = 10.25$) titrates carboxylic acid groups and lactones, and sodium acid carbonate ($pK_{NaHCO_3} = 6.37$) neutralizes only the carboxylic acid groups (Kim et al. 2012). The concentration of a specific surface group is calculated from the neutralization point usually determined as the point at pH 7 on the Boehm titration curve (Radovic et al. 2004).

5.1.1.2 Potentiometric Titrations

The potentiometric titration provides qualitative and quantitative information on the amount and strength of acid sites on the surface of a solid. The information obtained is complementary to that acquired by other techniques, such as Boehm titration and infrared spectroscopy (Bandosz and Ania 2006). In this method it is assumed that the study system is composed of acidic sites that are characterized by their acid constants, Ka, which are deprotonated at a certain pH (Contescu et al. 1995).

Functional groups present on the surface of a porous solid are exposed to a wide range of inter and intra-molecular interactions, such as inductive effect, resonance, tautomeric, steric hindrance and hydrogen bonding. These interactions can alter the acid-base behavior of Bronsted, and as a result a continuous distribution is obtained in terms of the protonic affinity of surface chemical properties such as acidity and basicity. The groups concentration can be described by a continuous pKa distribution, function f(pKa), in which the pK ranges can be overlapped by several chemically different groups (Jagiello et al. 1994; Bandosz and Ania 2006). The importance of differentiating the pKa values of the surface groups of the porous solids is to evaluate the effect of pH on the surface groups and to determine what type of mechanism is likely to occur in a process of adsorption of charged species. In the classification of groups it is assumed that values of $pKa < 8$ correspond to carboxylic acid groups and those with $pKa > 8$ are classified as phenols and quinones (Contescu et al. 1993).

Significant potentiometric titration data are obtained within a pH window between 3 and 11. Surface groups whose pKa values are outside the experimental pH window do not react during acid-base titration, although they can dissociate (or bind) protons spontaneously on contact of the solid with the aqueous solution. This has no consequence for the determination of the acidity distribution function for acid-base groups of moderate strength ($3 < pKa < 11$). However, when strongly acidic ($pKa < 3$) or basic ($pK > 11$) groups are present; they may not be detected in the function of pK, (pK). Its presence can only be inferred by comparing the adjusted proton-binding curve and experimental data (Bandosz et al. 1993; Contescu et al. 1997).

5.1.1.3 The pH of the Point Zero Charge

The pH of an aqueous suspension of a porous solid is a useful parameter that allows a first approximation of the functional groups present on the surface of a solid. The Bronsted acid groups tend to donate their protons to the water molecules and therefore the surface is negatively charged (and the pH decreases), while the Lewis base groups adsorb the protons in the solution and are positively charged (Bandosz et al. 1993). The charge is in function: of the type of ions present; of the characteristic of the surface; of the nature of the solid and of the pH solution.

Surface charge can be determined using several methods, including acid-base titration and mass titration; the methods are based on determining the maximum possible transfer of protons between the particle and the solution.

In the mass titration method, according to the procedure described by Noh and Schwarz (1990), the overall pH resulting from the maximum transfer of protons is measured when different amounts of solid are weighed and a given volume of an aqueous solution is added, from 0.1 N of NaCl until the pH of the solution does not change with the addition of the solid. According to the information above, by definition of the pH at the point of zero charge, pH_{PZC} is the pH at which the net surface charge is neutral; that is, when all the groups present on the surface reach their equilibrium of dissociation and association. The pH_{PZC} values change systematically with the modification of surface chemistry, for example an oxidized solid will have a lower pH_{PZC} value (Jagiello et al. 1995; Bandosz and Ania 2006).

The distribution of the surface charge with respect to the pH of the solution is an important piece of information, because it allows to explain the adsorption of ions and contributes in the elucidation of the adsorption mechanism. Thus: if the $pH > pH_{PZC}$, the acid functional groups dissociate, and they release the protons in the middle and a negatively charged surface is produced, which favors the interaction with cations. On the other hand, if the $pH < pH_{PZC}$, the basic sites combine with the protons of the medium to leave a positively charged surface, and interaction with anions is favored (Jagiello et al. 1995; Bandosz and Ania 2006).

5.1.1.4 Thermal Analysis – Mass Spectroscopy

Thermal analysis of porous solids allows to study the presence of surface functional groups due to their thermal stability (for example: carbonaceous solids). In the thermal analysis (TA) the solid must be heated at a programmed heating rate and in an inert atmosphere which serves at the same time as a gas carrier. The technique induces the thermal desorption of species adsorbed on the surface of the solid. The desorbed gaseous products are analyzed by techniques such as mass spectroscopy and IR, gas chromatography or gravimetric analysis. In general, oxygen-containing functional groups are mostly broken down in the form of water, carbon dioxide and carbon monoxide, while functional groups with nitrogen release nitrous oxide (Salame and Bandosz 2001; Figueiredo and Pereira 2012).

The interpretation of the thermal desorption profiles according to the number of peaks and the desorption temperature, provides useful information about the type of desorbed species and the nature of the interactions between those species and the solid. The concentration of the different surface groups can be estimated by the deconvolution of the TA profiles. It has been shown that a multiple Gaussian function fits the data appropriately (Haydar et al. 2000; Salame and Bandosz 2001; Bandosz and Ania 2006; Figueiredo and Pereira 2012). CO_2 desorption occurs mainly due to acidic groups.

The groups that produce CO_2 decompose at two different temperatures, corresponding to two types of chemical and energetically differentiable groups since the decomposition temperature is related to the strength of the group. According to this, desorption at low temperature (375 °C) is assigned to carboxylic acids, and at high temperature (625 °C) it is attributed to carboxylic anhydrides and lactones. In the same way, the groups that produce CO are assigned to different groups at 625 and 825 °C corresponding to phenols, ethers and carbonyls. Water peaks are recorded in the TA profiles at various temperatures. A peak at low temperatures is generally attributed to the desorption of physisorbed water, while at high temperatures it is assigned to the desorption of water bound to oxygen complexes by hydrogen bonds, to the condensation of phenolic groups, or to dehydration in reactions of the vicinal carboxylic groups to obtain carboxyl anhydrides (Haydar et al. 2000; Jagiello et al. 1995; Figueiredo and Pereira 2012).

5.2 Materials and Methods

5.2.1 *Activated Carbon*

Activated carbon was obtained from coconut shells and physical activation. The solid was sieved to a particle size of 1 mm and washed with a dilute solution of 0.01 M HCl, to remove impurities and part of the ashes. Afterwards, the sample was washed with distilled water to remove the excess of acid, dried for 24 h at 90 °C and stored in plastic containers under nitrogen.

Subsequently, the activated carbon was subjected to an oxidation process by impregnation with aqueous solution of different oxidants: HNO_3 6 M, at their boiling temperature, (GACoxN), H_2O_2 10 M (GACoxP) and NaClO 2 M (GACoxCl) at room temperature. One sample more was obtained from GAC and a thermal treatment at 900 °C. The samples were washed with distilled water until the pH of the water was constant.

5.2.2 *Physicochemical Characterization*

5.2.2.1 Textural Characterization

Textural parameters, such as area BET, pore volume and pore width of activated carbons, were evaluated by physical adsorption of N_2 at −196 °C and CO_2 at 0 °C in an automatic Autosorb 3B apparatus (Quantachrome). The apparent surface area, pore volume and width were determined by the BET and QSDFT model, respectively.

5.2.2.2 pH at the Point Zero Charge

The pH value at the point of zero charge, pH_{PZC}, was determined by the mass titration method, for which different weighed amounts of activated carbon (0.010–0.600 g) were placed in a series of 50-mL glass bottles, to each one of which was added 10 mL of 0.1 M NaCl. The bottles were stoppered and left under constant stirring for 48 h. Next, the pH value of each solution was measured using a CG 840B Schott pH meter (Figueiredo and Pereira 2012).

5.2.2.3 Boehm Titration

The surface chemistry of activated carbons was characterized by Boehm titrations. For this purpose, 0.500 g of each activated carbon was taken, to which 50 mL each of 0.1 M NaOH, Na_2CO_3, and $NaHCO_3$ was added to determine the acidity of groups or 50 mL of 0.1 M HCl was added to determine the basicity. The mixtures were kept at a constant temperature of 25 °C with constant stirring for 48 h. Subsequently, a 10-mL aliquot of each solution in contact with the activated carbon samples was titrated employing corresponding 0.1 M standard solutions of HCl or NaOH (Salame and Bandosz 2001; Figueiredo and Pereira 2012; Kim et al. 2012).

5.2.2.4 Potentiometric Titration

The potentiometric titrations were carried out in an automatic titrator (DMS Titrando 888, Metrohm). The titrations are performed on a suspension of 0.100 g of the solid and 50 ml of 0.01 M $NaNO_3$, and are constantly stirring and under N_2 atmosphere to prevent the influence of atmospheric CO_2 until obtaining a stable pH value. Standard solutions of NaOH and HCl 0.100 M were employed as titrant agents. The experiments were carried out in a pH range of 3 to 10. If the initial pH is different from 3 the volume of the corresponding reagent should be added and recorded. The experimental data were transformed into the proton binding curves (Q), which were deconvoluted into a continuous pKa distribution corresponding to the different surface groups by using the SAEIUS method (Bandosz et al. 1993).

5.2.2.5 Thermogravimetric Analysis/ Mass Spectroscopy

The thermogravimetric analysis was conducted on a TA (Thermal Analyzer) instrument (SDT Q600) with approximately 30 mg of the solid, which was heated at a rate of 10 °C/min to a maximum of 1000 °C. The experiments were performed under helium atmosphere (Flow: 100 mL/min). From the recorded weight loss, the thermogravimetric and the differential thermogravimetric curves were obtained. During the thermal analysis, the gases produced by the decomposition of the surface groups were collected for analysis by a mass spectrometer (ThermoStar Gas Mass Spectrometer, GSD; Pfeiffer Vacuum) and the thermal profiles m/z were obtained (Haydar et al. 2000).

5.3 Discussion and Analysis of Results

5.3.1 Thermogravimetric Analysis–Mass Spectroscopy

For carbonaceous porous materials from lignocellulosic precursors several pyrolysis processes may occur at a certain temperature range, even they may overlap with each other. Therefore, thermogravimetric analysis provides information of the total mass loss with the temperature rising and the thermal stability of the material, and differential thermogravimetric analysis may give a distinct signal representing each process. Figure 5.1 (a–e) correspond to the thermogravimetric analysis and

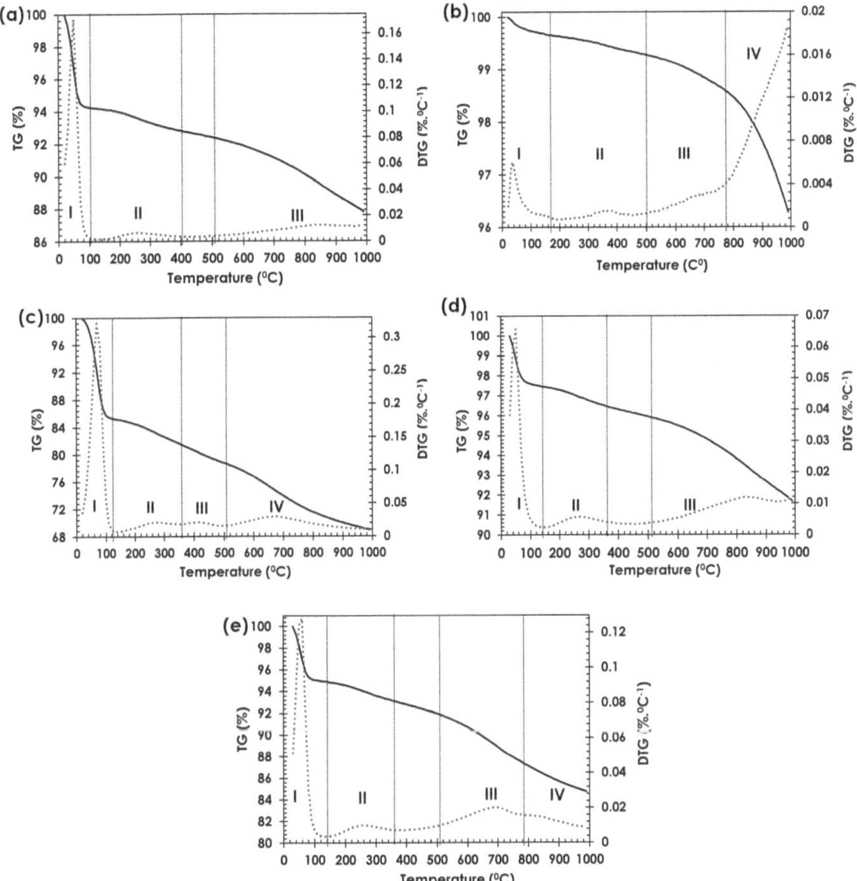

Fig. 5.1 Thermogravimetric and differential thermogravimetric curves. The thermal degradation of activated carbon samples have been investigated between room temperature and 1000° in nitrogen for (**a**) GAC, (**b**) GAC900, (**c**) GACoxN, (**d**) GACoxP, (**e**) GACoxCl

differential thermo-gravimetric curves for every sample Thermograms show an initial mass loss related to the removal of moisture. The moisture content of the solids fluctuates between 6.70 and 18.9%, being higher in the oxidized carbonaceous solids. In general, the moisture content is mainly due to the ability of porous solids to retain air humidity under ambient conditions. Taking this into account, the solids modified by the oxidation treatments present higher concentration of sites available to interact with the water molecules and consequently an increase in the hydrophilic character of the surface (Contescu et al. 1995, 1996).

All thermogravimetric curves show a continuous mass loss between 3% and 32% from room temperature to 1000 °C with shallow plateaus. But differential thermogravimetric curves demonstrate that the thermal processes consists of four regions of weight loss, at least, except GAC and GACoxP, for which only three are revealed. For activated carbons GAC, GAC 900, GACoxP and GACoxCl, initial mass loss varies between 2% and 6%, which is related to the loss of moisture. This first stage for GACoxN reflects greater weight loss, approximately 14.7%, which is related to the hydrophilic character of this surface, its location and a higher concentration of surface groups.

The gaseous products of the pyrolysis were monitored by mass spectrometry. In the analysis of the evolved gaseous products the following molecules were monitored: water (H_2O m/z = 18), carbon monoxide (CO, m/z = 28) and carbon dioxide (CO_2, m/z = 44). The mass spectra thermal profile of the gases evolved during pirólisis is given in Fig. 5.2 (a–c).

Figure 5.2(a), represents the water molecules desorption in the temperature range. As it can be seen, more water desorbs from the surface of the oxidized samples than from GAC. The lower signal was found for the GAC900 sample. The slight shift to higher temperatures for desorption of water indicates the need for more energy to remove water molecules from the porous structure. This behavior is more evident for the GACoxN, where stronger adsorption forces may occur owing to the heterogeneity of its surface chemistry.

The thermograms show two or three regions above 140 °C (see Fig. 5.1a–e), the change in the shape of thermogravimetric curve indicates desorption of thermally labile surface compounds at different temperatures; in other words, mass loss may be

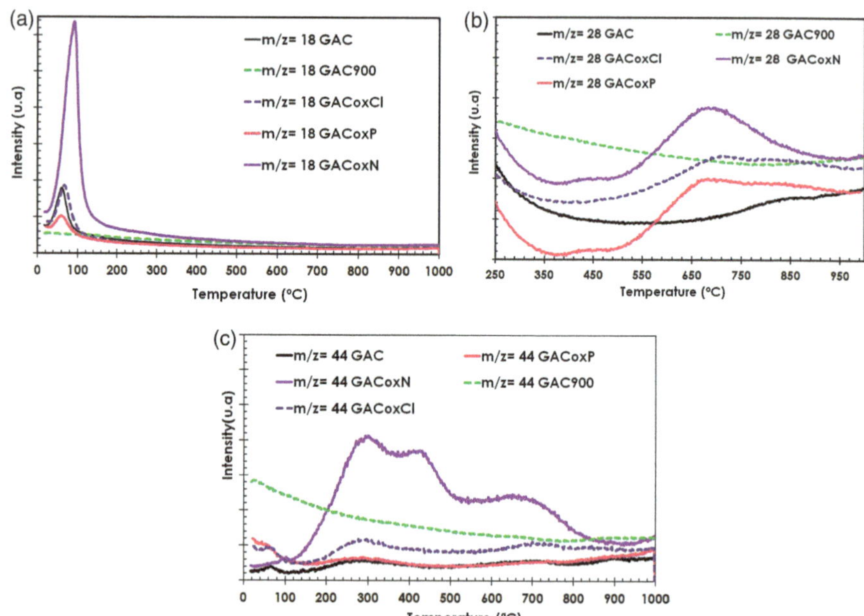

Fig. 5.2 Mass Spectrum thermal profile of gas release of pyrolysis process (**a**) m/z = 18, (**b**) m/z = 28, (**c**) m/z = 44 of activated carbons. (—= GAC; ---= GAC900; —= GACoxN; —= GACoxP; ---= GACoxCl)

related to desorption of different compounds from surface oxygen complexes. This desorption can be monitored in the thermal profile of the mass spectrum.

For carbonaceous materials the decomposition of the surface oxygenated groups in water, carbon dioxide and carbon monoxide, was analyzed on the m/z thermal profiles. The signals assignment of thermogravimetric curve is related to the characteristics of the experiment, such as the heating rate and the textural characteristics of the solid. However, some ranges have been established for the assignment of oxygenated surface groups. Carbon dioxide peaks correspond to carboxylic acid groups at low temperature, which is to say temperatures below 400 °C, and from lactone type groups at high temperatures, which means temperatures between 450 °C and 650 °C. Carboxylic anhydrides are decomposed into carbon monoxide and carbon dioxide at temperatures above 650 °C. Carbon monoxide signal at temperatures above 650 °C corresponds to ether groups and up to about 900 °C comes from phenols and carbonyl groups. Quinone groups have been associated with signals between 700 °C and 1000 °C (Haydar et al. 2000; Salame and Bandosz 2001; Almeida et al. 2016).

Profiles m /z = 28 and 44, are shown in Fig. 5.2(b, c), and their trend clearly indicates that surface oxidation treatments promote an increase in the amount of surface oxygen groups represented by an increase in carbon monoxide CO and carbon dioxide signals.

Carbon monoxide and carbon dioxide spectra obtained for oxidized activated carbons indicate that carbon dioxide desorption is more significant than that of carbon monoxide due to the formation of acidic groups. For the GACoxN samples two contributions are evident for carbon dioxide thermal profiles. They are centered on temperatures approximate to 270 °C (the same temperature for GAC, GAC900, GACoxP and GACoxCl) and 410 °C (no detected in the other samples) and these are attributed to the decomposition of two different types of carboxylic acid groups, as it is expected on a heterogeneous surface. The difference in decomposition temperature is related to the acid strength of the group, it is considered that at a lower temperature the stronger acid group is decomposed and at a higher temperature the carboxylic acids with lower strength are decomposed (Zielke et al. 1996; Figueiredo et al. 1999; Burg and Cagniant 2008; Vivo-Vilches et al. 2014; Almeida et al. 2016).

The carbon dioxide thermal profiles of GAC and GACoxP evidence a similar amount of acidic groups that decompose in the range of temperature, while for the GACoxCl sample an increase in the signal between 180 and 400 ° C was found. Carbon dioxide peaks above 600 °C are attributed to carboxylic anhydrides and lactones (Treviño-Cordero et al. 2013; Ashish et al. 2012; Burg and Cagniant 2008). These kinds of groups are useful to interact with ionic species dissolved in aqueous solution, by the formation of surface complexes and ion exchange (Zielke et al. 1996; Figueiredo et al. 1999; Burg and Cagniant 2008; Vivo-Vilches et al. 2014; Almeida et al. 2016).

The low concentration of surface groups because of the thermal treatment for the GAC 900 sample is evidenced by an almost constant signal throughout the temperature range except for a small contribution located between 800 and 1000 °C, associated with desorption of oxygenated surface groups at temperature of the treatment and the formation of higher thermal stability and decomposition of the solid (Burg and Cagniant 2008; Vivo-Vilches et al. 2014).

In the m/z thermal profiles of carbon monoxide (Fig. 5.2b) all samples exhibit signals between 430 and 1000 °C. Although the analysis of these profiles cannot distinguish different contributions above 500 °C, the signal may be associated with the thermal stability of each group: Carboxylic anhydride (400–630 °C), phenol (600–700 °C), ether (700 °C), carbonyl (800–900 °C) and Quinone (700–1000 °C). The signals are more intense for the GACoxN, GACoxCl and GACox P samples (Salame and Bandosz 2001; Burg and Cagniant 2008; Gorgulho et al. 2008; Figueiredo and Pereira 2012; Kohl et al. 2010).

5.3.2 Determination of Surface Groups by Titration Methods: Boehm Method, Point of Zero Charge and Potentiometric Titration

Tables 5.1 and 5.2 summarize the results obtained from the Boehm method, pH_{PZC}, and potentiometric titration. Boehm titrations data are presented as µmol/g.

Different pKa intervals are correlated with the change of acidity of the functional groups due to the nature of the groups near the interest group. This phenomenon is

Table 5.1 Superficial groups (µmol/g) determinate by Boehm titration and pH_{PZC}

	Sample				
Surface groups	GAC	GAC 900	GACoxN	GACoxP	GACoxCl
Carboxylic	62.23	0.00	217.3	133.5	33.90
Lactonic	34.73	10.50	43.02	30.22	54.23
Phenolic	73.00	31.50	59.55	93.20	339.0
Acidity total	169.9	41.99	319.9	257.0	427.1
Basicity total	77.80	250.6	39.70	91.94	102.6
Groups totals	247.7	292.5	359.5	349.0	529.7
pHpzc	5.4	8.9	3.4	6.2	7.2

Table 5.2 pKa distribution and surface groups density by pKa value (µmol/g) determined by potentiometric titration

	Sample				
pH	GAC	GAC 900	GACoxN	GACoxP	GACoxCl
pH 3–4	–	–	3.08 (301.8)	–	3.25 (21.95)
pH 4–5	4.91 (53.85)	–	4.98 (111.4)	4.58 (41.56)	4.28 (60.34)
pH 5–6	–	5.01 (27.56)	–	5.85 (30.22)	5.56 (37.62)
pH 6–7	6.53 (46.67)	6.50 (61.67)	6.40 (124.6)	–	6.57 (51.72)
pH 7–8	7.69 (32.31)	7.73 (51.17)	–	7.05 (46.61)	–
pH 8–9	–	–	8.22 (135.6)	8.84 (35.27)	8.20 (69.74)
pH 9–10	9.32 (55.04)	9.81 (143.0)	9.83 (152.2)	9.87 (78.09)	9.94 (105.8)
pH 10–11	–	–		–	11.1 (86.98)
Total	187.9	191.1	826.0	231.8	434.9

called the inductive effect, which is negative or positive if the closest group enhances or weakens the acid strength. The negative inductive effect may also be due to the electronegativity or field effect of the substituent, wherein the dipole of the substituent produces an electrostatic field that affects the carboxylic proton, favoring its ionization. Generally, electro-withdrawing groups increase the acidity and decrease the basicity, while the electro-donor groups act in opposite direction (Figueiredo et al. 1999; Bandosz and Ania 2006). In the case of an activated carbon the same effect can occur due to several groups that can be found on the surface and enhance the acid strength of the groups; among these, the carbonyl and hydroxyl groups. In the case of like-aromatic structures (graphenic layers), resonance also determines the acid strength of the adjacent groups by inductive effect (Jagiello et al. 1995; Bandosz and Ania 2006).

To assign pKa values obtained in potentiometric titrations to functional groups, it is assumed that the carboxylic acid groups generally have pKa values between 2 and 5, while Lactone groups (formed by cyclic esters, product of condensation of an alcohol group with a carboxylic acid group in the same molecule) tend to hydrolyze in acidic or basic media and have an acid dissociation constant that depends on the ring size, taking pKa between about 6 and 8.5, as in the case of butyrolactone and phenolphthalein, respectively. Phenolic groups have pKa values close to 9. As to pyrone type groups located on the edges of the graphenic layers, they may have pKa values higher to 9 and up to 13, depending on the position of the ketone and ether ring effect (Jagiello et al. 1995; Zielke et al. 1996; Bandosz and Ania 2006; Dominguez et al. 2013; Almeida et al. 2016).

The changes in surface chemistry detected for each of the GAC series solids are analyzed below:

As-received Activated Carbon, GAC The as received activated carbon, GAC, has a distribution of acidic and basic groups on the surface, in which total acidity is higher with respect to the total basicity; within the acid and phenolic groups, it has higher concentration. Its surface has overall acidic character also evidenced by a $pH_{PZC} = 5.4$. The potentiometric titration results show that the carboxylic acid groups ($pKa < 7$) on the surface have pKa values of 4.91 and 6.53, and phenols and quinone groups of 7.69 and 9.32, respectively. As based on the Boehm titration data, mainly these groups contribute to the surface acidity of this carbon.

As mentioned above in the introduction, the nature and concentration of the surface groups can be modified by heat or chemical treatments. Modification with oxidizing agents can increase the concentration of the oxygenated surface groups, while a heat treatment under inert atmosphere enables selective decomposition of some of these groups. In this research the surface chemistry of the activated carbons was modified by treatment with various solutions of oxidizing agents (nitric acid, hydrogen peroxide and sodium hypochlorite) and heat treatment. The effects of modification treatments are evaluated below:

Heat Treatment at 900 °C, GAC900 Decomposition of surface groups can be monitored from the $m/z = 18$, 28 and 44 thermal profiles (Fig. 5.2a–c), and from

Boehm and potentiometric titration. Generally, a decrease in the concentration of surface acid groups due to their thermal decomposition has been found. Thus carboxylic groups decompose between 250 and 400 °C, lactones between 400 and 650 °C and phenols between 600 and 800 °C, and at higher temperatures, above 1000 °C, the quinone and pyrone groups (Contescu et al. 1993; Bandosz and Ania 2006; Figueiredo and Pereira 2012). Therefore, the heat treatment at 900 °C effectively decreases the groups concentration on the surface, removing even totally carboxylic groups or species with a pKa value < 5, as in the case of the GAC900 sample.

Furthermore, the heat treatment under an inert atmosphere, such as nitrogen, is used to obtain activated carbons with basic and hydrophobic character, due to the elimination of interaction sites with water molecules, such as the oxygenated functional groups. The basicity of the activated carbons is attributed to chromene and pyrone type groups which decompose at high temperatures and to the electron density on the graphenic layers (π electrons delocalized, located away from the edges of the graphenic layers), according to the following equation:

$$C_\pi + H_3O^+ \rightarrow C_\pi \ldots \quad \ldots H_3O^+ \tag{5.1}$$

The electron density of activated carbons is influenced by surface groups at the edges of the graphenic layers; therefore a change in the basicity parameter is often associated to the removal of surface groups near these electron density planes. In the case of the GAC900, it is observed an increase of the pH_{PZC} moving to 8.9, it may be correlated with desorption of the thermo labile acid groups.

Furthermore it can observed that the total basicity increases as well as the concentration of surface groups with pKa values between 9 and 11, which are not necessarily related to the presence of phenol or carbonyls groups. This is also attributed to other protonated basic structures by the initial condition of the titration ($pH < 3$). These kinds of solids are suitable for acid adsorbate as well as organic molecules with a high charge density (π electrons delocalized) (Bandosz and Ania 2006).

Nitric Acid, GACoxN The treatment with nitric acid is used to oxidize the carbonaceous surface, to increase the total acidity by promoting the formation of functional groups, to remove part of the inorganic matter from the solid and to improve the surface hydrophilicity. Oxidation reactions occurring on the surface of activated carbons resemble the reactions of oxidation of polyaromatic molecules. Some of the mechanisms by which oxygenate surface groups are formed with the treatment with nitric acid have been previously reported (Eqs. 5.2, 5.3, and 5.4) (Figueiredo et al. 1999).

The oxidation reaction with nitric acid can occur through a radical mechanism as shown in Eq. 5.2(a–e) or by the rupture of a C–C bond to form the ketone group as shown in Eq. 5.2(f) and 5.4.

$$\begin{array}{lr}
HNO_3 + HNO_2 \rightleftarrows 2\dot{N}O_2 + H_2O & (a)\\
\dot{N}O_2 + H^+ \rightleftarrows H\dot{N}O_2^+ & (b)\\
Ar - CH_2 - CH_2 - Ar + H\dot{N}O_2^+ \rightleftarrows Ar - \dot{C}H - CH_2 - Ar + H\dot{N}O_2^+ & (c)\\
Ar - \dot{C}H - CH_2 - Ar + \dot{N}O_2 \rightleftarrows Ar - CHONO - CH_2 - Ar & (d)\\
Ar - CHONO - CH_2 - Ar + H_2O_2 \rightleftarrows Ar - CHOH - CH_2 - Ar + HNO_2 & (e)\\
Ar - CO - CH_2 - Ar + \dot{N}O_2 \rightleftarrows Ar - \dot{C}O - ONO - CH_2 - Ar & (f)
\end{array} \tag{5.2}$$

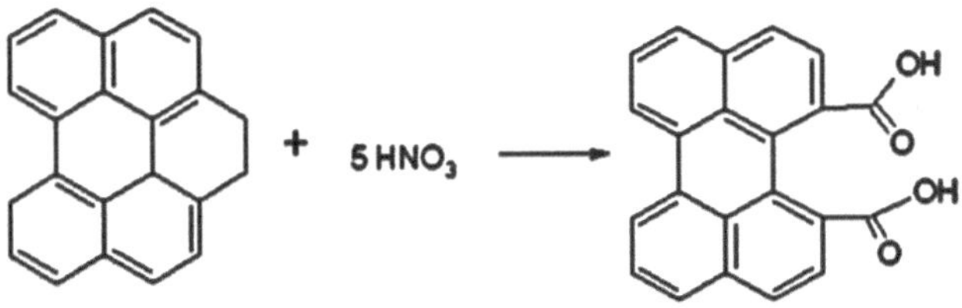

$$+ \ 5HNO_3 \longrightarrow \quad OH \quad O \quad OH \quad O \tag{5.3}$$

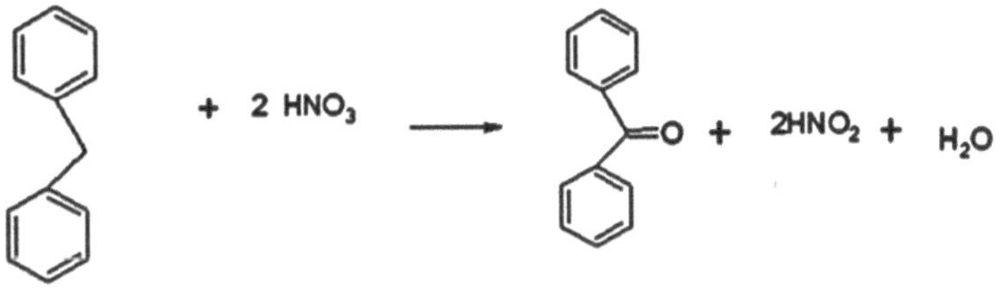

$$+ \ 2\ HNO_3 \longrightarrow \quad =O + 2HNO_2 + H_2O \tag{5.4}$$

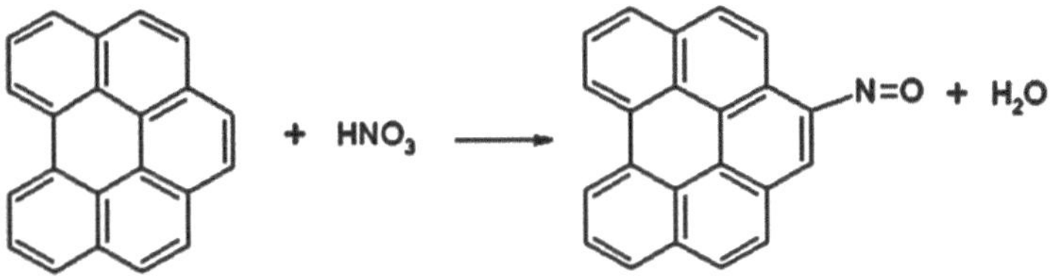

$$+ \ HNO_3 \longrightarrow \quad N{=}O + H_2O \tag{5.5}$$

Equation 5.2 represents the formation of dicarboxylic groups that can form a result of the rupture of an aliphatic bond of a site in a nonaromatic edge of the basal planes of the carbon. The reaction begins with the rupture of the single bond –C–C-at the alpha position of a Benzyl carbon atom. Equation 5.2 represents oxidation when there is a metilene group ($–CH_2–$) between two aromatic systems which may result in the formation of ketone type groups. Additionally, the nitrogen may be added to the carbon surface by a similar reaction to Benzene nitration. The mechanism involves the formation of the highly reactive nitronium ion (NO_2^-), which can form a nitrated product (Eq. 5.4) (Zielke et al. 1993; Almeida et al. 2016).

Among the generated groups, the amount of carboxylic acids increases of about four times compared to their concentration in the initial sample (217.3 μmol/g) (Table 5.1). By comparing Fig. 5.3 (a, c) the change and the increase in the intensity of peaks in the distribution of pKa for GACoxN sample. This trend represents the increase in the concentration of surface functional groups and the heterogeneity of the surface by inductive effect between groups which alters the range of acid dissociation constants that depend on other nearby groups and the size of the graphene layer. As it is expected after the oxidation process with nitric acid species with pKa values at 3.08 (301.8 μmol/g) and 4 (111.4 μmol/g) were obtained, two different acid carboxylic just like it was suggested by TA-MS studies. Additionally, the lactone and/or carboxylic anhydrides groups are represented by the peaks at pKa values between 6 (124.6 μmol/g) and 8 (135.6 μmol/.g) and the phenolic groups by

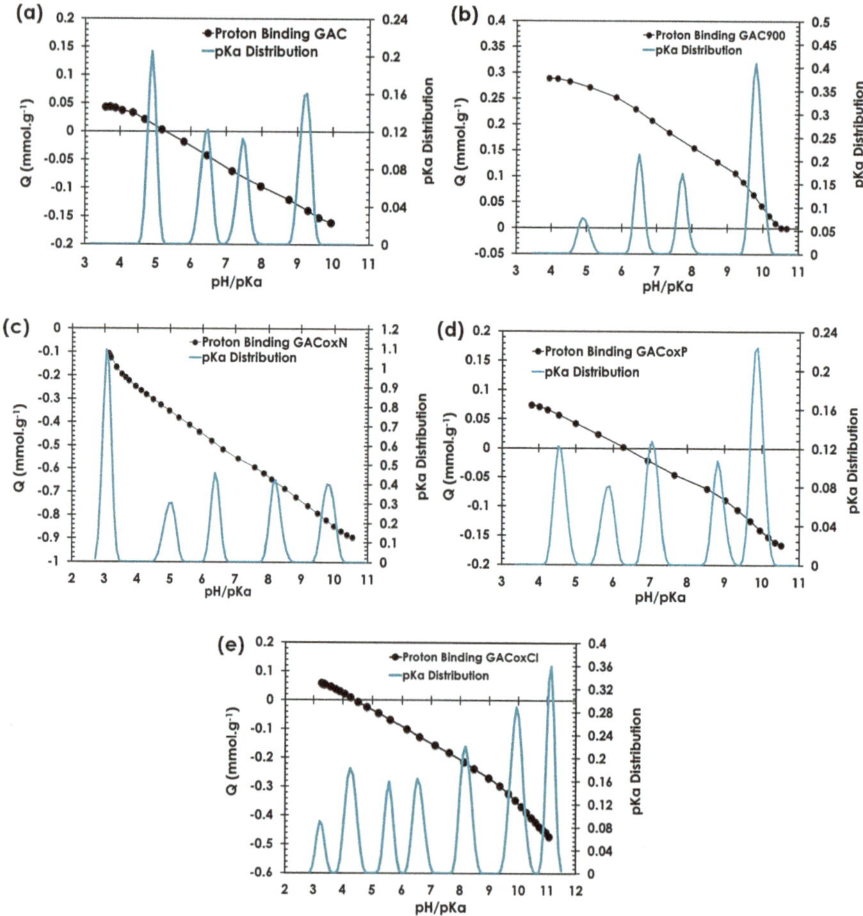

Fig. 5.3 Protons binding curves (blue line) and pKa distributions of the surface groups for the activated carbons, (**a**) GAC, (**b**) GAC900, (**c**) GACoxN, (**d**) GACoxP and (**e**) GACoxCl

pKa above 9(152.2 μmol/g) (Table 5.2). The change in the concentration of the acid surface groups (approximately 4.4 times with respect to GAC) is also evidenced by the decrease in pH_{PZC} to 3.4.

Together with the increase in the number of acidic groups on the carbon surface there is a decrease in the basic character of the surface. That decrease is a result of the neutralization of basic groups by the nitric acid used for oxidation.

Hydrogen Peroxide, GACoxP Like the nitric acid, the hydrogen peroxide (H_2O_2) is an oxidant which has been used for modification of carbonaceous solids. Hydrogen peroxide is a product of the addition of H_2 and O_2, or an adduct of an oxygen atom and one H_2O molecule, which has the ability of oxidizing organic compounds with an efficiency of 47% and with the generation of water as the only theoretical product.

Hydrogen peroxide is thermodynamically unstable and decomposes into oxygen and water, according to the Eq. 5.6:

$$H_2O_2 \quad + \quad H_2O_2 \quad \rightarrow \quad 2H_2O \quad + \quad O_2 \tag{5.6}$$

Activated carbon can promote the decomposition of hydrogen peroxide because of its graphite structure and the functional groups on its surface. Activated carbon degrades hydrogen peroxide to water and oxygen. The decomposition is favored by the presence of phenolic groups on the surface, which interact with the anion of the dissociation of the peroxide. The species formed on the surface interacts with other molecules of hydrogen peroxide which decompose and enable the release of O_2 and the regenerating of the carbon surface as described in the following equations (Long et al. 2008; Bach and Semiat 2011; Domínguez et al. 2013; Chen et al. 2014; Pinho et al. 2015):

$$GAC - OH \ + H^{+}OOH^{-} \quad \rightarrow \quad GAC - OOH \ + H_2O \tag{5.7}$$

$$GAC - OOH \ + \ H_2O_2 \quad \rightarrow \quad GAC - OH \ + H_2O \ + \ O_2 \tag{5.8}$$

The path taken by the decomposition of H_2O_2 involves a variety of reactive species that can interact with the surface; hence in the presence of electro-attractant acid groups in the activated carbon surface they will attract the delocalized electrons in the basal planes of the solid surface, which restricts the availability of these electrons to interact with the molecules of hydrogen peroxide.

On the other hand, the radical species from hydrogen peroxide are gradually formed by side reactions including electron transfer reactions and the production of organic species caused by the rupture of bonds on the carbonaceous surface. The presence of basic groups such as pyrones, chromenes, ethers and carbonyls increase the decomposition of hydrogen peroxide to radical species by accepting an electron from the activated carbon. Peroxide radical dissociation can be described as follows (Long et al. 2008; Bach and Semiat 2011; Chen et al. 2014; Pinho et al. 2015):

$$GAC^{+} + H_2O_2 \rightarrow GAC + OOH\cdot + H^{+} \quad (5.9)$$

$$OOH\cdot \leftrightarrow H^{+} + O_2^{-} \quad (5.10)$$

$$H_2O_2 + O_2^{-} \rightarrow OH^{-} + OH\cdot + O_2 \quad (5.11)$$

An activated carbon will act as an adsorbent of the different species formed during peroxide decomposition and therefore the formation of oxygenated complexes on the solid surface can take place. In the simplest case H_2O_2 oxidizes directly the surface compounds in a reaction without the formation of O_2, also by the addition of the species HO_2^{-} or oxygen atom to the carbon surface. The following reactions represent some possible reactions that may occur in the oxidation of active carbon with hydrogen peroxide (Gómez-Serrano et al. 1994; Ailing et al. 2014; Adedeji et al. 2014):

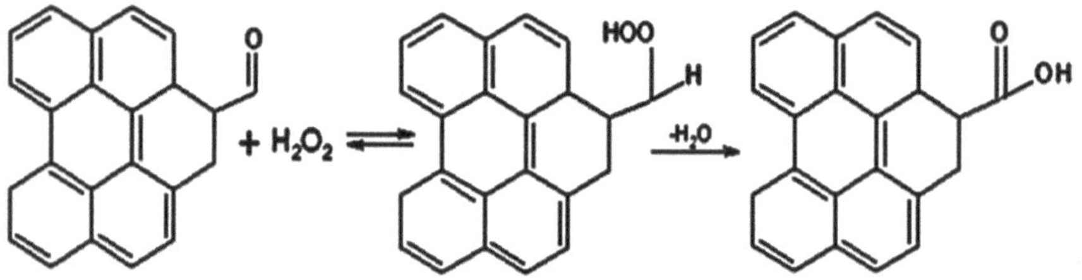

(5.12)

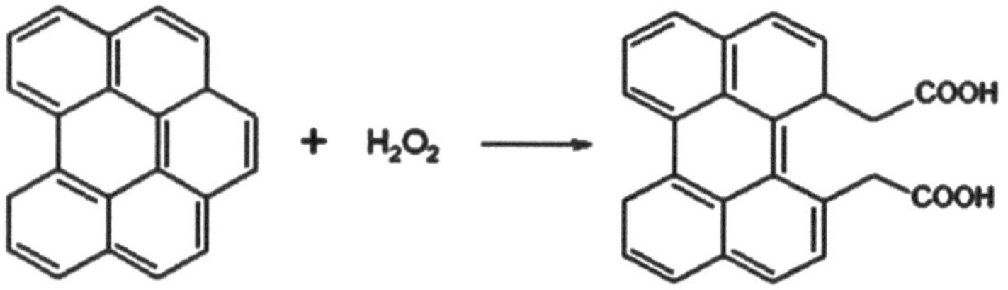

(5.13)

Owing to the effect of hydrogen peroxide on the carbonaceous surface, it is expected that the acidity parameter increases. In fact in the case of the GACoxP sample the increase in this parameter is close to 44% (Table 5.2). For this sample the highest contributions of carboxylic acid groups and phenolic groups are found. The potentiometric titration results also indicated the formation of species with pKa values centered on 5.85 (30.22 μmol/g) and between 7.05 (46.61 μmol/g) and 9.87 (78.09 μmol/g) for phenol groups. The increase in the number OH groups has been found when the pH of the solution depends only on the initial concentration of the mixture of activated carbon and hydrogen peroxide solution. The total basicity of the surface increases slightly compared to GAC, this leads to an increase in the pH_{PZC} to 6.2 for GACoxP.

Sodium Hypochlorite, GACoxCl Like the previous two solutions, sodium hypochlorite acts as an oxidizer of the carbonaceous surface. The oxidation reaction of a carbonaceous surface has already been described in literature (Vasu 2008); the pH of the solution that is in contact with the carbonaceous surface determines the species that act as oxidizing agent. In an acidic solution, the OCl decomposes more easily than in a basic solution; this is why NaOCl solutions are maintained at basic pH. In the NaOCl solutions there is a dynamic balance, which is represented by the following equation:

$$NaOCl + H_2O \rightarrow 2NaOH + HClO \Leftrightarrow Na^+ + OH^- + H^+ + OCl^- \quad (5.14)$$

Through oxidation with sodium hypochlorite, different species are obtained on the carbonaceous surface. The reaction depicted in Eq. 5.13 in which dicarboxylic acids are formed is also applicable in this case, as shown in Eqs. 5.15(b) and 5.16. Breaking bonds C=C and C–C eventually occurs, starting with the addition of hypochlorous acid to a double bond carbon as presented in Eq. 5.15. Another mechanism is oxidation of a methyl or metilene group to carboxylic acid (Eq. 5.16). As it is shown in the equations, chlorination is an intermediate step in the process of oxidation and rupture of C–C bond when occurs by attack of the ion OCl^-. It should also be noted that these reactions are slower than those occurring with nitric acid, because hypochlorite is less reactive than nitric acid (Vinke et al. 1994; Vasu 2008).

$$R-CH=CH-R + HClO \longrightarrow R-\underset{OH}{CH}-\underset{Cl}{CH}-R \quad (a)$$

$$R-\underset{OH}{CH}-\underset{Cl}{CH}-R + 3\,ClO^- \xrightarrow[-3\,H_2O]{+3\,OH^-} 2\,R-COO^- + 4\,Cl^- \quad (b)$$

$$R-CH_3 + 3\,ClO^- \xrightarrow[-2\,H_2O]{OH^-} R-COO^- + 3\,Cl^- \quad (c)$$

(5.15)

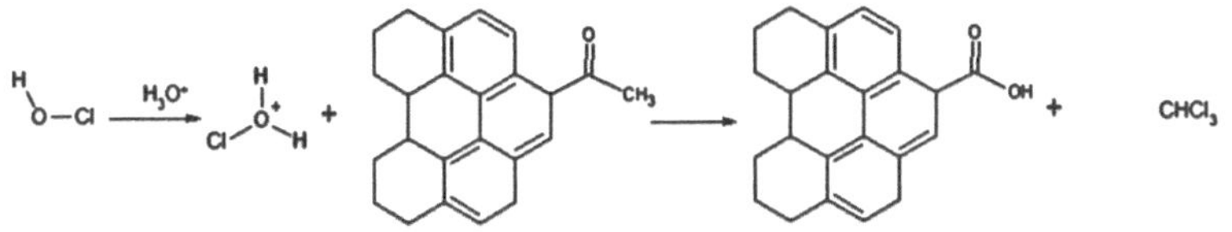

$$H-O-Cl \xrightarrow{H_3O^+} Cl-\overset{+}{O}H_2 + \text{Ar}-C(=O)-CH_3 \longrightarrow \text{Ar}-C(=O)-OH + CHCl_3 \quad (5.16)$$

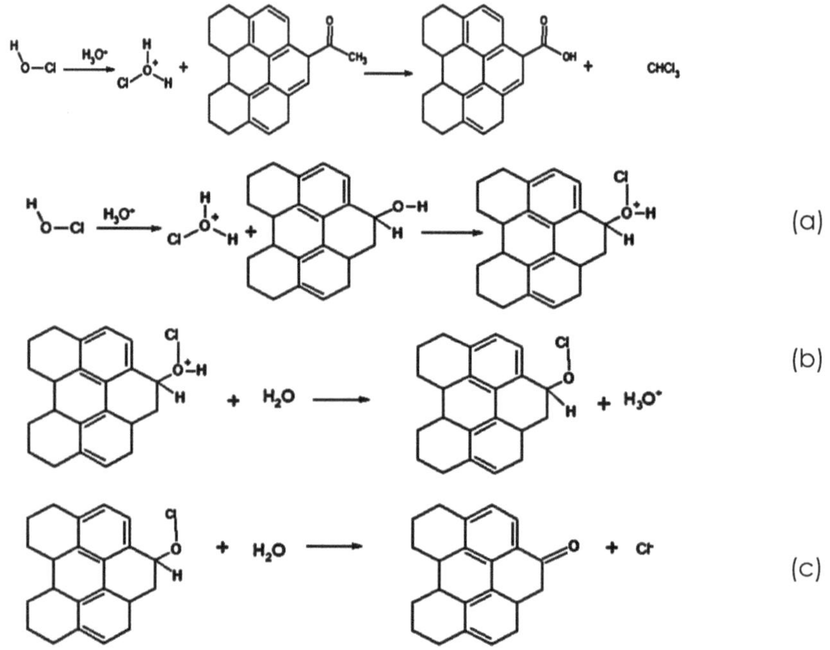

(a)

(b)

(c)

(5.17)

The formation of phenolic, carbonyl and quinones groups has also been reported for this type of treatments, Eq. 5.17 represents this type of mechanism.

According to above and Boehm titration data, treatment with NaOCl increased total acidity by promoting the formation of functional groups, especially lactonic (54.23 μmol/g) and phenolic groups (339.0 μmol/g), the last one show an increase of about six times compared to the initial sample (Table 5.1). Potentiometric titration evidences a pKa distribution different from GAC, in which three species of carboxylic acids with pKa values 3.25 (21.95 μmol/g), 4.28(60.34 μmol/g) and 5.56 (37.62 μmol/g) were observed. As for potentiometric titration, Lactone groups with values of pKa close to 6.5 double their concentration with respect to the solid GAC, whereas the species with pKa values of 8.2 are also formed after modification with sodium hypochlorite, as well as formation of species with pKa >10 values, these species can be attributed to weak acids such as phenols and carbonyl groups, as well as to the presence of basic groups such as pyrone; therefore the basicity parameter also increased compared to GAC. This change in concentration of the surface groups is also evidenced by an increase of pH_{PZC} to 7.2.

5.4 Effects of the Surface Oxidation over Textural Parameter: Apparent Area, Volume Pore and Characteristic Energy

The textural characteristics of the activated carbons such as apparent area, micropore volume and characteristic energy, parameters evaluated from the experimental data of the nitrogen isotherms and the Brunauer-Emmett-Teller (BET), Dubinin-Astakhov (DA) are presented in Table 5.3. Apparent areas (S_{BET}) are between 687–871 $m^2.g^{-1}$ and micropore volumes are between 0.26–0.36 $cm^3\ g^{-1}$, respectively. The differences in the textural characteristics evidence the effect of the oxidizing agents used on the textural properties of the activated carbons in function of the oxidizing force of the modifying agent used (HNO_3, H_2O_2 and NaOCl).

In general the attack of the oxidizing agents is carried out mainly on carbon atoms located in the opening of the pores or on the surface of the solid, because these atoms do not have the cohesive forces compensated, as in an atom located inner of the solid, which makes them labile for the interaction with the oxidizing agents. Decrease in textural parameters by oxidation may be due to formation of humic substances which are byproducts during the oxidation process, the formation of surface oxygen groups are located at the edges of the pore openings which limits the accessibility of the nitrogen molecule to porous structures, some pores can be destroyed or deepening by removal of the C atoms and the mechanical pore destruction by the surface tension of the oxidizing solution (Vivo-Vilches et al. 2014).

Figure 5.4 show the correlation between the densities of chemical groups evaluated by Boehm method and apparent area and characteristic energy. It is possible to observe the decrease of the apparent surface area as the number of superficial groups increases (Bansal and Goyal 2005; Marsh and Rodriguez-Reinoso 2005; Sing 2014; Thommes and Cychosz 2014).

On the other hand, the apparent surface area of the solid GACoxP increased with respect to the solid GAC, according to the pore volume after the oxidation process it could be observed that there is an increase in the quantity of pores with dimensions close to 30 Å (Fig. 5.5), so in the oxidation process not only the formation of oxygenated surface groups is expected but also the opening of porous structures (see Fig. 5.5) (Thommes and Cychosz 2014), which may be related to the rupture of the carbon grains due to the reaction between hydrogen peroxide and activated carbon.

About energy characteristic (Fig. 5.4), It is observed that the surface is more energetic for the oxidized solids, according to its more heterogeneous surface

Table 5.3 Textural properties of activated carbon

	Sample				
Textural	GAC	GAC 900	GACoxN	GACoxP	GACoxCl
S_{BET}	849	889	815	871	687
V_{mic}	0.35	0.36	0.35	0.36	0.26
E_0	7.64	7.38	8.45	7.66	9.30

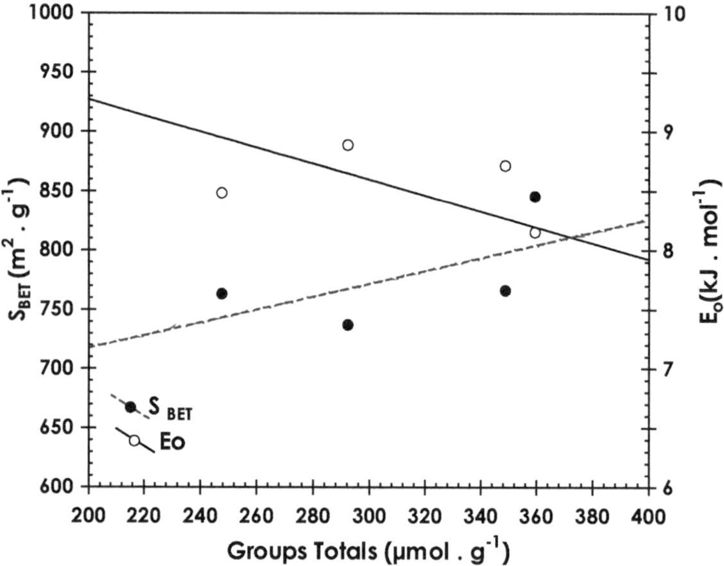

Fig. 5.4 Correlation of groups total determinate by Boehm titration and apparent surface area and characteristic energy

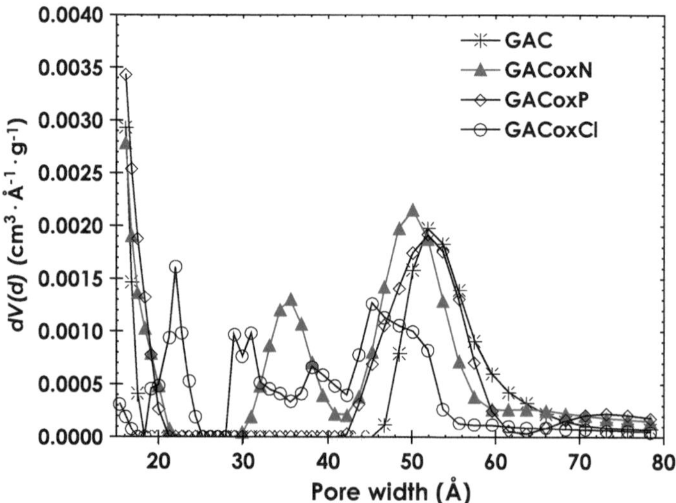

Fig. 5.5 Comparison of Pore size distribution obtained from QSDFT analysis of N_2 adsorption and of GAC and oxidized samples

chemistry, the solid GACoxP has the highest pore volume and its pores are less energetic than the other two oxidized solids. According to the effect of hydrogen peroxide on the carbonaceous surface it is expected that the acidity parameter increases, it can be corroborated with Boehm titration, the increase in this parameter is close to 44% (Table 5.1), where the largest contribution were of carboxylic acid and phenolic groups.

The decrease in textural parameters by oxidation with sodium hypochlorite is greater than after nitric acid and hydrogen peroxide treatments. This behavior is consistent with strongest oxidation of the surface. The reaction between sodium hypochlorite solution and activated carbon was very exothermic and causes the destruction of the carbon granules which affects the porosity of the solid.

For all the activated carbons, the pore dimensions range between 7 and 80 Å, and the highest contribution is by far in the micropore region (i.e., $d < 25$ Å). A secondary smaller contribution of larger pores between 30 and 70 Å can be observed in all the samples, which is retained upon all the surface modification of GAC but appears to be shifted also towards mesopores for GACoxN and towards micropores for GACoxCl.

The attack of the oxidizing agents is carried out mainly on carbon atoms located in the opening of the pores or on the outer surface of the solid, because for these atoms, the cohesive forces are not compensated, as in the atoms located in the inner surface of the solid. The agents used for the modification of surface chemistry mainly act by including oxygen atoms on the surface of the activated carbon. The inclusion of oxygenated surface groups can produce a decrease in the number of pores, which the nitrogen can access, hence determining a worsening of the textural parameters. Moreover, a significant effect on the surface chemistry is also determined because of oxygenated group insertion.

5.4.1 Mechanisms of Heavy Metal Adsorption from Aqueous Solution

As it was previously addressed, functional groups such as carboxylic acid, phenolic and lactone among others are generated during the oxidation process. The polar character of these groups decreases the hydrophobic character of the carbonaceous surface and promotes the ions adsorption in aqueous solution (Zhang et al. 2017).

The importance of oxygenated functional groups on the surface of adsorbent solids has been highlighted by several authors (González and Pliego-Cuervo 2014; Arcibar-Orozco et al. 2015; Hernández-Eudave et al. 2016; Peláez et al. 2015; Ribeiro et al. 2015; Zhang et al. 2017; Aguayo-Villarreal et al. 2017). For example, metal ions adsorption from aqueous solution is mainly established by specific interactions between functional groups and ions through different mechanisms.

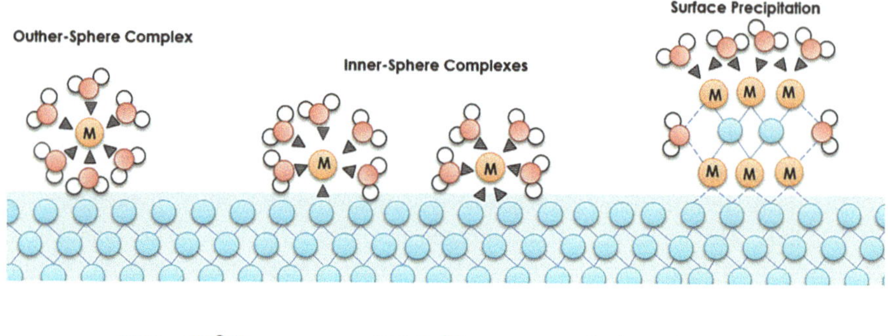

$\equiv \mathbf{SOH} + (\mathbf{M}^{2+})_{aq} \rightarrow \equiv \mathbf{SOH}(\mathbf{M}^{2+})$ Outer-sphere complex

$\equiv \mathbf{SOH} + (\mathbf{M}^{2+})_{aq} \rightarrow \equiv \mathbf{SOM}^{+} + \mathbf{H}^{+}$ Inner-sphere complex

$\equiv \mathbf{SOH} + (\mathbf{M}^{2+})_{aq} \rightarrow (\equiv \mathbf{SO})_2\mathbf{M}^{0} + \mathbf{2H}^{+}$ Bidentate Inner-sphere complex

Fig. 5.6 Inner and outer sphere complex and surface precipitation representation

The functional groups can have the same donor-acceptor character found in organic groups, which determine the mechanism of adsorption, like the formation of metal complexes, such as COOH-M and electron acceptor-donor reactions, so the increase of ion removal is attributed to the increase of surface sites available for adsorption (Aguayo-Villarreal et al. 2017).

Many mechanisms can be described from Bronsted-Lowry, Pearson and Lewis acid-basic theories. The interaction between ion and surface can be described according to surface complex models. These complexes can be of inner- or outer-sphere depending on the formation of chemical bond or the approach of the ion to the surface groups within a critical distance in which the ions are separated by one or more water molecules. Furthermore, a change of the pH zone given by the concentration and the kind of groups on the surface can lead to the surface precipitation (Stumm, 1995; Moreno-Tovara et al. 2014; Ribeiro et al. 2015; Serrano-Gómez et al. 2015; Peláez et al. 2015; Aguayo-Villarreal et al. 2017) (Fig. 5.6).

Surface complexation models should define specific surface sites, surface chemical reactions and their equilibrium constants, also mass and charge balances. Ion adsorption mechanisms and surface characteristics can be established from experimental observations such as chemical titrations, point of zero charge shifts, pH effects and calorimetry measurements. Also, the adsorbate properties in the case of ionic species, the ionic radius, the coordination sphere, solubility and electronegativity should be considered. For example, a bigger ionic radius between different ions can define the higher possibility to interact with the surface. This implies that the bigger ion is able to bind with a higher number of oxidized groups, and its coordination sphere is bigger too. Then the ion has more probabilities to form a

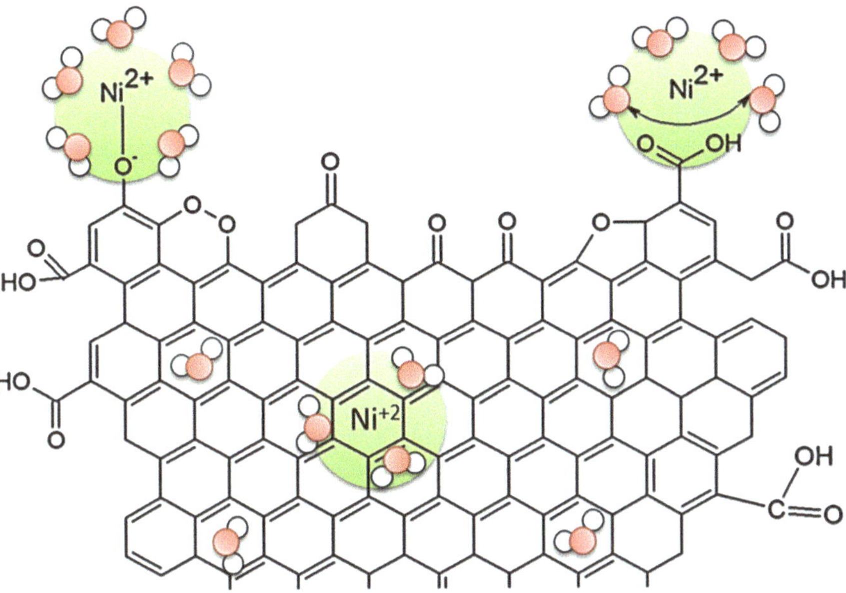

Fig. 5.7 Representation of ion adsorption mechanisms on carbonaceous surface

stable bond with distant oxidized groups through electrostatic interactions and Van der Walls forces (Ribeiro et al. 2015; Arcibar-Orozco et al. 2015; Hernandez-Eudave et al. 2016; Aguayo-Villarreal et al. 2017).

Figure 5.7 shows the possible mechanisms present in the ions adsorption process on carbonaceous surface. For this kind of surface, both ion exchange and electrostatic interactions occur. The intervention of the oxygenated surface groups in the ion exchange adsorption mechanism can be confirmed by an increase in the removal capacity of the modified solids by oxidation treatments. Also because of the dependence of the pH of the process, or proton ions concentration (González and Pliego-Cuervo 2014; Hernandez-Eudave et al. 2016). Probably there is a permanently basal/interlayer surfaces charge and it is responsible for cation exchange mechanism, also the surface complexation can be dependent on the variably charged edge surfaces at high pH. This behavior suggests that the adsorption of the metal ions is attributed to columbic forces between the positively charged ions and the negative surface charge of the adsorbent, based on attractive electrostatic interactions between donor oxygenated groups (Lewis Bases) on the surface of carbonaceous solids and the electron acceptor nature of heavy metal ions (Lewis acids). In addition, it is possible the formation of surface complexes by the interaction of the ions with the delocalized π electrons of the graphene layers for the activated carbons (González and Pliego-Cuervo 2014; Ribeiro et al. 2015; Arcibar-Orozco et al. 2015; Hernandez-Eudave et al. 2016; Aguayo-Villarreal et al. 2017).

5.5 Conclusions

A series of carbonaceous solids prepared from activated carbons and different oxidant agents for the modifications have been chemically characterized, which have altered the chemical characteristics of the starting solid, for example by analyzing the curves of TG-MS of the starting solid and its modifications it can be seen that the oxidized solids show a more pronounced loss of mass at temperatures below 200 °C, this loss is attributed to the hygroscopic properties of the solids, which increases with the inclusion of hydrophilic groups. That is, the presence of oxygenated surface groups may be related to the moisture content of the solid, increasing the hydrophilic character of the activated carbon surface.

The loss of mass above 200 °C corresponds to the degradation of the surface groups at different temperatures, evidencing the different stability for each group. At low temperatures desorption of CO_2 and CO is associated with the presence of carboxylic acid groups, lactones and carboxylic anhydrides, while at high temperatures indicate the presence of phenol and carbonyl groups.

The analysis of the data of both Boehm titration and potentiometric titration allowed observe the change in the distribution of the acidic and basic surface groups due to the modifications made to the solids. Oxidation with the different solutions employed results in an increase in the number of acid groups for the three solids (GACoxN, GACoxP and GACoxCl). On the other hand, it is evident the reduction in the basic sites, that can be justified because the increase of the content of oxygenated groups decreases the electronic density of the basal planes and therefore reduces the basicity of the surface of the coal.

The oxidation reactions occurring on the surface of the activated carbon resemble the oxidation reactions of polyaromatic molecules. The zero charge point of the solids is modified with the change in the concentration of the surface groups promoted by each treatment, i.e. a greater amount of acid groups as in the case of solids GACoxN will give way to a pH_{PZC} acid.

Solids with surface chemistry heterogeneous can be use in the removal of heavy metal ions from aqueous solution; it can be observed the importance of functional groups mainly by establishing specific interactions between functional groups and ions through different mechanisms including the formation metal complexes, such as COOH-M and acceptor reactions of electron donors. Such mechanisms are favored in the process of oxidation and functionalization.

Acknowledgments The authors wish to thank the framework agreement established between the Universidad Nacional de Colombia and Universidad de Los Andes (Bogotá, Colombia). Also they want express their acknowledgments to Dr. Teresa Bandosz scientific support to carry out this research. The authors also wish to thank the grant by the Faculty of Sciences of Universidad de los Andes (Colombia) in all of No. 28-11-2017 Call 2018-2019 for the funding of research programs for Professors of Associated, Full Professors and Emeritus Professors (classified in the Professoral order).

References

Adedeji A, Yun Hui L, Young Min J (2014) Surface oxidation of activated carbon pellets by hydrogen peroxide for preparation of CO_2 adsorbent. J Ind Eng Chem 20:2130–2137. https://doi.org/10.1016/j.jiec.2013.09.042

Aguayo-Villarreal IA, Bonilla-Petriciolet A, Muñiz-Valencia R (2017) Synthesis of activated carbons from pecan nutshell and their application in the antagonistic adsorption of heavy metal ions. J Mol Liq 230:686–695. https://doi.org/10.1016/j.molliq.2017.01.039

Ailing R, Peng H, Bin G, Jing H, Bei L (2014) Surface oxidation of activated carbon pellets by hydrogen peroxide for preparation of CO_2 adsorbent. J Ind Eng Chem 20:2130–2137. https://doi.org/10.1016/j.jiec.2013.09.042

Almeida LA, Castro A, Mendoça F, Mesquita J (2016) Characterization of acid functional groups of carbon dots by nonlinear regression data fitting of potentiometric titration curves. Appl Surf Sci 370:486–495. https://doi.org/10.1016/j.apsusc.2016.02.128

Arcibar-Orozco J, Rangel-Mendez J, Diaz-Flores PE (2015) Simultaneous adsorption of Pb(II)-Cd (II), Pb(II)-Phenol, and Cd(II)-Phenol by activated carbon cloth in aqueous solution. Water Air Soil Pollut 226:2197–2203. https://doi.org/10.1007/s11270-014-2197-1

Ashish S, Aniruddha M, Prathmesh S, Dattatraya P, Prakash R, Mansing A, Sanjay K (2012) Removal of Bi (III) with adsorption technique using coconut shell activated carbon. Chin J Chem Eng 20:768–775. https://doi.org/10.1016/S1004-9541(11)60247-4

Bach A, Semiat R (2011) The role of activated carbon as a catalyst in GAC/iron oxide/H_2O_2 oxidation process. Desalination 273:57–63. https://doi.org/10.1016/j.desal.2010.04.020

Bandosz T, Ania C (2006) Surface chemistry of activated carbons and its characterization. In: Activated carbon surfaces in environmental remediation activated carbon surfaces in environmental remediation. Elsevier, New York, pp 160–229 ISBN:9780080455952

Bandosz T, Jagiello J, Contescu C, Schwarz J (1993) Characterization of the surfaces of activated carbons in terms of their acidity constant distributions. Carbon 31:1193–1202. https://doi.org/10.1016/0008-6223(93)90072-I

Bansal RC, Goyal M (2005) Activated carbon and its surface structure. In: Activated carbon adsorption. Taylor & Francis Group, New York, pp 1–60 ISBN:9780824753443

Boehm HP (1966) Chemical identification of surface groups. Adv Catal 16:179–274. https://doi.org/10.1016/S0360-0564(08)60354-5

Burg P, Cagniant D (2008) Characterization of carbon surface chemistry. In: Chemistry and physiscs of carbon. Taylor & Francis Group, New York, pp 29–172 ISBN:9781420042986

Chen C, Li X, Tong Z, Li Y, Li M (2014) Modification process optimization, characterization and adsorption property of granular fir-based activated carbon. Appl Surf Sci 315:203–211. https://doi.org/10.1016/j.apsusc.2014.07.111

Contescu C, Jagiello J, Schwarz J (1993) Heterogeneity of proton binding sites at the oxide/solution Interface. Langmuir 9:1754–1765. https://doi.org/10.1021/la00031a024

Contescu C, Popa V, Miller J, Ko E, Schwarz J (1995) Proton affinity distribution of TiO_2-SiO_2 and ZrO_2-SiO_2 mixed oxides and their relationship to catalyst activities for1-butene isomerization. J Catal 157:244–258. https://doi.org/10.1006/jcat.1995.1285

Contescu C, Popa V, Schwarz J (1996) Heterogeneity of hydroxyl and deuteroxyl groups on the surface of TiO2 polymorphs. J Coll Interf Sci 180:149–161. https://doi.org/10.1006/jcis.1996.0285

Contescu A, Contescu C, Putyera K, Schwarz J (1997) Surface acidity of carbons characterized by their continuous pK distribution and Boehm titration. Carbon 35:83–94. https://doi.org/10.1016/S0008-6223(96)00125-X

Domínguez C, Ocón P, Quintanilla A, Casas J, Rodriguez J (2013) Highly efficient application of activated carbon as catalyst for wet peroxide oxidation. Appl Catal B Environ 140:663–670. https://doi.org/10.1016/j.apcatb.2013.04.068

Figueiredo J, Pereira MFR (2012) The role of surface chemistry in catalysis with carbons. Catal Today 150:2–10. https://doi.org/10.1016/j.cattod.2009.04.010

Figueiredo J, Pereira MFR, Freitas MMA, Orfao JJM (1999) Modification of the surface chemistry of activated carbons. Carbon 37:1379–1389. https://doi.org/10.1016/S0008-6223(98)00333-9

Goertzen S, Thériault K, Oickle A, Tarasuk A, Andreas H (2012) Standardization of the Boehm titration. Part I. CO_2 expulsion and end point determination. Carbon 48:1252–1261. https://doi.org/10.1016/j.carbon.2009.11.050

Gómez-Serrano V, Acedo-Ramos M, López-Peinado A, Valenzuela-Calahorro C (1994) Oxidation of activated carbon by hydrogen peroxide; Study of surface functional groups by FTIR Fuel 73:387–395. https://doi.org/10.1016/0016-2361(94)90092-2

González PG, Pliego-Cuervo YB (2014) Adsorption of Cd(II), Hg(II) and Zn(II) from aqueoussolution using mesoporous activated carbonproduced from *Bambusa vulgaris* striata. Chem Eng Res Des 92:2715–2724. https://doi.org/10.1016/j.cherd.2014.02.013

Gorgulho H, Mesquita J, Goncalves F, Pereira M, Figueiredo J (2008) Characterization of the surface chemistry of carbon materials by potentiometric titrations and temperature-programmed desorption. Carbon 46:1544–1555. https://doi.org/10.1016/j.carbon.2008.06.045

Haydar S, Moreno-Castilla C, Ferro-Garcia MA, Carrasco-Marin F, Rivera-Utrilla J, Perrard A, Joly JP (2000) Regularities in the temperature-programmed desorption spectra of CO and CO_2 from activated carbons. Carbon 38:1297–1308. https://doi.org/10.1016/S0008-6223(99)00256-0

Hernandez-Eudave MT, Bonilla-Petriciolet A, Moreno-Virgen MR, Rojas-Mayorga CK, Tovar-Gómez R (2016) Design analysis of fixed-bed synergic adsorption of heavy metals and acid blue 25 on activated carbon. Desalin Water Treat 57:9824–9836. https://doi.org/10.1080/19443994.2015.1031710

Jagiello J (1994) Stable numerical solution of the adsorption integral equation using splines. Langmuir 10:2778–2785. https://doi.org/10.1021/la00020a045

Jagiello J, Bandosz T, Putyera K, Schwarz J (1995) Determination of proton affinity distribution for chemical systems in aqueous environments using a stable numerical solution of the adsorption integral equation. J Colloid Interf Sci 172:341–346. https://doi.org/10.1006/jcis.1995.1262

Kim Y, Yang S, Lim H, Kim T, Park C (2012) A simple method for determining the neutralization point in Boehm titration regardless of the CO_2 effect. Carbon 50:3315–3323. https://doi.org/10.1016/j.carbon.2011.12.030

Kohl S, Drochner A, Vogel H (2010) Quantification of oxygen surface groups on carbon materials via diffuse reflectance FT-IR spectroscopy and temperature programmed desorption. Catal Today 150(1–2):67–70. https://doi.org/10.1016/j.cattod.2009.05.016

Long X, Cheng H, Xin Z, Xiao W, Li W, Yuan W (2008) Adsorption of ammonia on activated carbon from aqueous solutions. Environ Prog 27:225–233. https://doi.org/10.1002/ep.10252

Marsh H, Rodriguez-Reinoso F (2005) Activated carbon (origins). In: activated carbon. Elsevier Science Ltd, Oxford, pp 13–81 ISBN:9780080455969

Moreno-Tovara R, Terrés E, Rangel-Mendez JR (2014) Oxidation and EDX elemental mapping characterization of an orderedmesoporous carbon: Pb(II) and Cd(II) removal. Appl Surf Sci 303:373–380. https://doi.org/10.1016/j.apsusc.2014.03.008

Noh JS, Schwarz JA (1990) Effect of HNO_3 treatment on the surface acidity of activated carbons. Carbon 28:675–682. https://doi.org/10.1016/0008-6223(90)90069-B

Onida B, Fiorilli S, Camarota B, Perrachon D, Bruzzoniti MC (2008) Acidic functional groups incorporated in ordered mesoporous materials: a comparison among different host matrices. In: Gédéon A, Massiani P, Babonneau F (eds) Zeolites and related materials: trends, targets and challenges proceedings of 4th international FEZA conference. Elsevier, Amsterdam/London, p 67 ISBN:978-4-444-53297-8

Peláez AA, Herrera A, Bautista A, Salazar M (2015) Preparation and characterization of activated carbon from Agave tequilana Weber for the removal of textile dyes and heavy metals. Desalin Water Treat 57:21105–21117. https://doi.org/10.1080/19443994.2015.1111815

Pinho M, Silva A, Fathy N, Attia A, Gomes J (2015) Activated carbon xerogel–chitosan composite materials for catalytic wet peroxide oxidation under intensified process conditions. J Environ Chem Eng 3:1243–1251. https://doi.org/10.1016/j.jece.2014.10.020

Radovic L, Moreno-Castilla C, Rivera-Utrilla J (2004) Carbon materials as adsorbents in aqueous solutions. In: Chemistry and physics of carbon. A series of advances. Ed Marcel Dekker, New York, pp 293–297 ISBN:9780824740887

Ribeiro RFL, Soares VC, Costa LM, Nascentes CC (2015) Production of activated carbon from biodiesel solid residues: an alternative for hazardous metal sorption from aqueous solution. J Environ Manag 162:123–131. https://doi.org/10.1016/j.jenvman.2015.05.023

Saini VK, Andrade M, Pinto ML, Carvalho AP, Pires J (2010) How the adsorption properties get changed when going from SBA-15 to its CMK-3 carbon replica. Sep Purif Tech 75:366–376. https://doi.org/10.1016/j.seppur.2010.09.006

Salame I, Bandosz T (2001) Surface chemistry of activated carbons: combining the results of temperature-programmed desorption, Boehm, and potentiometric titrations. J Coll Interf Sci 240:252–258. https://doi.org/10.1006/jcis.2001.7596

Serrano-Gómez J, López-González H, Olguín MT, Bulbulian S (2015) Carbonaceous material obtained from exhausted coffee by an aqueous solution combustion process and used for cobalt (II) and cadmium (II) sorption. J Environ Manag 156:121–127. https://doi.org/10.1016/j.jenvman.2015.03.013

Silvestre-Albero A, Gonçalves M, Itoh T, Kaneko K, Endo M, Thommes M, Rodríguez-Reinoso F, Silvestre-Albero J (2012) Well-defined mesoporosity on lignocellulosic-derived activated carbons. Carbon 50:66–72. https://doi.org/10.1016/j.carbon.2011.08.007

Sing KSW (2014) Assessment of surface area by gas adsorption. In: Adsorption by powders and porous solids principles, methodology and applications, 2nd edn. Elsevier Ltd/Academic, Oxford/San Diego, pp 237–263 ISBN:978-0-12-598920-6

Smith JL, Herman RG, Klier K (2005) Ion exchange, core-level shifts, and bond strengths in mesoporous solid acid SBA-15. Stud Surf Sci Catal 158:797–804. https://doi.org/10.1016/S0167 2991(05)80415 X

Stumm W (1995) Chapter 1: The inner-sphere surface complex a key to understanding surface reactivity. In: Advances in chemistry. American Chemical Society, Washington, DC, pp 1–32. https://doi.org/10.1021/ba-1995-0244.ch001

Thommes M, Cychosz K.A (2014) Physical adsorption characterization of nanoporous materials: progress and challenges. Adsorption 20:233–250. https://doi.org/10.1007/s10450-014-9606-z

Treviño-Cordero H, Juárez L, Mendoza D, Hernández V, Bonilla A, Montes-Morán M (2013) Synthesis and adsorption properties of activated carbons from biomass of Prunus domestica and Jacaranda mimosifolia for the removal of heavy metals and dyes from water. Ind Crop Prods 42:315–323. https://doi.org/10.1016/j.indcrop.2012.05.029

Vasu E (2008) Surface modification of activated carbon for enhancement of nickel(II) adsorption. E-J Chem 5:814–819. https://doi.org/10.1155/2008/610503

Vinke P, Van Der Elik M, Verbree M, Voskamp AF, Van Bekkum H, Vinke P (1994) Modification of the surfaces of a gas activated carbon and a chemically activated carbon with nitric acid, hypochlorite, and ammonia. Carbon 32:675–686. https://doi.org/10.1016/0008-6223(94)90089-2

Vivo-Vilches JF, Bailón-García E, Pérez-Cadenas AF, Carrasco-Marín F, Maldonado-Hódar FJ (2014) Tailoring the surface chemistry and porosity of activated carbons: evidence of reorganization and mobility of oxygenated surface groups. Carbon 68:520–530. https://doi.org/10.1016/j.carbon.2013.11.030

Zhang C, Liu X, Lu X, He M, Meijer EJ, Wang R (2017) Surface complexation of heavy metal cations on clay edges: insights from first principles molecular dynamics simulation of Ni(II). Geochim Cosmochim Acta 203:54–68. https://doi.org/10.1016/j.gca.2017.01.014

Zielke U, Hüttinger K, Hoffman WP (1996) Surface-oxidized carbon fibers: I. Surface structure and chemistry. Carbon 34:983–998. https://doi.org/10.1016/0008-6223(96)00032-2

Chapter 6
Thermodynamic Properties of Heavy Metals Ions Adsorption by Green Adsorbents

Mohamed Nasser Sahmoune

Contents

Abstract There is growing interest in the use of cheap organic materials to clean heavy metal pollution by adsorption. This chapter presents the thermodynamic parameters of the adsorption of heavy metals ions adsorption by green adsorbents: the Gibbs free energy, entropy, and enthalpy. Research indicates that the temperature of the adsorption medium is the most important parameter influencing thermodynamic analysis. As a result, the adsorption of heavy metals ions by green adsorbents is spontaneous in most cases, with $\Delta G°$ lower than 0. Some researchers found that the uptake of heavy metals ions increased with temperature, an endothermic process. A more enhanced level of uptake in parallel with a temperature rise resembles the nature of a chemisorption mechanism, with $\Delta H°$ higher than 0. In contrast, other authors obtained the opposite trends, with $\Delta H°$ lower than 0, low temperature caused a high adsorption (exothermic process), and the mechanism was mainly physical adsorption. Some authors reported positive $\Delta S°$ values for adsorption of heavy metals ions by green adsorbents, suggesting the affinity of metals ions for adsorbents

M. N. Sahmoune (✉)
Department of Process Engineering, Faculty of Engineering Sciences, University of Boumerdes, Boumerdes, Algeria

Laboratory of Coatings, Materials and Environment, University of Boumerdes, Boumerdes, Algeria

G. Crini, E. Lichtfouse (eds.), *Green Adsorbents for Pollutant Removal*, Environmental Chemistry for a Sustainable World 18,
https://doi.org/10.1007/978-3-319-92111-2_6

used. On the other hand, negative $\Delta S°$ values were also reported by diverse authors, indicating a decrease in the randomness at the solid/solution interface during the adsorption process. No conclusion should be drawn based on corresponding values of thermodynamic parameters. Since the thermodynamic parameters were evaluated from very different adsorbent/adsorbate combinations, it was not possible to note a correlation between the corresponding enthalpy change and entropy change following adsorption.

6.1 Introduction

Green adsorbents are derived from agricultural sources and by-products (vegetables, fruits, foods), giving them a very low production cost when compared to commercial activated carbon (Kyzas and Kostoglou 2014). However, the green adsorbents materials have some negative sides such as low uptake capacity and releasing organic components in terms of high chemical oxygen demand (COD) and biological oxygen demand (BOD). Thus, they can cause oxygen reduction in water (Abdolali et al. 2014).

The main pollutants studied for the treatment of industrial effluents are dyes (Sahmoune and Yeddou 2016), heavy metals (Sahmoune et al. 2011) and others such as pesticides (Ouznadji et al. 2016) and phenols. Heavy metals can contaminate potable water source, groundwater, soil, and have tendency for bioaccumulation in the food chain.

Among several physical and chemical methods, adsorption process is one of the effective techniques that have been successfully employed for heavy metal removal from wastewater (Srivastava et al. 2015). There is an increasing demand for green adsorbents as alternative low-cost adsorbents. Heavy metal removal studies using green adsorbents with adsorption technique have recently been conducted by many researchers to elucidate the adsorption capacities and binding mechanisms in aqueous media (Ackacha and Meftah 2014; Ahmad et al. 2014; Ramaraju et al. 2014; Yagub et al. 2014; Ali et al. 2016; Gupta et al. 2016; Li et al. 2016).

Among the process parameters frequently investigated in the literature, temperature is shown to affect adsorption capacity (Sahmoune 2016). It indicates how adsorption can be an endothermic or exothermic process. When adsorption capacity decreased with temperature, the process was claimed to be exothermic, and vice versa. Thermodynamic analysis of the adsorption process is also essential to conclude whether the process is spontaneous or not. The feasibility and spontaneous nature of the adsorption process is reflected by the thermodynamic parameters (Doke and Khan 2013). Thermodynamic study and a good perception of temperature influence on adsorption process can help to understand the adsorption mechanism in heavy metals removal. Most studies estimated changes in free energy ($\Delta G°$), enthalpy ($\Delta H°$), and entropy ($\Delta S°$) under standard states based on a set of temperature-dependent equilibrium adsorption quantities (Liu and Lee 2014).

During the past 10 years, some reviews have been published that mainly focused on different aspects of adsorption thermodynamic (Milonjic 2007; Liu 2009; Liu and Lee 2014; Chang et al. 2016). Milonjic (2007) emphasized a correct calculation of thermodynamic parameters, especially the change of the free energy of adsorption at a solid/liquid interface. Liu (2009) also reviewed calculation of the free energy of adsorption in different ways in the literature. Liu and Lee (2014) directed their review towards the probable flaws in many adsorption studies when they assessed their thermodynamic parameters. Chang et al. (2016) recently published a review listing the free energy of adsorption, enthalpy and entropy change data for adsorption equilibrium reported in biosorption literature during January 2013–May 2016.

The present chapter briefly reviews the spontaneity of adsorption of heavy metals by selected green adsorbents based on thermodynamic parameters. We examined if the adsorption process is feasible or non-feasible, spontaneous or non spontaneous, favorable or not, followed by the presentation of adsorption mechanisms. This chapter can help all researchers working in the adsorption area, and especially those who may not be familiar with the adsorption thermodynamic.

6.2 Distinguish Between Chemisorption and Physisorption

The phenomenon of adsorption is essentially an attraction of adsorbate molecules to an adsorbent surface. Metals ions interact with solid adsorbents with forces originating either from the physical Van der Waals interaction (physisorption) or from the chemical adsorption (chemisorption) in which the adsorption caused by the formation of chemical bonds between the surfaces of solids (adsorbent) and heavy metals (adsorbate). Chemically adsorbed adsorbates are immobilized within the surface or on the surface. In physisorption the adsorbate is attached to the surface by weak van der Waals forces. However the adsorbate is less strongly attached to the surface compared to chemisorption (Yagub et al. 2014). In chemisorption, the forces operating in these cases are similar to those of a chemical bond. Physical adsorption occurs with low heat of adsorption (20–40 $kJmol^{-1}$). In contrast, in chemical adsorption, the heat of adsorption are high i.e. about 40–400 $kJmol^{-1}$. The process of physisorption is rapid, reversible i.e. desorption of the adsorbate occurs by increasing the temperature and does not require any activation energy. In chemisorption the process is may be slow, irreversible and specific in nature and occurs only when there is some possibility of compound formation between the heavy metal being adsorbed and the solid adsorbent (Aksakal and Ucun 2010). Physisorption is not specific in nature i.e. all heavy metals are adsorbed on all solids to some extent. In addition, in physical adsorption, equilibrium is established between the adsorbate and the liquid phase resulting multilayer adsorption. However, in chemisorption, when the surface is covered by the monomolecular layer (monolayer adsorption), the capacity of the adsorbent is essentially exhausted. Accordingly, based on the different reversibility and specific of chemical and physical adsorption processes, thermal desorption of the adsorbed sorbent could provide important information for the study of adsorption mechanism.

6.3 Calculation of Thermodynamic Parameters

Thermodynamic parameters of adsorption from solutions provide a great deal of information concerning the type and mechanism of the adsorption process. The Gibbs energy change (ΔG°), enthalpy change (ΔH°) and entropy change (ΔS°) thermodynamic parameters have been estimated to evaluate the feasibility of the adsorption process. The Gibbs energy change (ΔG°) indicates the degree of spontaneity of an adsorption process, and a higher negative value reflects a more energetically favorable adsorption (Din et al. 2014). Generally, the change of free energy for physisorption is between −20 and 0 kJ/mol, however, chemisorption is a range of −80 to −400 kJ/mol (Kula et al. 2008).

In the study of adsorption the Gibbs energy change ΔG° is calculated as follows:

$$\Delta G^{\circ} = -RT \ln K \tag{6.1}$$

In many studies of adsorption thermodynamics (Mohammadi et al. 2010; Ma et al. 2012; Guo et al. 2013; Yu and Luo 2014; Salman et al. 2015), the distribution constant K can be expressed as in relation (6.2):

$$K = \frac{C_{ad}}{C_e} \tag{6.2}$$

In which C_{ad} (mg/l) and C_e (mg/l) are the concentration of solute adsorbed at equilibrium and the solute concentration in solution at equilibrium, respectively.

R is the gas constant with a value of 8.314 $Jmol^{-1}K^{-1}$, and T is the absolute temperature in kelvins,

The relationship of (ΔG°) to enthalpy change (ΔH°) and entropy change (ΔS°) of adsorption is expressed as:

$$\Delta G^{\circ} = \Delta H^{\circ} - T\Delta S^{\circ} \tag{6.3}$$

Substituting Eq. 6.1 into Eq. 6.3 gives

$$\ln K = -\frac{\Delta H^{\circ}}{RT} + \frac{\Delta S^{\circ}}{R} \tag{6.4}$$

The values of ΔH° and ΔS° are determined from the slop and intercept of the linear plot of (*ln K*) vs. (1/T) of Eq. 6.4 (Fig. 6.1).

The positive value of change in enthalpy (ΔH°) indicates that the adsorption is an endothermic process, while positive value of change in entropy (ΔS°) reflects the increased randomness at the solid/solution interface.

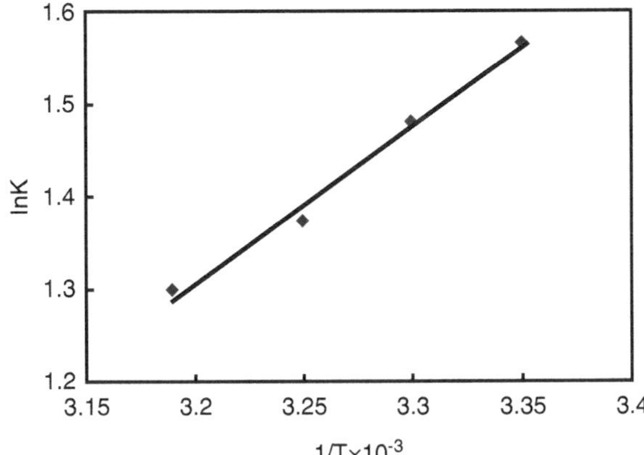

Fig. 6.1 Plot of ln K vs. 1/T for estimation of thermodynamic parameters for the adsorption The values of ΔH° and ΔS° are determined from the slop and intercept of the linear plot of (*ln K*) vs. (1/T). These parameters indicate that the adsorption process is spontaneous or not and endothermic or exothermic. The standard enthalpy change (ΔH°) for the adsorption process is:
(a) Positive value indicates that the process is endothermic in nature
(b) Negative value indicate that the process is exothermic in nature
The positive value of (ΔS°) reflects the affinity of metal ion for adsorbent used, while the negative value of (ΔS°) indicates the decrease randomness at the solid – solution interface during fixation of metal ion on the active sites of the adsorbent

The thermodynamic equilibrium constant of adsorption can be reasonably approximated by the Langmuir equilibrium constant (Liu 2009).

$$K = K_L \tag{6.5}$$

In this case, the Langmuir equilibrium constant can be applied for determination of ΔG°.

The general form of the Langmuir equation is given by the following expression (Sahmoune et al. 2009):

$$q_e = q_{\max} \frac{C_e K_L}{C_e K_L + 1} \tag{6.6}$$

In which q_e (mg/g) and $q_{\max}$ (mg/g) are the amount adsorbed at equilibrium and maximum adsorption capacity of adsorbent.

K_L is Langmuir equilibrium constant related to the energy of adsorption (lmg^{-1}).

The Langmuir model suggests that uptake occurs on a homogeneous surface by monolayer sorption without interaction between adsorbed molecules (Sahmoune ct al. 2011). In addition, thc modcl assumcs uniform cncrgics of adsorption onto the surface and no transmigration of the adsorbate (Sahmoune et al. 2011).

6.4 Isosteric Heat of Adsorption

Knowledge of the heat of adsorption is important for the characterization and optimization of an adsorption process (Ahmaruzzaman and Gupta 2011). Isosteric heat of adsorption (ΔH_x, $Kjmol^{-1}$) is defined as the heat of adsorption determined at constant amount of adsorbate adsorbed.

The isosteric heat of adsorption is calculated using the Clausius-Clapeyron equation:

$$\frac{d(\ln C_e)}{dT} = \frac{\Delta H_x}{RT^2} \tag{6.7}$$

Integrating the above equation gives the following equation:

$$\ln C_e = -\left(\frac{\Delta H_x}{R}\right)\frac{1}{T} + K \tag{6.8}$$

where K is a constant.

(ΔH_x) is calculated from the slope of the ln *Ce* versus ($1/T$) for different amount of heavy metal adsorbed on adsorbent.

The isosteric heat of adsorption is calculated from the slope of the plot of ln Ce versus $1/T$ for different amounts of adsorbate onto adsorbent.

The magnitude of ΔHx value gives information about the adsorption mechanism as physical sorption or chemical adsorption (Ahmaruzzaman and Gupta 2011).

6.5 Activation Energy

The activation energy is the minimum kinetic energy required for a particular reaction to carry out (Ali et al. 2016). Activation energy also provides a measure of the energetic barrier that the adsorbate ion have to ride previously to being fixed by the adsorption sites (Ali et al. 2016). The activation energy of a reaction is usually denoted by Ea, and given in units of $kJmol^{-1}$. It can be determined from experimental measurements of the adsorption rate constant at different temperatures according to the Arrhenius equation as follows:

$$\ln k = \ln A - \frac{E_a}{RT} \tag{6.9}$$

Where k is the adsorption rate constant, A is the Arrhenius constant or (frequency factor), which is a measure of the accessibility of the reaction sites to adsorbate. By plotting ln k versus 1/T and from the slope and the intercept, values of Ea and A can be obtained. The magnitude of activation energy may give an idea about the type of adsorption (Ali et al. 2016).

Some of the assigned values of Ea (kJ/mol) include 8–25 to physical adsorption, less than 21 to aqueous diffusion, 20–40 to pore diffusion and greater than 84 to ion exchange (Nuhoglu and Malkoc 2009).

6.6 Thermodynamic Investigations of Heavy Metals Adsorption on Green Adsorbents

As given in Table 6.1, the effect of temperature on the adsorption of heavy metals by selected green adsorbents is discussed in the following sub-sections.

6.6.1 Agricultural Sources and By-products

Sarada et al. (2017) carried out study of the adsorptive characteristics of cadmium from aqueous solution onto green plant biomass *Araucaria heterophylla*. Batch experiments were carried out. The % removal of Cd (II) on *Araucaria heterophylla* decreased from 92% to 86% with an increase in temperature from 25 to 40 °C at the initial cadmium concentration of 30 mg/l. Authors calculated various thermodynamic parameters, such as ΔG°, ΔH° and ΔS° that designated the biosorption to be a spontaneous and exothermic. The experimental ΔH° value is found to be less than 40 kJ/mol, which indicates that the adsorption of Cd^{2+} ions on *Araucaria heterophylla* is physisorption.

Jeyaseelan and Gupta (2016) reported that temperature remarkably influenced the equilibrium metal uptake. The adsorption of heavy metal increased as the temperature decreased, indicating that low temperature favored Cr (VI) removal by adsorption onto natural green tea leaves. Authors stated that the increase in adsorption capacity with decrease in temperature could be to a strong bonding between chromium and the adsorbent.

Borna et al. (2016) tried using *Hibiscus Cannabinus* kenaf herb as adsorbent for the treatment of wastewaters contaminated with Cr (VI) in batch and column system. A maximum adsorption of 582 μg/g Cr (VI) was achieved at temperature 23 °C. The isothermal data could be well described by the Langmuir equation. The thermodynamic parameters of the chromium ions uptake onto the sorbent indicated that the adsorption was feasible, spontaneous and dominated by physisorption process.

Ackacha and Meftah (2014) described adsorption study of Cd (II) onto *Acacia tortilis* seeds. Authors reported that cadmium uptake by this adsorbent increased from 706 to 1004 mg/g as the temperature of solution increased from 293 to 323 K. The data followed Freundlich adsorption isotherm. Gibbs free energy was found to be −4.77, −5.26 and −6.07 kJ/mol at temperature 293, 303 and 323 K respectively. The adsorption was endothermic having an enthalpy change of 8.1 kJ/mol. The small positive ΔS° value (0.044 kJ/mol) in the system under investigation indicates that no

Table 6.1 Effect of temperature on the adsorption of heavy metals by selected green adsorbents

Adsorbent	Metal	Temperature range (K)	Adsorption process	References
Acacia arabica sawdust	Hg (II)	293–333	Endothermic	Meena et al. (2008)
Acacia leucocephala bark	Cu (II)	303–323	Exothermic	Munagapati et al. (2010)
Acacia tortilis seeds	Cd (II)	293–323	Endothermic	Ackacha and Meftah (2014)
Bael leaves	Pb (II)	303–323	Endothermic	Chakravarty et al. (2010)
Bamboo sawdust	Zn (II)	303–323	Endothermic	Sulaiman et al. (2011)
Banana peel	Pb (II)	293–323	Endothermic	Li et al. (2016)
Banana peel	Cd (II)	293–323	Endothermic	Li et al. (2016)
Beech sawdust	Cr (III)	293–323	Endothermic	Witek-Krowiak (2013)
Cashew nut shell	Pb (II)	303–333	Exothermic	Senthil Kumar (2014)
Cashew nut shell	Ni(II)	303–333	Exothermic	Senthil Kumar et al. (2011)
Chestnut shell	Cu (II)	293- 313	Exothermic	Yao et al. (2010)
Coffee residues	Pb (II)	288–303	Endothermic	Wu et al. (2016)
Coffee residues	Zn (II)	288–303	Endothermic	Wu et al. (2016)
Daucus carota L.	Mn (II)	293–333	Endothermic	Güzel et al. (2008)
Daucus carota L.	Co (II)	293–333	Endothermic	Güzel et al. (2008)
Fiber of palm tree	Pb (II)	298–313	Exothermic	Hikmat et al. (2014)
Ficus glomerata	Cr (VI)	303–323	Endothermic	Rao and Rehman (2010)
Green tea leaves	Cr (VI)	298–323	Exothermic	Jeyaseelan and Gupta (2016)
Hibiscus rosa sinensis	Zn (II)	293–313	Endothermic	Vankar et al. (2012)
Hydrilla verticillata	Pb (II)	298–313	Exothermic	Huang et al. (2009)
Lemon peel	Co (II)	298–318	Exothermic	Bhatnagar et al. (2010)
Mangifera indica sawdust	Cr (VI)	298–308	Exothermic	Kapur and Mondal (2013)
Mango leaves	As (III)	283–318	Endothermic	Kamsonlian et al. (2012)
Oak wood char	Cr (VI)	298–318	Endothermic	Mohan et al. (2011)

(continued)

Table 6.1 (continued)

Adsorbent	Metal	Temperature range (K)	Adsorption process	References
Oak bark char	Cr (VI)	298–318	Endothermic	Mohan et al. (2011)
Olive oil waste	Ni(II)	298–333	Endothermic	Nuhoglu and Malkoc (2009)
Olive stone	Cd (II)	293–313	Exothermic	Kula et al. (2008)
Orange peels	Cr (III)	303–333	Exothermic	Ugbe et al. (2014)
Peanut hull	Cu (II)	298–338	Endothermic	Ali et al. (2016)
Petiole	Pb (II)	298–313	Exothermic	Hikmat et al. (2014)
Pine cone	As (III)	293–318	Endothermic	Van Vinh et al. (2015)
Pinus brutia leaf	Ce (III)	293–333	Exothermic	Kütahyalı et al. (2012)
Polygonum orientale Linn	Pb (II)	283–313	Endothermic	Wang et al. (2010)
Rice husk	As (III)	283–318	Endothermic	Kamsonlian et al. (2012)
Rice husk (modified)	Cu (II)	303–330	Endothermic	Jaman et al. (2009)
Rose waste	Cr (III)	303–333	Exothermic	Iftikhar et al. (2009)
Saccharum bengalense	Cd (II)	283–333	Endothermic	Din et al. (2014)
Sesame husk	Pb (II)	298–308	Endothermic	Surchi (2011)
Straw ash	Ni (II)	288–353	Endothermic	Arshadi et al. (2014)
Straw ash	Co (II)	288–353	Endothermic	Arshadi et al. (2014)
Sugarcane bagasse	As (III)	298–318	Endothermic	Gupta et al. (2015)
Sugarcane bagasse	As (V)	298–318	Endothermic	Gupta et al. (2015)
Tamarind fruit shell	Cu (II)	303–333	Endothermic	Anirudhan and Radhakrishnan (2008)
Tamrix articulata	Cd (II)	303–333	Exothermic	Al Othman et al. (2011)
Triticum durum Desf	Cu (II)	293–303	Endothermic	Aydin et al. (2008)

Some researchers found that the uptake of heavy metals ions increased with higher temperature (endothermic process). A more enhanced level of uptake in parallel with a temperature rise resembles the nature of a chemisorption mechanism ($\Delta H^\circ > 0$). In contrast, other authors obtained the opposite trends ($\Delta H^\circ < 0$), low temperature caused a high adsorption (exothermic process), and the mechanism was mainly physical adsorption significant structural change occurs in the adsorbent material. Authors claimed that Cd (II) ions were adsorbed onto the sorbent by ion exchange and surface adsorption mechanisms.

Ni (II), Cd (II), Cu (II), and Co (II) adsorptive removal from water by barley straw ash has been explored (Arshadi et al. 2014). Batch experiments were carried out. Authors noted that adsorption of metals increases with increasing temperature, indicating that the removal process was endothermic. Authors found that the entropy values are positive for all the experiments and presumed that the positive value of $\Delta S°$ revealed the increased randomness and an increase in the degrees of freedom at the adsorbent-solution interface.

Ugbe et al. (2014) examined batch trivalent chromium adsorption by untreated *Citrus sinensis* orange peels. The effect of the temperature on the adsorption of trivalent chromium has been investigated at 30, 40, 50 and 60 °C. It was observed that upon increasing the temperature from 30 to 60 °C, the amount of chromium adsorbed by orange peels decreased. Authors noted decrease in the $\Delta G°$ with decrease in temperature which indicated more efficient adsorption at low temperature. The $\Delta G°$ found are in the range −20 and 0 $kJmol^{-1}$, which indicates that the adsorption of Cr^{3+} ions onto orange peels is physisorption. All values of enthalpy change were found to be negative between −12.98 and −7.48 J/mol, indicating an exothermic adsorption process. The negative values of free energy indicated the spontaneous nature of adsorption.

Din et al. (2014) reported the sorption of Pb (II) onto plant *Saccharum bengalense* in batch experiments. The Gibbs free energy of adsorption ($\Delta G°$) was determined to be between −6.78 and −10.78 kJ/mol at temperature between 10 and 60 °C. The decrease in the $\Delta G°$ value with an increase in temperature favors the removal process of lead at high temperature. The values of $\Delta H°$ and $\Delta S°$ for *Saccharum bengalense* adsorption were reported to be 0.084 kJ/mol and 16.66 kJ/molK, respectively. The adsorption process was found to be spontaneous, feasible and endothermic. The energy of activation calculated by the Arrhenius equation (Ea = 5 kJ/mol) indicated an adsorption process of a physisorption.

Kamsonlian et al. (2012) reported batch sorption of aqueous As (III) using mango leaves powder (MLP) and rice husk (RH). Arsenic removal increased with increasing temperature. This means that the adsorption of As (III) onto MLP and RH is an endothermic process.

The free energy of adsorption for each adsorbent at all temperature was negative and decreased with the rise in temperature, which implies that the process is more favorable at higher temperature. Enthalpy change of adsorption $\Delta H°$ of As (III) by MLP and RH were found to be 23.89 and 56.26 kJ/mol, suggesting a chemisorption mechanism and indicative for an endothermic process. Entropy change of adsorption $\Delta S°$ of As (III) by MLP and RH were found to be 144.38 and 247.17 kJ/molK. This means that the adsorption of As (III) by MLP and RH occurs with increase in disorder at solid–liquid interface (Doke and Khan 2013).

Kütahyalı et al. (2012) examined cerium biosorption by modified *Pinus brutia* leaf powder. The sorption of Ce (III) is well described by the Langmuir isotherm. The maximum biosorption capacity was 109 mg/g of biomass at temperature 20 °C.

Thermodynamic parameters for the removal of Ce (III) are $\Delta G^\circ = -19.6$ and −20.2 kJ/mol at 20 and 60 °C respectively; $\Delta H^\circ = -15$ kJ/mol and $\Delta S^\circ = 15.6$ J/mol K. The negative free energy values indicate the feasibility of the process and spontaneous nature of biosorption. Enthalpy changes at different temperatures were found to be negative thereby indicating exothermic nature of the process. The positive value of entropy indicates the irreversibility and stability of the biosorption process.

The adsorption potential of *Polygonum orientale Linn*-based activated carbon (PLAC) to remove Pb (II) from aqueous solution was also investigated using a batch mode experiment (Wang et al. 2010). An increase in temperature is observed to induce a positive effect on the adsorption process, indicating the endothermic of adsorption. Authors concluded that the mechanism of adsorption was chemical ion exchange.

Removal of Cu^{2+} ions from aqueous solution using the garden grass was examined by Hossain et al. (2012). Authors found the adsorption was favorable at lower temperature. The adsorption was found to be non spontaneous and physisorption was the main mechanism of adsorption. A positive value of ΔS° revealed an increased randomness between solid-solution interfaces during the adsorption of copper on garden grass (Hossain et al. 2012).

Yao et al. (2010) evaluated the sorption potential of chestnut shell for the removal of copper ions. The removal of Cu (II) decreased with increasing temperature. Authors calculated the ΔH° and ΔS° values of Cu (II) sorption on chestnut shell to be −17.42 kJ/mol and −54.66 J/molK, respectively. Gibbs free energy change ΔG° was calculated to be −1.39, −0.88 and −0.30 kJ/mol at different temperatures in the range 293–313 K. The value of ΔG° increase with an increase of temperature and better adsorption is obtained at low temperature. The small negative free energy values also indicate that the adsorption of copper by chestnut shell was driven by physical sorption.

Bhatnagar et al. (2010) removed cobalt by using lemon peel. It was observed that with increase in temperature, adsorption capacity decreases. The negative ΔH° (−21.2 kJ/mol) means that the adsorption process was carried out as exothermic at 25–45 °C. Furthermore, the positive ΔS° (54.61 J/mol) indicates the affinity of lemon peel for cobalt. Gibbs free energy change (ΔG°) was calculated to be −37.47 and −38.56 kJ/mol for 25 and 45 °C, respectively. The negative ΔG° values indicated spontaneous nature of the adsorption. Authors claimed that the physical adsorption occurs during adsorption.

Huang et al. (2009) utilized untreated *Hydrilla verticillata* for aqueous lead removal. The effect of the temperature on the adsorption of lead has been investigated at 25, 30, 35 and 40 °C. Authors found that the most suitable temperature was 25 °C. When the temperature decreases from 40 to 25 °C, the magnitude of free energy change shifts to high negative value suggested that the biosorption was more spontaneous at low temperature. The biosorption process was exothermic. The positive value of entropy change (33.85 J/mol K) shows the increase randomness after the biosorption of lead on *H. verticillata*.

Nuhoglu and Malkoc (2009) investigated olive oil waste as adsorbent in the removal of nickel from aqueous solution. The effect of the temperature on the adsorption of the nickel ions has been investigated at 25, 45 and 60 °C. Results showed that the process was endothermic. The authors demonstrated that physisorption was the main mechanism of adsorption.

Distillation waste of rose petals was used to remove Cu (II) and Cr (III) from aqueous solutions (Iftikhar et al. 2009). The best performance for Cu (II) and Cr (III) removal was observed at a temperature of 303 K. Authors found the adsorption was favorable at lower temperature. Thermodynamic parameters indicated that the metal adsorption was spontaneous and exothermic in nature. Positive $\Delta S°$ values reflect the affinity of rose waste for copper and chromium ions, and the increase of randomness at the solid/liquid interface during the adsorption of Cu^{2+} and Cr^{3+} ions onto rose waste biomass. The activation energy for Cu (II) and Cr (III) were found to be 68.53 and 114.45 kJ/mol, respectively. These values (greater than 65 kJ/mol) indicate that ion exchange may play a significant role in the adsorption process.

Kula et al. (2008) studied the adsorption removal of Cd (II) using a low-cost adsorbent, prepared from olive stone, an agricultural solid by-product. Olive stone is impregnated with $ZnCl_2$ as activated agent and carbonized to produce low-cost adsorbent. The effect of temperature on the adsorption of the cadmium has been investigated at 293, 303 and 313 K.

Authors reported that cadmium uptake by treated olive stone decreased slightly from 0.75 to 0.6 mg/g as the temperature of solution increased from 293 to 313 K, indicated that cadmium adsorption is favored at low temperature. Authors calculated thermodynamic parameters ($\Delta G°$, $\Delta H°$, and $\Delta S°$) that designated the adsorption to be spontaneous exothermic and physical in nature.

Investigations were made to carry out the adsorption capacities of activated carbon derived from olive bagasse for the removal of Cr (VI) from aqueous solution (Demiral et al. 2008). Authors observed that upon increasing the temperature from 25 to 45 °C, the amount of hexavalent chromium adsorbed by olive bagasse increased from 88.59 to 109.89 mg/g. This implies that the adsorption process is endothermic in nature. The increase in $\Delta G°$ value with increasing temperature shows that the adsorption is more favorable at high temperature. The positive value of $\Delta S°$ (0.145 kJ/mol K) indicates that there is an increase in the randomness in the system solid/solution interface during the adsorption process.

6.6.2 *Agricultural Residues and Wastes*

Wu et al. (2016) performed a series of batch experiments for the removal of three heavy metals, i.e. Cu (II), Pb(II) and Zn(II) by waste coffee residues. With increasing temperature the metal ions adsorption capacity increased, which proves an increase in feasibility of adsorption at higher temperatures. Authors calculated the relevant thermodynamic parameters.

Their thermodynamic studies showed that the process was feasible and spontaneous at all temperatures and was endothermic in nature. Authors pointed out that the adsorption of Cu^{2+}, Pb^{2+} and Zn^{2+} on waste coffee residues was driven by a physisorption process.

Peanut hull is a local natural abundant agricultural waste in Egypt. The impact of temperature on the adsorption isotherm of copper ions by peanut hull at pH 4 was explored by Ali et al. (2016) using a series of batch experiments. They found that the adsorption is non spontaneous (positive Gibbs energy) and the percentage of Cu (II) removal increased slightly from 83.34 to 87.26 as the temperature increased from 25 to 65 °C. The adsorption was endothermic having an enthalpy change of 4.44 $kJmol^{-1}$ at 25 °C. Authors claimed that the adsorption mechanism of Cu (II) onto peanut hulls is a chemical adsorption process.

Banana peel, a common biological waste in the form of powder (BPCF) was used for adsorption of Cu (II), Pb (II), Cd (II) and Cr (VI) from aqueous solutions (Li et al. 2016). Carboxyl, hydroxyl, and K, Ca, Mg present in banana peel serve as sites for the metal cation exchange. The effect of the temperature on the adsorption of heavy metals by banana peel has been investigated at 20, 30 and 40 °C. The decrease in the $\Delta G°$ value for all metal ions with an increase in temperature indicates more efficient adsorption at high temperature. The positive values of $\Delta S°$ for all metal ions reflects the affinity of the banana peel adsorbent for heavy metals ions, and the increase of randomness at the solid/liquid interface during the adsorption of Cu^{2+},Pb^{2+}, Cd^{2+} and Cr^{6+} ions onto BPCF. All values of enthalpy change $\Delta H°$ were found to be positive indicating an endothermic adsorption process.

Sugarcane bagasse (SCB) a low-cost agro waste product was investigated for As (III) and As (V) removal (Gupta et al. 2015). The effect of temperature (25, 35 and 45 °C) on As (III)/As (V) adsorption was studied with 2.5 g/l of Sugarcane bagasse in 500 mg/l metal ion solution at pH 7. Thermodynamic studies suggested the spontaneous and endothermic nature of adsorption of As (III) and As (V) by Sugarcane bagasse. Authors suggested that adsorption of arsenic could take place by chemical ion exchange mechanism.

Van Vinh et al. (2015) used raw and treated pine cone for the adsorption of arsenic pollutant. Thermodynamic parameters, for the removal of As (III) by raw pine cone are: $\Delta G° = -3.74$ and –5.19 kJ/mol at 293 and 318 K respectively; $\Delta H° = 13.25$ kJ/mol and $\Delta S° = 0.058$ kJ/mol K and those for treated pine cone are: $\Delta G° = -10.21$ and –13.74 at 293 and 318 K respectively; $\Delta H° = 31.1$ kJ/mol and $\Delta S° = 0.141$ kJ/molK. The negative free energy indicated that the feasibility of the process and spontaneous nature of adsorption. Enthalpy changes at different temperatures were found to be positive thereby indicating endothermic nature of the process. A small positive entropy values in the system indicated that no structural change occurred in the adsorbent material.

Hikmat et al. (2014) explored the use of agricultural waste materials adsorbents such as petiole and fiber of the date palm for aqueous Pb (II) removal. An increase in temperature is observed to induce a negative effect on the adsorption process. The enthalpy change ($\Delta H°$) and the entropy change ($\Delta S°$) for Pb(II) adsorption on petiole and fiber are determined to be within the ranges of (−17.34) – (−12.25) kJ/mol and (−76.03) – (−67.03) J/molK, respectively.

The free energy for each adsorbent was negative and increased with the rise of temperature; therefore the adsorption is more favorable at low temperature.

Senthil Kumar (2014) described adsorption study of Pb(II) onto cashew nut shell. Author reported that maximum removal of lead ions was obtained at temperature 30 °C. As the temperature of the solution is changed from 30 to 60 °C the adsorption percentage of lead (II) adsorbed onto cashew nut shell decreased from 96% to 90% for the equilibrium time 30 min and initial concentration 10 mg/l. These results indicated the exothermic nature of adsorption onto this adsorbent. Thermodynamic data on cashew nut shell followed physical adsorption.

Kapur and Mondal (2013) used *Mangifera indica* sawdust to adsorb chromium (VI). Cr (VI) came down from 70% to 35% in the temperature range of 25–35 °C for initial Cr (VI) concentration of 50 mg/l, adsorbent dose of 6 g/l and pH 2.6 indicating that the adsorption is exothermic. The negative values of $\Delta G°$, $\Delta H°$ and activation energy Ea obtained from thermo dynamical analysis of the experimental data confirmed the spontaneous and exothermic nature of the adsorption. The experimental thermodynamic $\Delta S°$ is found near 0 kJ/mol K, which indicates that no significant structural change occurs in the adsorbent material.

Witek-Krowiak (2013) evaluated the sorption potential of beech sawdust for the removal of copper and chromium ions. Maximum removal of Cu (II) and Cr (III) were 36.04 and 50.04 mg/g, respectively, at temperature 50 °C. The effect of temperature was studied at temperature range 20–50 °C. Upon increasing temperature from 20 to 50 °C, the amount of copper adsorbed by beech sawdust increased from 31.38 to 36.04 mg/g and the amount of chromium adsorbed by the same adsorbent increased from 43.56 to 50.04 mg/g. The estimated thermodynamic parameters indicated that the metal biosorption was spontaneous, endothermic and chemical in nature.

Vankar et al. (2012) explored the use of natural dye waste of *Hibiscus rosa sinensis* for aqueous Zn (II) removal. The thermodynamic parameters indicated the feasibility, endothermic, and spontaneous nature of the biosorption process between 20–40 °C. The positive value of $\Delta S°$ (0.08 kJ/mol) show that there is an increase in randomness at the solid/solution interface during the adsorption process, and also reflects the affinity of the adsorbent for zinc ion. Physical adsorption was claimed as the dominant mechanism.

Cadmium metal ion was removed from water by *meranti* wood sawdust (Rafatullah et al. 2012). The effect of temperature showed decreased adsorption capacity with increasing temperature from 30 to 50 °C. The adsorption was exothermic having an enthalpy change of −37.17 kJ/mol at 30–50 °C.When temperature decreases from 50 to 30 °C, the magnitude of free energy change shifts to high negative value (from −6.15 to −9.02 kJ/mol) suggested that the adsorption was more spontaneous at low temperature. A negative entropy (−0.093 kJ/mol K) of adsorption, reflects that no structural change occurs in *meranti* wood sawdust when cadmium ions are taken up on this biomass.

Sulaiman et al. (2011) investigated Zn (II) removal from aqueous solution using modified bamboo sawdust. The effect of the temperature on the adsorption of the zinc has been investigated at 30, 40 and 50 °C. Authors noted that adsorption capacity increased as the temperature increased, thereby indicating the process to be endothermic. Gibbs free energy change ΔG° was calculated to be –6.92, −5.68 and –5.07 kJ/mol for Zn (II) adsorption. The negative ΔG° values indicated thermodynamically feasible and spontaneous nature of the adsorption. The decrease in ΔG° values with increase in temperature showed a decrease in feasibility of adsorption at higher temperature. Enthalpy change of adsorption ΔH° was found to be – 23.83 kJ/mol. The negative value of entropy change (−0.055 kJ/mol) suggested a decrease in the randomness at the solid/solution interface during the adsorption process.

The equilibrium adsorption characteristics of Cr (VI) ions from aqueous solutions having respective metal ion concentrations in the range of 1–100 mg/l for Oak wood char and Oak bark char were studied at different temperatures in the range of 25–45 °C (Mohan et al. 2011). An increase in temperature is observed to induce a positive effect on the adsorption process. Author found the adsorption to be temperature dependent. Oak wood char and Oak bark char adsorbed hexavalent chromium by chemical adsorption and the process was endothermic in nature.

Senthil Kumar et al. (2011) reported batch sorption of aqueous Ni (II) using an agricultural waste cashew nut shell. Authors found that the biosorption percentage of Ni (II) by cashew nut shell biomass decreased from 77% to 72% as temperature was increased from 30 to 60 °C, showing exothermic nature of the process. The small negative ΔG° values imply favorable and spontaneous sorption process and indicate physical adsorption of nickel onto cashew nut shell. The negative entropy values suggested a decrease in the randomness at the solid/solution interface during the biosorption process.

The adsorptions of lead from aqueous solutions by Brazilian sawdust samples (*Caryocar spp.*, *Manilkara spp.* and *Tabebuia spp.*) were investigated by Prado et al. (2010). The thermodynamic parameters for *Caryocar spp.* are ΔG° = −14.3 kJ/mol, ΔH° = −2 kJ/mol, and ΔS° = 41 J/molK, and those for *Manilkara spp.* are ΔG° = −14.7 kJ/mol, ΔH° = −3.2 kJ/mol, and ΔS° = 38 J/molK, and those for *Tabebuia spp.* are ΔG° = −15.6 kJ/mol, ΔH° = −2.3 kJ/mol, and ΔS° = 45 J/molK. This thermodynamic data showed spontaneous and exothermic nature of the sorption processes. The positive value of ΔS° reflects the affinity of Pb (II) for each sorbent used and reveal the increase in randomness at the solid-solution interface through the adsorption of Pb (II) metal on the active sites of selected adsorbent.

The impact of temperature on the adsorption of copper, chromium, nickel, and lead ions by meranti sawdust at pH 6 and adsorbent dose of 5 g/l was explored by Rafatullah et al. (2009). They found that the uptake of ions increased with higher temperature indicating the process is endothermic. Authors speculated that the ion exchange and hydrogen bonding may be the principal mechanism for the removal of heavy metals.

Pinus sylvestris Linn was used for Cr (VI) adsorption (Ucun et al. 2008). The experimental result by workers showed that, increasing temperature favors the adsorption. Authors found the adsorption to be temperature dependent. The adsorption was endothermic having an enthalpy change of 62.02 kJ/mol at 25–45 °C. The positive enthalpy of adsorption indicated chemical adsorption. The activation energy of reaction (Ea) was determined as 41.74 kJ/mol using the Arrhenius equation. Gibbs free energy and entropy were found to be –22.30 kJ/mol and 272.99 J/molK at 35 °C, respectively.

Djeribi and Hamdaoui (2008) removed Cu (II) from aqueous solution using *Cedar* sawdust by batch mode. A maximum sorption capacity of 294.12 mg/g was achieved at temperature 25 °C. The sorption of Cu (II) by sawdust was seen to be increasing with a decrease in temperature showing exothermic nature of the process. A decrease in the adsorption of Cu (II) with the rise in temperature may be due to the increasing tendency to desorb from the interface to the solution. The results from thermodynamic studies indicated that the adsorption was feasible, spontaneous and exothermic process in nature.

Meena et al. (2008) reported the adsorption of chromium, lead, mercury and copper metal ions on *Acacia arabica* sawdust. The removal of Cr (VI), Pb (II), Hg (II) and Cu (II), by sawdust has been found temperature dependent. The adsorption of metal ions has been found to increase with an increase in temperature from 20 to 60 °C. Enthalpy change ($\Delta H°$) was calculated to be 7.46 kJ/mol for Cr (VI), 26 kJ/mol for Hg (II), 21.62 kJ/mol for Pb (II) and 29.79 kJ/mol for Cu (II). The positive $\Delta H°$ indicates the endothermic nature of the adsorption process at 20–60 °C. Entropy change ($\Delta S°$) has been estimated to be very small under experimental conditions. Accordingly no significant structural change occurs in *Acacia arabica* sawdust. The negative ($\Delta G°$) values indicated thermodynamically feasible and spontaneous nature of adsorption.

6.7 Conclusion

For engineering practice, the thermodynamic parameters would be required to develop a full understanding of the overall adsorption. Indeed, the spontaneity of the process can be inferred. Thermodynamic parameters provide very useful information to the adsorption mechanisms and investigate the range of temperature at which the adsorption process is favorable or unfavorable. Unfortunately, the effect of temperature on the adsorption capacity is not always given in the literature. It should not be forgotten that the neglect of a possible temperature effect may lead to serious deviation of experimental results.

It is clear from the literature survey that green adsorbents are capable of adsorption of heavy metals. In most cases, in the present review, the adsorption process is thermodynamically feasible and spontaneous. Only two papers of the adsorption processes are non spontaneous.

The adsorption of heavy metals onto the active sites of green adsorbents is endothermic or exothermic. Adsorption occurs with decrease in disorder at adsorbent-adsorbate interface, or occurs with increase in disorder at adsorbent-adsorbate interface. Further, it was found that both physisorption and chemical adsorption played the chief role in the adsorption mechanisms of metal. No conclusion should be drawn based on corresponding values of thermodynamic parameters. Since the thermodynamic parameters were evaluated from very different adsorbent/adsorbate combinations, it is not possible to note a correlation between the corresponding enthalpy change and entropy change following adsorption.

Because of diverse experimental conditions and the influence of various environmental factors, it is very difficult to come to a consensus regarding the best green adsorbent suited for the adsorption of heavy metals.

References

Abdolali A, Guo WS, Ngo HH, Chen SS, Nguyen NC, Tung KL (2014) Typical lignocellulosic wastes and by-products for biosorption process in water and wastewater eatment: a critical review. Bioresour Technol 160:57–66. https://doi.org/10.1016/j.biortech.2013.12.037

Ackacha MA, Meftah SA (2014) *Acacia tortilis* seeds as a green chemistry adsorbent to clean up the water media from cadmium cations. Int J Environ Sci Dev 5:375–379. https://doi.org/10.7763/IJESD.2014.V5.513

Ahmad M, Rajapaksha AU, Lim JE, Zhang M, Bolan N, Mohan D, Vithanage M, Lee SS, Ok YS (2014) Biochar as a sorbent for contaminant management in soil and water: a review. Chemosphere 99:19–33. https://doi.org/10.1016/j.chemosphere.2013.10.071

Ahmaruzzaman M, Gupta VK (2011) Rice husk and its ash as low-cost adsorbents in water and wastewater treatment. Ind Eng Chem Res 50:13589–13613. https://doi.org/10.1021/ie201477c

Aksakal O, Ucun H (2010) Equilibrium, kinetic and thermodynamic studies of the biosorption of textile dye (reactive red 195) onto *Pinus sylvestris* L. J Hazard Mater 181:666–672. https://doi.org/10.1016/j.jhazmat.2010.05.064

Al Othman ZA, Hashem A, Habila MA (2011) Kinetic, equilibrium and thermodynamic studies of cadmium (II) adsorption by modified agricultural wastes. Molecules 16:10443–10456. https://doi.org/10.3390/molecules161210443

Ali RM, Hamad HA, Hussein MM, Malash GF (2016) Potential of using green adsorbent of heavy metal removal from aqueous solutions: adsorption kinetics, isotherm, thermodynamic, mechanism and economic analysis. Ecol Eng 91:317–332. https://doi.org/10.1016/j.ecoleng.2016.03.015

Anirudhan TS, Radhakrishnan PG (2008) Thermodynamics and kinetics of adsorption of Cu (II) from aqueous solutions onto a new cation exchanger derived from tamarind fruit shell. J Chem Thermodyn 40(2008):702–709. https://doi.org/10.1016/j.jct.2007.10.005

Arshadi M, Amiri MJ, Mousavi S (2014) Kinetic, equilibrium and thermodynamic investigations of Ni(II), Cd(II), Cu(II) and Co(II) adsorption on barley straw ash. Water Res Ind 6:1–17. https://doi.org/10.1016/j.wri.2014.06.001

Aydın H, Bulut Y, Yerlikaya C (2008) Removal of copper (II) from aqueous solution by adsorption onto low-cost adsorbents. J Environ Manag 87:37–45. https://doi.org/10.1016/j.jenvman.2007.01.005

Bhatnagar A, Minocha AK, Sillanpaa M (2010) Adsorptive removal of cobalt from aqueous solution by utilizing lemon peel as biosorbent. Biochem Eng J 48:181–186. https://doi.org/10.1016/j.bej.2009.10.005

Borna MO, Pirsaheb M, Niri MV, Mashizie KR, Kakavandi B, Zare MR, Asadi A (2016) Batch and column studies for the adsorption of chromium(VI) on low-cost *Hibiscus Cannabinus* kenaf, a green adsorbent. J Taiwan Inst Chem Eng 68:80–89. https://doi.org/10.1016/j.jtice.2016.09.022

Chakravarty S, Mohanty A, Sudha TN, Upadhyay AK, Konar J, Sircar JK, Madhukar A, Gupta KK (2010) Removal of Pb(II) ions from aqueous solution by adsorption using bael leaves (*Aegle marmelos*). J Hazard Mater 173:502–509. https://doi.org/10.1016/j.jhazmat.2009.08.113
Chang Y, Lai JY, Lee DJ (2016) Thermodynamic parameters for adsorption equilibrium of heavy metals and dyes from wastewaters: research updated. Bioresour Technol 222:513–516. https://doi.org/10.1016/j.biortech.2016.09.125
Demiral H, İ Demiral I, Tümsek F, Karabacakoğlu B (2008) Adsorption of chromium(VI) from aqueous solution by activated carbon derived from olive bagasse and applicability of different adsorption models. Chem Eng J 144:188–196. https://doi.org/10.1016/j.cej.2008.01.020
Din MI, Hussain Z, Mirza ML, Shah AT, Athar MM (2014) Adsorption optimization of lead (II) using *Saccharum bengalense* as non-conventional low cost biosorbent: isotherm and thermodynamic modeling. Int J Phytoremediation 16:889–908. https://doi.org/10.1080/15226514.2013.803025
Djeribi R, Hamdaoui O (2008) Sorption of copper (II) from aqueous solutions by cedar sawdust and crushed brick. Desalination 225:95–112. https://doi.org/10.1016/j.desal.2007.04.091
Doke KM, Khan EM (2013) Adsorption thermodynamics to clean up wastewater; critical review. Rev Environ Sci Biotechnol 12:25–44. https://doi.org/10.1007/s11157-012-9273-z
Guo L, Li G, Liu J, Meng Y, Tang Y (2013) Adsorptive decolorization of methylene blue by crosslinked porous starch. Carbohydr Polym 93:374–379. https://doi.org/10.1016/j.carbpol.2012.12.019
Gupta A, Vidyarthi SR, Sankararamakrishnan N (2015) Concurrent removal of As(III) and As (V) using green low cost functionalized biosorbent – *Saccharum officinarum* bagasse. J Environ Chem Eng 3:113–121. https://doi.org/10.1016/j.jece.2014.11.023
Gupta N, Kushwaha AK, Chattopadhyaya MC (2016) Application of potato (*Solanum tuberosum*) plant wastes for the removal of methylene blue and malachite green dye from aqueous solution. Arab J Chem 9:S707–S716. https://doi.org/10.1016/j.arabjc.2011.07.021
Güzel F, Yakut H, Topal G (2008) Determination of kinetic and equilibrium parameters of the batch adsorption of Mn(II), Co(II), Ni(II) and Cu(II) from aqueous solution by black carrot (*Daucus carota L.*) residues. J Hazard Mater 153:1275–1287. https://doi.org/10.1016/j.jhazmat.2007.09.087
Hikmat NA, Qassim BB, Khethi MT (2014) Thermodynamic and kinetic studies of lead adsorption from aqueous solution onto petiole and fiber of palm tree. Am J Chem 4:116–124. https://doi.org/10.5923/j.chemistry.20140404.02
Hossain MA, Ngo HH, Guo WS, Setiadi T (2012) Adsorption and desorption of copper(II) ions onto garden grass. Bioresour Technol 121:386–395. https://doi.org/10.1016/j.biortech.2012.06.119
Huang L, Zeng G, Huang D, Li L, Huang P, Xia C (2009) Adsorption of lead(II) from aqueous solution onto *Hydrilla verticillata*. Biodegradation 20:651–660. https://doi.org/10.1007/s10532-009-9252-4
Iftikhar AR, Bhatti HN, Hanif MA, Nadeem R (2009) Kinetic and thermodynamic aspects of Cu(II) and Cr(III) removal from aqueous solutions using rose waste biomass. J Hazard Mater 161:941–947. https://doi.org/10.1016/j.jhazmat.2008.04.040
Jaman P, Chakraborty D, Saha P (2009) A study of the thermodynamics and kinetics of copper adsorption using chemically modified Rice husk. Clean 2009(37):704–711. https://doi.org/10.1002/clen.200900138
Jeyaseelan C, Gupta A (2016) Green tea leaves as a natural adsorbent for the removal of Cr (VI) from aqueous solutions. Air Soil Water Res 9:13–19. https://doi.org/10.4137/ASWR.S35227
Kamsonlian S, Suresh S, Ramanaiah V, Majumder CB, Chand S, Kumar A (2012) Biosorptive behaviour of mango leaf powder and rice husk for arsenic(III) from aqueous solutions. Int J Environ Sci Technol 9:565–578. https://doi.org/10.1007/s13762-012-0054-6

Kapur M, Mondal MK (2013) Mass transfer and related phenomena for Cr (VI) adsorption from aqueous solutions onto *Mangifera indica* sawdust. Chem Eng J 218:138–146. https://doi.org/10.1016/j.cej.2012.12.054

Kula I, Uğurlu M, Karaoğlu H, Celik A (2008) Adsorption of Cd(II) ions from aqueous solutions using activated carbon prepared from olive stone by $ZnCl_2$ activation. Bioresour Technol 99:492–501. https://doi.org/10.1016/j.biortech.2007.01.015

Kütahyalı C, Sert Ş, Çetinkaya B, Yalçıntaş E, Acar MB (2012) Biosorption of Ce(III) onto modified *Pinus brutia* leaf powder using central composite design. Wood Sci Technol 46:721–736. https://doi.org/10.1007/s00226-011-0437-8

Kyzas GZ, Kostoglou M (2014) Green adsorbents for wastewaters: a critical review. Materials 7:333–364. https://doi.org/10.3390/ma7010333

Li Y, Liu J, Yuan Q, Tang H, Yu F, Lv X (2016) A green adsorbent derived from banana peel for highly effective removal of heavy metal ions from water. RSC Adv 6:45041–45048. https://doi.org/10.1039/C6RA07460J

Liu Y (2009) Is the free energy change of adsorption correctly calculated? J Chem Eng Data 54:1981–1985. https://doi.org/10.1021/je800661q

Liu X, Lee DJ (2014) Thermodynamic parameters for adsorption equilibrium of heavy metals and dyes from wastewaters. Bioresour Technol 160:24–31. https://doi.org/10.1016/j.biortech.2013.12.053

Ma J, Yu F, Zhou L, Jin L, Yang M, Luan J, Tang Y, Fan H, Yuan Z, Chen J (2012) Enhanced adsorptive removal of methyl orange and methylene blue from aqueous solution by alkali-activated multiwalled carbon nanotubes. Appl Mater Int 4:5749–5760. https://doi.org/10.1021/am301053m

Meena AK, Kadirvelu K, Mishra GK, Rajagopal C, Nagar PN (2008) Adsorptive removal of heavy metals from aqueous solution by treated sawdust (*Acacia arabica*). J Hazard Mater 150:604–611. https://doi.org/10.1016/j.jhazmat.2007.05.030

Milonjic SK (2007) A consideration of the correct calculation of thermodynamic parameters of adsorption. J Serb Chem Soc 72:1363–1367. https://doi.org/10.2298/JSC0712363M

Mohammadi M, Hassani AJ, Mohamed AR, Najafpour GD (2010) Removal of rhodamine B from aqueous solution using palm shell-based activated carbon: adsorption and kinetic studies. J Chem Eng Data 55:5777–5785. https://doi.org/10.1021/je100730a

Mohan D, Rajput S, Singh VK, Steele PH, Pittman CU Jr (2011) Modeling and evaluation of chromium remediation from water using low cost bio-char, a green adsorbent. J Hazard Mater 188:319–333. https://doi.org/10.1016/j.jhazmat.2011.01.127

Munagapati VS, Yarramuthi V, Nadavala SK, Alla SR, Abburi K (2010) Biosorption of Cu(II), Cd (II) and Pb(II) by *Acacia leucocephala* bark powder: kinetics, equilibrium and thermodynamics. Chem Eng J 157:357–365. https://doi.org/10.1016/j.cej.2009.11.015

Nuhoglu Y, Malkoc E (2009) Thermodynamic and kinetic studies for environmentaly friendly Ni (II) biosorption using waste pomace of olive oil factory. Bioresour Technol 100:2375–2380. https://doi.org/10.1016/j.biortech.2008.11.016

Ouznadji ZB, Sahmoune MN, Mezenner NY (2016) Adsorptive removal of diazinon: kinetic and equilibrium study. Desalin Water Treat 57(4):1880–1889. https://doi.org/10.1080/19443994.2014.978386

Prado AGS, Moura AO, Holanda MS, Carvalho TO, Andrade RDA, Pescara IC, de Oliveira AHA, Okino EYA, Pastore TCM, Silva DJ, Zara LF (2010) Thermodynamic aspects of the Pb adsorption using Brazilian sawdust samples: removal of metal ions from battery industry wastewater. Chem Eng J 160:549–555. https://doi.org/10.1016/j.cej.2010.03.066

Rafatullah M, Sulaiman O, Hashim R, Ahmad A (2009) Adsorption of copper (II), chromium (III), nickel (II) and lead (II) ions from aqueous solutions by *Meranti* sawdust. J Hazard Mater 170:969–977. https://doi.org/10.1016/j.jhazmat.2009.05.06

Rafatullah M, Sulaiman O, Hashim R, Ahmad A (2012) Removal of cadmium (II) from aqueous solutions by adsorption using *Meranti* wood. Wood Sci Technol 46:221–241. https://doi.org/10.1007/s00226-010-0374-y

Ramaraju B, Reddy PMK, Subrahmanyam C (2014) Low cost adsorbents from agricultural waste for removal of dyes. Environ Prog Sustain Energy 33:38–46. https://doi.org/10.1002/ep.11742

Rao RAK, Rehman F (2010) Adsorption studies on fruits of Gular (*Ficus glomerata*): removal of Cr (VI) from synthetic wastewater. J Hazard Mater 181:405–412. https://doi.org/10.1016/j.jhazmat.2010.05.025

Sahmoune MN (2016) The role of biosorbents in the removal of arsenic from water. Chem Eng Technol 39:1617–1628. https://doi.org/10.1002/ceat.201500541

Sahmoune MN, Yeddou AR (2016) Potential of sawdust materials for the removal of dyes and heavy metals: examination of isotherms and kinetics. Desalin Water Treat 57:24019–24034. https://doi.org/10.1080/19443994.2015.1135824

Sahmoune MN, Louhab K, Boukhiar A, Addad J, Barr S (2009) Kinetic and equilibrium models for the biosorption of Cr (III) on *Streptomyces rimosus*. Toxicol Environ Chem 91(7):1291–1303. https://doi.org/10.1080/02772240802613731

Sahmoune MN, Louhab K, Boukhiar A (2011) Advanced biosorbents materials for removal of chromium from water and wastewaters. Environ Prog Sustain Energy 30:284–293. https://doi.org/10.1002/ep.10473

Salman M, Athar M, Farooq U (2015) Biosorption of heavy metals from aqueous solutions using indigenous and modified lignocellulosic materials. Rev Environ Sci Biotechnol 14:211–228. https://doi.org/10.1007/s11157-015-9362-x

Sarada B, Krishna Prasad M, Kishore Kumar K, Murthy CVR (2017) Biosorption of Cd^{+2} by green plant biomass, *Araucaria heterophylla*: characterization, kinetic, isotherm and thermodynamic studies. Appl Water Sci 7:3483–3496. https://doi.org/10.1007/s13201-017-0618-1

Senthil Kumar P (2014) Adsorption of Lead(II) ions from simulated wastewater using natural waste: a kinetic, thermodynamic and equilibrium study. Environ Prog Sustain Energy 33:55–64. https://doi.org/10.1002/ep.11750

Senthil Kumar P, Ramalingam S, Kirupha SD, Murugesan A, Vidhyadevi T, Sivanesan S (2011) Adsorption behavior of nickel (II) onto cashew nut shell: equilibrium, thermodynamics, kinetics, mechanism and process design. Chem Eng J 167:122–131. https://doi.org/10.1016/j.cej.2010.12.010

Srivastava S, Agrawal SB, Mondal MK (2015) A review on progress of heavy metal removal using adsorbents of microbial and plant origin. Environ Sci Pollut Res 22:15386–15415. https://doi.org/10.1007/s11356-015-5278-9

Sulaiman O, Ghani NS, Rafatullah M, Hashim R (2011) Removal of zinc (II) ions from aqueous solutions using surfactant modified bamboo sawdust. Sep Sci Technol 46:2275–2282. https://doi.org/10.1080/01496395.2011.594846

Surchi KMS (2011) Agricultural wastes as low cost adsorbents for Pb removal: kinetics, equilibrium and thermodynamics. Int J Chem 3:103–112. https://doi.org/10.5539/ijc.v3n3p103

Ucun H, Bayhan YK, Kaya Y (2008) Kinetic and thermodynamic studies of the biosorption of Cr (VI) by *Pinus sylvestris* Linn. J Hazard Mater 153:52–59. https://doi.org/10.1016/j.jhazmat.2007.08.018

Ugbe FA, Pam AA, Ikudayisi AV (2014) Thermodynamic properties of chromium (III) Ion adsorption by sweet orange (*Citrus sinensis*) peels. Am J Anal Chem 5:666–673. https://doi.org/10.4236/ajac.2014.510074

Van Vinh N, Zafar M, Behera SK, Park HS (2015) Arsenic(III) removal from aqueous solution by raw and zinc loaded pine cone biochar: equilibrium, kinetics, and thermodynamics studies. Int J Environ Sci Technol 12:1283–1294. https://doi.org/10.1007/s13762-014-0507-1

Vankar PS, Sarswat R, Sahu R (2012) Biosorption of zinc ions from aqueous solutions onto natural dye waste of *Hibiscus rosa sinensis*: thermodynamic and kinetic studies. Environ Prog Sustain Energy 31:89–99. https://doi.org/10.1002/ep.10535

Wang L, Zhang J, Zhao R, Li Y, Li C, Zhang C (2010) Adsorption of Pb(II) on activated carbon prepared from Polygonum orientale Linn.: kinetics, isotherms, pH, and ionic strength studies. Bioresour Technol 101:5808–5814. https://doi.org/10.1016/j.biortech.2010.02.099

Witek-Krowiak A (2013) Application of beech sawdust for removal of heavy metals from water: biosorption and desorption studies. Eur J Wood Prod 71:227–236. https://doi.org/10.1007/s00107-013-0673-8

Wu CH, Kuo CY, Guan SS (2016) Adsorption of heavy metals from aqueous solutions by waste coffee residues: kinetics, equilibrium, and thermodynamics. Desalin Water Treat 57:1–9. https://doi.org/10.1080/19443994.2014.1002009

Yagub MT, Sen TK, Afroze S, Ang HM (2014) Dye and its removal from aqueous solution by adsorption: a review. Adv Coll Int Sci 209:172–184. https://doi.org/10.1016/j.cis.2014.04.002

Yao Z-Y, Qi J-H, Wang L-H (2010) Equilibrium, kinetic and thermodynamic studies on the biosorption of Cu(II) onto chestnut shell. J Hazard Mater 174:137–143. https://doi.org/10.1016/j.jhazmat.2009.09.027

Yu L, Luo Y (2014) The adsorption mechanism of anionic and cationic dyes by Jerusalem artichoke stalk-based mesoporous activated carbon. J Environ Chem Eng 2:220–229. https://doi.org/10.1016/j.jece.2013.12.016

Chapter 7
Biosorption and Biodegradation of Polycyclic Aromatic Hydrocarbons (PAHs) by Microalgae

Bhawana Pathak, Shalini Gupta, and Reeta Verma

Contents

Abstract Polycyclic aromatic hydrocarbons (PAHs) are ubiquitous organic pollutants, primarily generated during the process of incomplete combustion, extraction, exploitation and transportation of fossil fuel. PAHs are environmental and human health hazards due to recalcitrance, toxicity, carcinogenic and mutagenic nature. Therefore, a sustainable cleanup approach is required for the removal of PAHs from

---

B. Pathak (✉) · S. Gupta · R. Verma
School of Environment and Sustainable Development, Central University of Gujarat, Gandhinagar, Gujarat, India

G. Crini, E. Lichtfouse (eds.), *Green Adsorbents for Pollutant Removal*, Environmental Chemistry for a Sustainable World 18,
https://doi.org/10.1007/978-3-319-92111-2_7

contaminated sites. Efficiency of biosorption process for the removal of toxic pollutants has been thoroughly studied in the past. This chapter focuses on the application of microalgae green biosorbents for the removal of PAHs. Characteristics, environmental fate of PAHs and algal biochemistry are summarized. Algae cell structural constituents act as specific binding sites for removal of pollutant, and share enzymatic systems similar to bacteria. Major enzymes responsible for biodegradation of PAHs are described. Immobilization and co-culture technique for enhance biosorption are discussed.

7.1 Introduction

Polycyclic aromatic hydrocarbon (PAH) are ubiquitously distributed in the different forms of biosphere such as atmosphere, hydrosphere, lithosphere biota; due to its wide usage at industrial level. PAH compounds are release into environment by various sources which can be broadly classified into anthropogenic e.g. oil refinery, paint, chemical manufacturing, and carpet industry. Refinery industries are among one of the major source for release of PAH compound. Forest fire, volcano eruption among the natural sources of PAH compounds. PAH has long been of global concern due to its recalcitrance attributed to fused benzene ring and delocalized pi bonds resulting in toxicity, and carcinogenic potential. PAHs are lipophilic in nature therefore; get easily distributed into the aquatic food chains resulting in bioaccumulation and biomagnification of PAHs (Ke et al. 2010). Different technologies have been explored during the last decades for the decontamination of persistent PAHs from environment. Physical and chemical treatment techniques are common but not environment friendly. To overcome the negative impact of physical and chemical methods on the environment, biological methods such as biosorption or biodegradation are more suitable. Biological methods are cost effective and environment friendly to decontaminate soils and effluents. Presently much focus is drawn on biosorption which is one of the bioremediation techniques.

Biosorption is a physico-chemical process and includes several mechanisms such as absorption, adsorption, ion exchange, surface complexation and precipitation. Biosorption is a property of both living or dead organisms and their components, has been signaled as a promising biotechnology for pollutant removal from solution (Phang et al. 2015). The technology is efficient, simple, analogous operation to conventional ion exchange technology, and availability of biomass (Fomina and Gadd 2014). Industry surveys of refinery biological treatment plants have suggested that sorbed PAH do not accumulate in the biosolids, which indicates that biodegradation of sorbed PAHs occurs (Stringfellow and Cohen 1999). The most extensively used micro-organisms in the field of bioremediation are bacteria, yeast and fungi. Different bacterial and fungal strains exhibit different PAH removal efficiencies, which could be enhanced by infusing different variable factors. In recent years micro-algae has gained attention for its efficiency to remove various toxic pollutants in wastewater treatment systems by autotrophic growth (Semple et al. 1999; Lei et al. 2002, 2007). Algae are important producers of energy in both aquatic and terrestrial ecosystems. Despite major advantages of algal growth and its potential in uptake and degradation of diversified pollutants present in water systems, only few reports are available regarding their involvement in hydrocarbon biodegradation (Kaoutar et al. 2014).

Biosorption of organic contaminant on algae is more beneficial compared to other microorganisms spill clean-up sorbents. Algae does not require organic carbon source for growth, these small cells grow phototropically and utilize carbon dioxide from the atmosphere or inorganic carbon present in water ($H_2CO_3^-$, $HCO_3^-$and CO_3^{2-}). Some algae, primarily blue-green algae (cyanobacteria) can also grow mixotrophically using both organic and inorganic carbon sources. Cyanobacteria are also known to grow effectively under low nutrient conditions. Thus, a reliable supply of algal biomass can be generated at low cost and there is much scope for utilization of algae as biosorbent due to their high surface area to volume ratio and the nature of cell wall (Mishra and Mukherji 2012). A better understanding of the relative roles of biosorption and biodegradation and the algal physiology of those processes, contributes to a better understanding of the fate of PAHs in biological treatment systems. This chapter highlights biodegradation and biosorption mechanism, role of enzymes and other advance techniques for enhancement of biosorption process and biodegradadtion of PAHs simultaneously.

7.2 PAH: Physico-chemical Properties

It has been known that PAH compounds are persistent in the environment due to their physicochemical properties. United States Environment Protection Agency has designated 16 PAH compound as priority pollutants. Characteristic features of PAH compounds include very low aqueous solubility and vapor pressure, high hydrophobicity (high log Pow), high adsorption coefficient (Table 7.1). High thermodynamic stability of the aromatic ring makes them recalcitrant hence, cannot be degraded easily under natural conditions (Cao et al. 2009; Haritash and Kaushik 2009; Moscoso et al. 2012) (Fig. 7.1).

7.3 Environmental Fate of PAHs

Hydrophobicity and high boiling points of PAH compounds cause equilibriums between solid, aqueous and vapour phases slow and leads to persistence. Due to high octanol-water partition coefficients and low vapor pressure, PAHs are rapidly absorbed by living organisms and particulate matter. PAHs may be present in the vapor and/or dissolved phase, micelle form, sorbed to colloidal organic matter or to particles, and incorporated into biota in water (Takáčová et al. 2014). In aqueous or

Table 7.1 Physicochemical properties

S. No.	PAH compound	Number of ring	Melting point (°C)	Boiling point (°C)	Vapour pressure Pa at 25 °C	Density (g/cm^3)	N octanol water partition coefficient (log kow)	Solubility in water at 25 °C (mg/L)	Henry law constant at 25 °C (kpa)	Ionization potential (ev)
1.	Napthalene	2	78–80	217	–	1.16	–	31.6	4.4×10^{-4}	8.12
2.	Acenapthylene	2	92–93	–	$8.9 \times 10–^{1}$	0.899	4.07	–	1.14×10^{-3}	8.22
3.	Acenapthene	2	95	279	2.9×10^{-1}	1.024	3.92	3.93×10^{-3}	1.48×10^{-4}	7.68
4.	Fluorene	2	115–116	295	8×10^{-2}	1.203	4.18	1.98×10^{-3}	1.01×10^{-2}	7.88
5.	Antracene	3	216.4	342	8×10^{-4}	1.283	4.5	73	7.3×10^{-2}	7.439
6.	Phenanthrene	3	100.5	340	1.6×10^{-2}	0.98	4.6	1.29×10^{-3}	3.98×10^{-3}	8.19
7.	Flouroanthene	3	108.8	375	1.2×10^{-3}	1.252	5.22	260	6.5×10^{-4}	7.9
8.	Pyrene	4	150.4	393	6×10^{-4}	1.271	5.18	135	1.1×10^{-3}	7.5
9.	Benzo (A) Anthracene	4	160.7	400	2.8×10^{-5}	1.226	5.61	14	–	7.54
10.	Chrysene	4	253.8	448	8.4×10^{-5}	1.274	5.91	2	–	7.8
11.	Benzo (B) Fluoranthene	4	168.3	481	6.7×10^{-5}	–	6.12	1.2	5.1×10^{-5}	Na
12.	Benzo(K) Fluoranthene	4	215.7	480	1.3×10^{-8}	–	6.84	0.76	4.4×10^{-5}	Na
13.	Benzo(A) Pyrene	5	178.1	496	7.3×10^{-7}	1.351	6.5	3.8	3.54×10^{-5}	7.23
14.	Benzo(Ghi) Perylene	6	278.3	545	1.4×10^{-8}	1.329	7.1	0.26	2.7×10^{-5}	Na
15.	Indeno(1,2,3-Cd)Pyrene	5	163.6	536	1.3×10^{-8}	–	6.58	62	2.9×10^{-5}	Na
16.	Dibenzo(A,H) Anthracene	5	266.6	524	1.3×10^{-8}	1.282	6.5	0.5	7×10^{-6}	7.57

Source: Agency for toxic substances and disease registry (ATSDR), *Na* not available

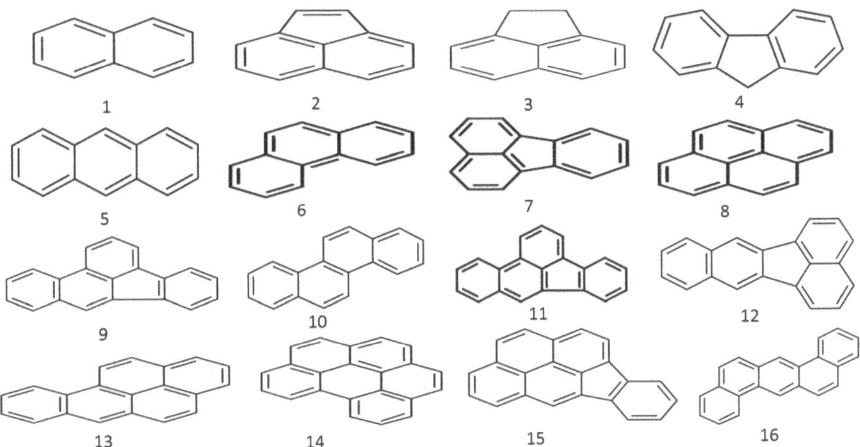

Fig. 7.1 Molecular structure of PAHs

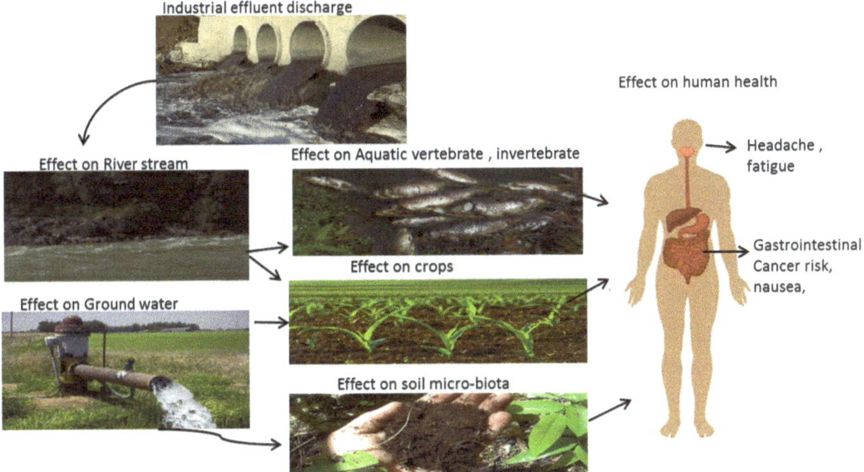

Fig. 7.2 Major routes of entry of PAHs in abiotic and biotic component of ecosystem

solid phase they get strictly bonded to organic matter hence, mobility in the soil-plant system is very slow. They are either trapped in the pores, fixed with covalent or hydrogen bonds, or bound during humification processes (Hatzinger and Alexander 1995). Migration of PAHs from the top soil is slow and seems to migrate bounded to particles (Jones et al. 1989; Pascale et al. 1997). The major ways of entry of PAHs in the water ways is by effluent discharge and in atmosphere by power plants, coal tar production industries, sludge as a raw industrial waste may contaminate the land (Fig. 7.2).

Table 7.2 Treatment technologies for PAH remediation

Containment	Thermal techniques	Physical techniques	Chemical techniques
Contaminated materials are enclosed in physical, chemical or hydraulic barriers which prevent the pollutants to migrate and contaminate groundwater	Incineration	Volatilization	Chemical oxidation
	Thermal desorption	Photolysis	Photocatalysis
		Adsorption	Coagulation
		Electro-remediation	Flocculation
		Electro-osmosis, filtration	

7.4 Treatment of PAH: Conventional Methods

The strategies of PAH remediation depend on the extent of contamination and the source of exposure. These methods are solely based on physical and chemical methods (Table 7.2) used at industrial level. These techniques do not provide any permanent solution for the removal of these hazardous compounds, also these current techniques are costly, require high mechanical inputs and are not environment friendly.

7.5 Biosorption

The technological breakthroughs must allow the integration of innovative biotechnology-based processes for recovery and/or removal of organic contaminants from natural sites or industrial waste treatment processes. Biotechnology covers all the aspects of the application of biooxidation, biosorption, bioreduction, bioaccumulation, bioprecipitation, bioflotation, bioflocculation, and biosensors (Macek and Mackova 2011). One important and widely studied alternative is biosorption, where certain types of biomass are able to bind and concentrate contaminants from even very dilute aqueous solutions. Biosorption is metabolism independent process which refers to the rapid chemical reaction between the surface and the adsorbate and ion exchange processes that occur at the cell surface (Ashkenazy et al. 1997). Microbial biomass provides a sink, by biosorption to cell walls, pigments and extracellular polysaccharides, intracellular accumulation, of organic compounds in and/or around cells, hyphae or other structures. The performance of a living biosorption can be affected by various factors like physiological state of the organism, age of the cells, availability of micronutrients during their growth and the environmental conditions such as pH, temperature, and the presence of certain co-ions are important parameters that may enhance or inhibit the phenomenon. Biosorbent can easily be produced using inexpensive growth media or

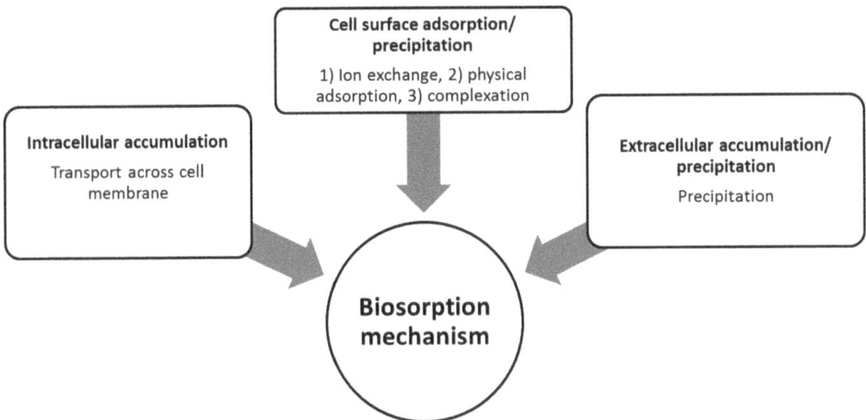

Fig. 7.3 Classification according to location of biosorption mechanism

obtained as a by-product from industry (Ahluwalia and Goyal 2007; Macek and Mackova 2011). Biosorption has many attractive features including removal of organic and inorganic contaminants over a broad range of pH and temperature, its rapid kinetics of adsorption and desorption and low capital and operation cost make this technology highly reliable over conventional techniques.

The mechanism of biosorption involves a solid phase sorbent or biosorbent; biological material and a liquid phase solvent, normally water containing a dissolved species to be sorbed sorbate, metal ions. It is a mass transfer process by which a substance is transferred from the liquid phase to the surface of a solid, and substance becomes bound by physical and/or chemical interactions. Due to higher affinity of the sorbent for the sorbate species, the latter is attracted and bounded by different methods. The process continues till equilibrium is established between the amount of pollutant adsorbed and its remaining concentration in the solution. The degree of sorbent affinity for the sorbate determines its distribution between the solid and liquid phases. Recent biosorption experiments have focused on the use of waste materials, which are by-products or the waste materials from large-scale industrial operations as biosorbent, for e.g. the waste mycelia available from fermentation processes, olive mill solid residues (Pagnanelli et al. 2002), activated sludge from sewage treatment plants (Hammaini et al. 2003), biosolids (Norton et al. 2004), macrophytes (Keskinkan et al. 2004).

Biosorption can be classified as on the basis of location in cells as presented in Fig. 7.3. Research on biosorption and biodegradation of organic pollutants has mainly concentrated on bacteria and fungi (Dean-Ross et al. 2002; Aksu 2005) with much less emphasis on algae. Nevertheless, the limited number of studies has demonstrated the capability of algae to accumulate and degrade wastewater-borne organic pollutants, such as phenol, dyes, PAHs i.e. napthalene, phenanthrene, benzo [α]pyrene and other aromatic compounds (Semple et al. 1999; Aksu and Akpinar 2001; Aksu 2005). Presently studies on microalgae, including *Chlamydomonas* sp.,

Chlorella miniata, *Scenedesmus platydiscus*, *Scenedesmus quadricauda*, *Selenastrum capricornutum* and *Synechocystis* sp. exhibit biosorption and biodegradation ability of individual and mixed PAH species (Chan et al. 2006; Lei et al. 2002, 2007). Among these algal species; *S. capricornutum*, a freshwater unicellular green alga, has received extensive attention for its ubiquitous occurrence, easy cultivation and, most importantly, for its high removal efficiency of high molecular weight (HMW) PAHs such as fluoranthene, pyrene and benzo α pyrene from aqueous solutions (Schoeny et al. 1988; Lei et al. 2002; Chan et al. 2006).

7.6 Cellular Structure and Biochemistry of Algae

Algae refer to a large and diverse group of organisms that contain chlorophyll and carry out oxygenic photosynthesis. It is important to note that algae are distinct from Cyanophyta. Class Cyanophyceae, the blue-green algae are also oxygenic phototrophs, but are eubacteria (true bacteria), and are therefore evolutionarily distinct from algae. The colonial forms of algae occur as aggregates of cells. Each cell shares common functions and properties, including the storage products they utilize as well as the structural properties of cell walls. Accordingly, algae divisions include Cyanophyta, Prochlorophyta, Phaeophyta, Chlorophyta, Charophyta, Euglenophyta, Chrysophyta, Pyrrhophyta, Cryptophyta and Rhodophyta. When comparing Phaeophyta (brown algae) to other common algal divisions such as the Chlorophyta (green algae), important differences are seen in the storage products utilized as well as in the cell wall chemistry. Biosorption in algae has mainly been attributed to the cell wall properties where both electrostatic attraction and complexation can play a role. Cryptophyta, for example, does not have a cell wall. Pyrrhophyta (dinoflagellates) can be "naked" or protected by cellulosic "thecal" plates (Bold and Wynne 1985; Lee 1989; Davis et al. 2003). The Chrysophyceae of the division Chrysophyta can be either "naked" or have scales, cellulosic walls or a cell envelope (Lee 1989). However, from biosorption point of view, two orders are of great importance, namely Laminariales and Fucales. Both orders are abundant in nature and include most structurally complex algae. Laminariales are collectively referred to as "kelps". The carboxylic groups are generally the most abundant acidic functional group in the brown algae. It also constitutes the highest percentage of titratable sites (typically greater than 70%) in dried brown algal biomass.

The bio-adsorption capacity of the algae is directly related to the presence of these sites on the alginate polymer, which itself comprises a significant component, up to 40% of the dry weight of the dried seaweed biomass (Percival and McDowell 1968; Davis et al. 2003). The second most abundant acidic functional group in brown algae is the sulfonic acid of fucoidan. Sulfonic acid groups typically play a secondary role, except when metal binding takes place at low pH. Hydroxyl groups are also present in all polysaccharides but they are less abundant. Thereby green algae consist of mainly cellulose and high percentage of the cell wall is protein (10–70%) bonded to polysaccharides forming glycoproteins (Romera et al. 2007). Proteins can contribute significantly to metal binding, offering the functional groups of amino acids (hydroxyl, carboxyl, sulfhydryl, amine, amide, imidazole (Romera et al. 2007; Dana Ivánová et al. 2012). From the present literature it can be concluded that the biochemistry in context with biosorption of hydrocarbons has been less explored. Hence it is a thrust area of research.

7.7 Biosorption and Biodegradation of PAHs by Algae

Organic molecules that are not biodegradable can still be removed from the wastewater by the microbial biomass via the process of biosorption. Biosorption is a promising alternative to replace or supplement the present removal processes of organic pollutants from wastewaters (Aksu 2005). Biosorption is an efficient technology for the removal of several micro-pollutants such as metals, phenols, and dyes from wastewater. For its implementation in a wastewater treatment plant, microbial cells such as bacteria, yeast, fungi, algae with biosorption ability can be separated from the water e.g., by sedimentation; after the biological process. Immobilization of the separated microbial cells can also be carried out during the treatment (Michalak et al. 2013; Sanches et al. 2017). Biosorption is less expensive than biodegradation and can be carried out with dead biomass (Sanches et al. 2017). Biodegradation is the most commonly reported biological mechanism for PAH removal. Biosorption by algal community may be extremely efficient for their removal (Sanches et al. 2017). List of algae species used for the degradation of PAHs has been given in Table 7.3. Algae are widely used as biosorbent for the biosorption of heavy metals; however biosorption of oil on algae has not been explored. Very limited research has been conducted for biosorption of organic pollutants on algae. Microalgae are one of the major primary producers in aquatic ecosystems, and play vital roles in the fate of PAHs in those environments. The mechanism involved in the removal of PAHs by microalgae is similar to that of heavy metals and other organic contaminants (Hong et al. 2008). The biosorption process involves two stages adsorption and absorption. Adsorption stage involves the adherence of contaminant species at the cell surface, attributed to physico-chemical forces independent of metabolism. The later stage involves slow active absorption, accumulation and degradation; this stage is totally species specific.

Biosorption of diesel and lubricating oil has been studied by Koelmans et al. (1995); Lei et al. (2007) using dead biomass of *Spirulina* sp. and *Scenedesmus abundans*. The rate and extent of sorption was studied in well mixed batch systems containing oil 0.1–2%, v/v and biomass 0.1% suspended in water. Sorption of diesel on *Spirulina* sp. was instantaneous. The Freundlich and Langmuir model provide adequate fit for diesel sorption on algae but not for lubricating oil. A three parameter model, the Sips model, provided good fit for all the experimentally generated isotherms and yielded maximum biosorption capacity of diesel and lubricating oil in the range of 12–14 g/g in 12 days. PAHs are widely recognized hydrophobic organic compounds which tend to bioconcentrate on algae, fish and other aquatic

Table 7.3 Algae species for PAH compound degradation

Algae sp.	PAH compound	References
Chlamydomonas reinhardtii	9H-Fluorene-9-one, Phenanthrene, 4-Methylbenzo[c]cinnoline, 1-Methylanthracene, 2-Methylanthracene, 1,2-dimethyl phenanthrene, 3,4-dimethyl phenanthrene	Semple et al. (1999)
Selenastrum capricornutum	Benzene,toluene,naphthalene, phenanthrene, pyrene, chlorobenzene, 1,2-dichlorobenzene, nitrobenzene, 2,6-dinitrotoluene, di-n-butylphthalate	Chekroun et al. (2014) and Semple et al. (1999)
Scenedesmus obliques	Naphthalene sulphonic acid	Cerniglia (1992)
Ochromonas danica	Phenols	Semple and Cain (1996)
Synechococcus PCC 7002	Phenol	Wurster et al. (2003) and Ghasemi et al. (2011)
Ochromonas danica	Phenol	Semple et al. (1999) and Ghasemi et al. (2011)
Phormidium valderianum	Phenol	Shashirekha et al. (1997) and Ghasemi et al. (2011)
Ankistrodesmus braunii & *S. quadricauda*	Low molecular phenol, catechol	Pinto et al. (2002) and Pinto et al. (2003)
Chlamydomonas sp., *Chlorella miniata*, *Chorella vulgaris*, *Scenedesmus platydiscus*, *S. quadricauda*, *S. capricornutum*, *Synechosystis* sp.	Pyrene, Fuoranthene	Lei et al. (2002, 2007)
Consortium of *Chlorella sorokiniana* and *Pseudomonas migulae*	Phenanthrene	Muñoz et al. (2003) and Chekroun et al. (2014)
Monoraphidium braunii	Bisphenol	Gattullo et al. (2012)
Agmenellum quadruplicatum	Bisphenol	Cerniglia et al. (1979) and Chekroun et al. (2014)
Anabaena cylindrical, *Chlamydomonas ulvaensis*, *Chlorella pyrenoidosa*, *Euglena gracilis*, *Phormidium foveolarum*, *Scenedesmus basiliensis*	Phenol, catechol	Ellis (1977)

organisms. Some researchers have reported biosorption of organic micropollutants, such as, chlorobenzenes on *Anabaena* and *Scenedesmus* and biosorption of low concentration of fluoranthene and pyrene on various microalgae. Several strains of microalgae are known to metabolize/transform naphthalene, phenanthrene, anthracene, benzo α pyrene and other PAHs (Cerniglia et al. 1979, 1980a, b; Narro et al. 1992a, b; Safonova et al. 2005; Chan et al. 2006; El-Sheekh et al. 2012). Mishra and Mukherji (2012) found that biosorption rate of lubricating oil is significantly lower than that of diesel due to, the additives present in lubricating oil. Algae can be used for development of low cost sorbents for removal of oil and can affect the fate and transport of spilled oil (Imran et al. 2012).

Efficiency of four microalgal species, namely, *Chlorella vulgaris*, *Scenedesmus platydiscus*, *Scenedesmus quadricauda*, and *Selenastrum capricornutum* to remove fluoranthene 1.0 mg/l, pyrene 1.0 mg/l, and a mixture of fluoranthene and pyrene; each at a concentration of 0.5 mg/l was evaluated. The study revealed that removal was algal species specific and toxicant-dependent. *Selenastrum capricornutum* was the most effective species while *C. vulgaris* was the least efficient species in removing and transforming polycyclic aromatic hydrocarbons (PAHs) with removal percentage of 78 and 48 respectively in 7 days. All species, except *S. platydiscus* exhibited higher *fluoranthene* removal efficiency than pyrene. The removal efficiency of fluoranthene and pyrene in a mixture was comparable, or higher than the respective single compound, suggesting that the presence of one PAH stimulated the removal of the other PAH. Most research work was on the uptake and/or metabolism of a single PAH contaminant, and very few reported the situation with two or more PAHs. In nature, it is common to have a mixture of PAH contamination, interacting with each other (Canet et al. 2001; Tang et al. 2005). However, the possible stimulatory, antagonistic, and competitive uptake and metabolism of the combined PAHs received little exploration (Dean-Ross et al. 2002; Herwijnen et al. 2003; Lei et al. 2007). Likewise *Oscillatoria* sp., strain JCM examined under photoautotrophic growth conditions has been reported to oxidize naphthalene to 1-naphthol (Fig. 7.4). Marine Cyanobacterium *Agmenellum quadruplicatum* strain PR-6 can convert phenanthrene to phenanthrene trans-9,10-dihydrodiol and 1-methoxyphenanthrene (Figs. 7.5 and 7.6) (Cerniglia et al. 1980a; Narro et al. 1992a, b). Oxidation of Benzo α pyrene by microalgae *Selanastum capricornutum* has been evaluated. Transformation of Benzo α pyrene resulted in the formation of cis-4,5-, 7,8-, 9,10- and 11,12-Benzo α pyrene-dihydrodiols. The transformation involves a dioxygenase system, similar to bacterial PAH degradation systems but unlike those of eukaryotic organisms (like fungi) which involve monooxygenase systems (Warshawsky et al. 1988; Ghosal et al. 2016). The effects of gold, white or UV-A fluorescent lights for the biotransformation of Benzo α pyrene and phototoxicity of carcinogenic PAHs in different algal systems was determined. It has been observed that algae for e.g. *S. capricornutum*, *Scenedesmus acutus* and *Ankistrodesmus braunii* are able to degrade Benzo α pyrene to dihydrodiols. Also, the degradation varies with the kind and intensity of light sources (Warshawsky et al. 1995). Microalgal strain *Prototheca zopfii* immobilized in polyurethane foam has been reported to accumulate a mixture of PAHs in the matrix, when incubated with mixture of aliphatic hydrocarbons and

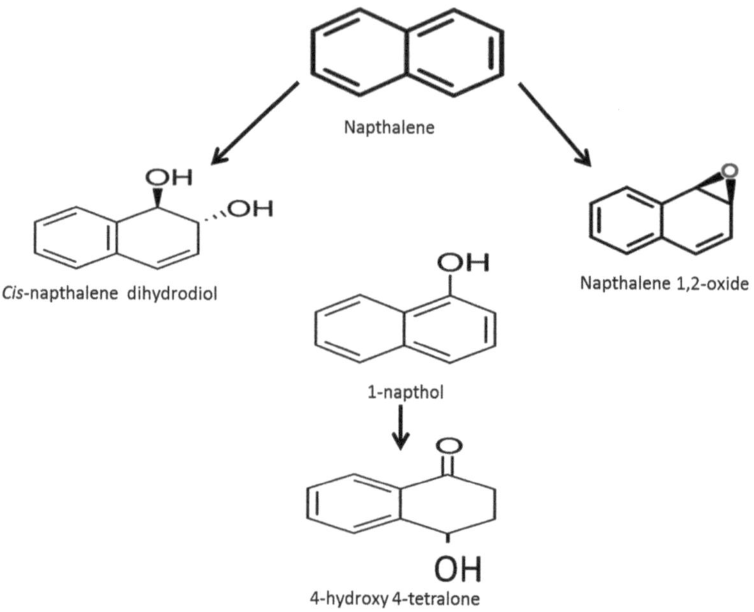

Fig. 7.4 Metabolism of naphthalene by the cyanobacterium *Oscillatoria* sp., strain JCM. (Cerniglia et al. 1980a, b, modified)

PAHs as substrates. Though the immobilized organism can degrade n-alkanes but PAH degradation rate was very less. However, PAHs accumulation did not impair the degradation of PAHs. In case of free-living cells, the organisms can reduce the concentration of both PAHs and n-alkanes satisfactorily (Ueno et al. 2006, 2008; Ghosal et al. 2016). The accumulation and biodegradation of phenanthrene and fluoranthene by the two diatoms *Skeletonema costatum* and *Nitzschia* sp., enriched from a mangrove aquatic ecosystem was studied by Hong et al. (2008). It was observed that the strains were capable of accumulating and degrading phenanthrene and fluoranthene simultaneously. PAHs accumulation and degradation capability of *Nitzschia* sp. were higher than those of *S. costatum*. Further, it had been observed that the degradation of fluoranthene by the two diatoms was slower, compared to phenanthrene. The strains also showed similar or higher efficiency in the removal of the phenanthrene–fluoranthene mixture than phenanthrene or fluoranthene alone (Hong et al. 2008). The efficiency of seven microalgal species to remove pyrene from solution was reported by Lei et al. (2002). In a recent study removal of benzo α pyrene by biosorption and degradation was determined by two microalgal species *Selenastrum capricornutum* and *Scenedesmus acutus*. It has been seen that *S. capricornutum* can remove 99% of BaP after 15 h of exposure, whereas *S. acutus* can remove 95% after 72 h of exposure.

The effects of metals on biosorption and biodegradation of fluorene, phenanthrene, fluoranthrene, pyrene and benzo α pyrene by *Selenastrum capricornutum* were investigated with the appearance of three metabolites 4,5 dihydrodiol-BaP;

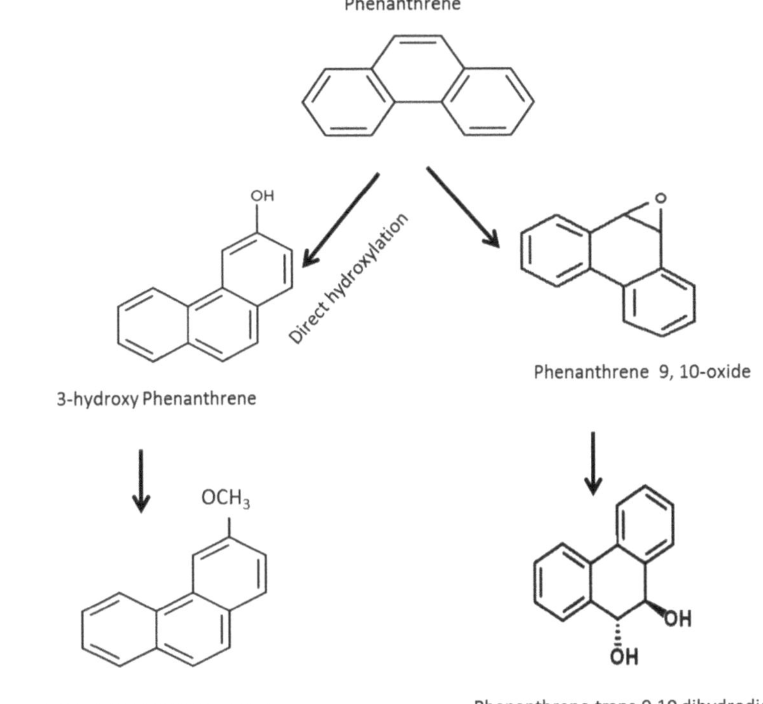

Fig. 7.5 Postulated pathways for phenanthrene metabolism by *Agmenellum quadruplicatum* PR-6. (Cerniglia et al. 1979, modified)

7,8-dihydrodiol-BaP; and 9,10 dihydrodiolBaP at short time periods from 0.25 to 72 h (Ke et al. 2010; Garcia de Llasera et al. 2016; Ghosal et al. 2016). Metal dosage and exposure time yielded a significant effect on the ability of removal of low molecular weight PAHs like fluorine and phenanthrene by the alga. For high molecular weight PAHs like fluoranthrene, pyrene and benzo α pyrene, the removal efficiency was not affected by the presence of metals (Patel et al. 2016; Ghosal et al. 2016). Takáčová et al. (2014) reported the biodegradation of benzo α pyrene by microalgae *Chlorella kessleri*. The removal efficiency of PAHs, and the effectiveness of live and dead cells, was found to be predominantly PAH dependent (Ke et al. 2010; Luo et al. 2014a, b; Ghosal et al. 2016). Furthermore, the significant impact of biosorption in bioremoval makes this approach even more attractive both in terms of technical and economic aspects.

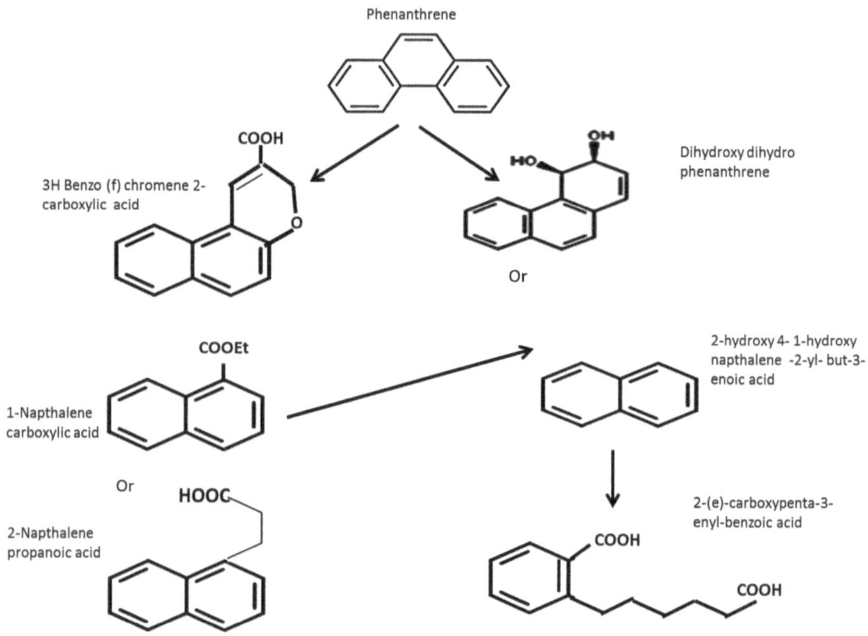

Fig. 7.6 Phenanthrene degradation by *Scedesmus obliquous* ES 55. (Safonova et al. 2005, modified)

7.8 Role of Enzyme During Biosorption and Biodegradation of PAH by Algae

Enzymes are the most important aspect of any contaminant degradation pathway and are microalgae species specific. Microalgae inhabiting at the real contaminated sites are capable of degrading organic pollutants. *Chlorella* ssp. *MM3*, a soil isolate from a former cattle dip site, was assessed in degrading pyrene both in aqueous medium and soil slurry. Study showed that differential protein expression in pyrene-grown cells of the microalga. Distinct accumulation of dihydrolipoamide acetyltransferase or dihydrolipoyl transacetylase, components of pyruvate dehydrogenase complex was observed. This indicated a possible role of this enzyme in microalgal degradation of pyrene. Strain MM3 can grow on pyrene in culture medium at concentrations as high as 250 μM. When grown in presence of 50 μM pyrene, the cell density increased from 1.1×10^{-5} cells mL/L to 16.45×10^{5} cells mL/L within 7 days (Subashchandrabose et al. 2017).

Similarly, a fresh water green alga, *Micractinium pusillum,* under nitrogen starvation conditions accumulated 100-fold of dihydrolipoamide dehydrogenase. Dihydrolipoamide dehydrogenase is also a component of the same pyruvate dehydrogenase complex that regenerates the lipoamide during pyruvate decarboxylation (Li et al. 2012; Subashchandrabose et al. 2017). Apparently, as a consequence of pyrene induced stress in algal cells, accumulation of dihyrolipoyl transscetylase clearly implicates role of the enzyme in biodegradation of pyrene by *Chlorella* sp. MM3. Under photoautotrophic growth conditions, *Agmenellum quadruplicatum* PR-6 used in toxicity study through algal lawn bioassay, indicated that 9-phenanthrol and 9,10-phenanthrenequinone inhibit the growth of algae. *Agmenellum quadruplicatum* must possess enzymes that catalyze the enzymatic hydration of phenanthrene 9,10-oxide, since it forms optically active phenanthrene trans-9,10-dihydrodiol as a metabolite of phenanthrene. The formation of an optically active trans-9,10-dihydrodiol in which only one atom of oxygen originates from O_2 strongly indicates that *A. quadruplicatum PR-6* oxidizes phenanthrene to phenanthrene trans-9,10-dihydrodiol via a monooxygenase-epoxide hydrolase-catalyzed reaction (Lu and Miwa 1980; Sato and Omura 1978; Narro et al. 1992a, b).

Glutathione S-transferase played an important role in biotransformation of pyrene by microalgal species and their activities varied from species to species (Lei et al. 2003). *S. platydiscus* consist higher efficiency in the removal of fluoranthene than pyrene; indicating pyrene as more recalcitrant PAH compound. The microalga also showed comparable or higher efficiency in the removal of fluoranthene and pyrene in a mixture than the respective single PAH, suggesting a stimulatory effect between fluoranthene and pyrene (Lei et al. 2007). The use of a cytochrome P-450 monooxygenase system in this alga is known as a mechanism of detoxification (Torres et al. 2008). High levels of monohydroxylated fluoranthene and 9-fluorenone, found in both living and heat-killed cells, indicated that these fluoranthene intermediates could be products of biodegradation or photodegradation or both. A small amount of fluoranthene/pyrene dihydrodiols was also identified in living cells. As the dihydroxylated PAHs could be either trans-dihydrodiols through the monooxygenase pathway or cis-dihydrodiols through the dioxygenase pathway (Heitkamp and Cerniglia 1989; Ke et al. 2010).

Warshawsky et al. (1988, 1995); Schoeny et al. (1988), examined the effects of the chlorophyte alga, *Selenastrum capricornutum*, on benzo α pyrene and observed dioxygenase system to oxidize the PAH compound to cis-dihydrodiols. Cis-dihydrodiols were subsequently, converted to sulfate ester and K and L-glucoside conjugates. The presence of the ring hydroxylating dioxygenase system is of particular importance as this mechanism is typically found only in bacteria. Trans dihydrodiols typically originate from epoxidation by the action of cytochrome P-450 monooxygenases and epoxide hydrolases on the PAH molecule (Ghosal et al. 2016).

Similarly Kirso and Irha (1998) studied the bioconcentration and transformation of a priority PAH, benzo α pyrene, by brown *Fucus vesiculosus* and *Chorda filum* (brown), *Furcellaria lumbricalis* (red), Enteromorpha *intestinalis*, *Cladophora glomerata* and *Chara aspera* (green). The results indicated that of all the benzo α pyrene consumed, 89–99% was found in the biomass of Fucus, an insignificant part was in the solution, and the remainder (up to 4%) was not recovered, i.e., was considered to have been metabolized. For green and chara algae the proportion of transformed PAHs was more essential, 42–49%. The transformation of benzo α pyrene in marine and freshwater algae is species specific and depends on the presence and activity of enzymes localized in the plant cells. The most important enzyme systems for detoxification of benzo α pyrene are o-diphenol oxidase, cytochrome P450, and peroxidase. The data obtained indicate the important role of marine and freshwater algae in the fate of carcinogenic PAHs in the environment (Table 7.4).

7.9 Advance Treatment Technologies in Biosorption

7.9.1 Immobilization

Information about advances in immobilization techniques and biotechnology with freshwater and marine microalgae is scattered. Current uses of immobilized microalgae include metabolite production, culture collection handling, energy procurement and removal of undesired or valuable substances from media (nutrients, metals and different pollutant agents). The use of immobilized algal cells in water purification processes has been reported long ago (Robinson et al. 1988). Microalgae form part of the organisms fixed in percolating filters of wastewater treatment plants. Immobilization can be a solution to major problems of employing microorganisms, including microalgae, to remove pollutants from wastewater is the difficulty in separating them after treatment. Immobilization techniques by embedding microbial cells into a matrix to make big particles could solve the harvesting problem. Growth of *C. vulgaris* immobilized in alginate was not affected and its efficiency to remove nitrogen and phosphorus was even higher than the free cell counterpart (Lau et al. 1997).

Knowledge on treating PAH-contaminated wastewater using immobilized microalgae is very limited. Use of blank alginate beads and immobilized dead algal cells for the removal of naphthalene from aqueous solutions was investigated by Ashour et al. (2008). The effects of contact time, solution pH, and naphthalene concentration on the biosorption of naphthalene on blank alginate beads or immobilized dead algal cells were studied. The study suggested that removal of naphthalene on both sorbents was pH dependent and significant removal of naphthalene was obtained at pH 4. Dynamic biosorption experiments revealed that the biosorption of naphthalene on either sorbent was rapid where the equilibrium uptake occurred within 10 min. Biosorption of naphthalene on either sorbent followed the pseudo-second order kinetics. Removal and degradation of a mixture of phenanthrene, fluoranthene and pyrene by alginate-immobilized algal beads by *Selenastrum capricornutum* was studied. The free cells of *S. capricornutum* were also found to have a higher ability to degrade PAHs than *Chlorella* (Lei et al. 2007). Another possible mean to enhance the removal and degradation of contaminants by immobilized microorganisms is to increase bead density (number of beads used). The removal of ammonium by alginate-immobilized *C. vulgaris* at 8 and 12 beads mL/L were significantly higher than that at other densities, 4, 16 and 20 beads mL/L, while the phosphate removal was not affected by bead density (Tam and Wong

Table 7.4 Enzymes released by algae for degradation of PAH compounds

PAH compound	Concentration	Incubation time	Algal strain	Possible enzyme	References
Pyrene	50 μM	21 days	*Chlorella* ssp. *MM3*	Dihydrolipoyl transacetylase	Subashchandrabose et al. (2017)
Pyrene	NA	NA	*Micractinium pusillum*	Dihydrolipoamide dehydrogenase	Li et al. (2012)
Phenanthrene	NA	NA	*Agmenellum quadruplicatum*	Monooxygenase-epoxide hydrolase	Lu and Miwa (1980), Sato and Omura (1978) and Narro et al. (1992b)
Naphthalene	4.8 mol	24 h	*Oscillatoria* sp., *strain JCM*	NA	Cerniglia et al. (1980b)
Pyrene	NA	NA	*S. capricornutum*	Glutathione S-transferase (GST)	Lei et al. (2003)
Bap	99%	15 h of exposure	*S. capricornutum*, *S. acutus*	Dioxygenase, are o-diphenol oxidase, cytochrome P450, and peroxidase	Ke et al. (2010) and Ghosal et al. (2016)

NA not available

2000). Dense microalgal beads would limit the light penetration and enhance the self-shading problem among beads which in turn reduced cell growth and metabolic activities (Garbayo et al. 2002). For the industrial application of biosorption, immobilization of biosorbent is necessary for solid/liquid separation. Immobilized biomass beads can then be packed in biosorption column, which is perhaps the most effective device for continuous operations (Aksu 2005). Lower growth rates of immobilized algal cells, probably due to restrictions in the diffusivity of nutrients to the immobilized cells, as well as protection of cells entrapped in immobilizing matrixes can also be exploited in order to facilitate handling of culture collections (Lukavsky 1988). One of the most recent reviews focused on the use of immobilized algae for wastewater, nitrogen, phosphorus and metal removal. Immobilization techniques can be primarily divided into two groups: "passive" and "active" immobilization (Mallick 2002).

7.9.2 Passive Immobilization

Many microorganisms, including some groups of microalgae have a natural tendency to attach to surfaces and grow on them (Robinson et al. 1986). This characteristic can be exploited in order to immobilize cells on carriers of different types (Codd 1987). Normally, those processes are easily reversible and contamination of effluents with unstuck cells is unavoidable. Adsorbent materials (carriers) for passive immobilization can be natural or synthetic. With respect to natural carriers, recently efforts have been made involving loofa biomass (Akhtar et al. 2004). In case of algae, *Scenedesmus obliquus* cells were immobilized in polyvinyl and polyurethane for the removal of nitrate from water. Survival of adsorbed cells was compared with entrapped cells i.e. cells immobilized by "active" immobilization, by mixing concentrated cells with one of the pre-polymers. Cellular growth is higher for adsorbed cells than that measured for entrapped cells; possibly due to the toxicity of the pre-polymers, although these authors reported that no toxic effects were found due to the residual presence of pre-polymer reagents (Urrutia et al. 1995). Yamaguchi et al. (1999) achieved a noticeable degradation of hydrocarbons by the colorless hydrophobic microalgae *Prototheca zopfii,* adsorbed to 8 mm-cubes of polyurethane foam in a bubble reactor.

7.9.3 Active Immobilization

Active immobilization techniques, includes the use of flocculants agents, chemical attachment and gel entrapment.

(a) Flocculant agents: Primarily used in order to avoid tedious and expensive centrifugation when algae are intended to be removed from a liquid medium.

Among the commonly used flocculants, chitosan has been the most widely employed. Chitosan is a linear amino polysaccharide of b-D-glucosamine (2-amino-2-deoxy-b-D-glucan) units joined by 1–4 linkages (Oungbho and Muller 1997). The inconvenience of chitosan in immobilizing techniques is its weak stability. Kaya and Picard (1996) tried to solve this problem using high viscosity chitosan and konjac flour (glucomannan, obtained from tuberous root of the konjac tree –*Amorphophalus konjac* in order to enhance the stability of floccules immobilizing viable cells of *Scenedesmus bicelularis* to use them in tertiary treatment of wastewaters, but concluded that konjac flour did not significantly modify the rheological properties of mixed chitosan solutions.

(b) Chemical attachment: Presents some great disadvantages when living cells are intended to be immobilized, because the chemical interaction, mainly due to covalent bonding, cross-linking – involving glutaraldehyde, for instance – or photo cross linkable resins cause damages in cellular surface and drastically reduces viability of cells (Codd 1987).

(c) Gel entrapment: This method is the most widely used technique for algal immobilization. Gel entrapment can be performed by the use of synthetic polymers such as acrylamide, photo cross linkable resins, polyurethanes, proteins such as gelatine, collagen or egg white or natural polysaccharides like agars, carrageenans or alginates (Codd 1987). Blanco et al. (1999) described the use of polysulphone, a thermoplastic material and epoxy resin entrapping cells of the cyanobacteria *Phormidium laminosum,* in order to check the capacity of biosorption and desorption, by the use of 0.1 M HCl of Cu(II), Fe(II), Ni (II) and Zn(II). Epoxy resins consist of two components that react with each other forming a hard, inert material. One is a bisphenol and the other is epichlorohydrin. A study by Chaillan et al. (2006) described the appearance of cyanobacterial mats (*Phormidium animale*) in petroleum-polluted site and observed no biodegradation of crude oil by cyanophytes, but from other organisms present in the mat formed by Phormidium. Hence, concluded that biofilm-forming bacteria covering macroalgae can also degrade oil (Radwan et al. 2002). *C. sorokinianain* aggregation with bacteria was able to successfully remove salicylate; intermediate compound of PAH from a photobioreactor in an experiment described by Muñoz et al. (2003). Hydrocarbons, nevertheless, have been reported to be degraded by Ca-alginate immobilized colorless *P. zopfii* (Suzuki et al. 1998; Yamaguchi et al. 1999).

7.10 Algal and Bacterial Microcosm

Effective synergism and excellent PAH degradation can be achieved with microalgal-bacterial consortium. Yet its success is highly dependent on the consortium assembly. Therefore, proper selection of the microbes for biodegradation is paramount to the success and efficiency of the degradation process. The bacterial assemblages are known to influence the development or decline of algal blooms

(Fukami et al. 1997). The molecular oxygen from algal photosynthesis is used as an electron acceptor by bacteria to degrade organic matter while the carbon dioxide from the bacterial mineralization completes the photosynthetic cycle. The symbiotic interactions of microalgae and bacteria form the basis of the biological oxygen demand removal in the wastewater treatment ponds, first reported by Oswald et al. (1953). Furthermore, microalgae can help in co-metabolic degradation or produce biosurfactants and extracellular matters to enhance bacterial activity for increasing pollutant bioavailability (Muñoz et al. 2003; Gods'gift and Fagade 2016).

Gods'gift and Fagade (2016) studied three species of bacteria and two species of microalgae isolated, characterized, screened and emerged as best hydrocarbon utilizers. The consortia comprised of *Chlorella minutissimma* and *Aphanocapsa* sp. as microalgae inoculants, while *Citrobacter* sp. SB9, *Pseudomonas aeruginosa SA3* and *Bacillus subtilis* SA7 as bacterial inoculants. Growth dynamics, pH and degradation of the effluents'PAH content were analyzed for a period of 10 days. Consortium BCC (bacteria inoculants and *Chlorella minutissimma*), BCCA (all inoculants) and BCA (bacteria inoculants and *Aphanocapsa* sp.) had PAH degradation percentage of 92.09%, 67.76% and 47.19% respectively. Only BCA had an acidic pH after 10 days. Salicylic acid (sodium salt), phenol and phenanthrene were tested as model contaminants for representing a wide range of aqueous solubility and toxicity. The fate of these products was studied in algae–bacteria combinations. Influence of microalgae on the degradation of environmental pollutants by bacterial strains was also studied. Berthe-Corti et al. (1998) demonstrated the use of algae in combination with bacteria for the treatment of volatile organic compounds. However, the algae were grown in a separate reactor to provide oxygen for biological oxygen demand removal in the main reactor. The potential of algal–bacterial microcosms was also studied for the biodegradation of salicylate, phenol and phenanthrene. The isolation and characterization of aerobic bacterial strains capable of mineralizing each pollutant were first conducted. *Ralstonia basilensis* was isolated for salicylate degradation, *Acinetobacter haemolyticus* for phenol and *Pseudomonas migulae* and *Sphingomon* for phenanthrene. The green alga *Chlorella sorokiniana* was then cultivated in the presence of the pollutants at different concentrations, showing increasing inhibitory effects in the following order: salicylate < phenol < phenanthrene. The synergistic relationships in the algal–bacterial microcosms were clearly demonstrated, since for the three contaminants tested a substantial removal of > 85% was recorded only in the systems inoculated with both algae and bacteria and incubated under continuous lighting. This study presents enhanced biodegradation of toxic aromatic pollutants by algal–bacterial microcosms in a one-stage treatment (Borde et al. 2003). The influence of the initial composition of an algal-bacterial microcosm constituted of *Chlorella sorokiniana* and *Ralstonia basilensis* was tested for the fed-batch degradation of salicylate at 5 mM. Salicylate degradation was always limited by the oxygen generation rate, which was initially proportional to the algal density, but rapidly became limited by the availability of light once the algae started to grow. The decrease of the salicylate removal rate observed at high algal densities was likely caused by mutual shading within the algal population and the increase of oxygen consumption due to algal dark respiration.

With repeated salicylate amendments, all systems converged towards the same characteristics, reaching an optimum rate of salicylate degradation at 1 mmoL/L/day.

Future work should focus on more toxic and/or recalcitrant compounds e.g. polycyclic aromatic hydrocarbons, since in such cases, algal inhibition, or degradation rate may be limited by bacterial growth rather than the photosynthetic activity (Millemann et al. 1984; Guieysse et al. 2002). Muñoz and Guieysse (2006) reviewed the interactions between algae and bacteria in processes designed for the treatment of hazardous contaminants. Production of oxygen by algae improves degradation of substances that must be degraded aerobically. Both bacteria and algae could produce defending substances against the other co-immobilized organism. When co-immobilized with algae, increase of pH and oxygen due to photosynthesis in the media can slow down bacterial growth. On the other hand, consumption of CO_2 and extracellular matter production such as exopolysaccharydes by algae can enhance bacterial growth rate. Conversely, CO_2 and growth promoter substances production by bacteria can enhance microalgal growth. Combinations of solar energy and algal immobilization technologies can be successfully used in industrial processes (Mallick 2002; Garrido 2008).

7.11 Factors Affecting Rate of Biosorption

7.11.1 *Surfactant*

Surfactants play a vital role in the solubilisation of PAHs in soil and water systems (Cheng and Wong 2006). Surfactants were successfully used for the remediation of sites contaminated with several organic chemicals (Mulligan et al. 2001). The impact of surfactants on phycoremediation of PAHs-contaminated soil in a slurry-phase has not been fully explored. The surfactant-mediated phycoremediation of organic contaminants in soil slurry was studied by Subashchandrabose et al. 2017. Microalgae were used for screening the toxicity of surfactants (Lewis 1990). Several lipophilic non-ionic surfactants are reported to enhance growth of microalgae (Ernst et al. 1983). Conversely, some non-ionic surfactants are degraded by certain microalgae (Davis and Gloyna 1969). Similarly, for desorption of pyrene in soil-water systems, Tween 80 is more efficient than Triton X-100 (Cheng and Wong 2006). When used in the bioslurry-based system non-ionic surfactant assists in desorption of sorbed PAHs in soil to aqueous medium. This mechanism ultimately results in the enhanced biodegradation of PAHs by the microorganisms (Kim et al. 2001).

7.11.2 Surface Area to Biovolume (S/V) Ratio

High S/V ratio of freshwater algae possesses high potential for biosorption after the examination of eight species in the accumulation of atrazine (Tang et al. 1998). Two microalgal species exhibited different degrees of cellular uptake and metabolism of PAHs, either singly or in a mixture. However, the relationship between the percentage PAH accumulation and the S/V ratio were insignificant ($P < 0.05$). The removal of PAHs by microalgae species was directly related to the initial cell density employed. In general, higher cell density or biomass provides more surface area and cell volume for adsorption and absorption of pollutants, leading to better removal in a shorter period of time (Hong et al. 2008).

7.11.3 pH

pH is an important parameter for biosorption of pollutant from aqueous solution because it may affect the solubility and the degree of ionization of the adsorbate during reaction. pH influences solution chemistry of the adsorbate, activity of the functional groups in the biosorbent and competition of sorbate ions. The variation of pH affects the effectiveness as hydrogen ion itself is a tough competing adsorbate (Puranik et al. 2005). Highly alkaline pH conditions (pH 10.0) led to a decline in biosorption efficiency. In general, the metal biosorption rate decreases to a large extent with an increase in alkalinity mainly at pH > 6.0–7.0 (Vannela and Verma 2006). Generally neutral pH value (pH 7) supports the biosorption rate (Al-Homaidan et al. 2014). Pérez-Gregorio et al. (2010) studied that increased pH values of aqueous solution may lead to increases in the concentration of hydroxide ion in n-hexane solution. So, phenanathrene compound are greater in their ionized form. Thus, maximum adsorption occurred at pH 8 of organic solution and in the lowest presence of ionized forms. Whereas, Davis et al. 2003, Ozer et al. 2006 reported that hydrocarbon and oil droplets often carry negative charge in aqueous solution. Thus, negative charge on the biosorbent surface at pH exceeding pH may hinder sorption through electrostatic repulsion and this may have an adverse impact on both absorption and adsorption processes. Moreover, pH can also impacts the surface charge on sorbates (Djerdjev and Beattie 2008).

7.11.4 Ionic Strength

The solution ionic strength is also reported to impact sorption of heavy metal ions on algae (Mishra and Mukherji 2012). Beolchini et al. (2006) showed the negative effect of ionic strength and the positive effect of pH on biosorption performances: the highest determined value for copper specific uptake was about 60 mg/g at pH 6 and about 15 mg/g at pH 4. Chung et al. 2007 reported maximum sorption was attained at low ionic strength, at neutral pH and at 30–35 °C. The expected trend for dissolved organic solute, i.e., increase in sorption with increase in ionic strength was not observed for oil sorption on algae. The trend in specific oil/diesel uptake variation with ionic strength was significantly affected by the type of algae. Sorption of both lubricating oil and diesel on *Spirulina* sp. was highest at the lowest ionic strength (background electrolyte, $NaNO_3$, concentration of 0.01 M), decreased up to $NaNO_3$ at 0.1 M, increased significantly at 0.5 M and subsequently decreased again as $NaNO_3$ at 1 M (Mishra and Mukherji 2012).

7.11.5 Temperature

Temperature has a vital effect on biosorption process, as increase or decrease in temperature can influence the extent of biosorption. Usually temperature enhances biosorptive removal of adsorptive pollutants when increased by increasing surface activity and kinetic energy of the adsorbate. Contrarily, this may also damage the physical structure of the biosorbent (Park et al. 2010). Metal uptake increased gradually with an increase in temperature. The rate of copper biosorption by the nonliving cells was rapid reaching a maximum of 90.61% at 37 °C temperature. The study indicated that removal of copper from aqueous to adsorbent is rapid at ambient temperatures compared to higher temperatures 45–60 °C. Biosorption was in the range of 82.3–85% at 45–60 °C and 78.8% at 26 °C (Al-Homaidan et al. 2014).

7.11.6 Metal

As a useful indicator, the application with microalgae for aquatic toxicity assays has been extensively conducted. Unique eco-niche in the aquatic food web and high sensitivity to a wide spectrum of pollutants, including PAHs and metals are the traits of microalgae. Heavy metals are commonly found contaminants along with the PAH compounds in wastewaters discharged from hydrocarbon processing, iron/steel manufacturing, electroplating, electronics manufacturing dye manufacturing and smelting (Mielke et al. 2004; Ke et al. 2010). Combined effect of cadmium, zinc, copper and nickel on the effectiveness of *S.capricornutum* to remove mixed PAHs from culture medium by biosorption and biodegradation was conducted by Mielke et al. 2004. Fluorene and phenenthrene, fluoranthene and pyrene, and benzo α pyrene were used. Removal efficiency of PAHs by algae was examined by live and dead algae cells. The study established that both live and dead algal (heat killed) cells are capable of PAHs removal from the medium in the order fluoranthene < phenenthrene < fluoranthene < pyrene < benzo α pyrene. Presence of metal significantly influenced the cell-PAHs interaction; which was PAH species dependent. In case of low molecular weight PAHs, metal dosage and exposure time has a positive effect on their removal (99% of fluoranthene and 89% of phenenthrene) in 7 day duration. In case of high molecular weight PAHs metal presence did not affect the removal efficiency whereas the uptake in ethyl acetate fraction of the biomass was increased. The effects of metals on the removal efficiency of PAHs by *S. Capricornutum* were PAH species dependent. Under metal stress, the removal of low molecular weight PAHs fluoranthene and phenenthrene from the medium by algal cells was enhanced. Metal stress also significantly increased the EA-extractable biomass uptake of high molecular weight PAHs. The enhancement effect of metals on the removal of flurene and phenethrene may contribute to oxidation processes by reactive oxygen species generated under stress conditions (Mittler 2002). Both metals and PAHs could act as reactive oxygen species inducers in algal cells (Babu et al. 2003). PAHs and metal ions can cause damage to microalgae structurally and functionally (Nassiri et al. 1996). Wang and Zheng (2008) reported the combined effects of fluoranthene and copper on a marine diatom *Phaeodactylum tricornutum*. Both fluoranthene and copper resulted in ultra-structural impairments including cytoplasmic vacuolization, organelle changes and the appearance of cells with multilayered cell walls and excretion of organic matter. The study demonstrated *S. capricornutumis* as a robust algal species with some tolerance to the toxicity posed by PAHs and metals. The enhancement effect of metals on the removal of flurene and phenenthrene may contribute to oxidation processes by reactive oxygen species generated under stress conditions (Mittler 2002). It might be of great significance, using metal-containing effluents generated from the surrounding chemical industries to dilute the strong oily wastewaters. Hence, the positive effect of the metals on PAH removal, could be utilized.

7.12 Industrial Perspective of Biosorption

Most of the biosorption studies were carried out in laboratory scale. Transfer of knowledge from laboratory scale to industrial applications is a relatively slow process. The first pilot plant installations of biosorption technology was established in USA and Canada (Tsezos 1999). Several commercial biosorbents have been proposed for theremoval of heavy metal ions from industrial or mining wastewaters. These sorbents incorporated biomass of cyanobacteria (*Spirulina*), yeasts, algae or plants (*Lemna* sp., *Sphagnum* sp.), immobilized in polymeric (polysulphone, polyethylene, polypropylene) porous beads (Tsezos et al. 2012; Michalak et al. 2013). Large scale study in South Africa, reported by Bosman and Hendricks (1980) concerning the removal of industrial nitrogenous wastes with high-rate algal ponds concluded that a multi-stage algal system is required for exerting the full removal potential of nitrogen by algal biomass incorporation followed by algal harvesting. New technologies and devices for cultivation and removal of contaminats from industrial wastewater have also been developed. Like waste water treatment using attached alagal based system, The Algal Turf Scrubbers (ATS) – Ecologically Engineered, Algal Based System for nutrient reduction. ATS was used in treatment of dairy manure effluent in central Maryland (USA) and agricultural wastewater in the Florida Everglades (Adey et al. 2011). Commercialization of this technology is developed by HydroMentia Inc., which builds and operates ATS mainly in Florida. Recently, research to improve the performance of ATS has continued, with tests involving new applications and evaluations of the harvested biomass (Adey et al. 2011; Valeta and Verdegem 2015; Sukačová and Červený 2017). Industrial application of biosorption necessitates immobilization of biosorbent for solid/liquid separation. Immobilized biomass beads can be used in packed sorption column, which is perhaps the most effective device for continuous operations. For the removal of PAHs, fungi, yeast and bacteria are represented the efficient classes of biosorbents relative to other biomass types. The biosorptive capacity and the time needed to reach the equilibrium have differed from one to another combination of microorganism-PAHs. Industrial effluents contain several pollutants simultaneously; little attention has been given to sorption of PAHs and heavy metal in a single system. More information on biosorption is required to determine the best combination of organics, biomass types and environmental conditions.

7.13 Conclusion

In the field of biosorption and biodegradation, microorganisms such as bacteria, fungi and yeast have gained immense attention for their abilities to accumulate and degrade organic and inorganic pollutants in their cell structures. But research studies exploring the utilization of microalgae in biosorption and degradation of organic contaminants such as PAHs are limited. In the present review potential of microalgae as a biosorbent for PAHs sorption and degradation has been discussed in detail. The study highlights biosorption; a two stage process and species specific. pH, temperature, metal species present, surface area to biovolume ratio and surfactant are important factors affecting the rate of biosorption. In literature it has been found that most of the research is limited to single PAH compound biosorption. Biosorption of mixed PAH is still in blooming stage, with huge unexplored aspects. This aspect needs to be addressed as in nature more than one PAH compound could be present that may affect the biosorption and degradation of one on the other. Research indicated that pH of biosorption medium is the most important parameter influencing the biosorption capacity. Enzymes released by organisms play a crucial role for the degradation of organic compound. Components of pyruvate dehydrogenase complex have been reported in various studies suggesting key role of the enzyme in degradation of PAH compounds. However, more elaborative research is required to understand other enzymes and mechanism responsible in the degradation of PAHs. Enzymatic behaviour of microalgae during the process and key enzymes responsible for the process are yet to be understood effectively. Further research should be done for enhancement of biosorbent potential of microalgae.

Much more information on algal-based processes are required to reliably predict the short- and long-term impact of pollutant stress on freshwater systems. It would be very interesting to find out if algae reported to degrade/complete removal of PAH and other hazardous compounds are able to protect the phytoplankton, from the adverse effects of these compounds. However, complete proteome analysis will provide more insight into the metabolism and non-target effects of PAHs towards microalgae. Genetic engineering offers a promising tool to improve the absorption and bioremediation of many organic compounds and increase microalgal tolerance to these compounds. In addition, the mechanisms underlying the enhanced degradation of pollutants by algal-bacterial interaction need to be elucidated (Semple et al. 1999). The use of immobilized biomass, rather than native biomass, has been recommended for large-scale application of biosorption process. Immobilization techniques increase the overall cost of biosorbents, and decrease their biosorptive rates and capacities. Thus, more research studies are required to overcome the drawbacks. Use of microalgae as a biosorbent is a blooming technique and is still limited to laboratories only. Efforts should be made to explore biosorption and biodegradation technique among industries for treatment for PAHs contaminants as an eco-friendly and cost effective technique to clean up the environment.

References

Adey WH, Kangas PC, Mulbry W (2011) Algal turf scrubbing: cleaning surface waters with solar energy while producing a biofuel. Bioscience 61:434–441. https://doi.org/10.1525/bio.2011.61.6.5

Ahluwalia SS, Goyal D (2007) Microbial and plant derived biomass for removal of heavy metals from wastewater. Bioresour Technol 98:2243–2257. https://doi.org/10.1016/j.biortech.2005.12.006

Akhtar N, Iqbal J, Iqbal M (2004) Removal and recovery of nickel(II) from aqueous solution by loofa sponge immobilized biomass of *Chlorella sorokiniana*: characterization studies. J Hazard Mater 108:85–94. https://doi.org/10.1016/j.jhazmat.2004.01.002

Aksu Z (2005) Application of biosorption for the removal of organic pollutants: a review. Process Biochem 40:997–1026. https://doi.org/10.1016/j.procbio.2004.04.008

Aksu Z, Akpinar D (2001) Competitive biosorption of phenol and chromium(VI) from binary mixtures onto dried anaerobic activated sludge. Biochem Eng J 7:183–193. https://doi.org/10.1016/S1369-703X(00)00126-1

Al-Homaidan AA, Al-Houri HJ, Al-Hazzani AA, Moubayed MS (2014) Biosorption of copper ions from aqueous solutions by *Spirulina Platensis* biomass. Arab J Chem 7:57–62. https://doi.org/10.1016/j.arabjc.2013.05.022

Ashkenazy R, Gottlieb L, Yannai S (1997) Characterization of acetone-washed yeast biomass functional groups involved in lead biosorption. Biotechnol Bioeng 55:1–10. https://doi.org/10.1002/(SICI)1097-0290(19970705)55:1<1::AID-BIT1>3.0.CO;2-H

Ashour I, Abu Al-Rub FA, Sheikha D, Volesky B (2008) Biosorption of naphthalene from refinery simulated waste-water on blank alginate beads and immobilized dead algal cells. Sep Sci Technol 43:2208–2224. https://doi.org/10.1080/01496390801887351

Babu TS, Akhtar TA, Lampi MA, Tripuranthakam S, Dixon DG, Greenberg BM (2003) Similar stress responses are elicited by copper and ultraviolet radiation in the aquatic plant *Lemna gibba*: implication of reactive oxygen species as common signals. Plant Cell Physiol 44:1320–1329. PMID: 14701927

Beolchini F, Pagnanelli F, Toro L, Veglio F (2006) Ionic strength effect on copper biosorption by *Sphaelotilus natans*: equilibrium study and dynamic modeling in membrane reactor. Water Res 40:144–152. https://doi.org/10.1016/j.watres.2005.10.031

Berthe-Corti L, Conradi B, Hulsch R, Sinn B, Wiesehan K (1998) Microbial cleaning of waste gas containing volatile organic compounds in a bioreactor system with a closed gas circuit. Acta Biotechnol 18:291–304

Blanco A, Sanz B, Llama MJ, Serra JL (1999) Biosorption of heavy metals to immobilized *Phormidium Laminosum* biomass. J Biotechnol 69:227–240

Bold HC, Wynne MJ (1985) Introduction to the algae. Structure and reproduction, 2nd edn. Prentice Hall, Inc, Englewood Cliffs, 720 pp

Borde X, Guieysse B, Delgado O, Munoz R, Hatti-Kaul R, Nugier CC (2003) Synergistic relationships in algal-bacterial microcosms for the treatment of aromatic pollutants. Bioresour Technol 86:293–300. https://doi.org/10.1016/S0960-8524(02)00074-3

Bosman J, Hendricks F (1980) The development of an algal pond system for the removal of nitrogen from an inorganic industrial; effluent. In: Proceedings of international symposium on aquaculture in wastewater NIWP. CSIR, Pretoria, pp 26–35

Canet R, Birnstingl JG, Malcolm DG, Lopez-Real JM, Beck AJ (2001) Biodegradation of polycyclic aromatic hydrocarbons (PAHs) by native microflora and combinations of white-rot fungi in a coal-tar contaminated soil. Bioresour Technol 76(2):113–117. https://doi.org/10.1016/S0960-8524(00)00093-6

Cao B, Nagarajan K, Loh KC (2009) Biodegradation of aromatic compounds: current status and opportunities for biomolecular approaches. Appl Microbiol Biotechnol 85:207–228. https://doi.org/10.1007/s00253-009-2192-4

Cerniglia CE (1992) Biodegradation of polycyclic aromatic hydrocarbons. Biodegradation 3:351–368. https://doi.org/10.1007/BF00129093

Cerniglia CE, Gibson DT, Van Baalen C (1979) Algal oxidation of aromatic hydrocarbons: formation of 1-naphthol from naphthalene by *Agmenellum Quadruplicatum*, strain PR-6. Biochem Biophys Res Commun 88:50–58. https://doi.org/10.1016/0006-291X(79)91695-4

Cerniglia CE, Baalen CV, Gibson DT (1980a) Metabolism of naphthalene by cyanobacterium *Oscillatoria* sp. Strain JCM. J Gen Microbiol 116:485–494 0022-1287/80/0000-8717

Cerniglia CE, Gibson DT, Van Baalen C (1980b) Oxidation of naphthalene by cyanobacteria and microalgae. J Gen Microbiol 116:495–500 0022-1287/80/0000-871

Chaillan F, Gugger M, Saliot A, Couté A, Oudot J (2006) Role of cyanobacteria in the biodegradation of crude oil by a tropical cyanobacterial mat. Chemosphere 62:1574–1582. https://doi.org/10.1016/j.chemosphere.2005.06.050

Chan SMN, Luan T, Wong MH, Tam NFY (2006) Removal and biodegradation of polycyclic aromatic hydrocarbons by *Selenastrum Capricornutum*. Environ Toxicol Chem 25:1772–1779. https://doi.org/10.1897/05-354R.1

Chekroun KB, Sánchez E, Baghour M (2014) The role of algae in bioremediation of organic pollutants. Int Res J Pub Environl Health 1(2):19–32 ISSN 2360-8803. http://www.journalissues.org/irjpeh/

Cheng KY, Wong JWC (2006) Effect of synthetic surfactants on the solubilization and distribution of PAHs in water/soil-water systems. Environ Technol Vol 27(8):835–844. https://doi.org/10.1080/09593332708618695

Chung MK, Tsui MTK, Cheung KC, Tam NFY, Wong MH (2007) Removal of aqueous phenanthrene by brown seaweed *Sargassum hemiphyllum*: sorption-kinetic and equilibrium studies. Sep Purif Technol 54:355–362. https://doi.org/10.1016/j.seppur.2006.10.008

Codd GA (1987) Immobilized micro-algae and cyanobacteria. Br Phycol Soc Newslett 24:1–5

Davis E, Gloyna E (1969) Anionic and nonionic surfactant sorption and degradation by algae cultures. J Am Oil Chem Soc 46:604–608. https://doi.org/10.1007/BF02544977

Davis TA, Bohumil V, Alfonso M (2003) A review of the biochemistry of heavy metal biosorption by brown algae. Water Res 37:4311–4330. https://doi.org/10.1016/S0043-1354(03)00293-8

Dean-Ross D, Moody J, Cerniglia CE (2002) Utilization of mixtures of polycyclic aromatic hydrocarbons by bacteria isolated from contaminated sediment. FEMS Microbiol Ecol 41:17. https://doi.org/10.1111/j.1574-6941.2002.tb00960.x

Djerdjev AM, Beattie JK (2008) Electroacoustic and ultrasonic attenuation measurements of droplet size and f-potential of alkane-in-water emulsions: effects of oil solubility and composition. Phys Chem 10:4843–4852. https://doi.org/10.1039/b807623e

Ellis BE (1977) Degradation of phenolic compounds by freshwater algae. Plant Sci Lett 8:213–216. https://doi.org/10.1016/0304-4211(77)90183-3

El-Sheekh MM, Ghareib MM, Abou-El-Souod GW (2012) Biodegradation of phenolic and polycyclic aromatic compounds by some algae and cyanobacteria. J Bioremed Biodegr 3:133. https://doi.org/10.4172/2155-6199.1000133

Ernst R, Gonzales CJ, Arditti V (1983) Biological effects of surfactants: Part 6 – Effects of anionic, non-ionic and amphoteric surfactants on a green alga (*Chlamydomonas*). Environ Pollut (Ser A) 31:159–175

Fomina M, Gadd GM (2014) Biosorption: current perspectives on concept, definition and application. Bioresour Technol 160:3–14. https://doi.org/10.1016/j.biortech.2013.12.102

Fukami K, Nishijima T, Ishida Y (1997) Stimulative and inhibitory effects of bacteria on the growth of microalgae. Hydrobiologia 358:185–199. https://doi.org/10.1023/A:1003139402315

Garbayo I, Leon R, Vigara J, Vilchez C (2002) Inhibition of nitrate consumption by nitrite in entrapped *Chlamyodomonas Reinhardtii* cells. Bioresour Technol 81:207–215. https://doi.org/10.1016/S0960-8524(01)00138-9

Garcia de Llasera MP, Olmos-Espejel Jde J, Diaz-Flores G, MontanoMontiel A (2016) Biodegradation of benzo(a)pyrene by two freshwater microalgae *Selenastrum Capricornutum* and *Scenedesmus Acutus*: a comparative study useful for bioremediation. Environ Sci Pollut Res Int 23:3365–3375. https://doi.org/10.1007/s11356-015-5576-2

Garrido Moreno I (2008) Microalgae immobilization: current techniques and uses. Bioresour Technol 99:3949–3964. https://doi.org/10.1016/j.biortech.2007.05.040

Gattullo EC, Hanno B, Steinberg Christian EW, Elisabetta L (2012) Removal of bisphenol A by the freshwater green alga *Monoraphidium braunii* and the role of natural organic matter. Sci Total Environ 416:501–506. https://doi.org/10.1016/j.scitotenv.2011.11.033

Ghasemi Y, Sara R-A, Elham F (2011) The biotransformation, biodegradation, and bioremediation of organic compounds by microalgae. J Phycol 47:969–980. https://doi.org/10.1111/j.1529-8817.2011.01051.x

Ghosal D, Ghosh S, Dutta TK, Ahn Y (2016) Current state of knowledge in microbial degradation of polycyclic aromatic hydrocarbons (PAHs): a review. Front Microbiol 7:1369. https://doi.org/10.3389/fmicb.2016.01369

Gods'gift OE, Fagade OE (2016) Microalgal-bacterial consortium in polyaromatic hydrocarbon degradation of petroleum – based effluent. J Bioremed Biodegr 7:359. https://doi.org/10.4172/2155-6199.1000359

Guieysse B, Borde X, Muñoz R, Hatti-Kaul R, Nugier-Chauvin C, Patin H, Mattiasson B (2002) Influence of the initial composition of algal-bacterial microcosms on the degradation of salicylate in a fed-batch culture. Biotechnol Lett 24:531–538. https://doi.org/10.1023/A:1014847616212

Hammaini A, Gonzalez F, Ballester A, Blazquez ML, Munoz JA (2003) Simultaneous uptake of metals by activated sludge. Miner Eng 16:723–729. https://doi.org/10.1016/S0892-6875(03)00166-3

Haritash AK, Kaushik CP (2009) Biodegradation aspects of polycyclic aromatic hydrocarbons (PAHs): a review. J Hazard Mater 169:1–15. https://doi.org/10.1016/j.jhazmat.2009.03.137

Hatzinger PB, Alexander M (1995) Effect of aging of chemicals in soil on their biodegradability and extractability. Environ Sci Technol 29:537–545. https://doi.org/10.1021/es00002a033

Heitkamp MA, Cerniglia CE (1989) Polycyclic aromatic hydrocarbon degradation by a *Mycobacterium sp.* in microcosms containing sediment and water from a pristine ecosystem. Appl Environ Microbiol 55:1968–1973 0099-2240/89/081968-06

Herwijnen R, van de Sande BF, van der Wielen FWM, Govers HAJ, Parsons JR (2003) Infuence of phenanthrene and fluoranthene on the degradation of fluorene and glucose by *Sphingomonas sp.* strain LB126 in chemostat cultures. FEMS Microbiol Ecol 46:105–111

Hong YW, Yuan DX, Lin QM, Yang TL (2008) Accumulation and biodegradation of phenanthrene and fluoranthene by the algae enriched from a mangrove aquatic ecosystem. Mar Pollut Bull 56:1400–1405. https://doi.org/10.1016/j.marpolbul.2008.05.003

Imran A, Asim M, Khan Tabrez A (2012) Low cost adsorbents for the removal of organic pollutants from wastewater. J Environ Manag 113:170–183. https://doi.org/10.1016/j.jenvman.2012.08.028

Ivánová D, Kaduková J, Kavuličová J, Horváthová H (2012) Determination of the functional groups in algae *Parachlorella Kessleri* by potentiometric titrations. Nova Biotechnologica Et Chimica 11:93–99. https://doi.org/10.2478/v10296-012-0010-3

Jones KC, Stratford JA, Tidridge P, Waterhouse KS, Johnston AE (1989) Polynuclear aromatic hydrocarbons in an agricultural soil: long-term changes in profile distribution. Environ Pollut 56:337–351

Kaoutar Ben Chekroun, Esteban Sánchez, Mourad Baghour (2014) The role of algae in bioremediation of organic pollutants, International Research Journal of Public and Environmental Health Vol.1 (2), pp. 19–32, ISSN 2360–8803, http://www.journalissues.org/irjpeh/

Kaya VM, Picard G (1996) Stability of chitosan gel as entrapment matrix of viable *Scenedesmus bicellularis* cells immobilized on screens for tertiary treatment of wastewater. Bioresour Technol 56:147–155. https://doi.org/10.1016/0960-8524(96)00013-2

Ke L, Luo L, Wang P, Luan T, Tam NF (2010) Effects of metals on biosorption and biodegradation of mixed polycyclic aromatic hydrocarbons by a freshwater green alga *Selenastrum Capricornutum*. Bioresour Technol 101:6961–6972. https://doi.org/10.1016/j.biortech.2010.04.011

Keskinkan O, Goksu MZ, Basibuyuk M, Forster CF (2004) Heavy metal adsorption characteristics of a submerged aquatic plant (*Ceratophyllum demersum*). Bioresour Technol 92(2):197–200. https://doi.org/10.1016/j.biortech.2003.07.011

Kim IS, Park JS, Kim KW (2001) Enhanced biodegradation of polycyclic aromatic hydrocarbons using nonionic surfactants in soil slurry. Appl Geochem 16:1419–1428

Kirso U, Irha N (1998) Role of algae in fate of carcinogenic polycyclic aromaitc hydrocarbons in the aquatic environment. Ecotoxicol Environ Saf 41:83–89. https://doi.org/10.1006/eesa.1998.1671

Koelmans A, Anzion SF, Lijklema L (1995) Dynamics of organic micropollutant biosorption to cyanobacteria and detritus. Environ Sci Technol 29:933–940 0013-936x/95/0929-0933$09.00/0

Lau PS, Tam NFY, Wong YS (1997) Wastewater nutrients (N and P) removal by carrageenan and alginate immobilized *Chlorella vulgaris*. Environ Technol 18:945–951. https://doi.org/10.1080/09593331808616614

Lee RE (1989) Phycology. Cambridge University Press, Cambridge, 561 pp. ISBN-13 978-0-511-38669-5. www.cambridge.org/9780521864084

Lei AP, Wong YS, Tam NFY (2002) Removal of pyrene by different microalgal species. Water Sci Technol 46:195–201

Lei AP, Wong YS, Tam NFY (2003) Pyrene-induced changes of glutathione-S-transferase (GST) activities in different microalgal species. Chemosphere 50(3):293–301. https://doi.org/10.1016/S0045-6535(02)00499-X

Lei AP, Hu ZL, Wong YS, Tam NFY (2007) Removal of fluoranthene and pyrene by different microalgal species. Bioresour Technol 98:273–280. https://doi.org/10.1016/j.biortech.2006.01.012

Lewis MA (1990) Chronic toxicities of surfactants and detergent builders to algae: a review and risk assessment. Ecotoxicol Environ Saf 20:123–140 0 147-65 13/90

Li Y, Fei X, Deng X (2012) Novel molecular insights into nitrogen starvation-induced triacylglycerols accumulation revealed by differential gene expression analysis in green algae *Micractinium Pusillum*. Biomass Bioenergy 42:199–211. https://doi.org/10.1016/j.biombioe.2012.03.010

Lu AYH, Miwa GT (1980) Molecular properties and biological functions of microsomal epoxide hydrase. Annu Rev Pharmacol Toxicol 20:513–531. https://doi.org/10.1146/annurev.pa.20.040180.002501

Lukavsky J (1988) Long-term preservation of algal strains by immobilization. Arch Protistenkd 135:65–68. https://doi.org/10.1016/S0003-9365(88)80054-X

Luo L, Wang P, Lin L, Luan T, Ke L, Tam NFY (2014a) Removal and transformation of high molecular weight polycyclic aromatic hydrocarbons in water by live and dead microalgae. Process Biochem 49:1723–1732. https://doi.org/10.1016/j.procbio.2014.06.026

Luo S, Chen B, Lin L, Wang X, Tam NF, Luan T (2014b) Pyrene degradation accelerated by constructed consortium of bacterium and microalga: effects of degradation products on the microalgal growth. Environ Sci Technol 48:13917–13924. https://doi.org/10.1021/es503761j

Macek T, Mackova M (2011) Potential of biosorption technology. Microbial biosorption of metal. Springer, Dordrecht, ISBN: 978-94-007-0442-8. https://doi.org/10.1007/978-94-007-0443-5-2

Mallick N (2002) Biotechnological potential of immobilized algae for wastewater N, P and metal removal: a review. BioMetals 15:377–390. https://doi.org/10.1023/A:1020238520948

Michalak I, Chojnacka K, Witek-Krowiak A (2013) State of the art for the biosorption process – a review. Appl Biochem Biotech 170:1389–1416. https://doi.org/10.1007/s12010-013-0269-0

Mielke HW, Wang G, Gonzales CR, Powell ET, Le B, Quach VN (2004) PAHs and metals in the soils of inner-city and suburban New Orleans, Louisiana, USA. Environ Toxicol Pharmacol 18:243–247

Millemann RE, Birge WJ, Black JA, Cushman RM, Daniels KL, Franco PJ, Giddings JM, McCarthy JF, Stewart AJ (1984) Comparative acute toxicity to aquatic organisms of components of coal-derived synthetic fuels. Trans Am Fish Soc 113:74–85. https://doi.org/10.1577/1548-8659(1984)113<74:CATTAO>2.0.CO;2

Mishra PK, Mukherji S (2012) Biosorption of diesel and lubricating oil on algal biomass. 3 Biotech 2:301–310. https://doi.org/10.1007/s13205-012-0056-6

Mittler R (2002) Oxidative stress, antioxidants and stress tolerance. Trends Plant Sci 7:405–410. https://doi.org/10.1016/S1360-1385(02)02312-9

Moscoso F, Deive FJ, Longo MA, Sanromán MA (2012) Technoeconomic assessment of phenanthrene degradation by Pseudomonas stutzeri CECT 930 in a batch bioreactor. Bioresour Technol 104:81–89. https://doi.org/10.1016/j.biortech.2011.10.053

Mulligan CN, Yong RN, Gibbs BF (2001) Surfactant-enhanced remediation of contaminated soil: a review. Eng Geol 60:371–380

Muñoz R, Guieysse B (2006) Algal–bacterial processes for the treatment of hazardous contaminants: a review. Water Res 40:2799–2815. https://doi.org/10.1016/j.watres.2006.06.011

Muñoz R, Guieysse B, Mattiasson B (2003) Phenanthrene biodegradation by an algal–bacterial consortium in two-phase partitioning bioreactors. Appl Microbiol Biotechnol 61:261–267. https://doi.org/10.1007/s00253-003-1231-9

Narro ML, Cerniglia CE, Van Baalen C, Gibson DT (1992a) Evidence for an NIH shift in oxidation of naphthalene by the marine cyanobacterium Oscillatoria sp. strain JCM. Appl Environ Microbiol 58:1360–1363. PMC195598

Narro ML, Cerniglia CE, Van Baalen C, Gibson DT (1992b) Metabolism of phenanthrene by the marine cyanobacterium Agmenellum quadruplicatum PR-6. Appl Environ Microbiol 58:1351–1359. PMC195597

Nassiri Y, Ginsburger-Vogel T, Mansot JL, Wry J (1996) Effects of heavy metals on Tetraselmis suecica: ultrastructural and energy-dispersive X-ray spectroscopic studies. Biol Cell 86:151–160. https://doi.org/10.1016/0248-4900(96)84779-4

Norton L, Baskaran K, McKenzie T (2004) Biosorption of zinc from aqueous solutions using biosolids. Adv Environ Res 8:629–635. https://doi.org/10.1016/S1093-0191(03)00035-2

Oswald W, Gotaas H, Ludwig H, Lynch V (1953) Algae symbiosis in oxidation ponds: III. Photosynthetic oxygenation. Sew Ind Waste 25:692–705 www.jstor.org/stable/25032197

Oungbho K, Muller BW (1997) Chitosan sponges as sustained release drug carriers. Int J Pharm 156:229–237. https://doi.org/10.1016/S0378-5173(97)00201-9

Ozer A, Akkaya G, Turabik M (2006) Biosorption of Acid Blue 290 (AB 290) and Acid Blue 324 (AB 324) dyes on Spirogyra rhizopus. J Hazard Mater B135:355–364. https://doi.org/10.1016/j.jhazmat.2005.11.080

Pagnanelli F, Toro L, Veglio F (2002) Olive mill solid residues as heavy metal sorbent material: a preliminary study. Waste Manag 22:901–907. https://doi.org/10.1016/S0956-053X(02)00086-7

Park D, Yun Y-S, Park JM (2010) The past, present, and future trends of biosorption. Biotechnol Bioprocess Eng 15:86–102. https://doi.org/10.1007/s12257-009-0199-4

Pascale H, Schiavon M, Morel J-L, Lichtfouse E (1997) Polycyclic aromatic hydrocarbon (PAH) occurrence and remediation methods. Analusis EDP Sci 25:M56–M59 hal-00193277

Patel JG, Nirmal Kumar JI, Kumar RN, Khan SR (2016) Biodegradation capability and enzymatic variation of potentially hazardous polycyclic aromatic hydrocarbons—anthracene and pyrene by Anabaena fertilissima. Polycycl Aromat Compd 36:72–87. https://doi.org/10.1080/10406638.2015.1039656

Percival EGV, McDowell RH (1968) Chemistry and enzymology of marine algal polysaccharides. Academic, London. https://doi.org/10.1002/ange.19680802022

Pérez-Gregorio M, García-Falcón M, Martínez-Carballo E, Simal-Gándara J (2010) Removal of polycyclic aromatic hydrocarbons from organic solvents by ashes wastes. J Hazard Mater 178:273–281. https://doi.org/10.1016/j.jhazmat.2010.01.073

Phang SM, Chu WL, Rabiei R (2015) Phycoremediation. In: Sahoo D, Seckbach J (eds) The algae world. Cellular origin, life in extreme habitats and astrobiology, vol 26. Springer, Dordrecht. https://doi.org/10.1007/978-94-017-7321-8_13

Pinto G, Pollio A, Previtera L, Temussi F (2002) Biodegradation of phenols by microalgae. Biotechnol Lett 24:2047–2051. https://doi.org/10.1023/A:1021367304315

Pinto G, Pollio A, Previtera L (2003) Removal of low molecular weight phenols from olive oil mill wastewater using microalgae. Biotechnol Lett 25:1657. https://doi.org/10.1023/A:1025667429222

Puranik P, Modak J, Paknikar K (2005) A comparative study of the mass transfer kinetics of metal biosorption by microbial biomass. Hydrometallurgy 52:189–197. https://doi.org/10.1016/S0304-386X(99)00017-1

Radwan SS, Al-Hasan RH, Salamah S, Al-Dabbous S (2002) Bioremediation of oily sea water by bacteria immobilized in biofilms coating macroalgae. Int Biodeter Biodegr 50:55–59. https://doi.org/10.1016/S0964-8305(02)00067-7

Robinson PK, Mak AL, Trevan MD (1986) Immobilized algae: a review. Process Biochem 21:122–126

Robinson PK, Reeve JO, Goulding KH (1988) Kinetics of phosphorus uptake by immobilized Chlorella. Biotechnol Lett 10:17–20. https://doi.org/10.1007/BF01030017

Romera E, González F, Ballester A, Blázquez ML, Muñoz JA (2007) Comparative study of biosorption of heavy metals using different types of algae. Bioresour Technol 98:3344–3353. https://doi.org/10.1016/j.biortech.2006.09.026

Safonova E, Kvitko K, Kuschk P, Moder M, Reisser W (2005) Biodegradation of phenanthrne by the green alga Scendesmus obliquus Es-55. Eng Life Sci 5:234–239. https://doi.org/10.1002/elsc.200520077

Sanches S, Martins M, Silva AF, Galinha CF, Santos MA, Pereira IAC, Crespo MTB (2017) Bioremoval of priority polycyclic aromatic hydrocarbons by a microbial community with high sorption ability. Environ Sci Pollut Res 24:3550–3561. https://doi.org/10.1007/s11356-016-8014-1

Sato R, Omura T (1978) Cytochrome P-450. Academic, New York, pp 16–18

Schoeny R, Cody T, Warshawsky D, Radike M (1988) Metabolism of mutagenic polycyclic aromatic hydrocarbons by photosynthetic algal species. Mutat Res 197:289–302. https://doi.org/10.1016/0027-5107(88)90099-1

Semple KT, Cain RB (1996) Biodegradation of phenols by the alga Ochromonas danica. Appl Environ Microb 62:1265–1273. PMC167892

Semple KT, Cain RB, Schmidt S (1999) Biodegradation of aromatic compounds by microalgae. FEMS Microbiol Lett 170:291–300. https://doi.org/10.1111/j.1574-6968.1999.tb13386.x
Shashirekha S, Uma L, Subramanian G (1997) Phenol degradation by the marine cyanobacterium Phormidium valderianum BDU-30501. J Ind Microbiol Biotechnol 19:130–133. https://doi.org/10.1038/sj.jim.2900438
Stringfellow WTM, Cohen LA (1999) Evaluating the relationship between the sorption of PAHs to bacterial biomass and biodegradation. Wat Res 33:2535–2544. https://doi.org/10.1016/S0043-1354(98)00497-7
Subashchandrabose SR, Logeshwaran P, Venkateswarlu K, Naidu R, Megharaj M (2017) Pyrene degradation by Chlorella sp. MM3 in liquid medium and soil slurry: possible role of dihydrolipoamide acetyltransferase in pyrene biodegradation. Algal Res 23:223–232. https://doi.org/10.1016/j.algal.2017.02.010
Sukačová K, Červený J (2017) Can algal biotechnology bring effective solution for closing the phosphorus cycle? Use of algae for nutrient removal. Eur J Environ Sci 7:63–72. https://doi.org/10.14712/23361964.2017.6
Suzuki T, Yamaguchi T, Ishida M (1998) Immobilization of Prototheca zopfii in calcium alginate beads for the degradation of hydrocarbons. Process Biochem 33:541–546. https://doi.org/10.1016/S0032-9592(98)00022-3
Takáčová A, Smolinská M, Ryba J, Mackul'ak T, Jokrllová J, Hronec P (2014) Biodegradation of Benzo[a]Pyrene through the use of algae. Cent Eur J Chem 12:1133–1143. https://doi.org/10.2478/s11532-014-0567-6
Tam NFY, Wong YS (2000) Effect of immobilized microalgal bead concentrations on wastewater nutrient removal. Environ Pollut 107:145–151. https://doi.org/10.1016/S0269-7491(99)00118-9
Tang JX, Hoagland KD, Siegeried BD (1998) Uptake and bioconcentration of atrazine by selected freshwater algae. Environ Toxicol Chem 17:1085–1090 https://digitalcommons.unl.edu/entomologyfacpub/134
Tang L, Tang XY, Zhu YG, Zheng MH, Miao QL (2005) Contamination of polycyclic aromatic hydrocarbons (PAHs) in urban soils in Beijing, China. Environ Int 31:822–828. https://doi.org/10.1016/j.envint.2005.05.031
Torres MA, Barros MP, Campos SCG, Pinto E, Rajamani S, Sayre RT, Colepicolo P (2008) Biochemical biomarkers in algae and marine pollution: a review. Ecotoxicol Environ Saf 71:1–15. https://doi.org/10.1016/j.ecoenv.2008.05.009
Tsezos M (1999) Process Metallurgy 9:171–173
Tsezos M, Hatzikioseyian A, Remoudaki E (2012) Biofilm reactors in mining and metallurgical effluent treatment: biosorption, bioprecipitation, bioreduction processes. http://www.metal.ntua.gr/uploads
Ueno R, Ureno N, Wada S (2006) Synergistic effect of cell immobilization in polyurethane foam and use of thermo tolerant strain on mixed hydrocarbon substrate by Prototheca zopfii. Fish Sci 72:1027–1033. https://doi.org/10.1111/j.1444-2906.2006.01252.x
Ueno R, Wada S, Urano N (2008) Repeated batch cultivation of the hydrocarbon-degrading, microalgal strain Prototheca zopfii RND16 immobilized in polyurethane foam. Can J Microbiol 54:66–70. https://doi.org/10.1139/w07-112
Urrutia I, Serra JL, Llama MJ (1995) Nitrate removal from water by Scenedesmus obliquus immobilizes in polymeric foams. Enzym Microb Technol 17:200–205. https://doi.org/10.1016/0141-0229(94)00008-F
Valeta J, Verdegem M (2015) Removal of nitrogen by algal turf scrubber technology in recirculating aquaculture systém. Aquac Res 46:945–951
Vannela R, Verma SK (2006) Co^{2+}, Cu^{2+}, and Zn^{2+} accumulation by cyanobacterium *Spirulina platensis*. Biotechnol Prog 22:1282–1293. https://doi.org/10.1021/bp060075s
Wang L, Zheng B (2008) Toxic effects of fluoranthene and copper on marine diatom Phaeodactylum tricornutum. J Environ Sci 20:1363–1372. https://doi.org/10.1016/S1001-0742(08)62234-2

Warshawsky D, Radike M, Jayasimhulu K, Cody T (1988) Metabolism of benzo(a)pyrene by a dioxygenase enzyme system of the freshwater green alga Selenastrum capricornutum. Biochem Biophys Res Commun 152:540–544. https://doi.org/10.1016/S0006-291X(88)80071-8

Warshawsky D, Cody T, Radike M, Reilman R, Schumann B, LaDow K, Schneider J (1995) Biotransformation of benzo[a]pyrene and other polycyclic aromatic hydrocarbons and heterocyclic analogs by several green algae and other algal species under gold and white light. Chem Biol Interact 97:131–148. https://doi.org/10.1016/0009-2797(95)03610-X

Wurster M, Mundt S, Hammer E (2003) Extracellular degradation of phenol by the cyanobacterium *Synechococcus PCC 7002*. J Appl Phycol 15:171. https://doi.org/10.1023/A:1023840503605

Yamaguchi T, Ishida M, Suzuki T (1999) An immobilized cell system in polyurethane foam for the lipophilic micro-alga Prototheca zopfii. Process Biochem 34:167–171. https://doi.org/10.1016/S0032-9592(98)00084-3

Chapter 8
Adsorption and Oxidation Techniques to Remove Organic Pollutants from Water

Mustapha Mohammed Bello and Abdul Aziz Abdul Raman

Contents

Abstract The presence of recalcitrant organic pollutants in the environment has become a source of concern due to their detrimental effects. Effective technologies are therefore necessary. The combination of adsorption and advanced oxidation processes is emerging as a green technology for removing organic pollutants. Here, we review the applications of adsorption-advanced oxidation processes to remove organic pollutants from waters. We discuss the fundamentals and synergistic performance of adsorption-Fenton, adsorption-photocatalysis, ultrasound-adsorption and adsorption-Ozonation. There is a growing trend in using bifunctional materials that can act both as adsorbents and catalysts in photocatalysis and heterogeneous Fenton oxidation. Ultrasound irradiation is commonly combined with adsorption to enhance mass transfer through cavitation. However, the contribution of the cavitation-generated radicals has not received much attention. While previous

M. M. Bello (✉) · A. A. A. Raman
Department of Chemical Engineering, Faculty of Engineering, University of Malaya, Kuala Lumpur, Malaysia
e-mail: mmbello.cda@buk.edu.ng

G. Crini, E. Lichtfouse (eds.), *Green Adsorbents for Pollutant Removal*, Environmental Chemistry for a Sustainable World 18,
https://doi.org/10.1007/978-3-319-92111-2_8

studies have given more emphasis on the synergy of the combined processes in pollutants removal, we highlight the potential of the oxidation processes to simultaneously regenerate adsorbents in the combined processes.

8.1 Introduction

The rapid increase in the World's population and the consequent industrial activities have put enormous pressure on the environment. Many industrial processes generate large quantities of wastewater containing various pollutants, which are mostly discharged into water bodies. These pollutants include recalcitrant organic pollutants such as synthetic dyes, pesticides, pharmaceuticals, personal care products and other emerging pollutants. Although wastewater treatment plants have been well established, it is now being recognized that many of these pollutants could not be easily removed by the existing technologies (Perdigo et al. 2010; Comero et al. 2013; Luo et al. 2014). Therefore, more effective technologies must be developed for the removal of recalcitrant pollutants from the environment.

A lot of efforts have been devoted into developing effective treatment technologies that can address the challenges of removing recalcitrant pollutants. To this end, technologies such as adsorption and advanced oxidation processes have received wide attention as alternative methods for the treatment of recalcitrant wastewaters (Kyriakopoulos and Doulia 2006; Poyatos et al. 2010; Chowdhury and Balasubramanian 2014; Dewil et al. 2017). Adsorption involves the use of suitable material to sequester pollutants from wastewater while advanced oxidation processes rely on the generation of reactive radicals that can oxidize organic pollutants. While interest on adsorption is mainly due to its cost-effectiveness and availability of wide range of adsorbents, advanced oxidation processes can effectively mineralize recalcitrant organic pollutants (Dewil et al. 2017). Both adsorption and advanced oxidation process are considered green technologies for wastewater treatment. However, these technologies are not without limitations. Adsorption is rather a slow process and the need to continuously regenerate spent adsorbents is considered a disadvantage. On the other hand, some advanced oxidation processes may require the input of costly chemicals while others are energy intensive or limited by some environmental conditions. For example, Fenton oxidation requires the input of a relatively costly H_2O_2 (Asghar et al. 2015), produces large amount of iron sludge (Lima et al. 2017) and requires a strict acidic environment (Guo et al. 2017).

Generally, integrating different wastewater treatment processes can improve overall process performance and in some cases, lower treatment cost (Elsellami et al. 2009; Lau et al. 2016). Thus, combining adsorption with advanced oxidation processes can offer many advantages such as higher synergistic performance, reduced treatment time and chemical/energy inputs (Thankappan et al. 2015). Additionally, some of the limitations of individual process can be addressed through process integration (Cataldo et al. 2016). For example, since adsorption faces

Table 8.1 Types and description of the commonly combined adsorption-advanced oxidation processes

Process	Description
Adsorption-Fenton	Adsorption combined with homogeneous Fenton oxidation
	Adsorption combined with heterogeneous Fenton using bifunctional material
	Various composites can act as adsorbent as well as Fenton-like catalyst
Adsorption-photocatalysis	Involves photocatalyst degradation and adsorption on the surface of heterogeneous catalysts
	Bifunctional composites with photocatalytic and adsorption capability are utilized
Adsorption-ultrasound	Ultrasound-assisted adsorption is widely investigated
	Ultrasound sound cavitation promotes mixing and mass transfer
	Degradation is achieved through generation of radicals and pyrolytic effect
	Ultrasound can also be used in adsorbent regeneration
Adsorption-Ozonation	Activated carbon is commonly combined with Ozonation
	Various bifunctional materials are used in catalytic Ozonation

challenges in the exhaustion and the need for regeneration of spent adsorbent, advanced oxidation processes can be employed to degrade the adsorbed organic pollutants, leading to the regeneration of the adsorbent (Kim et al. 2015). In this context, advanced oxidation processes can be designed to occur simultaneously with the adsorption, ensuring the continues service of the adsorbent. Advanced oxidation processes such as ultrasound irradiation can be used to enhance mass transfer in adsorption process as well as to provide further degradation of pollutants by reactive radicals (Bazrafshan et al. 2017; Khataee et al. 2017).

The synergistic performance of adsorption-advanced oxidation processes in the treatment of recalcitrant wastewater have been widely reported (Cataldo et al. 2016; Alipanahpour et al. 2017; Fakhri et al. 2017; Maryam et al. 2017; Mojiri et al. 2017; Mu et al. 2017; Sharifpour et al. 2017). The advanced oxidation processes that are commonly combined with adsorption include Fenton oxidation, photocatalysis, ultrasound irradiation and Ozonation (Table 8.1). Some of the advanced oxidation processes involved the use of heterogeneous catalysts to drive the process. In this regard, materials with dual-function as catalysts and adsorbents can be employed. The advances in material science is playing a significant role in the development of such bifunctional materials. Graphene-based materials, for example, are being explored in the development of bifunctional composites for simultaneous adsorption-oxidation of organic pollutants (Chen et al. 2017; Luo et al. 2017; Mu et al. 2017). Many studies have reported the possibility of using such bifunctional materials under visible light irradiation in heterogeneous photo-Fenton and photocatalysis, signifying wider applications with possible reduction in the treatment cost.

Despite the opportunities in integrating adsorption with advanced oxidation processes, literature on the topic is rather sporadic and a review on the synergistic effects of the combined processes is yet to be presented. In this chapter, we present an overview on the application of combined adsorption-advanced oxidation processes for the removal of recalcitrant organic pollutants from water. The discussion is based on the classification in Table 8.1. First, a brief introduction on adsorption, including adsorbent types is offered. Advanced oxidation processes are then discussed, highlighting various types of processes and their performances. The discussions on the combined adsorption-advanced oxidation processes, which include adsorption-Fenton, adsorption-photocatalysis, adsorption-ultrasound and adsorption-Ozonation, are then offered. In each case, emphasis has been given to the mechanisms, synergistic performance as well as challenges faced by the combined processes.

8.2 Adsorption

8.2.1 Background

Adsorption is a well-established physical separation and purification process with numerous industrial applications. Generally, adsorption is a surface phenomenon, involving the use of solid materials to remove some substances from liquid or gas. The substance to be removed can adhere to the surface of the adsorbent through various mechanisms. Although, adsorption is used generally to also include the removal of pollutants from gaseous medium, its applications in wastewater treatment deals only with the removal of pollutants from aqueous solution using solid materials. Adsorption has proven to be an effective technology for removing various organic and inorganic pollutants from wastewater.

The basic concept of adsorption is depicted in Fig. 8.1. The surface for the adsorption is provided by the solid material referred to as *adsorbent*. The substance to be removed is termed as the *adsorbate*. The surface of the adsorbent possesses certain characteristics that allow the attachment of the adsorbate. As the adsorption occurs under certain conditions, a reversible phenomenon, termed *desorption*, is also possible. In desorption, the adsorbates are released from the surface of the adsorbent and moved back to the liquid phase (Fig. 8.1). Since adsorption is a surface phenomenon, the surface characteristics of the adsorbent is an important consideration (Worch 2012).

Generally, the term adsorption is often used loosely to also include the *absorption* of substances into the solid material. However, absorption is distinct from adsorption as absorption goes beyond surface interaction and involves the bulk transfer of the substance into the adsorbent. Often it is difficult to delineate between adsorption and absorption, especially in natural systems where complex materials can bind substances both on their surfaces and in the bulk structures (Worch 2012). In such cases, *sorption* is used to denote the combination of adsorption and absorption.

Adsorption has various applications in liquid/gas purification and other separation processes. In wastewater treatment, adsorption is among the most cost-effective

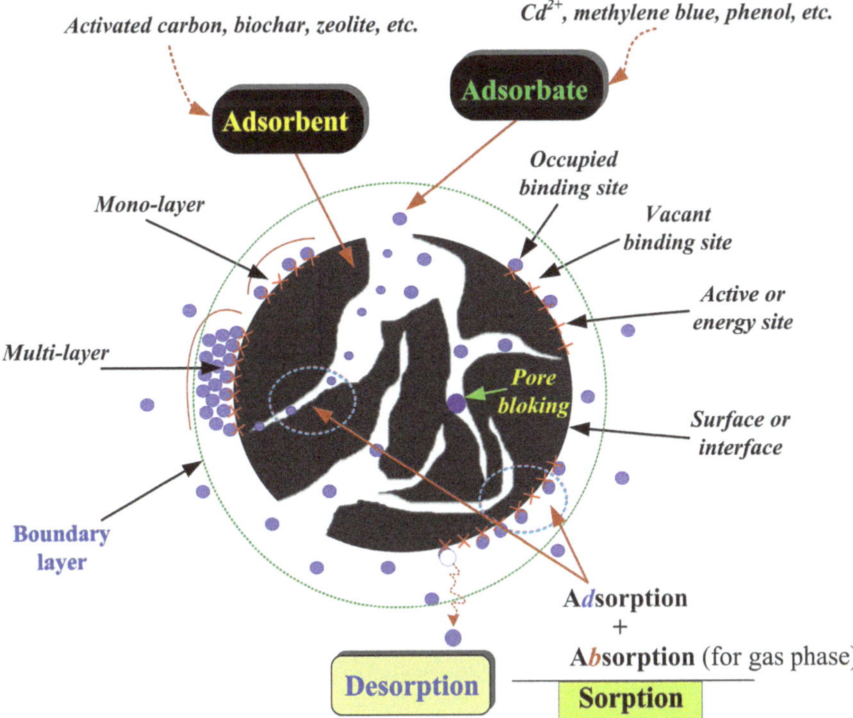

Fig. 8.1 Basic concept of adsorption. The adsorbate from liquid/gas phase are attached onto the surface of the adsorbent. This attachment can be a monolayer or multi-layer depending on the surface characteristics of the adsorbent and the nature of the adsorbate. Desorption can also occur where the adsorbate is released back into the liquid/gas phase. A distinction is also made between adsorption and absorption which goes beyond surface and involve bulk transfer of adsorbate into the adsorbent. (Adopted from Tran et al. (2017b) with permission from Elsevier)

treatment technologies and has been applied in the removal of various pollutants such as heavy metals (Peng et al. 2017; Uddin 2017), dyes (Tan et al. 2015; Ngulube et al. 2017), pharmaceuticals (Al-Khateeb et al. 2014; Ahmed 2017b), pesticides (Kyriakopoulos et al. 2005; Cederlund et al. 2016) and other recalcitrant pollutants. Some of the advantages of adsorption include its simplicity, cost-effectiveness and availability of numerous materials that can be used as adsorbents (Demiral and Güngör 2016; Ihsanullah et al. 2016; Aguayo-Villarreal et al. 2017).

8.2.2 Adsorbents

The intrinsic tendency of some natural porous materials to adsorb other substances was recognized as far back as the eighteenth century (Ruthven 1984). However, interest has been mostly on the engineered adsorbents, where materials with specific properties are developed to meet certain treatment objectives. A lot of progress has been recorded in the development of adsorbents, from the time when silica gel and commercial activated carbon were the major adsorbents, to date, where various engineered materials such as graphene-based composites, carbon nanotubes, chitosan-based composites and zeolite-based composites are being used. Some of the important considerations in selecting adsorbents include adsorption capacity, cost, stability and reusability, as well as the environmental implications.

8.2.2.1 Natural and Low-cost Adsorbents

Some solid materials have inherent abilities to adsorb and desorb other substances under natural conditions. These materials can be from natural sources such as clay minerals, biomass, agricultural wastes or industrial by-products. The basic characteristics of these materials are that they are low-cost and exhibit relatively low adsorptive capacities when used without any modification. However, it is possible to enhance the performances of these materials through physical and chemical modifications.

Among the various natural adsorbents, clay minerals are among the most widely investigated because of their abundance, low-cost and non-toxicity (Abidi et al. 2015; Shaban et al. 2017). Due to their negatively charged surface and colloidal nature, clay minerals are excellent adsorbents for cations and organic compounds. However, some form of surface modification may be necessary to enhance their adsorption capacity in many applications (Duc et al. 2006; Wang et al. 2016b). Other commonly used clay-based adsorbents include kaolin, red clay, montmorillonite and bentonite. Reviews on the applications of clay-based adsorbents have been presented recently (Zhu et al. 2016; Pandey 2017).

Agricultural wastes and by-products have been extensively investigated as alternative adsorbents in wastewater treatment (Amran et al. 2011; Anastopoulos and Kyzas 2014; Zhou et al. 2015). Agricultural wastes are lignocellulosic materials, consisting of lignin, cellulose and hemicelluloses components (Amran et al. 2011). Applications of agricultural wastes as adsorbents offer many advantages including availability, low cost, ease of regeneration and environmental friendliness (Nasuha and Hameed 2011; Hashemian et al. 2014). Details on agricultural wastes, including their applications and performances in treating recalcitrant wastewater have been presented earlier (Ioannidou and Zabaniotou 2007; Ahmaruzzaman 2008; Anastopoulos and Kyzas 2014; Zhou et al. 2015). However, despite the widely reported studies, agricultural wastes exhibit low adsorption capacities (Fig. 8.2) and required some form of modification for effective applications.

Byproducts and wastes from industrial activities such as fly ash, sludge, red mud and sawdust have also been used as adsorbents for the removal of recalcitrant organic pollutants (Ahmaruzzaman 2008). Although the applications of low-cost adsorbents in wastewater treatment is widely reported in the literature, their reported performances have been lower than engineered materials. However, their low costs and availability, coupled with development of various modification methods, have

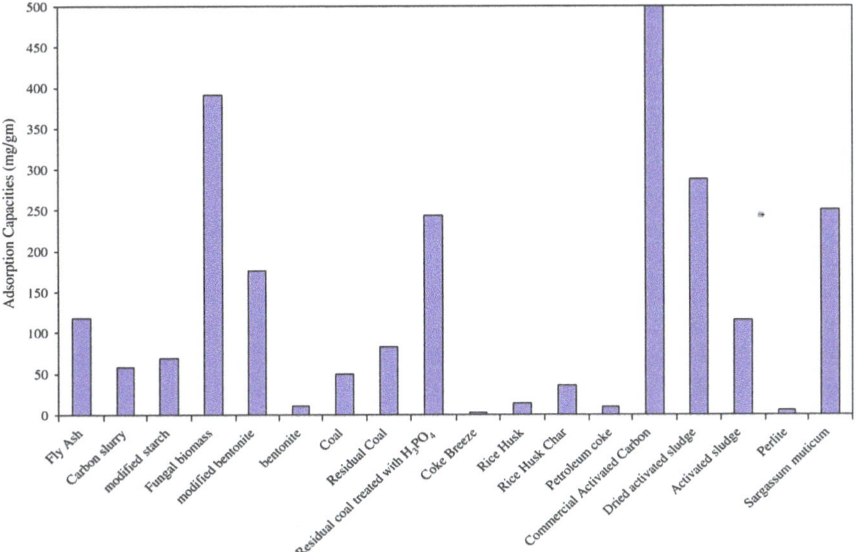

Fig. 8.2 Adsorption capacities of various low-cost adsorbents compared to commercial activated carbon in the adsorption of phenol. The adsorption capacity of commercial activated carbon is higher than all the adsorbents. The fungal biomass and dried activated sludge have relatively high adsorption capacities and may offer a more cost-effective treatment due to their low costs. (Adapted from Ahmaruzzaman (2008) with permission by Elsevier)

placed them at par with the engineered adsorbents. Nevertheless, these adsorbents are yet to be developed into industrial applications.

8.2.2.2 Engineered Adsorbents

Engineered adsorbents are those that are developed specifically for pollutants removal. These adsorbents are characterized by high surface area, adsorption capacity and stability. Commonly used engineered adsorbents include carbonaceous materials, polymeric materials and other composites. These materials are discussed here.

Carbon-based materials are the most widely used adsorbents for environmental applications. The commonly available carbon-based adsorbents include commercial activated carbons, graphene and carbon nanotubes. Among these, activated carbons are the most commonly used adsorbents for the removal of organic pollutants from the environment (Ahmaruzzaman 2008). Activated carbons are microporous particles with high surface area and internal porosity, which make them suitable adsorbents for environmental applications (Ioannidou and Zabaniotou 2007). The characteristics and performance of activated carbons depend on the precursor material and the activation methods employed. Generally, activated carbons are produced through different activation methods involving thermochemical treatment of the precursor material. Activated carbons can be produced in granular form or powdered form, each possessing certain advantages and disadvantages as discussed elsewhere (Kyriakopoulos and Doulia 2006). The literature on the applications of activated carbon in the treatment of recalcitrant organic pollutants is quite extensive. However, reviews on some of these applications have recently been presented (Mezohegyi et al. 2012; Ahmed 2017a; Kah et al. 2017).

Graphene, the newest member of the carbon family, has shown a great potential as an adsorbent. Graphene is a two-dimensional hexagonal network of carbon atoms, which possesses excellent structural, mechanical, electrical and thermal properties (Singh et al. 2011). These properties have made graphene a suitable candidate in various applications such electronics, power, medicine and environment. The properties of graphene and graphene-based composites have been discussed recently (Papageorgiou et al. 2017). Despite the excellent properties of graphene, it possesses some drawbacks such stacking and agglomeration during usage, which can hinder its effectiveness in certain applications (Singh et al. 2011). Consequently, graphene has been commonly modified to enhance its effectiveness in many applications (Kotov and Kelly 2006). Among the graphene derivatives, graphene oxide, a reduced form of graphene, has received interest as an excellent adsorbent owing to its abundant surface functional groups such as carboxyl, hydroxyl and epoxy groups (Mkhoyan et al. 2009). Details on graphene/graphene oxide, its preparation and composites have been reviewed previously (Rao et al. 2009; Singh et al. 2011; Randviir et al. 2014; Sherlala et al. 2018).

Besides graphene oxide, different modifications of graphene have been reported, especially the development of graphene nanocomposites using various organic and inorganic materials (Sherlala et al. 2018). With these modifications, additional functional groups are introduced into the composites, enhancing their adsorption capacities. Numerous graphene-based composites have been investigated as adsorbents in the removal of recalcitrant pollutants from wastewater. For example, the removal of a reactive dye from aqueous solution using a magnetic graphene-hydrogels was reported (Halouane et al. 2017). The composite achieved an adsorption capacity of about 114 mg/g and could be recycled five times while maintaining its capacity. In another study, the removal of a pharmaceutical pollutant using a magnetic-graphene oxide sponge was attempted, where the composite achieved an adsorption capacity of 473 mg/g (Yu et al. 2017). In another study, Du and Pan (2014) prepared a cellulose-graphene oxide composite for the adsorption of triazine pesticides from aqueous solution. The composite exhibited high performance, removing more than 95% of the initial concentrations of the pesticides. The applications of graphene for sequestration of organic pollutants have been reviewed recently (Ersan et al. 2017).

Carbon nanotubes are carbon materials that have been widely studied as adsorbents. Although the interests on carbon nanotubes preceded that of graphene, it has been overshadowed by graphene due to the comparative superior performance and versatility of the latter. Carbon nanotubes are tubular-structured carbons, with large surface areas, porous structure and strong affinity towards pollutants (Bhanjana et al.

2017). The surfaces of carbon nanotubes can be easily modified to accommodate other materials, forming hybrids/composites for various applications. Carbon nanotubes have been widely investigated as alternative adsorbents in the treatment of recalcitrant organic pollutants such as dyes (Kumar et al. 2013; Eskandarian et al. 2014), pesticides (Pyrzynska 2011; Hadi et al. 2017; Hua et al. 2017), pharmaceuticals (Gauden et al. 2011; Strachowski and Bystrzejewski 2015; Diaz-Flores et al. 2017) and phenolic compounds (Gauden et al. 2011; Strachowski and Bystrzejewski 2015; Diaz-Flores et al. 2017).

Carbon-based materials are among the most effective adsorbents for the removal of recalcitrant organic pollutants from the environment. Their excellent properties and adsorption capacities are unrivalled by other adsorbents. However, these materials are relatively expensive, which limit their large-scale applications. In addition, the applications of graphene and carbon nanotubes in adsorption are still at laboratory stage. Nevertheless, beside technical issues on the use of these nano-adsorbents, the overall treatment cost will be a deciding factor in their industrial applications.

Various synthetic polymers have been utilized as alternative adsorbents in wastewater treatment. Adsorption polymers are traditionally based on ion-exchange resins, which have lower surface area and consequently lower adsorption capacities compared to activated carbons. However, the advances in polymer science have resulted in the production of polymers with comparative surface areas to activated carbons (Kyriakopoulos and Doulia 2006). Polymeric adsorbents are low-energy demanding and can be easily regenerated at lower cost compared to activated carbons (Frimmel et al. 1999). There is a large body of literature on the application of polymeric adsorbents in the removal of organic pollutants. The use of polymeric adsorbents in wastewater treatment have been reviewed (Pan et al. 2009; Zhao et al. 2011; Sajid et al. 2018).

Recently, there is a growing effort in enhancing the performance of polymeric adsorbents through surface modification, copolymerization and other strategies. For example, Ying et al. (2017) prepared a poly-ampholytic polymer based on sodium alginate and polyacrylic acid-co-polymethacrylothyethyltrimethyl ammonium chloride for the adsorption of anionic dye, Food Yellow 3. The developed copolymer exhibited high adsorption capacity, removing up to 655 mg/g of the dye, which is among the highest reported for the removal of Food Yellow 3.

8.3 Advanced Oxidation Processes

8.3.1 Background

Advanced oxidation processes is a term used generally to refer to wastewater treatment processes that are characterized by the generation of hydroxyl radicals ($OH^{\cdot}$) capable of degrading organic pollutants. The $OH^{\cdot}$ is a very reactive radical with a redox potential of 2.8 ev that can react with organic pollutant in the order of $10^9\ M^{-1}\ S^{-1}$, converting them to carbon dioxide, water and intermediates (Neyens

Table 8.2 Types and descriptions of the common advanced oxidation processes used for the removal of organic pollutants from water

Advanced oxidation process	Description
Fenton oxidation	Homogeneous reaction between Fe^{2+}/Fe^{3+} and H_2O_2 to generate hydroxyl radicals
	Heterogeneous reaction between iron oxide or other metal oxides and H_2O_2 to generate hydroxyl radicals
Photocatalysis	Semiconductor is illuminated with ultraviolet irradiation to generate electron-hole pair
	Recombination of the electron and hole can occur, decreasing the efficiency of the process
Ozonation	Ozone is used as a direct oxidant but is selective in its reaction
	Ozone can decompose, especially driven by a catalyst, to generate hydroxyl radicals
Ultrasound cavitation	Ultrasound cavitation is used to generate hydroxyl radicals through bubbles formation and collapse
	Degradation can also occur through pyrolytic effect
Electrochemical oxidation	Processes that utilized electrolyte cells to generate hydroxyl radicals
	Most common is anodic oxidation using metal oxide as anode
	Electro-Fenton oxidation involves homogeneous Fenton and anodic oxidation

and Baeyens 2003). The discussion on advanced oxidation processes was initially limited to technologies involving ozone, hydrogen peroxide and ultraviolet irradiation as discussed by Glaze and co-workers (Glaze et al. 1987). Although the discussion of Glaze et al. was mainly on ozone, hydrogen peroxide and ultraviolet irradiation, the field of advanced oxidation processes has grown widely since then.

Nowadays, the term is used broadly, involving many processes such as Fenton oxidation, photocatalysis, Ozonation, ultraviolet irradiation, ultrasound cavitation, electrochemical oxidation, supercritical water oxidation and other emerging technologies. Table 8.2 shows the common advanced oxidation processes for wastewater treatment. Advanced oxidation processes have proven effective in the treatment of recalcitrant organic pollutants. A lot of developments have been recorded in the field of advanced oxidation processes, which are largely fueled by the advances in material science and the concerted efforts of researchers working in the field. Dewil et al. (2017) have recently offered some new perspectives on advanced oxidation processes, discussing their integration with other treatment technologies, solar-driven processes and the emerging electrochemical oxidation.

The effectiveness of advanced oxidation processes has led to their wide investigations by researchers working in developing treatment technologies for recalcitrant wastewaters. Due to the high oxidation power of $OH^{\cdot}$, organic pollutants are effectively degraded. However, complete mineralization is not always feasible because of the intensive energy or chemical requirements of some of the processes. Although some processes such as Fenton and Ozonation have been applied at industrial scales, many of the emerging technologies are still under laboratory investigations. Also, the mechanisms and effects of operational parameters of some of the processes are still being discussed.

8.3.2 *Types of Advanced Oxidation Processes*

Several approaches can be used in classifying advanced oxidation processes. While the common approach is to classify them based on the processes used to generate $OH^{\cdot}$ (Table 8.2), other classifications can be based on whether the process is homogeneous or heterogeneous and whether a source of energy is required or not. However, classifying advanced oxidation processes based on the processes used in generating the oxidant seems more practical. Discussion on other classification methods has been presented earlier (Poyatos et al. 2010). In the context of $OH^{\cdot}$ generating methods, Fenton-based processes, photocatalysis, Ozonation and ultrasound irradiation are discussed here.

8.3.2.1 Fenton Oxidation

Fenton oxidation involves the use ferrous iron (Fe^{2+}) to catalyze the decomposition of H_2O_2 to $OH^{\cdot}$ as shown in Eq. 8.1. The generated $OH^{\cdot}$ can then react with organic pollutants, oxidizing them to some organic radicals that can be further oxidized (Eq. 8.2). Fenton oxidation is an effective technology for the treatment of recalcitrant organic pollutants. The Fe^{2+} rapidly decomposes the H_2O_2, generating the $OH^{\cdot}$. The Fe^{2+} is reduced to Fe^{3+}, which further react with H_2O_2 at slower rate and produced hydroperoxyl radical ($^{\cdot}O_2H$) (Eq. 8.3). The hydroperoxyl radicals are also capable of reacting with organic pollutants, though at slower rate than $OH^{\cdot}$. Although the mechanism of Fenton oxidation is presented in these simple steps, the actual process is more complex, involving chain reactions that are still subject to discussion (Hu et al. 2015). Details on the mechanism of Fenton oxidation have been presented earlier (Neyens and Baeyens 2003; Babuponnusami and Muthukumar 2014; Wang et al. 2016a).

$$Fe^{2+} + H_2O_2 \rightarrow Fe^{3+} + OH^{\cdot} + OH^{-} \quad (8.1)$$

$$RH + OH^{\cdot} \rightarrow H_2O + R^{\cdot} \rightarrow \text{further oxidation} \quad (8.2)$$

$$Fe^{3+} + H_2O_2 \rightarrow Fe^{2+} + HO_2^{\cdot} + OH^{+} \quad (8.3)$$

Fenton oxidation can be a homogeneous or heterogeneous process, depending on the type of catalyst employed. Homogeneous Fenton oxidation uses soluble iron salts, such as $FeSO_4$, as the catalyst for decomposing H_2O_2. Since Fe^{2+} are readily available in the solution, there is little mass transfer resistance and the process proceeds rapidly. However, homogeneous Fenton oxidation possesses some drawbacks. The process requires the working pH to be maintained under acidic condition, in the range of 2.8–3.5. This is necessary to ensure that Fe^{2+} and Fe^{3+} maintain their active role and to avoid their precipitations as oxyhydroxides (Clarizia et al. 2017). Thus, pH values outside this range will impair the effectiveness of the process. Another drawback of process is the production of excessive sludge from the dissolved iron (Lima et al. 2017). At the end of the process, a neutralization stage is usually required to stop the Fenton reaction. This is achieved by adding a base to raise the pH of the solution to about 10, quenching the H_2O_2 and at the same time, precipitating iron sludge. A lot of efforts have been devoted towards addressing these limitations.

One of the approaches toward addressing the limitations of homogeneous Fenton oxidation is the use of heterogeneous catalysts. In heterogeneous Fenton oxidation, oxides of iron or other metals are employed as the catalysts for the decomposition of H_2O_2. Various forms of iron oxide, such as magnetite, maghemite, goethite and hematite have been used as catalysts in heterogeneous Fenton oxidation (Pouran et al. 2014). Heterogeneous Fenton oxidation offers the advantages of using wider operational pH since there is no concern for Fe^{2+}/Fe^{3+} precipitation. However, mass transfer limitation is seen as a disadvantage of heterogeneous Fenton oxidation compared to homogeneous Fenton (Cruz et al. 2017). Besides iron oxide, other metal catalysts have been used in heterogeneous Fenton oxidation, the so-called Fenton-like processes. In Fenton-like processes, the basic concept of Fenton oxidation is adopted with some modifications, usually in the catalyst used. Various metal oxides, and of recent, nanocomposites, are used as catalysts in Fenton-like processes. Another variation of Fenton, the photo-Fenton, uses a source of light such as ultraviolet irradiation to enhance the process performance. The possibility of using solar energy in the process makes it attractive due to the potential lower cost of treatment (Dewil et al. 2017; Villegas-Guzman et al. 2017).

Besides using heterogeneous catalysts, other approaches have been explored to overcome the limitations of homogeneous Fenton. One of these strategies involves the use of various chelating agents that allow the application of homogeneous Fenton oxidation at circumneutral pH. The basic mechanism is the formation of complexes between the chelating agents and Fe^{3+}, which prevent the precipitation of Fe^{3+}. Compounds such as oxalate, citrate, ethylenediaminetetraacetic acid, nitrilotriacetic acid can be used as chelating agents to form Fe^{3+} complexes. Clarizia et al. (2017) have recently reviewed the various strategies to conduct homogeneous Fenton oxidation at circumneutral pH.

Electro-Fenton oxidation is another approach that can enhance the performance of homogeneous Fenton oxidation. It involves combining anodic oxidation with homogeneous Fenton oxidation to generate the hydroxyl radicals. Anodic oxidation is an electrochemical process where hydroxyl radicals are generated directly through the use of high O_2 evolutions anode, such as Pt and PbO_2 anodes (Vasudevan and Oturan 2014). In electro-Fenton oxidation, soluble ferrous salt is introduced into the anodic oxidation process, which generate additional homogeneous hydroxyl radicals (Barhoumi et al. 2017). Electro-Fenton offers additional advantages such as the in-situ generation of H_2O_2 and cathodic regeneration of ferrous ions (Sirés et al. 2014; Paramo-Vargas et al. 2016).

Fluidized bed Fenton process is another strategy towards addressing the limitations of the homogeneous Fenton oxidation. In the process, homogeneous Fenton is conducted in a fluidized bed reactor, employing solids such as SiO_2 and Al_2O_3 as carriers. The fluidized carries provide surface for the precipitation of iron oxide, thus reducing the generation of iron sludge in the solution (Bello et al. 2017). The basic mechanism involves the crystallization of the ferric hydrolysis product on the surface of the carriers. Additionally, the fluidization provides excellent mixing between the Fenton's reagent and the target pollutant, enhancing process performance (Chen et al. 2016). Some studies have reported the superior performance of fluidized bed Fenton process compared to conventional homogeneous Fenton oxidation (Liu et al. 2014; Matira et al. 2015). Besides the Fenton's reagent, other important parameters include the types of carrier, particle size, loading, fluidization velocity and reactor geometry (Bello et al. 2017).

8.3.2.2 Photocatalysis

Photocatalysis is another form of advanced oxidation processes where semiconductor catalysts and a source of light are utilized to generate the oxidizing agent. Generally, photocatalysis is based on the ability of the photocatalyst to simultaneously adsorb reactants and absorb photons efficiently (Herrmann 2010). As a wastewater treatment process, photocatalysis offers many advantages such as complete mineralization of pollutants, operating under ambient condition and low operational cost. The fundamentals and mechanism of photocatalysis have been previously reviewed (Herrmann 2010; Meng et al. 2010; Mohamed and Bahnemann 2012; Bora and Mewada 2017).

Briefly, photocatalysis involves illuminating the surface of a semiconductor with photon energy equal to or greater than the bandgap energy of the material. This excites an electron from the valence band to the conducting band, leaving behind an empty band, which will lead to the creation of electron-hole pair (e^- -h^+). Considering TiO_2 as the photocatalyst, Eqs. 8.4, 8.5, 8.6, 8.7, 8.8 and 8.9 show the typical reactions while Fig. 8.3 shows the basic mechanism of photocatalysis. The degradation of pollutants occurs by the $OH^{\cdot}$ (Eq. 8.8) and superoxide radical anions ($O_2^{\cdot -}$) (Eq. 8.9). One of the major challenges in photocatalysis is the prevention of the recombination reaction that occurs in Eq. 8.5. Recombination of the generated charge carriers reduces the overall quantum efficiency and prevent them from reacting with the target pollutants (Choi et al. 1994).

$$TiO_2 + hv \rightarrow h^+ + e^- \quad (8.4)$$

$$e^- + h^- \rightarrow \text{energy} \quad (8.5)$$

$$H_2O + h^+ \rightarrow OH^{\cdot} + H^+ \quad (8.6)$$

$$O_2 + e^- \rightarrow O_2^{\cdot -} \quad (8.7)$$

$$OH^{\cdot} + \text{pollutant} \rightarrow\rightarrow\rightarrow H_2O + CO_2 \quad (8.8)$$

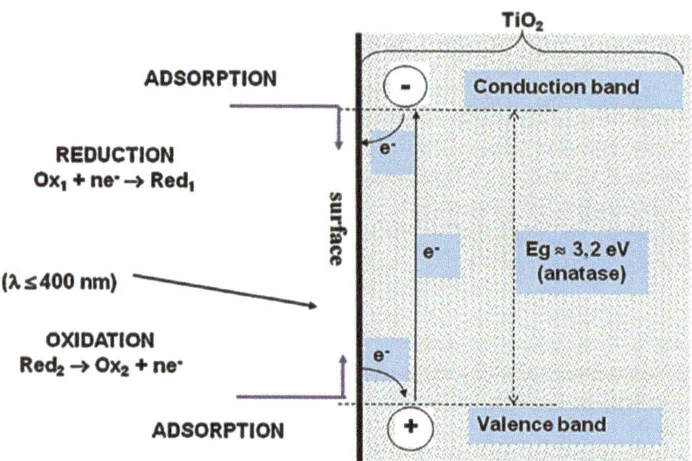

Fig. 8.3 Basic mechanism of photocatalysis using TiO_2. Once the surface of the photocatalyst is illuminated, the electron is excited and moved up to the conduction bad, leaving a + ve hole in the valence bad. The overall process involves reduction-oxidation and adsorption at the surface of the photocatalyst. (Adapted from Herrmann (2010) with permission from Elsevier)

$$O_2^{\cdot -} + \text{pollutant} \rightarrow\rightarrow\rightarrow H_2O + CO_2 \tag{8.9}$$

Although the work of Fujishima and Honda (1972) on photo-electrolysis of water on the surface of TiO_2-based electrode is generally considered to be the pioneering work on photocatalysis, the works of Barry and Stone (1960) and Doerffler and Hauffe (1964) on heterogeneous photocatalysis using ZnO were published earlier. However, photocatalysis was first applied for the removal of environmental pollutants by Frank and Bard in 1977, where they studied the photocatalytic degradation of cyanide ions (Frank and Bard 1977). A lot of progress has been recorded in the field of photocatalysis since the publications of these pioneering works. Among the semiconductors used, TiO_2 has been the most widely applied. This is largely due to the excellent properties of TiO_2 and its commercial availability. However, in the last few years, there has been a growing shift to the use of other metal oxides, especially those that are capable of high activity under visible-light.

Photocatalysis has been successfully used in the degradation of various organic pollutants such as dyes (Zhang et al. 2010; Yu et al. 2012; Sudrajat and Babel 2016; Saeed et al. 2017), pharmaceuticals (Keen and Linden 2013; Kanakaraju et al. 2014; He et al. 2016) pesticides (Laxma and Kim 2015; Kosera et al. 2017) and other recalcitrant pollutants. However, despite the effectiveness of photocatalysis using TiO_2, it suffers some limitations. These include wide band gap which limits its activity to specific wavelength, recombination of the photo-generated charge carriers and the need for downstream separation (Meng et al. 2010; Etacheri et al. 2015). Thus, addressing these limitations have received wide attention. Strategies such as heterogeneous coupling, doping of TiO_2 with suitable materials, and immobilization of TiO_2 can be explored in addressing these limitations (Choi et al. 1994; Bora and

Mewada 2017). Heterogeneous coupling can delay or prevent the recombination of the charge carriers, doping can extend the activity of the TiO_2 while immobilization can address the challenges of post-treatment separation.

Another active research area in photocatalysis is the development of visible-light active photocatalysts that can be used in solar energy-driven treatment process. Since visible-light contributes for 45% of the solar energy spectrum in contrast to 5% of ultraviolet, visible-light-driven photocatalysis appears very promising. However, the commonly used semiconductors are only responsive to a narrow band of the ultraviolet irradiation. Consequently, their photocatalytic activity under visible light is limited. To this end, some strategies have been adopted to extend the photocatalytic activity of semiconductors to the visible-light spectrum. These include doping with noble metals (Luo et al. 2015) and heterogeneous coupling (Ren et al. 2014).

8.3.2.3 Ozonation

Ozone (O_3) is one of the strongest chemical oxidants, having a very high thermodynamic oxidation potential (2.07 eV). Ozonation is a well-established advanced oxidation process for water treatment that is capable of oxidizing recalcitrant organic pollutants. Although ozone is a powerful oxidant, its oxidation reactions are rather selective and may be considered slow for wastewater treatment applications (Glaze et al. 1987). However, catalytic Ozonation and integrating Ozonation with other processes have been pursued as possible ways to improve its effectiveness. Some of the strategies being adopted to enhance the performance of Ozonation have been presented recently (Gomes et al. 2017).

The mechanism of Ozonation has been extensively discussed in the literature (Glaze et al. 1987; Umar et al. 2013; Gomes et al. 2017). In a classical Ozonation, organic pollutants are oxidized through direct reaction with molecular ozone and/or indirect reaction with $OH^{\cdot}$ that is generated from the decomposition of ozone (Broséus et al. 2009). While oxidation by direct molecular ozone may be limited to the degradation of multiple bonds and aromatic compounds, degradation by the $OH^{\cdot}$ is very fast and nonselective (Glaze et al. 1987; Broséus et al. 2009).

The pH condition and wastewater matrix determine whether direct or indirect oxidation will dominate. Direct oxidation dominates under acidic conditions or in the presence of radical scavengers that can hinder the chain reaction necessary to decompose ozone (Umar et al. 2013). On the other hand, indirect oxidation dominates under basic conditions or in the presence of compounds that promote radical-type chain reactions and $OH^{\cdot}$ formation (Staehelin and Hoigne 1985; Irmak et al. 2005). However, in a real wastewater, radical scavengers are always present and therefore, oxidation by molecular ozone may be the dominant mechanism (Nakada et al. 2007; Umar et al. 2013).

Equations 8.10, 8.11, 8.12, 8.13, 8.14 and 8.15 show the chain reactions that occur in the decomposition of ozone in aqueous medium. The reactions occur in three stages of initiation, propagation and termination as outlined below (Irmak et al. 2005; Umar et al. 2013).

Initiation stage: this involves the decomposition of ozone initiated by OH^-, resulting in the formation of hydroperoxide ion (HO_2^-) (Eq. 8.10). The HO_2^- further reacts with ozone to form $H_2^{\cdot}$ and $O_3^{\cdot-}$(Eq. 8.11).

$$O_3 + OH^- \rightarrow HO_2^- + k_1 = 70\,M^{-1}S^{-1} \quad (8.10)$$

$$HO_2^- + O_3 \rightarrow H_2^{\cdot} + O_3^{\cdot-}, k_2 = 2.2 \times 10^6 M^{-1}S^{-1} \quad (8.11)$$

Propagation stage: in this stage, ozone is further decomposed into ozonide ion $(O_3^{\cdot-})$ by reacting with $O_2^{\cdot-}$ (Eq. 8.12). The ozonide ion then reacts with $H^+$to form $HO_3^{\cdot}$ (Eq. 8.13), which is further converted to hydroxyl radical ($OH^{\cdot}$) (Eqs. 8.14 and 8.15). Further reactions may occur, producing $HO_4^{\cdot}$ and $HO_2^{\cdot}$ (Eqs. 8.16 and 8.17).

$$O_3 + O_2^{\cdot-} \rightarrow O_3^{\cdot-}, k_2 = 1.6 \times 10^9 M^{-1}S^{-1} \quad (8.12)$$

$$O_3^{\cdot-} + H^+ \rightarrow HO_3^{\cdot} \;\; k_3 = 5.0 \times 10^{10} M^{-1}S^{-1} \quad (8.13)$$

$$HO_3^{\cdot} \rightarrow O_3^{\cdot-} + H^+, \;\; pKa = 6.2 \quad (8.14)$$

$$HO_3^{\cdot} \rightarrow OH^{\cdot} + O_2, k_3 = 1.4 \times 10^5 M^{-1}S^{-1} \quad (8.15)$$

$$O_3 + OH^{\cdot} \rightarrow HO_4^{\cdot} \;\; k_4 = 2.0 \times 10^9 M^{-1}S^{-1} \quad (8.16)$$

$$HO_4^{\cdot} \rightarrow HO_2^{\cdot} + O_2 \;\; k_5 = 2.8 \times 10^4 M^{-1}S^{-1} \quad (8.17)$$

Termination stage: This stage may involve the recombination of $OH^{\cdot}$, $HO_2^{\cdot}$ and O_2.

Ozonation has been widely applied in the treatment of recalcitrant organic pollutants such as dyes (Panda and Mathews 2014; Hu et al. 2016), pharmaceuticals and personal care products (Hansen et al. 2016; Gomes et al. 2017) and pesticides (Chelme-Ayala et al. 2010; Sans and Esplugas 2017). As mentioned earlier, direct oxidation by ozone is quite selective and may not be able to decompose complex organic compounds. To speed up the process and ensure the production of $OH^{\cdot}$, various catalysts are applied in Ozonation, the so-called catalytic Ozonation. The catalyst acts as a radical promoter, ensuring the completion of the chain reactions and the consequent generation of $OH^{\cdot}$.

Metal oxides and activated carbon-based materials have been traditionally used as heterogeneous catalysts in catalytic Ozonation. However, with the increasing progress in material science, various composites have been developed for catalytic Ozonation such as graphene-based composites (Ahn et al. 2017; Yin et al. 2017), zeolite-based composites (Gümüs 2017; Ikhlaq and Kasprzyk-hordern 2017), carbon nanotubes (Wang et al. 2018), and alumina-based composites (Roshani et al. 2014; Ikhlaq et al. 2015). With the use of these composites, potential exist for combining adsorption with heterogeneous catalytic Ozonation.

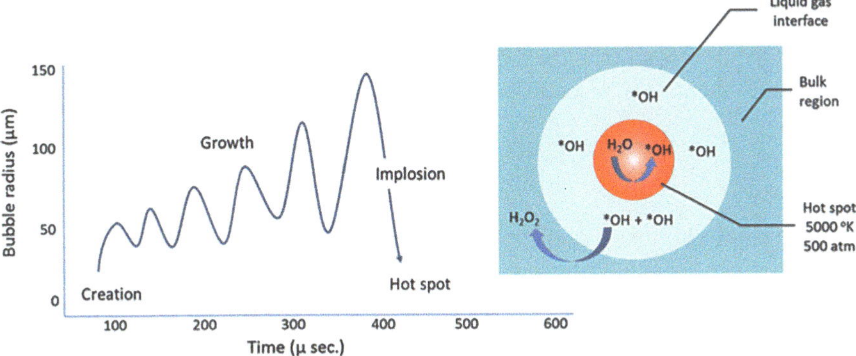

Fig. 8.4 Cavitation bubble formation and collapse which is the underlaying principle of acoustic cavitation. Two products from the bubble-collapse, hydroxyl radicals and heat, are utilized in the degradation of pollutants. The growth of the cavitation bubble depends on the intensity of the sound wave. (Adapted from Tran et al. (2015) with permission from Springer Nature)

8.3.2.4 Ultrasound

Ultrasound is a form of advanced oxidation processes where ultrasound irradiation is used to induce an acoustic cavitation capable of generating $OH^{\cdot}$. Although the generation of $OH^{\cdot}$ is emphasized, the cavitation phenomenon also results in the generation of local hot spots that can induce pyrolytic degradation of the pollutants. Thus, pollutants removal in ultrasound consists of the degradation by $OH^{\cdot}$ radicals and the pyrolytic destruction by the generated heat (Chowdhury and Viraraghavan 2009). In the literature, different names are used to denote the use of ultrasound irradiation for wastewater treatment, such as sonolysis, sonication and sonochemistry. However, the basic process remains the same.

In general, cavitation is induced in a liquid whenever the intensity of the ultrasound exceeds the cavitation thresholds of the liquid. This results in the formation, growth and collapse of bubbles in the liquid as depicted in Fig. 8.4. By these growth and collapse cycles, the sound waves exert negative and positive pressures respectively on the liquid (Chowdhury and Viraraghavan 2009). When the negative pressure reaches a certain threshold level, the average distance between the liquid molecules would exceed the critical molecular distance required to keep the liquid together (Mason and Lorimer 1988; Suslick 1989; Chowdhury and Viraraghavan 2009). At this point, the liquid would break down, forming cavitation bubbles under high temperature that can reach 5000 K and pressure up to 1000 atmosphere (Suslick 1990). These bubbles consist of compressed gas and vapor, which generate localized hot spots capable of pyrolyzing water molecules into hydroxyl and hydrogen radicals (Eqs. 8.18, 8.19 and 8.20) (Wang and Xu 2012). Details on the principles and theories of sonochemistry have been extensively discussed in the literature (Mason and Lorimer 1988; Suslick 1990; Adewuyi 2001; Adewuyi 2005; Gogate 2008; Wang and Xu 2012).

$$H_2O \rightarrow OH^{\cdot} + H^{\cdot} \tag{8.18}$$

The $OH^{\cdot}$ then react with the target organic pollutant (8.19). A side reaction may also occur, where $OH^{\cdot}$ recombines to form H_2O_2 (Eq. 8.20).

$$R + OH^{\cdot} \rightarrow RO + CO_2 + H_2O + \text{intermediates} \tag{8.19}$$

$$OH^{\cdot} + OH^{\cdot} \rightarrow H_2O_2 \tag{8.20}$$

The use of ultrasound cavitation as an advanced oxidation process is relatively new. However, there has been a growing interest in its applications in the degradation of organic pollutants as evident in the number of publications on the topic. Ultrasound offers some advantages in wastewater treatment such as high mass transfer and short reaction cycles, energy conservation, non-requirement of chemical input and non-generation of secondary pollution (Adewuyi 2001; Chowdhury and Viraraghavan 2009). Additionally, it can be easily integrated with other treatment technologies, making it versatile.

Numerous studies have been reported on the applications of ultrasound in the degradation of recalcitrant pollutants such as phenolic compounds (Gogate 2008), dyes (Das 2016; Khataee et al. 2017), pharmaceuticals (Rayaroth et al. 2016; Yang et al. 2016; Tran et al. 2017a), pesticides (Matouq et al. 2008; Agarwal et al. 2016) and other organic pollutants (Dietrich et al. 2017; Jawale et al. 2017). In most cases, ultrasound is combined with other treatment technologies, particularly adsorption and other advanced oxidation processes. However, despite the growing interest in the applications of ultrasound in wastewater treatment, aspects such as process kinetics, reactor design and scale-up of the technology are yet to be fully established (Chowdhury and Viraraghavan 2009).

8.4 Combined Adsorption-Advanced Oxidation Processes

The possibility of combining different wastewater treatment processes to achieve a set of treatment objectives is an important consideration in wastewater treatment. Process integration in wastewater treatment is an attractive approach that can be used to increase process performance, overcome some drawbacks and possibly lower operational cost (Cataldo et al. 2016). For example, since adsorption is a low-energy demanding technology while some advanced oxidation processes are energy intensive, integrating the two technologies may lower the overall energy requirement.

Combining the cost-effectiveness of adsorption with the effective degradation of advanced oxidation processes offers several benefits. Technologies such as Fenton and photocatalysis require the input of chemicals, which can be very high when treating high-strength wastewaters. Integrating adsorption with advanced oxidation processes, especially when low-cost adsorbents are employed, can lower the amount of chemicals required to achieve the treatment objective. Although advanced oxidation processes can degrade complex organic pollutants, complete mineralization is not always feasible. Consequently, the process may end up producing some intermediate pollutants. Thus, combining advanced oxidation processes with adsorption could ensure that these intermediates are removed from the treated wastewater (Dwivedi et al. 2017). Advanced oxidation processes such as Fenton oxidation and photocatalysis can be designed to enhance the regeneration of adsorbents by degrading the adsorbed pollutants.

8.4.1 *Adsorption-Fenton*

As discussed in Sect. 8.2.1, Fenton oxidation involves the use of Fe^{2+} to catalyze the decomposition of H_2O_2 for the generation of $OH^{\cdot}$. Many studies have reported the applications of combined adsorption-Fenton oxidation for the treatment of recalcitrant pollutants (Table 8.3). In this section, we discussed such applications, highlighting the fundamentals and synergistic performance of the combined process.

There are two possible routes for combining adsorption with Fenton oxidation (Fig. 8.5). The first involves the use of heterogeneous bifunctional materials which can act both as adsorbent and catalyst for the decomposition of H_2O_2 (Li et al. 2017). Although adsorption is an inherent part of heterogeneous Fenton oxidation, investigations into its contribution to the pollutant removal is scarcely reported (Akbar et al. 2017). Iron oxides, the commonly used catalysts in heterogeneous Fenton oxidation, are good adsorbents and can contribute toward the removal of target pollutants through adsorption. The second route of adsorption-Fenton oxidation involves the use of adsorbents, such as activated carbon, and homogeneous Fenton's reagent. In this case, a distinct adsorption process is integrated with the Fenton oxidation. Thus, the discussion on adsorption-Fenton oxidation are based on the two routes highlighted above and depicted in Fig. 8.5.

8.4.1.1 Synergistic Effect

The synergistic effect between adsorption and oxidation can be explored toward process improvement and addressing the limitations of conventional Fenton oxidation. For example, Hadjltaief et al. (2017) investigated the removal of 2-chlorophenol from aqueous solution using natural clay as both adsorbent and catalyst for Fenton-like oxidation. When the clay was used alone as adsorbent, only 22.1% removal was achieved. This removal increased to 61.72% when H_2O_2 was introduced into the system, indicating the catalytic activity of the natural clay and the synergy of adsorption and oxidation. In another study using adsorption-Fenton oxidation for removal of carbamazepine, the degradation efficiency increased from 49.39% under Fenton oxidation to 99.51% under combined adsorption-Fenton (Dwivedi et al. 2017). Thus, it is evident that combining adsorption with Fenton oxidation results in treatment performances higher than individual performances (Wang et al. 2014b; Hu et al. 2015; Li et al. 2015). Figure 8.6 shows the performance of combined

Table 8.3 Reported studies on Fenton-adsorption for the treatment of recalcitrant wastewater

Process description	Target pollutant	Operational conditions	Performance	References
Homogeneous Fenton + Granular activated carbon	Pharmaceutical: carbamazepine	Adsorbent dosage: 300 g/L	Fenton: 49.39%	Dwivedi et al. (2017)
		H_2O_2: 8.5 g/L	Fenton + adsorption: 99.51%	
		pH: 3.5		
		Time: 30–180 min		
Heterogeneous Fenton: Fe_3O_4/CeO_2	Dye: Methylene blue	Catalyst/adsorbent dosage: 1 g/L	Adsorption: 40%	Li et al. (2017)
		H_2O_2: 163.7 mM	Fenton + adsorption: 100%	
		pH: 4–6		
		Time: 120 min		
Heterogeneous Fenton: Magnetic nanoparticles +activated carbon	Dye: Direct red 16	Catalyst/adsorbent dosage: 0.2 g/L	Adsorption: 22.6%	Akbar et al. (2017)
		H_2O_2: 6 mM	Adsorption + Fenton: 88.8%	
		pH: 2		
		Time: 30–180 min		
Heterogeneous Fenton: Activated carbon-Fe_3O_4	Pharmaceutical: Tetracycline	Catalyst/adsorbent dosage: 0.2 g/L	Adsorption: 56%	Jafari et al. (2017)
		PS: 30 mM	Adsorption + Fenton: 80.77%	
		pH: 5.5		
		Time: 180 min		
Heterogeneous Fenton: Activated carbon+γ-Fe_2O_3	Dye: Methylene blue	Catalyst/adsorbent dosage: 2 g/L	Adsorption: 73%	Fayazi et al. (2016)
		H_2O_2: 12 mM	Adsorption + Fenton: 100%	
		pH: 5		
		Time: 15–30 min		
Homogeneous Fenton + activated carbon/ceramsite	Dyes	Adsorbent dosage: 175 mm	Adsorption: 65%	Lyu et al. (2016)
		[Dye]/[Fe^{2+}]/[H_2O_2]/: 1:1:10	Adsorption + Fenton: 94%	
		pH: 3		
		Time: 10 min		
Heterogeneous Fenton: $CuMnO_2$	Benzotriazole	Catalyst/adsorbent dosage: 1 g/L	Adsorption + Fenton: 90%	Zhang et al. (2016)
		H_2O_2/Benzotriazole: 14:1		
		pH: 3–8		
		Time: 180 min		

(continued)

Table 8.3 (continued)

Process description	Target pollutant	Operational conditions	Performance	References
Heterogeneous Fenton: Fe-zeolite	Dye: Methylene blue	Catalyst/adsorbent dosage: 0.3 g/L	Adsorption: 53.8%	Zhang et al. (2017b)
		H_2O_2: 25 g/L	Fenton + adsorption: 87.7%	
		pH: 4		
		Time: 180 min		
Heterogeneous Fenton: Iron mining residue	Mercaptobenzothiazole	Catalyst/adsorbent dosage: 3 g/L	Adsorption: 10%	Lau et al. (2016)
		H_2O_2: 25 g/L	Adsorption + Fenton: 100%	
		pH: 3		
		Time: 60 min		
Heterogeneous Fenton: Fe-Montmorillonite	Dye: Rhodamine blue	Catalyst/adsorbent dosage: 0.5 g/L	Adsorption: 80%	Guo and Zhang (2015)
		H_2O_2: 10 mM	Adsorption + Fenton: 99%	
		pH: 2–10		
		Time: 60 min		
Heterogeneous Fenton: Zeolite-Cobalt ferrite nanoparticles	Phenol & paracetamol	Catalyst/adsorbent dosage: 0.2 g/L	Adsorption: 40%	Irani et al. (2015)
		H_2O_2. 50 mM	Adsorption + Fenton: 99.5%	
		pH: 3		
		Time: 90 min		
Heterogeneous Fenton: Fe-granular activated carbon	Bisphenol A	Catalyst/adsorbent dosage: 0.7 g/L	Adsorption + Fenton: 99%	Kim et al. (2015)
		H_2O_2/Bisphenol A: 36–108		
		pH: 3		
Heterogeneous Fenton: Fe-activated carbon	Atrazine	Catalyst/adsorbent dosage: 2 g/L	Adsorption: 70%	Morales-Perez et al. (2015)
		H_2O_2: 284 μL	Adsorption + Fenton: 96%	
		pH: 6		
		Time: 120 min		
Heterogeneous Fenton: Zero valent iron residue	Furfural	Catalyst/adsorbent dosage: 50 g/L	Adsorption: 97%	Li et al. (2015)
		H_2O_2: 176 mM	Adsorption + Fenton: 100%	
		pH: 2		
		Time: 21 hr		

(continued)

Table 8.3 (continued)

Process description	Target pollutant	Operational conditions	Performance	References
Heterogeneous Fenton: nano-zerovalent iron +granular activated carbon	Nitrobenzene	Catalyst/adsorbent dosage: 0.4 g/L	Adsorption: 37%	Hu et al. (2015)
		H_2O_2: 5 mM	Adsorption + Fenton: 88%	
		pH: 4		
		Time: 100 min		
Homogeneous Fenton + granular activated carbon	Leather tannin effluent: Syntan	Catalyst/adsorbent dosage: 0.5 g/L	Fenton: 60%	Thankappan et al. (2015)
			Adsorption + Fenton: 98%	
		H_2O_2: 300 mg/L		
		pH: 3		
Homogeneous Fenton + activated carbon adsorption	Methyl methacrylate	Catalyst/adsorbent dosage: 0.019 mM	Adsorption:	Almazán-sánchez et al. (2014)
			Color: 67%	
		H_2O_2: 22.9 mM	COD: 42%	
		pH: 2	Adsorption + Fenton:	
		Time: 130 min	Color: 96%	
			COD: 60%	
Heterogeneous Fenton: sewage-derived carbon	Naphthalene dye	Catalyst/adsorbent dosage: 0.5 g/L	Adsorption: 40%	Gu et al. (2013)
		H_2O_2: 15 mM	Adsorption + Fenton: 94%	
		pH: 5		
		Time: 120 min		
Homogeneous Fenton + Fly ash adsorption	Dye: Rhodamine blue	Catalyst/adsorbent dosage: 80 g/L	Fenton: 72%	Chang et al. (2009)
			Adsorption + Fenton: 98%	
		H_2O_2: 6 mM		
		pH: 3		
		Time: 30 min		

adsorption-Fenton oxidation in the treatment of leather tannin effluent. The removal efficiency of the combined process is higher than either of the processes alone.

8.4.1.2 Simultaneous Adsorption-Fenton Using Heterogeneous Bifunctional Materials

A lot of efforts have been reported on the development of various composites with bifunctional capabilities as heterogeneous Fenton catalysts and adsorbents. The basic mechanism involved in the heterogeneous Fenton-adsorption include the

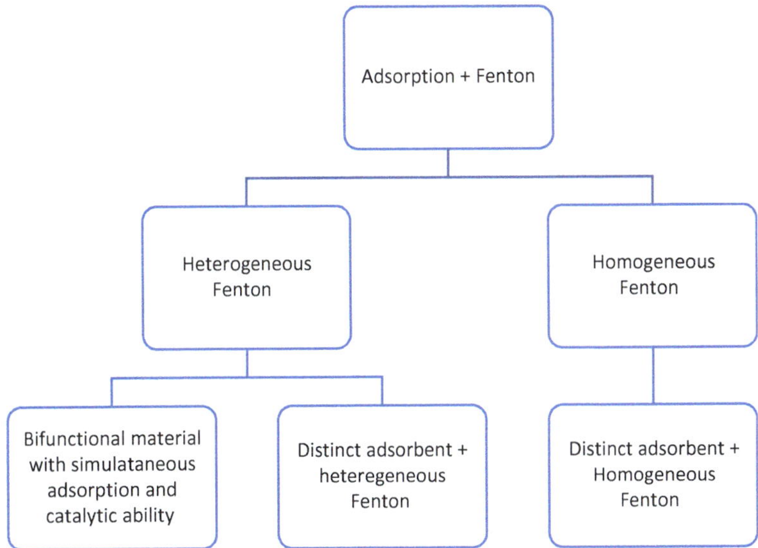

Fig. 8.5 Possible routes of combining adsorption with Fenton oxidation. For heterogeneous Fenton oxidation, the process may consist of a single bifunctional material with adsorption and catalytic ability. It may also involve a distinct heterogeneous Fenton and an adsorption stage. In homogeneous Fenton, a distinct adsorbent such as activated carbon is combined with Fenton's reagent

oxidation reactions between ferric/ferrous ions and H_2O_2, and the adsorption of both pollutants and oxidant on the surface of the composite (Ahmadi et al. 2017; Jafari et al. 2017). The Fenton reactions occur in the solution while the adsorbed oxidant further initiate some heterogeneous reaction at the surface of the composite. Figure 8.7 shows the mechanism of adsorption-oxidation using magnetic nanoparticles-functionalized carbon which is commonly employed as a heterogeneous Fenton catalyst.

Various heterogeneous composites have been utilized in adsorption-oxidation processes. These composites consist mainly of iron oxides and other compounds such as natural polymers (Gao et al. 2016; Cruz et al. 2017), clay minerals (Guo and Zhang 2015; Fida et al. 2017), carbon (Ahmadi et al. 2017; Jafari et al. 2017) and other metal oxides (Nguyen et al. 2011; Li et al. 2017). Iron oxides can be replaced by other metal oxides, such CuO and MnO_2, which act as Fenton-like catalysts (Zhang et al. 2016).

Carbonaceous materials are used to develop Fe-based composites with adsorption-oxidation capability. It is generally considered that the adsorption sites are mainly provided by the carbon while the catalytic effect is largely due to the iron. However, some studies have demonstrated that carbon materials, such as activated carbon and graphene, have intrinsic catalytic abilities toward decomposing H_2O_2 (Lucking et al. 1998; Maldonado-Hódar et al. 1999; Zhang et al. 2012). Among the carbon materials, activated carbon fibers are excellent support materials for developing heterogeneous catalyst. Their large surface areas and microporous structures,

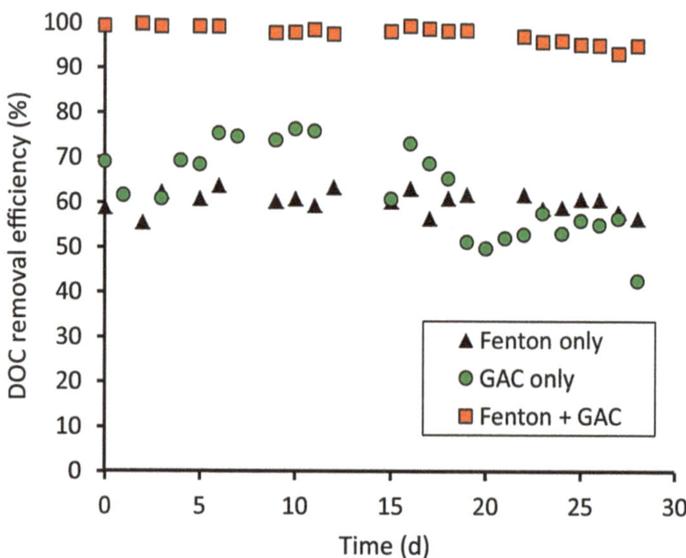

Fig. 8.6 Synergy of adsorption and Fenton oxidation in the removal of dissolve organic carbon from leather tannin effluent (experimental condition: syntan = 500 mg/L, $FeSO_4$ = 500 mg/L, H_2O_2 = 300 mg/L; pH = 3). The combined process was applied in sequence, with Fenton oxidation as the initial treatment. The combined process resulted in total removal of the dissolved organic carbon in contrast to either process alone. GAC: Granular activated carbon; DOC: Dissolved organic carbon. (Adapted from Thankappan et al. (2015) with permission by Elsevier)

high adsorption capacities and ease of modification are ideal in developing composites with dual adsorption-oxidation functions (Yao et al. 2013; Sun et al. 2014; Wang et al. 2014a). Wang et al. (2014b) used different activated carbon fibers in developing heterogeneous Fenton catalysts with dual function as adsorbent and catalyst. The composite exhibited high adsorptive and catalytic performance in the removal of various dyes from aqueous solution.

Jafari et al. (2017) prepared a heterogenous composite consisting of activated carbon and Fe_3O_4 for the removal of antibiotic tetracycline through adsorption-Fenton processes. While the composite could remove up to 56% of the initial concentration of the tetracycline through adsorption, more than 80% removal was achieved in the combined adsorption-Fenton process. Similarly, Fayazi et al. (2016) integrated activated carbon with γ-Fe_2O_3 for the removal of methylene blue from aqueous solution. The removal of the dye through adsorption reached 70%, which increased to 100% with the activation of Fenton-like process.

Metal oxides are also used to develop bifunctional composites for adsorption-Fenton process. In this regard, various transition metals such as copper, manganese, cerium, etc. are commonly used, with or without iron oxide in the composite. For example, Zhang et al. (2016) developed a mesoporous CuO/MnO_2 composite with adsorption-oxidation capability for the removal of emerging recalcitrant pollutant, benzotriazole. Figure 8.8 shows the reaction mechanism that occur when the

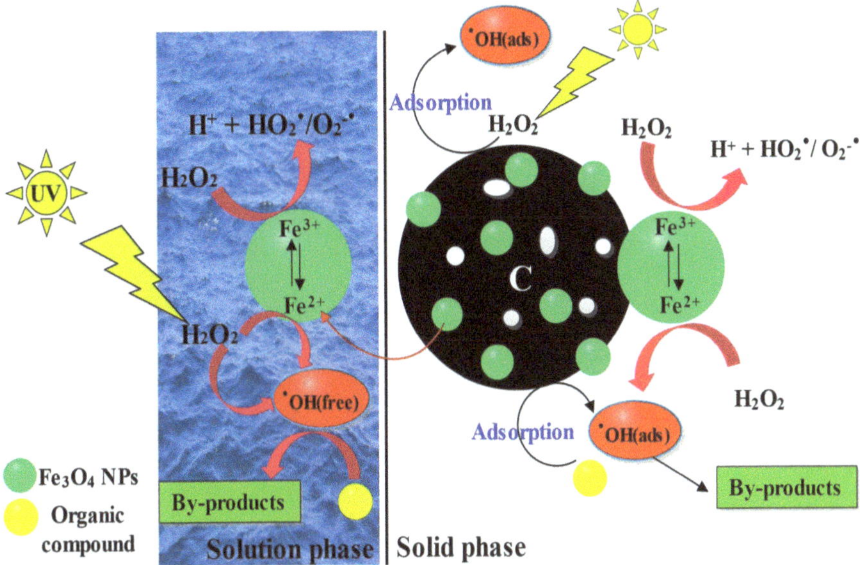

Fig. 8.7 Basic mechanism of adsorption-oxidation using magnetic nanoparticles-functionalized carbon as a heterogeneous Fenton catalyst. The oxidation occurs both in the solution phase and solid phase while the adsorption occurs at the surface of the composite. Both free OH· and adsorbed OH· participate in the homogeneous/heterogeneous Fenton reactions. Ultraviolet/visible irradiation enhances the decomposition of H_2O_2. (Adapted from Akbar et al. (2017) with permission from Elsevier)

composite was applied in the removal of benzotriazole. Both adsorption and catalytic oxidation occurred simultaneously. However, the experimental results indicated that the catalytic oxidation is the dominant mechanism for the removal of benzotriazole by the composite.

Another study utilized a mesoporous Fe_3O_4/CeO_2 composite with dual-function as adsorbent and Fenton-like catalyst for the removal of methylene blue from aqueous solution (Li et al. 2017). Under adsorption alone, more than 50% of the initial dye concentration was removed. When Fenton-like reaction was activated through the introduction of H_2O_2, almost 100% removal was achieved. While the adsorption capacity was due to the large surface area, pore size and oxygen vacancies on the composite, the catalytic ability was attributed to the synergy of Fe^{2+} and Ce^{4+} which can activate H_2O_2.

Heterogeneous minerals such as zeolite and clays have been used to developed bifunctional composites for adsorption-Fenton oxidation. For example, Zhang et al. (2017b) developed Fe-zeolite composite via molecular imprinting process for the removal of methylene blue from aqueous solution. About 54% removal was attributed to the adsorption by the Fe-Zeolite. On the other hand, the combined process was able to remove about 88% of the initial dye concentration.

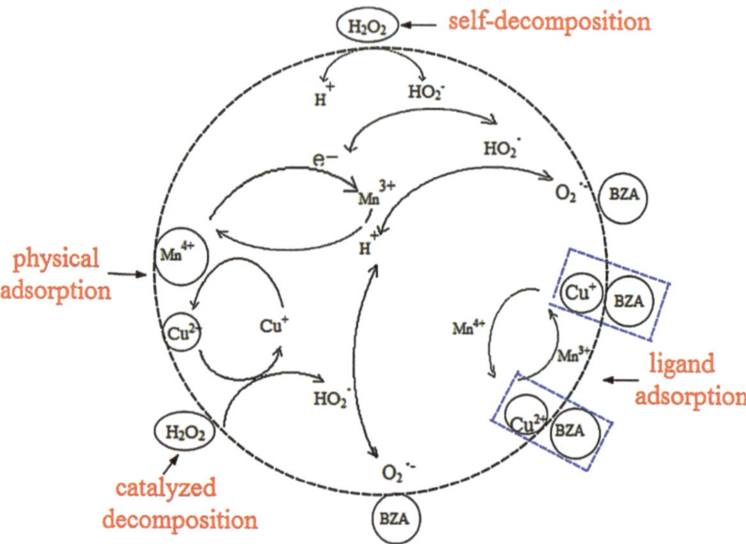

Fig. 8.8 Mechanism of adsorption-Fenton oxidation by CuO/MnO_2 composite as Fenton-like heterogeneous catalyst for the removal of benzotriazole. The adsorption occurs via chemisorption and complexation between Cu^{2+} and benzotriazole. Catalytic oxidation occurs when H_2O_2 undergo self-decomposition or is decomposed by Cu^{2+} to generate the superoxide radical ($H_2O^{\cdot}$). (Adapted from Zhang et al.(2016) with permission from Elsevier)

8.4.1.3 Adsorption Integrated with Fenton

As pointed out earlier, adsorption can be integrated with Fenton oxidation and many of its variants such as photo-Fenton and electro-Fenton. In such cases, each process is distinct, and their synergistic performance is evaluated and compared with either process standing alone. The process mainly involves a homogeneous Fenton oxidation and an adsorption stage, where activated carbon is commonly employed. The major advantage here is that the oxidation stage occurs very fast due to the homogeneous Fenton's reaction. The presence of activated carbon, or other adsorbents, can induced crystallization of iron oxide and hence, reduces the iron sludge generation that is associated with homogeneous Fenton oxidation (Garcia-Segura et al. 2016; Bello et al. 2017). Additionally, adsorption can serve as a filtration stage when applied in post-Fenton treatment, removing residual pollutants and sludge from the wastewater (Ramírez-sosa et al. 2013).

One of the popular approaches in combining homogeneous Fenton and adsorption is through the application of a fluidized bed reactor. In fluidized bed Fenton, solid particles are used as support materials, which also serve as adsorption sites for the target pollutants. The description of the mechanism that occurs in fluidized bed Fenton has been presented previously (Chen et al. 2016).

Many studies have reported the use of fluidized bed Fenton to remove recalcitrant organic pollutants. However, only few have considered the adsorption part of the process. For example, Lyu et al. (2016) reported the treatment of multi-dye wastewater using an integrated adsorption-Fenton oxidation in a fluidized bed reactor. In the study, the authors considered the two distinct processes of homogeneous Fenton and adsorption using activated carbon and ceramsite as adsorbents. The performance of the integrated processes was higher than individual process alone. Adsorption alone led to 65% COD reduction from the while 94% reduction was achieved when adsorption was integrated with the Fenton oxidation.

Adsorption-Fenton can be designed to occur simultaneously in one stage or in two distinct stages of adsorption followed by Fenton oxidation or vice-versa. Although many studies have reported the one-stage process, some researchers favored the two-stage (Almazán-sánchez et al. 2014; Candal 2014). In most cases, Fenton oxidation is applied first, followed by the adsorption stage. The advantage here is that the high concentration of the pollutant is rapidly degraded by the Fenton oxidation, while the residual pollutant is removed at the adsorption stage. For example, Almazán-sánchez et al. (2014) integrated a sequential Fenton oxidation-adsorption for the treatment of wastewater containing methacrylate. Fenton oxidation was applied first, which resulted in the removal of 67%, 42% and 41% of the initial colour, COD and TOC respectively. However, when the adsorption stage was introduced, the colour, COD and TOC removals rose to 96%, 60% and 58% respectively. A similar study by Meijide et al. (2017) considered the integration of electro-Fenton process with a fixed-bed activated carbon adsorption for the removal of pesticide, acetamiprid. Although the electro-Fenton was able to degrade the pesticide, total removal was only achieved when the adsorption was introduced.

Integrating adsorption with Fenton oxidation is an effective way of removing recalcitrant pollutants from wastewater as shown in Table 8.3. The use of bifunctional composite that can act as heterogeneous Fenton catalyst and adsorbent is quite attractive and is being widely pursued. Graphene-based composites are receiving attention and will play a significant role in developing adsorption-Fenton oxidation processes. Although many studies have utilized composites with both catalytic and adsorption ability, more attention has been given to the degradation process. Thus, the adsorption potentials of these composites need to be studied as well.

8.4.2 *Adsorption-Photocatalysis*

8.4.2.1 Background

The application of combined adsorption-photocatalysis is receiving wide attention as an alternative approach to remove recalcitrant organic pollutants from wastewater streams (Lin et al. 2010; Lin et al. 2017; Yang et al. 2017). By their heterogeneous nature, semiconductor photocatalysts have an inherent ability to adsorb pollutants on their surfaces. Thus, photocatalytic degradation of pollutants always involved some form of adsorption process. However, the photocatalytic degradation is normally the dominant mechanism, though this may be different when some photocatalyst composites are employed (Zhang et al. 2017a). Therefore, adsorption-photocatalysis treatment can occur simultaneously where the photocatalyst is designed to have high adsorption and catalytic capacities (Yang et al. 2017; Wang et al. 2017b). Alternatively, a distinct adsorption stage can be coupled to photocatalysis (Dai et al. 2015; Cataldo et al. 2016). However, using semiconductor-based composites with simultaneous adsorption-photocatalytic ability is the commonly employed route. Table 8.4 shows some reported studies on the applications of adsorption-photocatalysis for the treatment of recalcitrant organic pollutants. The synergistic performance of the combined processes can be clearly seen in the overall removal efficiencies of the pollutants.

8.4.2.2 Synergistic Effect

Heterogeneous photocatalysts provide active sites for the adsorption of both pollutants and the photo-generated reactants. Some photocatalysts may possess ion-exchange properties, making them excellent adsorbents (Kubo and Nakahira 2008; Camposeco et al. 2016; López-mu et al. 2016). Additionally, since composite materials possess active functional groups, adsorption of pollutants can occur even under dark condition. However, once the source of light is supplied, the photoelectron and holes are generated and adsorbed on the surface of the catalyst. Thus, both adsorption and photocatalytic degradation occurs on the surface of the composite (Zhang et al. 2017a). Figure 8.9 shows a typical mechanism in photocatalysis using vanadium-loaded titania nanotubes for the removal of Basic Violet 3 and perchloroethylene from wastewater under visible irradiation. Both pollutants are adsorbed on the nanotubes and oxidized by $OH^{\cdot}$ and $O_2^{\cdot -}$.

The commonly used semiconductors are faced with challenges of low quantum efficiency due to the rapid recombination of photoelectrons and holes, low adsorption capacities and wide band gaps. Some of the strategies that can overcome these challenges have been highlighted in Sect. 8.3.2.2. In this context, Zhang et al. (2017a) synthesized a Ag-TiO_2-reduced graphene oxide nanocomposite with bifunctional ability of adsorption and photocatalysis under visible light. While the Ag can assist in preventing the recombination of electron-hole pairs, the reduced graphene oxide possesses large surface area, enhancing the adsorption capacity of the composite. Interestingly, the composite can remove more than 99.5% of Rhodamine Blue compared to 13% removed by pristine TiO_2.

Iron-based composites have also been used in adsorption-photocatalysis for recalcitrant wastewater treatment. Fakhri et al. (2017) prepared magnetite quantum dots anchored tin dioxide nano-fibers for removal of ethyl methanesulfonate from aqueous solution. The composite was able to remove about 57% of the pollutant via adsorption and 94% under combined adsorption-photocatalysis. This clearly shows the synergy of the combined process. However, Di et al. (2017) observed that the contribution of adsorption was negligible in the removal of ibuprofen using Zn-Fe mixed metal oxides. This was perhaps due to the dominance of ZnO in the composite, which depends largely on photoactivation.

Table 8.4 Some recent studies on the application of adsorption-photocatalysis for the treatment of recalcitrant pollutants

Process description	Target pollutant	Operational conditions	Performance	References
Ag-TiO_2-graphene oxide	Rhodamine blue	Adsorbent/photocatalyst dosage: 0.6 g/L	Adsorption + degradation: 99.5%	Zhang et al. (2017a)
		Energy: A 300 W-420 nm visible light lamp		
		Time: 70 min		
Ag-Br-Graphene oxide hydrogels	Bisphenol A	Adsorbent/photocatalyst dosage: 2 g/L	Adsorption + degradation: 100%	Chen et al. (2017)
		Energy: Visible light lamp		
		Time: 90 min		
Fe_3O_4/Quantum Dots/SnO_2	Ethyl methanesulfonate	Adsorbent/photocatalyst dosage: 4 g/L	Adsorption: 56.85%	Fakhri et al. (2017)
		Energy: Visible light lamp	Adsorption + degradation: 93.85%	
		Time: 40 min		
ZnO-Graphene oxide nanorod	Dyes	Adsorbent/photocatalyst dosage: 0.5 g/L	Adsorption: 10–16%	Shanmugam et al. (2017)
		Energy: UV/Visible light	Adsorption + degradation: 75–95%	
		Time: 40 min		
V-loaded titania nanotubes	Dye	Adsorbent/photocatalyst dosage: 0.01 g/L	Adsorption: 70%	Lin et al. (2017)
		Energy: Visible light	Adsorption + degradation: 95%	
		Time: 180 min		
3D Ag_3-PO_4-graphene hydrogel	Bisphenol A	Adsorbent/photocatalyst dosage: 0.5 g/L	Adsorption: 75%	Mu et al. (2017)
		Energy: UV/visible light	Adsorption + degradation: 100%	
		Time: 12 min		
Hydroxyethyl acrylate-CuS hydroxyl	Sulfamethoxazole	Adsorbent/photocatalyst dosage: 20 g/L	Adsorption: 15%	Yang et al. (2017)
		Energy: Visible light	Adsorption + degradation: 90%	
		Time: 720 min		

(continued)

Table 8.4 (continued)

Process description	Target pollutant	Operational conditions	Performance	References
Bi_2WO_6-3D	Rhodamine blue	Adsorbent/ photocatalyst dosage: 1 g/L	Adsorption: 60%	Huang et al. (2017a)
		Energy: Visible light	Adsorption + degradation: 93.1%	
		Time: 130 min		
TiO_2-MoS_2-Graphene oxide	Bisphenol A	Adsorbent/ photocatalyst dosage: 0.5 g/L	Adsorption: 22.1%	Luo et al. (2017)
		Energy: UV light	Adsorption + degradation: 62.4%	
		Time: 300 min		
Carbon-TiO_2	Pesticide: Isoproturon	Adsorbent/ photocatalyst dosage: 2 g/L	Adsorption: 30%	García-díaz and Martín (2016)
		Energy: Ultraviolet light	Adsorption + degradation: 86%	
		Time: 180 min		

Due to the excellent properties of graphene oxide, there is growing interest in using graphene-based nanocomposites for adsorption-photocatalysis for the removal of recalcitrant pollutants. The adsorption sites are provided by the graphene sheet while the degradation is driven by the photocatalyst in the composite. In addition, graphene can enhance the photocatalytic activity of the composite through reducing the recombination of the photo-generated electron-hole pairs (Luo et al. 2017). For example, a three-dimensional graphene hydrogel consisting of Ag, Br and reduced graphene oxide was developed for the removal of bisphenol A through adsorption-visible light driven-photocatalysis (Chen et al. 2017). The adsorbent was sequestered by the 3-D graphene nanosheets and degraded by the Ag-Br nanoparticles under visible light irradiation. The composite performed 1.5-fold higher than the Ag-Br, indicating the synergy of adsorption and photocatalysis. Many studies have reported the excellent performance of graphene-based bifunctional composite such as reduced graphene oxide-TiO_2 (Cheng et al. 2016) 3D Ag_3PO_4-graphene hydrogel (Mu et al. 2017) and TiO_2-MoS_2-reduced graphene oxide (Luo et al. 2017) for recalcitrant wastewater treatment.

It can be seen from Table 8.4 that most of the bifunctional materials exhibited excellent performance under the visible light irradiation, in contrast to the conventional photocatalysis where ultraviolet light is normally utilized. The use of visible light-driven processes is more attractive since it can lower the treatment cost (Bello and Raman 2017).

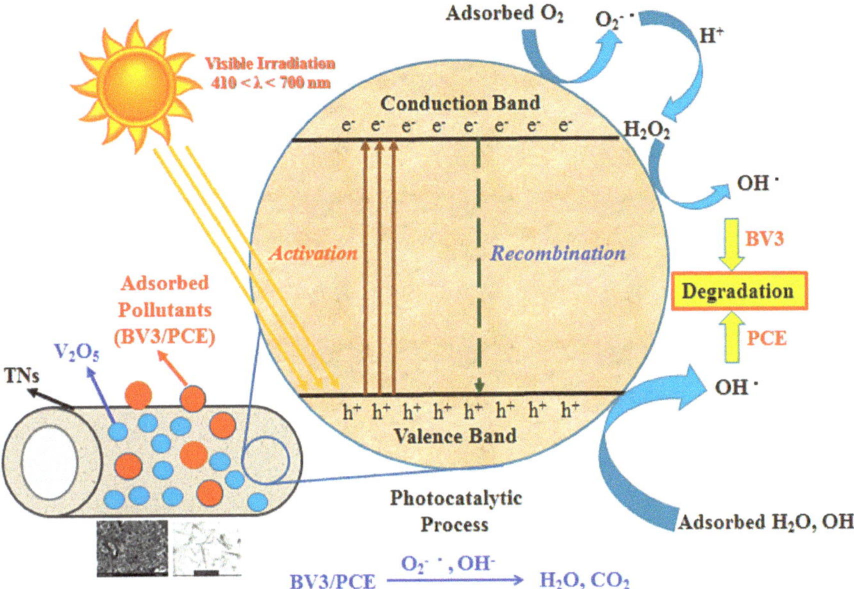

Fig. 8.9 Basic description of adsorption-photocatalysis process. Vanadium-loaded titania nanotubes were used in removing Basic Violet 3 and perchloroethylene from wastewater under visible irradiation. Both pollutants are adsorbed on the surface of titania nanotube while the degradation was driven by the hydroxyl radical ($OH^{\cdot}$) and superoxide ($O_2^{\cdot -}$). (BV3: Basic Violet 3, PCE: Perchloroethylene, TN: Titania nanotubes). (Adapted from Lin et al. (2017) with permission from Elsevier)

8.4.3 Adsorption-Ultrasound

8.4.3.1 Background

Ultrasound cavitation is commonly combined with adsorption for the removal of various pollutants, the so-called ultrasound-assisted adsorption. Combining adsorption with ultrasound has become a popular approach as it can increase efficiency and lower treatment time (Sayan and Edecan 2008; Roosta et al. 2014). The advantages of ultrasound, which have been highlighted in Sect. 8.3.2.4, could be explored in enhancing the adsorptive removal of recalcitrant pollutants from wastewater. One of the major challenges in adsorption is the adsorbent-solution mass transfer limitation. To this end, for example, cavitation can be used to enhance the mass transfer during adsorption process. Besides enhancing mass transfer, ultrasound can provide additional degradation of organic pollutants through hydroxyl radicals and pyrolytic effects. Consequently, ultrasound has been extensively coupled with adsorption for the removal of various recalcitrant pollutants.

The basic mechanism of ultrasound-adsorption, which include the ultrasound-driven adsorption and degradation by $OH^{\cdot}$, has been highlighted previously (Khataee et al. 2017). It includes (i) direct generation of $OH^{\cdot}$ and localized hot spots by the cavitation effect, and the subsequent degradation of pollutants by these products, (ii) ultrasound-enhanced adsorption: the ultrasound generated-energy enhances mass transfer and drives the adsorption through convection pathway. In some cases, heterogeneous catalysts are involved in the adsorbents/composites and a third effect will be introduced into the process. The heterogeneous catalyst can enhance the oxidation efficiency of the process through providing more nucleation sites for cavitation (Ghows and Entezari 2011; Pang and Abdullah 2013). Besides these, cavitation-associated phenomena such as micro-turbulence, acoustic waves and micro-jets are believed to enhance the interparticle collisions, improving the reactivity of the particles in the solution (Khataee et al. 2017). In their mechanistic studies in ultrasound-assisted adsorption for removal of aromatic compounds, Midathana and Moholkar (2009) have discussed the contributions of microstreaming, microturbulence, and acoustic waves in enhancing the adsorption process.

8.4.3.2 Synergistic Effect

Combining ultrasound with adsorption results in a synergistic treatment performance higher than either of the processes. For example, Khataee et al. (2017) investigated the removal of an azo dye, Acid Red 17, using a combined ultrasound-adsorption by Fe_3O_4-modified coffee waste hydrochar. Figure 8.10 shows the comparison of the treatment performances of the separate processes and the synergy of the combined processes. Adsorption alone could only remove about 15% of the initial concentration of the pollutant while degradation through cavitation was able to remove up to 34% of the pollutant after 80 min. The higher removal of the ultrasound is due to the higher degradation rate of the cavitation compared to adsorption. However, when the processes were combined, the removal could reach 100% using the Fe_3O_4-modified coffee waste hydrochar. Many studies have reported the synergy of ultrasound and adsorption in the removal of recalcitrant organic pollutants such as phenolic compounds (Li et al. 2002; Sonawane et al. 2008; Hamdaoui and Naffrechoux 2009; Dietrich et al. 2017), dyes (Das 2016; Asfaram et al. 2017; Dastkhoon et al. 2017; Sharifpour et al. 2017) and pharmaceuticals (Tran et al. 2017a).

Combining adsorption with ultrasound can also reduce the necessary time to reached equilibrium due to the enhanced mass transfer. Wang et al. (2017a) studied the application of MoS_2/graphene quantum dot for the removal of dyes under ultrasound-assisted adsorption. Figure 8.11 shows the removal rates and the adsorption capacities as function of time under different conditions. The removal rate reached almost 100% within 2 seconds under ultrasound-adsorption (Fig. 8.11a). On the other hand, the adsorption capacities using different initial concentrations of Rhodamine Blue reached maximum equilibrium stage within 4 seconds under ultrasound-adsorption (Fig. 8.11b). In the absence of ultrasound, the maximum adsorption capacities were achieved after about 30 min (Fig. 8.11c). This shows that combining ultrasound with adsorption can reduce treatment time.

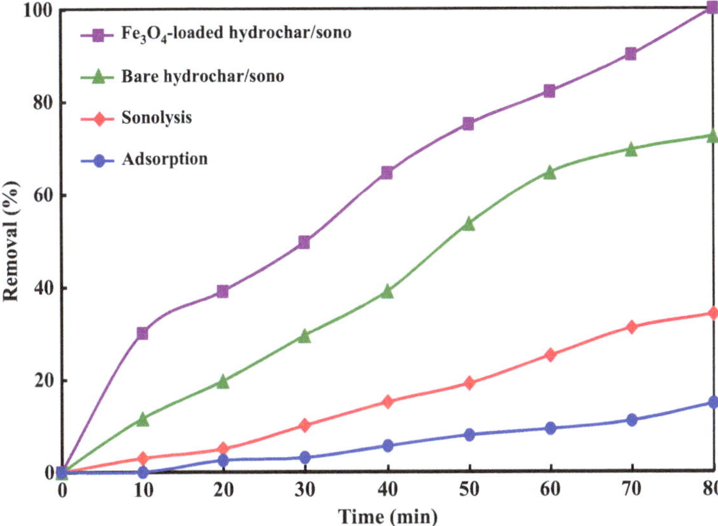

Fig. 8.10 Comparison of the performances of adsorption, ultrasound and ultrasound-assisted adsorption in the removal of Acid Red 17 from aqueous solution (Experimental conditions: [Dye] = 10 mg/L, [Fe_3O_4-CHC] = 1.0 g/L, ultrasound power = 300 W/L and pH = 6). The adsorption on bare hydro-char resulted in the lowest removal while the ultrasound-assisted adsorption using Fe_3O_4-loaded hydro-char gave the highest removal. The higher removal of ultrasound compared to adsorption is due to the hydroxyl and pyrolytic degradation ability of the ultrasound. (Adapted from Khataee et al. (2017) with permission from Elsevier)

The performance of ultrasound-adsorption depends largely on ultrasound power as well as the sonication time. These two factors determine the power requirement of an ultrasound-adsorption process. However, sonication time is more frequently discussed in the literature. Although conventional adsorption processes are characterized by long contact times due to mass transfer limitation, the contact time of ultrasound-adsorption is usually short because of the enhanced mass transfer brought about by the cavitation effects. Generally, there is increase in the process efficiency with increase in both sonication time and power. Figure 8.12 shows the effect of ultrasound power on the removal efficiency of a dye. The removal efficiency increased with increased in the ultrasound power. Besides giving the highest removal, the 400 W/L reduced the treatment time by about 25%. However, high ultrasound power and subsequent intense cavitation may erode the surface of adsorbents (Schueller and Yang 2001).

While emphasis is mostly placed on the effect of ultrasound on adsorption, it can also play a significant role in desorption of the adsorbed pollutants. This is of great interest in the regeneration of spent adsorbents. Ultrasound-assisted regeneration will be more attractive compared to chemical or thermal regeneration that are commonly employed. In ultrasound-assisted desorption, the cavitation-generated acoustic waves induce surface diffusion, which enhances the desorption process (Schueller and Yang 2001). Additionally, since an increase in temperature can

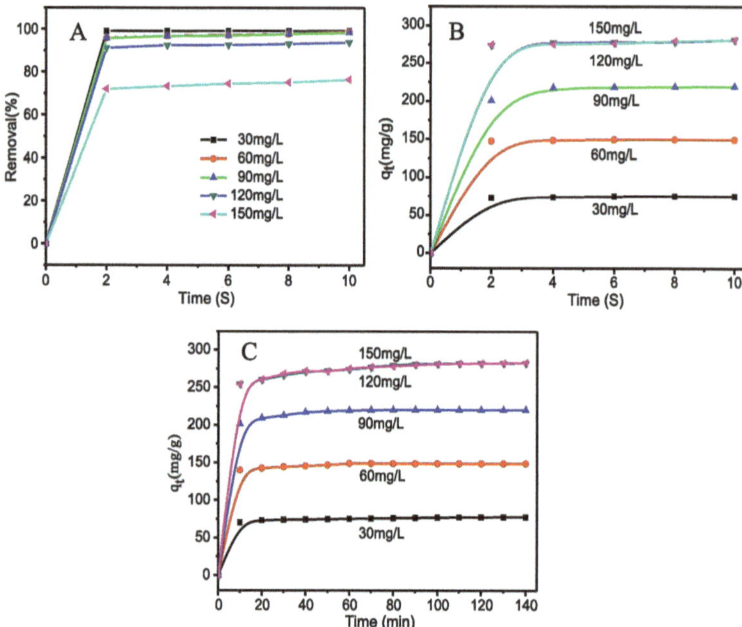

Fig. 8.11 (**a**) Removal rate and (**b**) adsorption capacity of MoS_2/graphene quantum dot as function of Rhodamine Blue concentrations under ultrasound (**c**) adsorption capacity of MoS_2/graphene quantum dot as function of Rhodamine Blue concentrations without ultrasound. Although the adsorption capacities are nearly the same, the equilibrium time was much shorter in case of ultrasound-adsorption (**b**) compared to adsorption alone (**c**). The reduction in process time is due to the enhanced mass transfer caused by the ultrasound cavitation. (Adapted from Wang et al. (2017a) with permission of Elsevier)

promote desorption, the localized hot spots near the adsorbent may further improve the desorption process. Figure 8.13 shows a typical ultrasound-assisted desorption of phenol from activated carbon. The ultrasound-assisted process exhibited higher desorption rate compared to other desorption processes shown. Details on ultrasound-assisted desorption can be found in previous literature (Schueller and Yang 2001; Breitbach et al. 2003; Wang and Yang 2007).

Most of the reported literature have focused on the contribution of ultrasound in effective mixing and the consequent enhanced mass transfer. However, the contribution of ultrasound through $OH^{\cdot}$ and pyrolytic degradation has not been commonly studied. It is well established that ultrasound-assisted adsorption exhibited higher performance than adsorption alone. However, the contribution of ultrasound is mostly considered in general, without evaluating the specific effect of each associated mechanism. Consequently, this will have to be addressed for proper understanding and development of ultrasound-assisted adsorption processes.

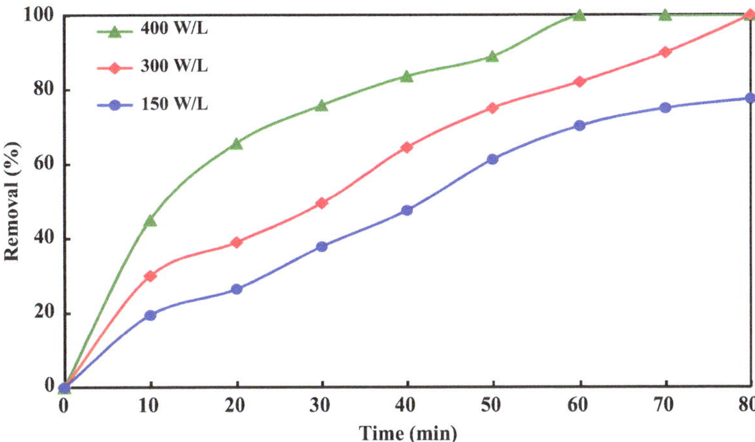

Fig. 8.12 Effect of ultrasound-power on the performance of ultrasound-assisted adsorption for the removal of Acid Red 15 (experimental conditions: [Dye] = 10 mg/L, [Fe_3O_4CHC] = 1.0 g/L and pH = 6). The removal efficiency increased with increasing ultrasound power. The 400 W/L gave the highest removal and reduced treatment time by about 25%. (Adapted from Khataee et al. (2017) with permission from Elsevier)

8.4.4 Adsorption-Ozonation

8.4.4.1 Background

The fundamentals of Ozonation have been discussed in Sect. 8.3.2.3. The combination of adsorption and Ozonation has not received much attention compared to other processes such as ultrasound and photocatalysis. Adsorption-Ozonation can be designed through the integration of distinct adsorption process and Ozonation process or through catalytic Ozonation using bifunctional composites. However, catalytic Ozonation using composites with adsorption-catalysis ability appears more attractive. Although various composites that can act as catalysts and adsorbents are being used in catalytic Ozonation, the discussion in the literature have given more emphasis on their catalytic function. However, there is strong possibility of adsorption between the surface functional groups of many composites and the pollutants (Maryam et al. 2017). For example, a reactive complex can be formed between ozone and the surface functional groups, which can then react with the organic pollutants in the solution (Nawrocki and Kasprzyk-hordern 2010; Huang et al. 2017b).

8.4.4.2 Synergistic Effect

The synergy of adsorption and catalytic Ozonation in wastewater treatment has been pointed earlier (Faria et al. 2008). While many studies have reported high adsorption

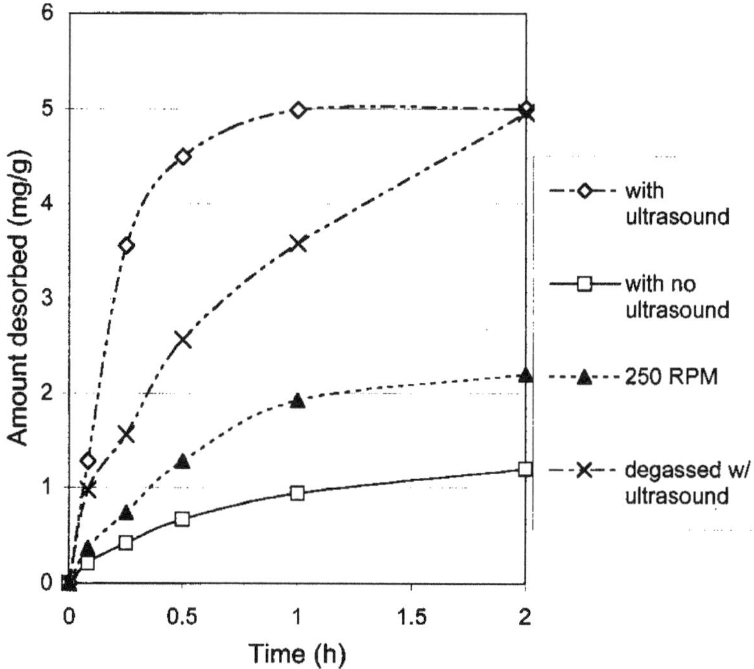

Fig. 8.13 Ultrasound-assisted desorption of phenol on activated carbon at 1 MHz, 60 W and 22 °C. Desorption in the absence of ultrasound can only remove about 2 mg/g of phenol after 2 h. Ultrasound cavitation improved the desorption process, desorbing about 5 mg/g under the same condition. (Adapted from Schueller and Yang (2001) with permission from American Chemical Society)

of organic molecules on the composites, others have indicated less contribution of adsorption (Nawrocki and Kasprzyk-hordern 2010). Some of the controversies surrounding the mechanism of catalytic Ozonation, including adsorption, have been discussed (Nawrocki 2013).

Carbon materials, particularly activated carbon, are commonly utilized as heterogeneous catalysts in catalytic Ozonation (Garcı and Beltra 2007; Gümüs 2017; Maryam et al. 2017; Huang et al. 2017b). Figure 8.14 shows a typical experimental set-up using simultaneous activated carbon catalyzed Ozonation for wastewater treatment. The performance of the combined process was higher than Ozonation alone due to the contribution of activated carbon, both as catalyst and adsorbent. Interestingly, Cataldo et al. (2016) studied the combination of Ozonation, photocatalysis and activated carbon adsorption for the treatment of saline wastewater. Although the authors reported an increase in reaction rate by 15% when the processes were combined, the individual contributions were not quantified. One interesting part of the work is that the possibility of regeneration of the spent activated carbon by the Ozonation was investigated, where a partial regeneration of about 40% was achieved.

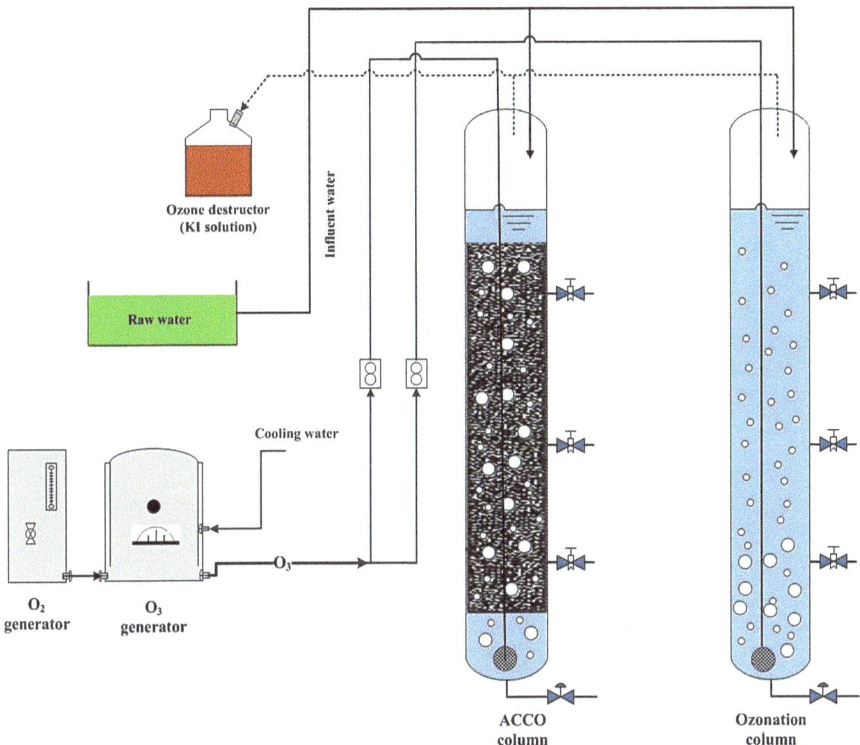

Fig. 8.14 A typical set-up for the combined adsorption-Ozonation using activated carbon as a heterogeneous catalyst cum adsorbent. The wastewater and ozone are introduced into the column containing activated carbon. The two processes can be separated by introducing a second column. ACCO: Activated carbon catalyzed Ozonation. (Adopted from Maryam et al. (2017) with permission by Elsevier)

Recently, the use of carbon-based bifunctional composites is receiving attention. For example, Huang et al. (2017b) studied the combined adsorption-catalytic Ozonation for the removal of Bisphenol A using multi-walled carbon nanotubes-Fe_3O_4 composites. Ozonation alone could remove about 50% of Bisphenol A after 40 min, while adsorption using the composite removed 21% of the pollutant. However, when Ozonation was combined with the composite, the removal increased to 91%, indicating the synergy of the processes (Fig. 8.15). The excellent adsorption and catalytic performance of the composite was attributed to its large surface area, hollow tube channels and abundant surface functional groups.

The work of Huang et al. (2017b) has also pointed out the surface reactions between activated carbon-based composite and organic pollutants, which involves some form of complex formation. The authors investigated the performance of combined a Ozonation and MnO_x/sewage sludge-derived activated carbon. While

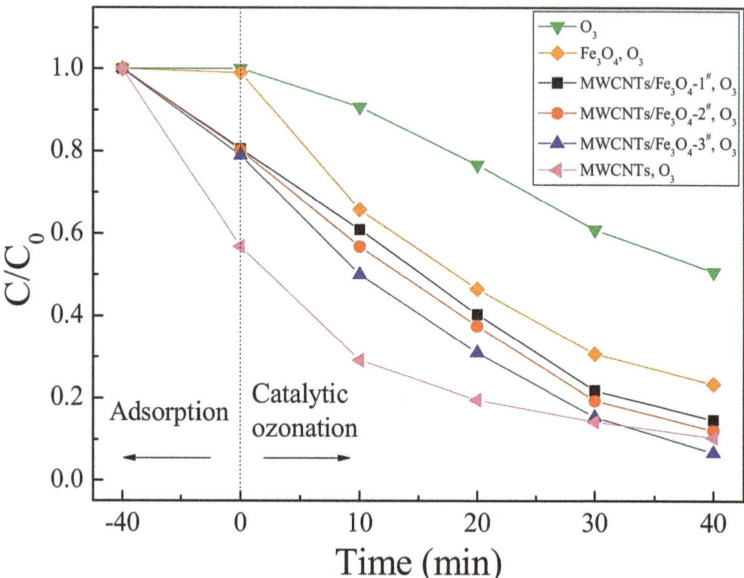

Fig. 8.15 Bisphenol A removal efficiency under different combinations of processes based on O_3/multi-walled carbon nanotubes-Fe_3O_4 (reaction conditions: initial pH = 7.0, initial Bisphenol A concentration = 50 ppm; O_3 flow rate = 0.3 L min^{-1}; ozone gas concentration = 3.0 mg L^{-1}; catalyst dose = 0.5 g L^{-1}). The designation of the composite as 1,2,3 reflects differences in preparation steps. The contributions of adsorption and catalytic Ozonation are clearly depicted. MWCNT: multiwalled carbon nanotubes. (Adopted from Huang et al. (2017c) with permission by Elsevier)

treatment by Ozonation alone resulted in about 10.3% removal efficiency, the combined process achieved more than 90% removal efficiency. The high performance was due to the degradation by $OH^{\cdot}$ and surface reactions. In another work, Mojiri et al. (2017) investigated the treatment of landfill leachate using electro-Ozonation and adsorption by zeolite/bentonite/cement composite. The leachate was first treated with Ozonation and then transferred to the powdered composite adsorbent reactor. The adsorption stage improved the overall COD removal from 64.8% to 88.2%.

Sakulthaew et al. (2015) studied the removal of polycyclic aromatic hydrocarbons through a two-stage process consisting of initial Ozonation, followed by adsorption using carbon nano-onions. The process involved a fast transformation of the pollutants by Ozonation and the removal of Ozonation product by the adsorption stage. The residence time in the Ozonation chamber was about 26 min, which resulted in the transformation of the polycyclic aromatic hydrocarbons to intermediates that were subsequently removed in the adsorption stage.

8.5 Conclusion

Combined adsorption-advanced oxidation processes are effective technologies for the removal of recalcitrant organic pollutants. Numerous studies have reported the synergy between adsorption and oxidation processes such as Fenton, photocatalysis, ultrasound and Ozonation. The higher performance of the combined processes is due to the enhanced adsorption and further degradation of pollutants via oxidation by hydroxyl radicals. The development of bifunctional materials with adsorption-catalytic ability has facilitated the combination of adsorption and oxidation processes such as photocatalysis and Fenton oxidation. The combination of adsorption and ultrasound, the so-called ultrasound assisted-adsorption, can greatly speed up adsorption through the effects of cavitation. However, control and optimization of the combined processes are quite challenging and have not been fully established. Although the discussion in the literature has placed more emphasis on the synergistic performance of adsorption and advanced oxidation processes, the potential of using advanced oxidation processes in desorption and regeneration of spent adsorbents is attractive. Advanced oxidation processes can regenerate spent adsorbents by degrading the adsorbed pollutants as demonstrated by many studies. Although the synergistic performance of the combined processes has been widely reported, the cost implications have not been discussed. Consequently, the extent of the reduction in the treatment cost, is not yet clear. Some of the advantages that may be derived from combining adsorption with advanced oxidation processes include higher performance, shorter treatment time and simultaneous regeneration of spent adsorbents.

References

Abidi N, Errais E, Duplay J, Berez A, Jrad A, Schafer G, Ghazi M, Semhi K, Trabelsi-Ayadi M (2015) Treatment of dye-containing effluent by natural clay. J Clean Prod 86:432–440. https://doi.org/10.1016/j.jclepro.2014.08.043

Adewuyi YG (2001) Sonochemistry: environmental science and engineering applications. Ind Eng Chem Res 40:4681–4715. https://doi.org/10.1021/ie010096l

Adewuyi YG (2005) Sonochemistry in environmental remediation. 1. Combinative and hybrid sonophotochemical oxidation processes for the treatment of pollutants in water. Environ Sci Technol 39:3409–3420. https://doi.org/10.1021/es049138y

Agarwal S, Tyagi I, Kumar V, Hadi M, Bagheri A (2016) Degradation of azinphos-methyl and chlorpyrifos from aqueous solutions by ultrasound treatment. J Mol Liq 221:1237–1242. https://doi.org/10.1016/j.molliq.2016.04.076

Aguayo-Villarreal IA, Bonilla-Petriciolet A, Muñiz-Valencia R (2017) Preparation of activated carbons from pecan nutshell and their application in the antagonistic adsorption of heavy metal ions. J Mol Liq 230:686–695. https://doi.org/10.1016/j.molliq.2017.01.039

Ahmadi M, Kakavandi B, Jorfi S, Azizi M (2017) Oxidative degradation of aniline and benzotriazole over PAC@$Fe^{II}Fe_2^{III}O_4$: a recyclable catalyst in a heterogeneous photo-Fenton-like system. J Photochem Photobiol A Chem 336:42–53. https://doi.org/10.1016/j.jphotochem.2016.12.014

Ahmaruzzaman M (2008) Adsorption of phenolic compounds on low-cost adsorbents: a review. Adv Colloid Interf Sci 143:48–67. https://doi.org/10.1016/j.cis.2008.07.002

Ahmed MJ (2017a) Adsorption of quinolone, tetracycline, and penicillin antibiotics from aqueous solution using activated carbons: review. Environ Toxicol Pharmacol 50:1–10. https://doi.org/10.1016/j.etap.2017.01.004

Ahmed MJ (2017b) Adsorption of non-steroidal anti-inflammatory drugs from aqueous solution using activated carbons: review. J Environ Manag 190:274–282. https://doi.org/10.1016/j.jenvman.2016.12.073

Ahn Y, Oh H, Yoon Y, Kyu W, Seok W, Kang J (2017) Effect of graphene oxidation degree on the catalytic activity of graphene for ozone catalysis. J Environ Chem Eng 5:3882–3894. https://doi.org/10.1016/j.jece.2017.07.038

Akbar A, Kakavandi B, Ra M, Kalantar F, Purkaram I, Ahmadi E, Esmaeili S (2017) Comparative treatment of textile wastewater by adsorption, Fenton, UV-Fenton and US-Fenton using magnetic nanoparticles-functionalized carbon (MNPs@C). J Ind Eng Chem 56:163–174. https://doi.org/10.1016/j.jiec.2017.07.009

Alipanahpour E, Ghaedi M, Asfaram A, Hajati S, Mehrabi F (2017) Preparation of nanomaterials for the ultrasound-enhanced removal of Pb^{2+} ions and malachite green dye: chemometric optimization and modeling. Ultrason Sonochem 34:677–691. https://doi.org/10.1016/j.ultsonch.2016.07.001

Al-Khateeb LA, Almotiry S, Salam MA (2014) Adsorption of pharmaceutical pollutants onto graphene nanoplatelets. Chem Eng J 248:191–199. https://doi.org/10.1016/j.cej.2014.03.023

Almazán-sánchez PT, Linares-hernández I, Martínez-miranda V, Lugo-lugo V (2014) Wastewater treatment of methyl methacrylate (MMA) by Fenton's reagent and adsorption. Catal Today 220–222:39–48. https://doi.org/10.1016/j.cattod.2013.09.006

Amran M, Salleh M, Khalid D, Azlina W, Abdul W, Idris A (2011) Cationic and anionic dye adsorption by agricultural solid wastes: a comprehensive review. Desalination 280:1–13. https://doi.org/10.1016/j.desal.2011.07.019

Anastopoulos I, Kyzas GZ (2014) Agricultural peels for dye adsorption: a review of recent literature. J Mol Liq 200:381–389. https://doi.org/10.1016/j.molliq.2014.11.006

Asfaram A, Ghaedi M, Hossein M, Azqhandi A, Goudarzi A, Hajati S (2017) Ultrasound-assisted binary adsorption of dyes onto Mn@CuS/ZnS-NC-AC as a novel adsorbent: application of chemometrics for optimization and modeling. J Ind Eng Chem 54:377–388. https://doi.org/10.1016/j.jiec.2017.06.018

Asghar A, Raman AAA, Daud WMAW (2015) Advanced oxidation processes for in-situ production of hydrogen peroxide/hydroxyl radical for textile wastewater treatment: a review. J Clean Prod 87:826–838. https://doi.org/10.1016/j.jclepro.2014.09.010

Babuponnusami A, Muthukumar K (2014) A review on Fenton and improvements to the Fenton process for wastewater treatment. J Environ Chem Eng 2:557–572. https://doi.org/10.1016/j.jece.2013.10.011

Barhoumi N, Oturan N, Ammar S, Gadri A (2017) Enhanced degradation of the antibiotic tetracycline by heterogeneous electro-Fenton with pyrite catalysis. Environ Chem Lett 15:689–693. https://doi.org/10.1007/s10311-017-0638-y

Barry TI, Stone FS (1960) The reactions of oxygen at dark and irradiated zinc oxide surface. Proc R Soc A Math Phys Eng Sci 255:124–144. https://doi.org/10.1098/rspa.1960.0058

Bazrafshan AA, Ghaedi M, Hajati S, Naghiha R, Asfaram A (2017) Synthesis of ZnO-nanorod-based materials for antibacterial, antifungal activities, DNA cleavage and efficient ultrasound-assisted dyes adsorption. Ecotoxicol Environ Saf 142:330–337. https://doi.org/10.1016/j.ecoenv.2017.04.011

Bello MM, Raman AAA (2017) Trend and current practices of palm oil mill effluent polishing: application of advanced oxidation processes and their future perspectives. J Environ Manag 198:170–182. https://doi.org/10.1016/j.jenvman.2017.04.050

Bello MM, Abdul Raman AA, Purushothaman M (2017) Applications of fluidized bed reactor in wastewater treatment – a review of the major design and operational parameters. J Clean Prod 141:1492–1514. https://doi.org/10.1016/j.jclepro.2016.09.148

Bhanjana G, Dilbaghi N, Kim K, Kumar S (2017) Carbon nanotubes as sorbent material for removal of cadmium. J Mol Liq 242:966–970. https://doi.org/10.1016/j.molliq.2017.07.072

Bora LV, Mewada RK (2017) Visible/solar light active photocatalysts for organic effluent treatment: fundamentals, mechanisms and parametric review. Renew Sust Energ Rev 76:1393–1421. https://doi.org/10.1016/j.rser.2017.01.130

Breitbach M, Bathen D, Schmidt-traub H (2003) Effect of ultrasound on adsorption and desorption processes. Ind Eng Chem Res 42:5635–5646. https://doi.org/10.1021/ie030333f

Broséus R, Vincent S, Aboulfadl K, Daneshvar A, Sauvé S, Barbeau B, Prévost M (2009) Ozone oxidation of pharmaceuticals, endocrine disruptors and pesticides during drinking water treatment. Water Res 43:4707–4717. https://doi.org/10.1016/j.watres.2009.07.031

Camposeco R, Castillo S, Navarrete J, Gomez R (2016) Synthesis, characterization and photocatalytic activity of TiO_2 nanostructures: nanotubes, nanofibers, nanowires and nanoparticles. Catal Today 266:90–101. https://doi.org/10.1016/j.cattod.2015.09.018

Candal R (2014) Adsorption of crystal violet on montmorillonite (or iron modified montmorillonite) followed by degradation through Fenton or photo-Fenton type reactions. J Environ Chem Eng 2:2344–2351. https://doi.org/10.1016/j.jece.2014.02.007

Cataldo S, Iannì A, Loddo V, Mirenda E, Palmisano L, Parrino F, Piazzese D (2016) Combination of advanced oxidation processes and active carbons adsorption for the treatment of simulated saline wastewater. Sep Purif Technol 171:101–111. https://doi.org/10.1016/j.seppur.2016.07.026

Cederlund H, Börjesson E, Lundberg D, Stenström J (2016) Adsorption of pesticides with different chemical properties to a wood biochar treated with heat and iron. Water Air Soil Pollut 227:1–12. https://doi.org/10.1007/s11270-016-2894-z

Chang S, Wang K, Li H, Wey M, Chou J (2009) Enhancement of Rhodamine B removal by low-cost fly ash sorption with Fenton pre-oxidation. J Hazard Mater 172:1131–1136. https://doi.org/10.1016/j.jhazmat.2009.07.106

Chelme-Ayala P, El-Din MG, Smith DW (2010) Kinetics and mechanism of the degradation of two pesticides in aqueous solutions by ozonation. Chemosphere 78:557–562. https://doi.org/10.1016/j.chemosphere.2009.11.014

Chen T-C, Matira EM, Lu M-C, Dalida MLP (2016) Degradation of dimethyl sulfoxide through fluidized-bed Fenton process. Int J Environ Sci Technol 300:1–12. https://doi.org/10.1016/j.jhazmat.2015.06.069

Chen F, An W, Liu L, Liang Y, Cui W (2017) Highly efficient removal of bisphenol a by a three-dimensional graphene hydrogel-AgBr@rGO exhibiting adsorption/photocatalysis synergy. Appl Catal B Environ 217:65–80. https://doi.org/10.1016/j.apcatb.2017.05.078

Cheng G, Xu F, Xiong J, Tian F, Ding J, Stadler FJ, Chen R (2016) Enhanced adsorption and photocatalysis capability of generally synthesized TiO_2-carbon materials hybrids. Adv Powder Technol 27:1949–1962. https://doi.org/10.1016/j.apt.2016.06.026

Choi W, Termin A, Hoffmann MR (1994) The role of metal ion dopants in quantum-sized TiO_2: correlation between photoreactivity and charge carrier recombination dynamics. J Phys Chem 98:13669–13679. https://doi.org/10.1021/j100102a038

Chowdhury S, Balasubramanian R (2014) Recent advances in the use of graphene-family nanoadsorbents for removal of toxic pollutants from wastewater. Adv Colloid Interf Sci 204:35–56. https://doi.org/10.1016/j.cis.2013.12.005

Chowdhury P, Viraraghavan T (2009) Sonochemical degradation of chlorinated organic compounds, phenolic compounds and organic dyes – a review. Sci Total Environ 407:2474–2492. https://doi.org/10.1016/j.scitotenv.2008.12.031

Clarizia L, Russo D, Di Somma I, Marotta R, Andreozzi R (2017) Homogeneous photo-Fenton processes at near neutral pH: a review. Appl Catal B Environ 209:358–371. https://doi.org/10.1016/j.apcatb.2017.03.011

Comero S, Loos R, Carvalho R, Anto DC, Locoro G, Tavazzi S, Paracchini B, Ghiani M, Lettieri T, Blaha L, Jarosova B, Voorspoels S, Servaes K, Haglund P, Fick J, Lindberg RH, Schwesig D, Gawlik BM (2013) EU-wide monitoring survey on emerging polar organic contaminants in wastewater treatment plant effluents. Water Res 7:6475–6487. https://doi.org/10.1016/j.watres.2013.08.024

Cruz A, Couto L, Esplugas S, Sans C (2017) Study of the contribution of homogeneous catalysis on heterogeneous Fe(III)/alginate mediated photo-Fenton process. Chem Eng J 318:272–280. https://doi.org/10.1016/j.cej.2016.09.014

Dai Y, Song Y, Tu X, Jiang Y, Yuan Y (2015) Sequential shape-selective adsorption and photocatalytic transformation of acrylonitrile production wastewater. Water Res 85:216–225. https://doi.org/10.1016/j.watres.2015.08.034

Das P (2016) Ultrasound assisted mixed azo dye adsorption by chitosan – graphene oxide nanocomposite. Chem Eng Res Des 117:43–56. https://doi.org/10.1016/j.cherd.2016.10.009

Dastkhoon M, Ghaedi M, Asfaram A (2017) Simultaneous removal of dyes onto nanowires adsorbent use of ultrasound assisted adsorption to clean waste water: chemometrics for modeling and optimization, multicomponent adsorption and kinetic study. Chem Eng Res Des 124:222–237. https://doi.org/10.1016/j.cherd.2017.06.011

Demiral H, Güngör C (2016) Adsorption of copper(II) from aqueous solutions on activated carbon prepared from grape bagasse. J Clean Prod 124:103–113. https://doi.org/10.1016/j.jclepro.2016.02.084

Dewil R, Mantzavinos D, Poulios I, Rodrigo MA (2017) New perspectives for advanced oxidation processes. J Environ Manag 195:93–99. https://doi.org/10.1016/j.jenvman.2017.04.010

Di G, Zhu Z, Zhang H, Zhu J, Lu H, Zhang W, Qiu Y, Zhu L, Küppers S (2017) Simultaneous removal of several pharmaceuticals and arsenic on Zn-Fe mixed metal oxides: combination of photocatalysis and adsorption. Chem Eng J 328:141–151. https://doi.org/10.1016/j.cej.2017.06.112

Diaz-Flores P, Arcibar-Orozco J, Perez-Aguilar N, Rangel-Mendez J, Medina VMO, Alcala-Jauegui (2017) Adsorption of organic compounds onto multiwall and nitrogen-doped carbon nanotubes: insights into the adsorption mechanisms. Water Air Soil Pollut 228:2–13. https://doi.org/10.1007/s11270-017-3314-8

Dietrich M, Franke M, Stelter M, Braeutigam P (2017) Degradation of endocrine disruptor bisphenol A by ultrasound-assisted electrochemical oxidation in water. Ultrason Sonochem 39:741–749. https://doi.org/10.1016/j.ultsonch.2017.05.038

Doerffler W, Hauffe K (1964) Heterogeneous photocatalysis I. The influence of oxidizing and reducing gases on the electrical conductivity of dark and illuminated zinc oxide surfaces. J Catal 3:156–170. https://doi.org/10.1016/0021-9517(64)90123-X

Du YL, Pan CP (2014) Preparation of cellulose/graphene composite and its applications for triazine pesticides adsorption from water. ACS Sustain Chem Eng 3:396–340. https://doi.org/10.1021/sc500738k

Duc M, Thomas F, Gaboriaud F (2006) Coupled chemical processes at clay/electrolyte interface: a batch titration study of Na-montmorillonites. J Colloid Interf Sci 300:616–625. https://doi.org/10.1016/j.jcis.2006.04.081

Dwivedi K, Morone A, Chakrabarti T, Pandey RA (2017) Evaluation and optimization of Fenton pretreatment integrated with granular activated carbon (GAC) filtration for carbamazepine removal from complex wastewater of pharmaceuticals industry. J Environ Chem Eng. https://doi.org/10.1016/j.jece.2016.12.054

Elsellami L, Chartron V, Vocanson F, Conchon P, Felix C, Guillard C, Retailleau L, Houas A (2009) Coupling process between solid-liquid extraction of amino acids by calixarenes and photocatalytic degradation. J Hazard Mater 166:1195–1200. https://doi.org/10.1016/j.jhazmat.2008.12.033

Ersan G, Apul OG, Perreault F, Karan T (2017) Adsorption of organic contaminants by graphene nanosheets: a review. Water Res 126:385–398. https://doi.org/10.1016/j.watres.2017.08.010

Eskandarian L, Pajootan E, Arami M (2014) Novel super adsorbent molecules, carbon nanotubes modified by dendrimer miniature structure, for the removal of trace organic dyes. Ind Eng Chem Res 53:14841–14853. https://doi.org/10.1021/ie502414t

Etacheri V, Di Valentin C, Schneider J, Bahnemann D, Pillai SC (2015) Visible-light activation of TiO_2 photocatalysts: advances in theory and experiments. J Photochem Photobiol C: Photochem Rev 25:1–29. https://doi.org/10.1016/j.jphotochemrev.2015.08.003

Fakhri A, Naji M, Afshar P (2017) Adsorption and photocatalysis efficiency of magnetite quantum dots anchored tin dioxide nano fibers for removal of mutagenic compound: toxicity evaluation and antibacterial activity. J Photochem Photobiol B Biol 173:204–209. https://doi.org/10.1016/j.jphotobiol.2017.05.041

Faria PCC, Orfao JJ, Pereira MF (2008) Activated carbon catalytic ozonation of oxamic and oxalic acids. Appl Catal A Gen 79:237–243. https://doi.org/10.1016/j.apcatb.2007.10.021

Fayazi M, Ali M, Afzali D, Mostafavi A (2016) Enhanced Fenton-like degradation of methylene blue by magnetically activated carbon/hydrogen peroxide with hydroxylamine as Fenton enhancer. J Mol Liq 216:781–787. https://doi.org/10.1016/j.molliq.2016.01.093

Fida H, Zhang G, Guo S, Naeem A (2017) Heterogeneous Fenton degradation of organic dyes in batch and fixed bed using La-Fe montmorillonite as catalyst. J Colloid Interf Sci 490:859–868. https://doi.org/10.1016/j.jcis.2016.11.085

Frank SN, Bard AJ (1977) Heterogeneous photocatalytic oxidation of cyanide ion in aqueous solutions at TiO2 powder. J Am Chem Soc 99:303–304. https://doi.org/10.1021/ja00443a081

Frimmel FH, Assenmacher M, So M, Abbt-braun G (1999) Removal of hydrophilic pollutants from water with organic adsorption polymers – part I. Adsorption behaviour of selected model compounds. Chem Eng Process 38:601–610

Fujishima A, Honda K (1972) Electrochemical photolysis of water at a semiconductor electrode. Nature 238:37–38. https://doi.org/10.1038/238037a0

Gao M, Zhang D, Li W, Chang J, Lin Q, Xu D (2016) Degradation of methylene blue in a heterogeneous Fenton reaction catalyzed by chitosan crosslinked ferrous complex. J Taiwan Inst Chem Eng 67:355–361. https://doi.org/10.1016/j.jtice.2016.08.010

Garcı JF, Beltra FJ (2007) Activated carbon promoted ozonation of polyphenol mixtures in water: comparison with single ozonation. Ind Eng Chem Res 46:8241–8247. https://doi.org/10.1021/ie0708881

García-díaz E, Martín MD (2016) Assessment of the effectiveness of combined adsorption and photocatalysis for removal of the herbicide isoproturon. Phys Chem Earth Parts A/B/C 91:77–86. https://doi.org/10.1016/j.pce.2015.09.011

Garcia-Segura S, Bellotindos LM, Huang Y-H, Brillas E, Lu M-C (2016) Fluidized-bed Fenton process as alternative wastewater treatment technology—a review. J Taiwan Inst Chem Eng 67:211–225. https://doi.org/10.1016/j.jtice.2016.07.021

Gauden PA, Pacholczyk A, Terzyk AP, Wis M, Wesołowski RP, Furmaniak S, Szczes A (2011) Phenol adsorption on closed carbon nanotubes. J Colloid Interf Sci 361:288–292. https://doi.org/10.1016/j.jcis.2011.05.032

Ghows N, Entezari MH (2011) Exceptional catalytic efficiency in mineralization of the reactive textile azo dye (RB5) by a combination of ultrasound and core-shell nanoparticles (CdS/TiO_2). J Hazard Mater 195:132–138. https://doi.org/10.1016/j.jhazmat.2011.08.049

Glaze WH, Kang J, Douglas H (1987) The chemistry of water treatment processes involving ozone, hydrogen peroxide and ultraviolet radiation. Ozone Sci Eng 9:335–352. https://doi.org/10.1080/01919518708552148

Gogate PR (2008) Treatment of wastewater streams containing phenolic compounds using hybrid techniques based on cavitation: a review of the current status and the way forward. Ultrason Sonochem 15:1–15. https://doi.org/10.1016/j.ultsonch.2007.04.007

Gomes J, Costa R, Quinta-ferreira RM, Martins RC (2017) Application of ozonation for pharmaceuticals and personal care products removal from water. Sci Total Environ 586:265–283. https://doi.org/10.1016/j.scitotenv.2017.01.216

Gu L, Zhu N, Guo H, Huang S, Lou Z, Yuan H (2013) Adsorption and Fenton-like degradation of naphthalene dye intermediate on sewage sludge derived porous carbon. J Hazard Mater 246–247:145–153. https://doi.org/10.1016/j.jhazmat.2012.12.012
Gümüs D (2017) A comparative study of ozonation, iron coated zeolite catalyzed ozonation and granular activated carbon catalyzed ozonation of humic acid. Chemosphere 174:218–231. https://doi.org/10.1016/j.chemosphere.2017.01.106
Guo S, Zhang G (2015) Green synthesis of a bifunctional Fe–montmorillonite composite during the Fenton degradation process and its enhanced adsorption. RSC Adv 6:2537–2545. https://doi.org/10.1039/C5RA25096J
Guo S, Yuan N, Zhang G, Yu JC (2017) Graphene modified iron sludge derived from homogeneous Fenton process as an efficient heterogeneous Fenton catalyst for degradation of organic pollutants. Microporous Mesoporous Mater 238:62–68. https://doi.org/10.1016/j.micromeso.2016.02.033
Hadi M, Shariati Z, Reza M, Shayeghi M (2017) Optimizing the removal of organophosphorus pesticide malathion from water using multi-walled carbon nanotubes. Chem Eng J 310:22–32. https://doi.org/10.1016/j.cej.2016.10.057
Hadjltaief BH, Sdiri A, Ltaief W, Costa DP, Galvez ME, Zina BM (2017) Efficient removal of cadmium and 2-chlorophenol in aqueous systems by natural clay: adsorption and photo-Fenton degradation processes. C R Chim. https://doi.org/10.1016/j.crci.2017.01.009
Halouane F, Oz Y, Meziane D, Barras A, Juraszek J, Singh SK, Kurungot S, Shaw PK, Sanyal R, Boukherroub R, Sanyal A, Szunerits S (2017) Magnetic reduced graphene oxide loaded hydrogels: highly versatile and efficient adsorbents for dyes and selective Cr(VI) ions removal. J Colloid Interf Sci 507:360–369. https://doi.org/10.1016/j.jcis.2017.07.075
Hamdaoui O, Naffrechoux E (2009) Adsorption kinetics of 4-chlorophenol onto granular activated carbon in the presence of high frequency ultrasound. Ultrason Sonochem 16:15–22. https://doi.org/10.1016/j.ultsonch.2008.05.008
Hansen KMS, Spiliotopoulou A, Kumar R, Escolà M, Bester K, Andersen HR (2016) Ozonation for source treatment of pharmaceuticals in hospital wastewater – ozone lifetime and required ozone dose. Chem Eng J 290:507–514. https://doi.org/10.1016/j.cej.2016.01.027
Hashemian S, Salari K, Yazdi ZA (2014) Preparation of activated carbon from agricultural wastes (almond shell and orange peel) for adsorption of 2-pic from aqueous solution. J Ind Eng Chem 20:1892–1900. https://doi.org/10.1016/j.jiec.2013.09.009
He Y, Sutton NB, Rijnaarts HHH, Langenhoff AAM (2016) Degradation of pharmaceuticals in wastewater using immobilized TiO_2 photocatalysis under simulated solar irradiation. Appl Catal B Environ 182:132–141. https://doi.org/10.1016/j.apcatb.2015.09.015
Herrmann JM (2010) Fundamentals and misconceptions in photocatalysis. J Photochem Photobiol A Chem 216:85–93. https://doi.org/10.1016/j.jphotochem.2010.05.015
Hu S, Yao H, Wang K, Lu C (2015) Intensify removal of nitrobenzene from aqueous solution using nano-zero valent iron/granular activated carbon composite as Fenton-like catalyst. Water Air Soil Pollut 226:1–13. https://doi.org/10.1007/s11270-015-2421-7
Hu E, Shang S, ming TX, Jiang S, lok CK (2016) Regeneration and reuse of highly polluting textile dyeing effluents through catalytic ozonation with carbon aerogel catalysts. J Clean Prod 137:1055–1065. https://doi.org/10.1016/j.jclepro.2016.07.194
Hua S, Gong J, Zeng G, Yao F, Guo M (2017) Remediation of organochlorine pesticides contaminated lake sediment using activated carbon and carbon nanotubes. Chemosphere 177:65–76. https://doi.org/10.1016/j.chemosphere.2017.02.133
Huang H, Cao R, Yu S, Xu K, Zhang Y (2017a) Single-unit-cell layer established $Bi_2WO_6$3D hierarchical architectures: efficient adsorption, photocatalysis and dye-sensitized photoelectrochemical performance. Appl Catal B Environ 219:526–537. https://doi.org/10.1016/j.apcatb.2017.07.084
Huang Y, Sun Y, Xu Z, Luo M, Zhu C, Li L (2017b) Removal of aqueous oxalic acid by heterogeneous catalytic ozonation with MnO_x/sewage sludge-derived activated carbon as catalysts. Sci Total Environ 575:50–57. https://doi.org/10.1016/j.scitotenv.2016.10.026

Huang Y, Xu W, Hu L, Zeng J, He C, Tan X, He Z (2017c) Combined adsorption and catalytic ozonation for removal of endocrine disrupting compounds over MWCNTs/Fe_3O_4 composites. Catal Today 297:143–150. https://doi.org/10.1016/j.cattod.2017.05.097

Ihsanullah AA, Al-amer AM, Laoui T, Al-marri MJ, Nasser MS, Khraisheh M, Ali M (2016) Heavy metal removal from aqueous solution by advanced carbon nanotubes: critical review of adsorption applications. Sep Purif Technol 157:141–161. https://doi.org/10.1016/j.seppur.2015.11.039

Ikhlaq A, Kasprzyk-hordern B (2017) Catalytic ozonation of chlorinated VOCs on ZSM-5 zeolites and alumina: formation of chlorides. Appl Catal B Environ 200:274–282. https://doi.org/10.1016/j.apcatb.2016.07.019

Ikhlaq A, Brown DR, Kasprzyk-hordern B (2015) Catalytic ozonation for the removal of organic contaminants in water on alumina. Appl Catal B Environ 165:408–418. https://doi.org/10.1016/j.apcatb.2014.10.010

Ioannidou O, Zabaniotou AÃ (2007) Agricultural residues as precursors for activated carbon production—a review. Renew Sust Energ Rev 11:1966–2005. https://doi.org/10.1016/j.rser.2006.03.013

Irani M, Roshanfekr L, Pourahmad H, Haririan I (2015) Optimization of the combined adsorption/photo-Fenton method for the simultaneous removal of phenol and paracetamol in a binary system. Microporous Mesoporous Mater 206:1–7. https://doi.org/10.1016/j.micromeso.2014.12.009

Irmak S, Erbatur O, Akgerman A (2005) Degradation of 17β-estradiol and bisphenol A in aqueous medium by using ozone and ozone/UV techniques. J Hazard Mater 126:54–62. https://doi.org/10.1016/j.jhazmat.2005.05.045

Jafari JA, Kakavandi B, Jaafarzadeh N, Rezaei Kalantary R, Ahmadi M, Akbar Babaei A (2017) Fenton-like catalytic oxidation of tetracycline by AC@Fe_3O_4 as a heterogeneous persulfate activator: adsorption and degradation studies. J Ind Eng Chem 45:323–333. https://doi.org/10.1016/j.jiec.2016.09.044

Jawale RH, Tandale A, Gogate PR (2017) Novel approaches based on ultrasound for treatment of wastewater containing potassium ferrocyanide. Ultrason Sonochem 38:402–409. https://doi.org/10.1016/j.ultsonch.2017.03.032

Kah M, Sigmund G, Xiao F, Hofmann T (2017) Sorption of ionizable and ionic organic compounds to biochar, activated carbon and other carbonaceous materials. Water Res 124:673–692. https://doi.org/10.1016/j.watres.2017.07.070

Kanakaraju D, Glass BD, Oelgemo M (2014) Titanium dioxide photocatalysis for pharmaceutical wastewater treatment. Environ Chem Lett 12:27–47. https://doi.org/10.1007/s10311-013-0428-0

Keen OS, Linden KG (2013) Degradation of antibiotic activity during UV/H_2O_2 advanced oxidation and photolysis in wastewater effluent. Environ Sci Technol 47:13020–13030. https://doi.org/10.1021/es402472x

Khataee A, Kayan B, Kalderis D, Karimi A, Akay S (2017) Ultrasound-assisted removal of acid red 17 using nanosized Fe_3O_4 -loaded coffee waste hydrochar. Ultrason Sonochem 35:72–80. https://doi.org/10.1016/j.ultsonch.2016.09.004

Kim JR, Huling SG, Kan E (2015) Effects of temperature on adsorption and oxidative degradation of bisphenol A in an acid-treated iron-amended granular activated carbon. Chem Eng J 262:1260–1267. https://doi.org/10.1016/j.cej.2014.10.065

Kosera VS, Cruz TM, Chaves ES, Tiburtius ERL (2017) Triclosan degradation by heterogeneous photocatalysis using ZnO immobilized in biopolymer as catalyst. J Photochem Photobiol A Chem 344:184–191. https://doi.org/10.1016/j.jphotochem.2017.05.014

Kotov NA, Kelly JW (2006) Carbon sheet solutions proteins downhill all the way. Nature 442:254–255. https://doi.org/10.1002/anie.200601677

Kubo T, Nakahira A (2008) Local structure of TiO_2-derived nanotubes prepared by the hydrothermal process. J Phys Chem C 112:1658–1662. https://doi.org/10.1021/jp076699d

Kumar V, Kumar R, Nayak A, Saleh A, Barakat MA (2013) Adsorptive removal of dyes from aqueous solution onto carbon nanotubes: a review. Adv Colloid Interf Sci 193–194:24–34. https://doi.org/10.1016/j.cis.2013.03.003

Kyriakopoulos G, Doulia D (2006) Adsorption of pesticides on carbonaceous and polymeric materials from aqueous solutions: a review. Sep Purif Rev 35:97–191. https://doi.org/10.1080/15422110600822733

Kyriakopoulos G, Doulia D, Anagnostopoulos E (2005) Adsorption of pesticides on porous polymeric adsorbents. Chem Eng Sci 60:1177–1186. https://doi.org/10.1016/j.ces.2004.09.080

Lau A, Alberto L, Teixeira C, Valéria F, Yokoyama L (2016) Evaluation of the mercaptobenzothiazole degradation by combined adsorption process and Fenton reaction using iron mining residue. Environ Technol 38:2032–2039. https://doi.org/10.1080/09593330.2016.1244571

Laxma PV, Kim K (2015) A review of photochemical approaches for the treatment of a wide range of pesticides. J Hazard Mater 285:325–335. https://doi.org/10.1016/j.jhazmat.2014.11.036

Li Z, Li X, Xi H, Hua B (2002) Effects of ultrasound on adsorption equilibrium of phenol on polymeric adsorption resin. Chem Eng J 86:375–379. https://doi.org/10.1016/S1385-8947(01)00301-1

Li F, Bao J, Zhang TC, Lei Y (2015) A combined process of adsorption and Fenton-like oxidation for furfural removal using zero-valent iron residue. Environ Technol 36:3103–3111. https://doi.org/10.1080/09593330.2015.1054317

Li K, Zhao Y, Song C, Guo X (2017) Magnetic ordered mesoporous Fe_3O_4/CeO_2 composites with synergy of adsorption and Fenton catalysis. Appl Surf Sci 425:526–534. https://doi.org/10.1016/j.apsusc.2017.07.041

Lima MJ, Silva CG, Silva AMT, Lopes JCB, Dias MM, Faria JL (2017) Homogeneous and heterogeneous photo-Fenton degradation of antibiotics using an innovative static mixer photoreactor. Chem Eng J 310:342–351. https://doi.org/10.1016/j.cej.2016.04.032

Lin K, Cheng H, Chen W, Wu C (2010) Synthesis, characterization, and adsorption kinetics of titania nanotubes for basic dye wastewater treatment. Adsorption 16:47–56. https://doi.org/10.1007/s10450-010-9216-3

Lin K, Lin Y, Cheng H, Haung Y (2017) Preparation and characterization of V-loaded titania nanotubes for adsorption/photocatalysis of basic dye and environmental hormone contaminated wastewaters. Catal Today 307:119–130. https://doi.org/10.1016/j.cattod.2017.05.075

Liu J, Li J, Mei R, Wang F, Sellamuthu B (2014) Treatment of recalcitrant organic silicone wastewater by fluidized-bed Fenton process. Sep Purif Technol 132:16–22. https://doi.org/10.1016/j.seppur.2014.04.050

López-mu M, Arencibia A, Cerro L, Pascual R, Melgar Á (2016) Adsorption of Hg(II) from aqueous solutions using TiO_2 and titanate nanotube adsorbents. Appl Surf Sci 367:91–100. https://doi.org/10.1016/j.apsusc.2016.01.109

Lucking F, Koser H, Jank M, Ritter A (1998) Iron powder, graphite and activated carbon as catalysts for the oxidation of 4-chlorophenol with hydrogen peroxide in aqueous solution. Water Res 32:2607–2614. https://doi.org/10.1016/S0043-1354(98)00016-5

Luo Y, Guo W, Hao H, Duc L, Ibney F, Zhang J, Liang S, Wang XC (2014) A review on the occurrence of micropollutants in the aquatic environment and their fate and removal during wastewater treatment. Sci Total Environ 473–474:619–641. https://doi.org/10.1016/j.scitotenv.2013.12.065

Luo B, Xu D, Li D, Wu G, Wu M, Shi W, Chen M (2015) Fabrication of a Ag/Bi_3TaO_7 plasmonic photocatalyst with enhanced photocatalytic activity for degradation of tetracycline. ACS Appl Mater Interfaces 7:17061–17069. https://doi.org/10.1021/acsami.5b03535

Luo L, Li J, Dai J, Xia L, Barrow CJ, Wang H, Jegatheesan J, Yang M (2017) Bisphenol A removal on TiO_2–MoS_2–reduced graphene oxide composite by adsorption and photocatalysis. Process Biochem 112:274–279. https://doi.org/10.1016/j.psep.2017.04.032

Lyu C, Zhou D, Wang J (2016) Removal of multi-dye wastewater by the novel integrated adsorption and Fenton oxidation process in a fluidized bed reactor. Environ Sci Pollut Res 23:20893–20903. https://doi.org/10.1007/s11356-016-7272-2

Maldonado-Hódar F, Madeira L, Portela M (1999) The use of coals as catalysts for the oxidative dehydrogenation of n-butane. Appl Catal A Gen 178:49–60. https://doi.org/10.1016/S0926-860X(98)00273-7

Maryam S, Mousavi S, Dehghanzadeh R, Ebrahimi SM (2017) Comparative analysis of ozonation (O_3) and activated carbon catalyzed ozonation (ACCO) for destroying chlorophyll a and reducing dissolved organic carbon from a eutrophic water reservoir. Chem Eng J 314:396–405. https://doi.org/10.1016/j.cej.2016.11.159

Mason TJ, Lorimer JP (1988) Sonochemistry: theory, applications and uses of ultrasound in chemistry. Ellis Horwood Ltd., New York

Matira EM, Chen T-C, Lu M-C, Dalida MLP (2015) Degradation of dimethyl sulfoxide through fluidized-bed Fenton process. J Hazard Mater 300:218–226. https://doi.org/10.1016/j.jhazmat.2015.06.069

Matouq MA, Al-anber ZA, Tagawa T, Aljbour S (2008) Degradation of dissolved diazinon pesticide in water using the high frequency of ultrasound wave. Ultrason Sonochem 15:869–874. https://doi.org/10.1016/j.ultsonch.2007.10.012

Meijide J, Rodríguez S, Sanromán MA, Pazos M (2017) Comprehensive solution for acetamiprid degradation: combined electro- Fenton and adsorption process. J Electroanal Chem 808:446–454. https://doi.org/10.1016/j.jelechem.2017.05.012

Meng CN, Jin B, Chow CWK, Saint C (2010) Recent developments in photocatalytic water treatment technology: a review. Water Res 44:2997–3027. https://doi.org/10.1016/j.watres.2010.02.039

Mezohegyi G, Van Der ZFP, Font J, Fortuny A, Fabregat A (2012) Towards advanced aqueous dye removal processes: a short review on the versatile role of activated carbon. J Environ Manag 102:148–164. https://doi.org/10.1016/j.jenvman.2012.02.021

Midathana VR, Moholkar VS (2009) Mechanistic studies in ultrasound-assisted adsorption for removal of aromatic pollutants. Ind Eng Chem Res 48:7368–7377. https://doi.org/10.1021/ie900049e

Mkhoyan KA, Contryman AW, Silcox J, Stewart DA, Eda G, Mattevi C, Miller S (2009) Atomic and electronic structure of graphene-oxide. Nano Lett 9:1058–1063. https://doi.org/10.1021/nl8034256

Mohamed HH, Bahnemann DW (2012) The role of electron transfer in photocatalysis: fact and fictions. Appl Catal B Environ 128:91–104. https://doi.org/10.1016/j.apcatb.2012.05.045

Mojiri A, Ziyang L, Hui W, Ahmad Z, Tajuddin R, Abu Amr S, Kindaichi T, Aziz HA, Farraji H (2017) Concentrated landfill leachate treatment with a combined system including electro-ozonation and composite adsorbent augmented sequencing batch reactor process. Process Saf Environ Prot 111:253–262. https://doi.org/10.1016/j.psep.2017.07.013

Morales-Perez AA, Arias C, Ramirez-Zamora R-M (2015) Removal of atrazine from water using an iron photo catalyst supported on activated carbon. Adsorption 22:49–58. https://doi.org/10.1007/s10450-015-9739-8

Mu C, Zhang Y, Cui W, Liang Y, Zhu Y (2017) Removal of bisphenol A over a separation free 3D Ag_3PO_4-graphene hydrogel via an adsorption-photocatalysis synergy. Appl Catal B Environ 212:41–49. https://doi.org/10.1016/j.apcatb.2017.04.018

Nakada N, Shinohara H, Murata A, Kiri K, Managaki S, Sato N, Takada H (2007) Removal of selected pharmaceuticals and personal care products (PPCPs) and endocrine-disrupting chemicals (EDCs) during sand filtration and ozonation at a municipal sewage treatment plant. Water Res 41:4373–4382. https://doi.org/10.1016/j.watres.2007.06.038

Nasuha N, Hameed BH (2011) Adsorption of methylene blue from aqueous solution onto NaOH-modified rejected tea. Chem Eng J 166:783–786. https://doi.org/10.1016/j.cej.2010.11.012

Nawrocki J (2013) Catalytic ozonation in water: controversies and questions. Discussion paper. Appl Catal B Environ 142–143:465–471. https://doi.org/10.1016/j.apcatb.2013.05.061
Nawrocki J, Kasprzyk-hordern B (2010) The efficiency and mechanisms of catalytic ozonation. Appl Catal B Environ 99:27–42. https://doi.org/10.1016/j.apcatb.2010.06.033
Neyens E, Baeyens J (2003) A review of classic Fenton's peroxidation as an advanced oxidation technique. J Hazard Mater 98:33–50. https://doi.org/10.1016/S0304-3894(02)00282-0
Ngulube T, Gumbo JR, Masindi V, Maity A (2017) An update on synthetic dyes adsorption onto clay based minerals: a state-of-art review. J Environ Manag 191:35–57. https://doi.org/10.1016/j.jenvman.2016.12.031
Nguyen TD, Phan NH, Do MH, Ngo KT (2011) Magnetic Fe_2MO_4 (M:Fe, Mn) activated carbons: fabrication, characterization and heterogeneous Fenton oxidation of methyl orange. J Hazard Mater 185:653–661. https://doi.org/10.1016/j.jhazmat.2010.09.068
Pan B, Pan B, Zhang W, Lv L, Zhang Q, Zheng S (2009) Development of polymeric and polymer-based hybrid adsorbents for pollutants removal from waters. Chem Eng J 151:19–29. https://doi.org/10.1016/j.cej.2009.02.036
Panda KK, Mathews AP (2014) Ozone oxidation kinetics of reactive blue 19 anthraquinone dye in a tubular in situ ozone generator and reactor: modeling and sensitivity analyses. Chem Eng J 255:553–567. https://doi.org/10.1016/j.cej.2014.06.071
Pandey S (2017) A comprehensive review on recent developments in bentonite-based materials used as adsorbents for wastewater treatment. J Mol Liq 241:1091–1113. https://doi.org/10.1016/j.molliq.2017.06.115
Pang YL, Abdullah AZ (2013) Effect of carbon and nitrogen co-doping on characteristics and sonocatalytic activity of TiO_2 nanotubes catalyst for degradation of Rhodamine B in water. Chem Eng J 214:129–138. https://doi.org/10.1016/j.cej.2012.10.036
Papageorgiou DG, Kinloch IA, Young RJ (2017) Mechanical properties of graphene and graphene-based nanocomposites. Prog Mater Sci 90:75–127. https://doi.org/10.1016/j.pmatsci.2017.07.004
Paramo-Vargas J, Granados SG, Maldonado-Rubio MI, Peralta-hernandez JM (2016) Up to 95% reduction of chemical oxygen demand of slaughterhouse effluents using Fenton and photo-Fenton oxidation. Environ Chem Lett 14:149–154. https://doi.org/10.1007/s10311-015-0534-2
Peng W, Li H, Liu Y, Song S (2017) A review on heavy metal ions adsorption from water by graphene oxide and its composites. J Mol Liq 230:496–504. https://doi.org/10.1016/j.molliq.2017.01.064
Perdigo A, Petre A, Rosal R, Rodrı A, Agu A, Ferna AR (2010) Occurrence of emerging pollutants in urban wastewater and their removal through biological treatment followed by ozonation. Water Res 44:578–588. https://doi.org/10.1016/j.watres.2009.07.004
Pouran SR, Abdul Raman AA, Wan Daud WMA (2014) Review on the application of modified iron oxides as heterogeneous catalysts in Fenton reactions. J Clean Prod 64:24–35. https://doi.org/10.1016/j.jclepro.2013.09.013
Poyatos JM, Muñio MM, Almecija MC, Torres JC, Hontoria E, Osorio F (2010) Advanced oxidation processes for wastewater treatment: state of the art. Water Air Soil Pollut 205:187–204. https://doi.org/10.1007/s11270-009-0065-1
Pyrzynska K (2011) Carbon nanotubes as sorbents in the analysis of pesticides. Chemosphere 83:1407–1413. https://doi.org/10.1016/j.chemosphere.2011.01.057
Ramírez-sosa DR, Castillo-borges ER, Méndez-novelo RI, Sauri-riancho MR, Barceló-quintal M, Marrufo-gómez JM (2013) Determination of organic compounds in landfill leachates treated by Fenton – adsorption. Waste Manag 33:390–395. https://doi.org/10.1016/j.wasman.2012.07.019
Randviir EP, Brownson DAC, Banks CE (2014) A decade of graphene research: production, applications and outlook. Mater Today 17:426–432. https://doi.org/10.1016/j.mattod.2014.06.001
Rao CNR, Sood AK, Subrahmanyam KS, Govindaraj A (2009) Graphene: the new two-dimensional nanomaterial. Angew Chem Int Ed 48:7752–7777. https://doi.org/10.1002/anie.200901678

Rayaroth MP, Aravind UK, Aravindakumar CT (2016) Degradation of pharmaceuticals by ultrasound-based advanced oxidation process. Environ Chem Lett 14:259–290. https://doi.org/10.1007/s10311-016-0568-0

Ren A, Liu C, Hong Y, Shi W, Lin S, Li P (2014) Enhanced visible-light-driven photocatalytic activity for antibiotic degradation using magnetic $NiFe_2O_4/Bi_2O_3$ heterostructures. Chem Eng J 258:301–308. https://doi.org/10.1016/j.cej.2014.07.071

Roosta M, Ghaedi M, Shokri N, Daneshfar A, Sahraei R, Asghari A (2014) Molecular and biomolecular spectroscopy optimization of the combined ultrasonic assisted/adsorption method for the removal of malachite green by gold nanoparticles loaded on activated carbon: experimental design. Spectrochim Acta A Mol Biomol Spectrosc 118:55–65. https://doi.org/10.1016/j.saa.2013.08.082

Roshani B, Mcmaster I, Rezaei E, Soltan J (2014) Catalytic ozonation of benzotriazole over alumina supported transition metal oxide catalysts in water. Sep Purif Technol 135:158–164. https://doi.org/10.1016/j.seppur.2014.08.011

Ruthven DM (1984) Principles of adsorption and adsorption processes. A Wiley-Interscience, New York

Saeed M, Ahmad A, Boddula R (2017) Ag@Mn_xO_y: an effective catalyst for photo-degradation of rhodamine B dye. Environ Chem Lett. https://doi.org/10.1007/s10311-017-0661-z

Sajid M, Khaled M, Baig N (2018) Removal of heavy metals and organic pollutants from water using dendritic polymers based adsorbents: a critical review. Sep Purif Technol 191:400–423. https://doi.org/10.1016/j.seppur.2017.09.011

Sakulthaew C, Comfort SD, Chokejaroenrat C, Li X, Harris CE (2015) Removing PAHs from urban runoff water by combining ozonation and carbon nano-onions. Chemosphere 141:265–273. https://doi.org/10.1016/j.chemosphere.2015.08.002

Sans C, Esplugas S (2017) Pesticides abatement by advanced water technologies: the case of acetamiprid removal by ozonation. Sci Total Environ 599–600:1454–1461. https://doi.org/10.1016/j.scitotenv.2017.05.065

Sayan E and Edecan ME (2008) An optimization study using response surface methods on the decolorization of reactive blue 19 from aqueous solution by ultrasound. Ultrason Sonochem 15:530–538. https://doi.org/10.1016/j.ultsonch.2007.07.009

Schueller BS, Yang RT (2001) Ultrasound enhanced adsorption and desorption of phenol on activated carbon and polymeric resin. Ind Eng Chem Res 40:4912–4918. https://doi.org/10.1021/ie010490j

Shaban M, Abukhadra MR, Shahien MG, Ibrahim SS (2017) Novel bentonite/zeolite-NaP composite efficiently removes methylene blue and Congo red dyes. Environ Chem Lett. https://doi.org/10.1007/s10311-017-0658-7

Shanmugam K, Manivel P, Thangavel R, Uyar T (2017) Multifunctional ZnO nanorod-reduced graphene oxide hybrids nanocomposites for effective water remediation: effective sunlight driven degradation of organic dyes and rapid heavy metal adsorption. Chem Eng J 325:588–600. https://doi.org/10.1016/j.cej.2017.05.105

Sharifpour E, Haddadi H, Ghaedi M (2017) Optimization of simultaneous ultrasound assisted toxic dyes adsorption conditions from single and multi-components using central composite design: application of derivative spectrophotometry and evaluation of the kinetics and isotherms. Ultrason Sonochem 36:236–245. https://doi.org/10.1016/j.ultsonch.2016.11.011

Sherlala AIA, Raman AAA, Bello MM, Asghar A (2018) A review of the applications of organo-functionalized magnetic graphene oxide nanocomposites for heavy metal adsorption. Chemosphere 193:1004–1017. https://doi.org/10.1016/j.chemosphere.2017.11.093

Singh V, Joung D, Zhai L, Das S, Khondaker SI, Seal S (2011) Graphene based materials: past, present and future. Prog Mater Sci 56:1178–1271. https://doi.org/10.1016/j.pmatsci.2011.03.003

Sirés I, Brillas E, Oturan MA, Rodrigo MA, Panizza M (2014) Electrochemical advanced oxidation processes: today and tomorrow. A review. Environ Sci Pollut Res 21:8336–8367. https://doi.org/10.1007/s11356-014-2783-1

Sonawane S, Chaudhari P, Ghodke S, Ambade S, Gulig S, Mirikar A, Bane A (2008) Combined effect of ultrasound and nanoclay on adsorption of phenol. Ultrason Sonochem 15:1033–1037. https://doi.org/10.1016/j.ultsonch.2008.03.006
Staehelin J, Hoigne J (1985) Decomposition of ozone in water in the presence of organic solutes acting as promoters and inhibitors of radical chain reactions. Environ Sci Technol 19:1206–1213. https://doi.org/10.1021/es00142a012
Strachowski P, Bystrzejewski M (2015) Comparative studies of sorption of phenolic compounds onto carbon-encapsulated iron nanoparticles, carbon nanotubes and activated carbon. Colloids Surf A Physicochem Eng Asp 467:113–123. https://doi.org/10.1016/j.colsurfa.2014.11.044
Sudrajat H, Babel S (2016) Rapid photocatalytic degradation of the recalcitrant dye amaranth by highly active N-WO_3. Environ Chem Lett 14:243–249. https://doi.org/10.1007/s10311-015-0538-y
Sun L, Yao Y, Wang L, Mao Y, Huang Z, Yao D, Lu W, Chen W (2014) Efficient removal of dyes using activated carbon fibers coupled with 8-hydroxyquinoline ferric as a reusable Fenton-like catalyst. Chem Eng J 240:413–419. https://doi.org/10.1016/j.cej.2013.12.009
Suslick KS (1989) The chemical effects of ultrasound. Sci Am 260:80–86
Suslick KS (1990) Sonochemistry. Science 247:1439–1445. https://doi.org/10.1126/science.247.4949.1439
Tan KB, Vakili M, Horri BA, Poh PE, Abdullah AZ, Salamatinia B (2015) Adsorption of dyes by nanomaterials: recent developments and adsorption mechanisms. Sep Purif Technol 150:229–242. https://doi.org/10.1016/j.seppur.2015.07.009
Thankappan R, Nguyen TV, Srinivasan SV, Vigneswaran S (2015) Removal of leather tanning agent syntan from aqueous solution using Fenton oxidation followed by GAC adsorption. J Ind Eng Chem 21:483–488. https://doi.org/10.1016/j.jiec.2014.03.008
Tran N, Drogui P, Brar SK (2015) Sonochemical techniques to degrade pharmaceutical organic pollutants. Environ Chem Lett 13:251–268. https://doi.org/10.1007/s10311-015-0512-8
Tran N, Drogui P, Brar SK, De CA (2017a) Synergistic effects of ultrasounds in the sonoelectrochemical oxidation of pharmaceutical carbamazepine pollutant. Ultrason Sonochem 34:380–388. https://doi.org/10.1016/j.ultsonch.2016.06.014
Tran NH, You S, Hosseini-bandegharaei A, Chao H-P (2017b) Mistakes and inconsistencies regarding adsorption of contaminants from aqueous solutions: a critical review. Water Res 120:88–116. https://doi.org/10.1016/j.watres.2017.04.014
Uddin MK (2017) A review on the adsorption of heavy metals by clay minerals, with special focus on the past decade. Chem Eng J 308:438–462. https://doi.org/10.1016/j.cej.2016.09.029
Umar M, Roddick F, Fan L, Abdul H (2013) Application of ozone for the removal of bisphenol A from water and wastewater – a review. Chemosphere 90:2197–2207. https://doi.org/10.1016/j.chemosphere.2012.09.090
Vasudevan S, Oturan MA (2014) Electrochemistry: as cause and cure in water pollution—an overview. Environ Chem Lett 12:97–108. https://doi.org/10.1007/s10311-013-0434-2
Villegas-Guzman P, Giannakis S, Torres-Palma RA, Pulgarin C (2017) Remarkable enhancement of bacterial inactivation in wastewater through promotion of solar photo-Fenton at near-neutral pH by natural organic acids. Appl Catal B Environ 205:219–227. https://doi.org/10.1016/j.apcatb.2016.12.021
Wang JL, Xu LJ (2012) Advanced oxidation processes for wastewater treatment: formation of hydroxyl radical and application. Crit Rev Environ Sci Technol 42:251–325. https://doi.org/10.1080/10643389.2010.507698
Wang Y, Yang RT (2007) Desulfurization of liquid fuels by adsorption on carbon-based sorbents and ultrasound-assisted sorbent regeneration. Langmuir 23:3825–3831. https://doi.org/10.1021/la063364z
Wang L, Yao Y, Sun L, Mao Y, Lu W, Huang S, Chen W (2014a) Rapid removal of dyes under visible irradiation over activated carbon fibers supported Fe(III)-citrate at neutral pH. Sep Purif Technol 122:449–455. https://doi.org/10.1016/j.seppur.2013.11.029

Wang L, Yao Y, Zhang Z, Sun L, Lu W, Chen W, Chen H (2014b) Activated carbon fibers as an excellent partner of Fenton catalyst for dyes decolorization by combination of adsorption and oxidation. Chem Eng J 251:348–354. https://doi.org/10.1016/j.cej.2014.04.088

Wang N, Zheng T, Zhang G, Wang P (2016a) A review on Fenton-like processes for organic wastewater treatment. J Environ Chem Eng 4:762–787. https://doi.org/10.1016/j.jece.2015.12.016

Wang S, Han Y, Cao X, Zhao D (2016b) Enhanced degradation of trichloroethylene using bentonite-supported nanoscale Fe/Ni and humic acids. Environ Chem Lett 14:237–242. https://doi.org/10.1007/s10311-015-0548-9

Wang C, Jin J, Sun Y, Yao J, Zhao G, Liu Y (2017a) In-situ synthesis and ultrasound enhanced adsorption properties of MoS_2/graphene quantum dot nanocomposite. Chem Eng J 327:774–782. https://doi.org/10.1016/j.cej.2017.06.163

Wang X, Wang J, Zhang J, Louangsouphom B, Song J, Wang X, Zhao J (2017b) Synthesis of expanded graphite C/C composites (EGC) based Ni-N-TiO_2 floating photocatalysts for in situ adsorption synergistic photocatalytic degradation of diesel oil. J Photochem Photobiol A Chem 347:105–115. https://doi.org/10.1016/j.jphotochem.2017.07.015

Wang J, Chen S, Quan X, Yu H (2018) Fluorine-doped carbon nanotubes as an efficient metal-free catalyst for destruction of organic pollutants in catalytic ozonation. Chemosphere 190:135–143. https://doi.org/10.1016/j.chemosphere.2017.09.119

Worch E (2012) Adsorption technology in water treatment: fundamentals, processes, and modeling. Walter de Gruyter GmBH & Co., Berlin

Yang B, Zuo J, Li P, Wang K, Yu X, Zhang M (2016) Effective ultrasound electrochemical degradation of biological toxicity and refractory cephalosporin pharmaceutical wastewater. Chem Eng J 287:30–37. https://doi.org/10.1016/j.cej.2015.11.033

Yang J, Li Z, Zhu H (2017) Adsorption and photocatalytic degradation of sulfamethoxazole by a novel composite hydrogel with visible light irradiation. Appl Catal B Environ 217:603–614. https://doi.org/10.1016/j.apcatb.2017.06.029

Yao Y, Wang L, Sun L, Zhu S, Huang Z, WangyangLu YM, Chen W (2013) Efficient removal of dyes using heterogeneous Fenton catalysts based on activated carbon fibers with enhanced activity. Chem Eng Sci 101:424–431. https://doi.org/10.1016/j.ces.2013.06.009

Yin R, Guo W, Du J, Zhou X, Zheng H, Wu Q, Chang J, Ren N (2017) Heteroatoms doped graphene for catalytic ozonation of sulfamethoxazole by metal-free catalysis: performances and mechanisms. Chem Eng J 317:632–639. https://doi.org/10.1016/j.cej.2017.01.038

Ying Z, Yu C, Jian Z, Zongrui T, Shaohua J (2017) Preparation of SA-g-(PAA-co-PDMC) polyampholytic superabsorbent polymer and its application to the anionic dye adsorption removal from effluents. Sep Purif Technol 188:329–340. https://doi.org/10.1016/j.seppur.2017.07.044

Yu K, Yang S, Liu C, Chen H, Li H, Sun C, Boyd SA (2012) Degradation of organic dyes via bismuth silver oxide initiated direct oxidation coupled with sodium bismuthate based visible light photocatalysis. Environ Sci Technol 46:7318–7326. https://doi.org/10.1021/es3001954

Yu B, Bai Y, Ming Z, Yang H, Chen L, Hu X, Feng S, Yang S (2017) Adsorption behaviors of tetracycline on magnetic graphene oxide sponge. Mater Chem Phys 198:283–290. https://doi.org/10.1016/j.matchemphys.2017.05.042

Zhang HAO, Chen DA, Lv X, Wang Y, Chang H, Li J (2010) Energy-efficient photodegradation of azo dyes with TiO_2 nanoparticles based on photoisomerization and alternate UV-visible light. Environ Sci Technol 44:1107–1111. https://doi.org/10.1021/es9029123

Zhang G, Wang S, Yang F (2012) Efficient adsorption and combined heterogeneous/homogeneous Fenton oxidation of amaranth using supported nano-FeOOH as cathodic catalysts. J Phys Chem 116:3623–3634. https://doi.org/10.1021/jp210167b

Zhang Y, Liu C, Xu B, Qi F, Chu W (2016) Degradation of benzotriazole by a novel Fenton-like reaction with mesoporous Cu/MnO_2: combination of adsorption and catalysis oxidation. Appl Catal B Environ 199:447–457. https://doi.org/10.1016/j.apcatb.2016.06.003

Zhang L, Ni C, Jiu H, Xie C, Yan J, Qi G (2017a) One-pot synthesis of Ag-TiO_2/reduced graphene oxide nanocomposite for high performance of adsorption and photocatalysis. Ceram Int 43:5450–5456. https://doi.org/10.1016/j.ceramint.2017.01.041

Zhang Y, Shang J, Song Y, Rong C, Wang Y, Huang W, Yu K (2017b) Selective Fenton-like oxidation of methylene blue on modified Fe-zeolites prepared via molecular imprinting technique. Water Sci Technol 75:659–669. https://doi.org/10.2166/wst.2016.525

Zhao X, Lv L, Pan B, Zhang W, Zhang S, Zhang Q (2011) Polymer-supported nanocomposites for environmental application: a review. Chem Eng J 170:381–394. https://doi.org/10.1016/j.cej.2011.02.071

Zhou Y, Zhang L, Cheng Z (2015) Removal of organic pollutants from aqueous solution using agricultural wastes: a review. J Mol Liq 212:739–762. https://doi.org/10.1016/j.molliq.2015.10.023

Zhu R, Chen Q, Zhou Q, Xi Y, Zhu J, He H (2016) Adsorbents based on montmorillonite for contaminant removal from water: a review. Appl Clay Sci 123:239–258. https://doi.org/10.1016/j.clay.2015.12.024

Chapter 9
Surface Engineered Magnetic Biosorbents for Water Treatment

Sofia F. Soares, Tiago Fernandes, Tito Trindade, and Ana L. Daniel-da-Silva

Contents

Abstract Water pollution is a matter of concern because of the adverse impact of contaminants on environment and human health. Common polymer-based sorbents are difficult to separate from treated water, a limitation that has restrained their use. Nanomaterials with magnetic features appear as advantageous alternatives to these conventional biosorbents offering the advantage of fast and easy magnetically-assisted separation. Moreover owing to reduced dimensions magnetic nanomaterials possess large specific surface area that favours adsorption. The surface modification with biopolymers enhances the adsorptive capabilities of magnetic nanoparticles without compromising the low-cost. However, in order to attain high-performance, a rational design of the surface of the magnetic biosorbents is essential.

In this chapter we present an overview of the most recent developments on magnetic biosorbents for water treatment. Primary attention is given to the chemical

---

Author contributed equally with all other contributors. Sofia F. Soares and Tiago Fernandes

S. F. Soares · T. Fernandes · T. Trindade · A. L. Daniel-da-Silva (✉)
Department of Chemistry and CICECO-Aveiro Institute of Materials, University of Aveiro, Aveiro, Portugal
e-mail: sofiafsoares@ua.pt; jtfernandes@ua.pt; tito@ua.pt; ana.luisa@ua.pt

G. Crini, E. Lichtfouse (eds.), *Green Adsorbents for Pollutant Removal*, Environmental Chemistry for a Sustainable World 18,
https://doi.org/10.1007/978-3-319-92111-2_9

strategies used for the surface modification of magnetic nanoparticles with biopolymers aiming to obtain highly effective, robust and reusable biosorbents with magnetic properties. Two different strategies are distinguished, the *in situ* functionalization and the post-synthesis surface functionalization. The later comprises two distinct stages, the synthesis of the magnetic nanoparticles and the surface functionalization, and allows a better control of each stage individually. Surface functionalization can involve the simple coating of magnetic nanoparticles with biopolymers or the covalent attachment of the biopolymer chains to the surface. Overall covalent immobilization of the biopolymer onto the particles surface, either by crosslinking or grafting approach, is recommended to ensure successful recycling and reuse of the biosorbents without significant loss of adsorption capacity.

Appropriate selection of the biopolymer and the route for surface modification of magnetic nanoparticles are crucial to obtain effective magnetic biosorbents optimized and specialized in the targeted pollutants. The performance of several magnetic biosorbents in the uptake of heavy metal species and organic pollutants from water is discussed in this chapter.

9.1 Introduction

Magnetic nanomaterials have attracted increased attention in the last decade for application as biosorbents in water remediation processes (Sousa et al. 2015; Mehta et al. 2015; Simeonidis et al. 2016; Adeleye et al. 2016; Reddy and Yun 2016). Owing to reduced size, nanomaterials possess large surface area to volume ratio available, which is a desirable feature for adsorptive applications. Nanomaterials possessing magnetic features are easily and quickly separated from treated water in the presence of an external magnetic field, which clearly represents an advantage in relation to non-magnetic biosorbents. Iron oxides nanoparticles are the most used in research for treatment of polluted water owing to low cost and moderate environmental impact (Tang and Lo 2013; Su 2017). Enhancement of adsorptive capacity of magnetic nanoparticles (MNPs) and selectivity towards target pollutants can be achieved via the chemical functionalization of the particles surfaces.

The search for eco-friendly and low-cost biosorbents has prompted the interest for new biopolymer-based nanomaterials and their use in water decontamination (Crini 2005; Daniel-da-Silva et al. 2013; Dehabadi and Wilson 2014; Carpenter et al. 2015). Surface modification with biopolymers occurring in nature, provides to the particles surfaces novel functional groups that may confer affinity towards a wide diversity of pollutants (Xu et al. 2012; Boamah et al. 2015; Vandenbossche et al. 2015). A vast number of biopolymers are produced in natural environment during the growth cycles of living organisms, like green plants, animals, bacteria, fungi and algae. Besides being available on a sustainable basis, biopolymers present

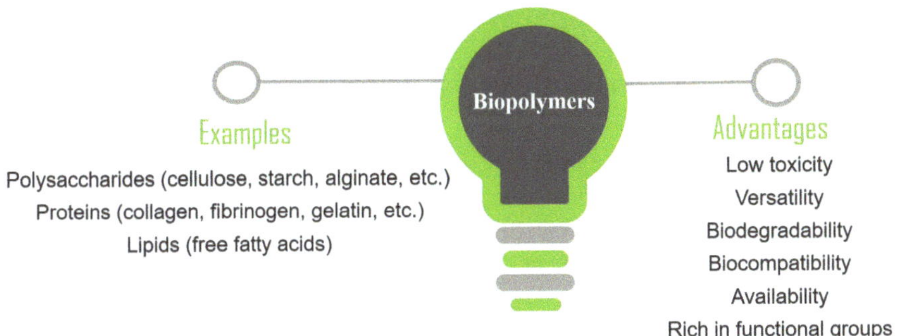

Fig. 9.1 Examples of biopolymers of different classes, such as polysaccharides, proteins and lipids. Polysaccharides include cellulose, starch and alginate among others. Collagen, fibrinogen and gelatin belong to the group of proteins and lipids include the free fatty acids. Biopolymers present several interesting environmentally friendly features such as low toxicity, versatility, biodegradability, biocompatibility, availability and they are rich in functional groups (Nair et al. 2017; Rebelo et al. 2017; Resch-Fauster et al. 2017)

the advantages of low cost, biodegradability and reduced toxicity (Nair et al. 2017; Rebelo et al. 2017; Resch-Fauster et al. 2017; Divya and Jisha 2017) (Fig. 9.1).

Ideally, a biosorbent for water remediation should fulfill the following requirements: specificity to target pollutants, high adsorptive performance, rapid adsorption, cost-effective, environmentally non-toxic, reusability and easy separation from treated water. Low-toxicity and easy magnetic separation can be in principle met by simple combination of biopolymers with magnetic iron oxides. However a rationale design of the surface of the MNPs is needed to attain specificity, high adsorption capacity and reusability. Advances in nanotechnology and in the field of colloidal science have extended the ability to tailor the surface of magnetic nanoparticles and to tune their physical-chemical properties to suit specific applications (Wu et al. 2015; Bohara et al. 2016). Appropriate selection of the biopolymer and the route for surface modification of magnetic nanoparticles are crucial to obtain effective biosorbents that meet the above mentioned requirements. For example, covalent immobilization of the biopolymer onto the particles surface, either by crosslinking or grafting approach, might be necessary to ensure successful recycling and reuse of the biosorbents without significant loss of adsorption capacity.

This chapter aims to provide a critical overview of the most recent developments in the field of magnetic biosorbents for application in water treatment. Primary attention is given to the chemical strategies used for the surface modification of magnetic nanoparticles with biopolymers aiming to obtain highly effective and robust biosorbents with magnetic properties. The main synthetic routes to prepare magnetic nanoparticles with controlled size and morphology are briefly enumerated. The performance of the magnetic biosorbents in the uptake of heavy metal species and organic pollutants from water is discussed. As a final note, the fate of the post-sorption magnetic biosorbents and its potential impact on the environment is put in perspective.

9.2 Magnetically Assisted Water Treatment

9.2.1 Magnetic Nanoparticles for Water Treatment

Magnetic separation relies on the fact that magnetic nanoparticles can be manipulated by an external magnetic field gradient. For the effective magnetic separation of micrometer or nanosized materials from a viscous flow, the magnetic force produced by the external magnetic field must overcome the drag force associated to the carrier fluid. The difference between the velocity of the magnetic nanoparticle and the velocity of the fluid ($\Delta \vec{v}$) can be expressed by Eq. (9.1) (Pankhurst et al. 2003; Sousa et al. 2015)

$$\Delta \vec{v} = \frac{R^2 \Delta\chi}{9\mu_0\eta} \nabla \left(\vec{B} \cdot \vec{B}\right) \tag{9.1}$$

where R is the radius of the particle, $\Delta\chi$ is the difference between the magnetic susceptibility of the particle and of the diamagnetic fluid and η is the viscosity of the fluid. The constant μ_0 is the magnetic permeability of free space and B is the magnetic flux density. From the above equation it follows that larger magnetic nanoparticles with higher magnetic susceptibility yield larger velocities. In other words larger particles are more efficient for magnetic separation. On the other hand, for adsorptive applications, smaller magnetic particles are desirable, owing to high surface area to volume ratio.

Iron oxide nanoparticles, more specifically magnetite (Fe_3O_4) and maghemite (γ-Fe_2O_3), are by far the most investigated magnetic nanoparticles for water treatment due to their adequate magnetic properties, low cost, chemical inertness and low toxicity (Xu et al. 2012; Kaur et al. 2014; Mehta et al. 2015; Su 2017). The easy synthesis, coating or surface functionalization, and the facility of tuning the size and particle shape, provide huge versatility to these materials. Both magnetic phases present inverse spinel crystal structure where the large oxygen ions are closely packed in a cubic order and the iron cations occupy the interstitial sites. Magnetite comprises both ferric (Fe^{3+}) and ferrous (Fe^{2+}) ions in its structure. Half of the Fe^{3+} ions occupy tetrahedral sites, while the Fe^{2+} ions together with remaining ferric cations, occupy the octahedral sites. Maghemite results from the topotactic oxidation of magnetite. Other spinel ferrites have been also reported for magnetically assisted water treatment (Reddy and Yun 2016).

Magnetite nanoparticles exhibit unique size-dependent magnetic properties (Fig. 9.2). Bulk magnetite is a ferrimagnetic material composed by multiple magnetic domains. Due to such magnetic domains, bulk magnetite exhibits a hysteresis curve when magnetization (M) is plotted versus magnetic field (H) and a permanent magnetization in the absence of a magnetic field is observed. The external field needed to bring this remanent magnetization to zero is designated by coercivity and

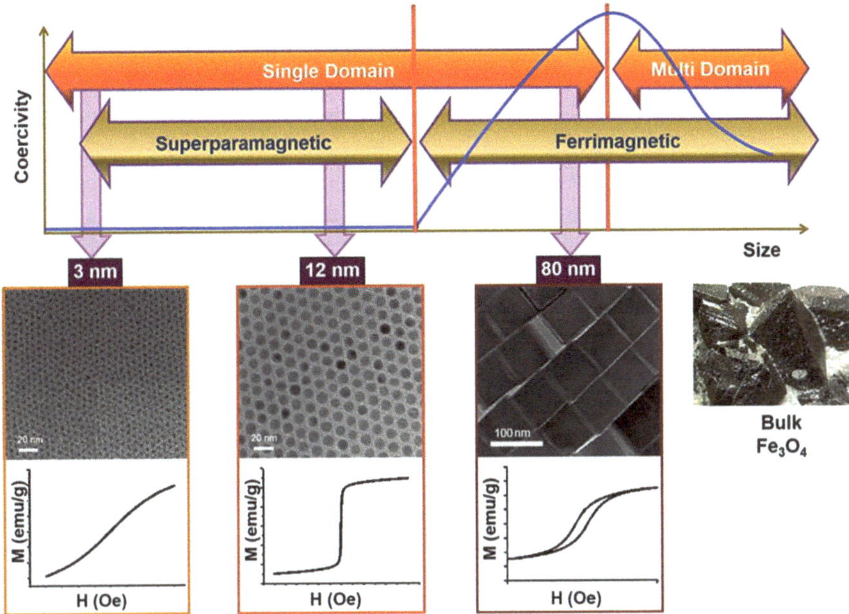

Fig. 9.2 Size-dependent magnetic properties of iron oxide nanoparticles. Inset TEM images and magnetization curves of magnetic nanoparticles. Bulk magnetite is a ferrimagnetic material composed by multiple magnetic domains. Magnetite particles become a single domain as the particle size decreases below 100 nm, where coercivity is maximized. When the particle size is smaller than 20 nm, the magnetization of magnetite nanoparticles is randomized so that they become superparamagnetic. (Adapted with permission from ref. Ling et al. (2015). *TEM* transmission electron microscopy)

in magnetite is maximized for sizes around 100 nm (Ling et al. 2015). The decrease of the particle size to the nanoscale brings consequences in terms of magnetic properties. Below a critical particle size (~25 nm for Fe_3O_4), it is energetically more favorable for magnetite particles to be composed by single magnetic domains and therefore exhibit superparamagnetic behavior. The magnetization curve of superparamagnetic nanoparticles does not exhibit hysteresis loop which means that in the absence of an external magnetic field these particles have zero magnetization, no coercivity and less tendency to agglomerate, an important feature for adsorptive applications.

For applications in magnetic separation, ferromagnetic particles are usually preferable over superparamagnetic nanoparticles because they show higher magnetophoretic response thus leading to faster separations. Owing to their notable properties, superparamagnetic magnetite and maghemite nanoparticles have been widely investigated for bio-applications, including drug delivery, magnetic resonance imaging and magnetic hyperthermia (Laurent et al. 2008; Wu et al. 2015; Ling et al. 2015).

9.2.2 Preparative Methods of Magnetic Iron Oxides

As described in the previous section, the properties of magnetic nanoparticles are strongly dependent on the particle size and shape. Thus, a number of synthetic strategies have been developed for the synthesis of magnetic nanoparticles with uniform morphology, narrow size distribution and tailored properties, as extensively reviewed elsewhere (Laurent et al. 2008; Wu et al. 2015; Ling et al. 2015). Examples of methods for the synthesis of colloidal magnetite nanoparticles include the co-precipitation method (Roth et al. 2015; Pušnik et al. 2016; Lin et al. 2017), the oxidative hydrolysis (Girginova et al. 2010; Reguyal et al. 2017), the hydrothermal treatment (Cheng et al. 2016; Gyergyek et al. 2017; Bhavani et al. 2017) and the thermal decomposition of iron containing molecular precursors (Jiang et al. 2014a; Glasgow et al. 2016; Bartůněk et al. 2016). These methods will be briefly reviewed in this section.

Co-precipitation

This is a simple and widely used method to synthesize Fe_3O_4 nanoparticles from mixtures of ferric (Fe^{3+}) and ferrous (Fe^{2+}) ions (molar ratio 2:1) in aqueous alkaline environment (pH 8–14) and non-oxidizing atmosphere. The global chemical reaction is depicted in Eq. (9.2).

$$Fe^{2+} + 2Fe^{3+} + 8OH^{-} \rightarrow Fe_3O_4 + 4H_2O \tag{9.2}$$

The size, shape and composition of the magnetic iron oxides prepared via co-precipitation depend on the type of salts used (e.g. chlorides, sulfates, nitrates), the Fe^{3+}/Fe^{2+} ratio, the reaction temperature, the pH value and ionic strength of the medium (Laurent et al. 2008; Su 2017). For example, samples prepared at hypostoichiometric ratios of hydroxide ions with respect to iron ions in combination with a high Fe^{3+}/Fe^{2+} ratio tend to lead to a mixture of the iron oxyhydroxides lepidocrocite (γ-FeO(OH)) and goethite (α-FeO(OH)) (Roth et al. 2015).

The main advantage of the co-precipitation process is that a large amount of nanoparticles can be synthesized. By varying the reaction conditions particles with size ranging from 2 to 17 nm can be obtained (Laurent et al. 2008). However this method does not provide a tight control of the particle size and usually polydisperse size distribution is obtained. The formation of the nanoparticles involves two stages, the formation of nuclei (nucleation) and the growth of the nuclei by diffusion of the species in solution to the surface of the crystal. In the co-precipitation method these stages are not separated, and nucleation can occur during the stage of growth, thus leading to size polydispersivity. Molecules or polymers containing chelating groups, e.g. carboxylate, can be added during the formation of the nanoparticles to facilitate the control of the particle size and colloidal stabilization (Bee et al. 1995). Because co-precipitation process occurs at mild temperatures in aqueous conditions, from room temperature (RT) to water boiling temperature, this method is compatible with the presence of biopolymers.

Oxidative Hydrolysis

Another aqueous-based synthesis route to prepare magnetite is the oxidative hydrolysis of iron (II) salts, a method that was firstly reported by Sugimoto and Matijevic (1980). It consists of partially oxidizing, ferrous hydroxide suspensions with mild oxidizing agents, typically nitrate ions. In contrast with the co-precipitation method, the oxidative hydrolysis can be used to prepare particles in a wide range of sizes, from nanometric up to dimensions as large as 1 micrometer (Zhang et al. 2009). The manipulation of the synthesis conditions, such as the ratio between the concentrations of the oxidant and the iron precursor and the initial pH, defines the nanoparticle size. Using this approach our group has prepared spherical and cubic magnetite particles with mean size of 50 nm and 100 nm, respectively that were subsequently functionalized to obtain magnetic sorbents (Girginova et al. 2010; Pinheiro et al. 2014; Carvalho et al. 2016; Tavares et al. 2016). As in the co-precipitation, this method does not provide a tight control over the size distribution of the nanoparticles.

Hydrothermal Reaction

In the hydrothermal process the synthesis is performed for several hours and in autoclaves where the pressure can be higher than 2000 psi and the temperature can be above 200 °C (Laurent et al. 2008). The reactants are premixed before being transferred to the autoclave, to ensure complete solvation. The formation of ferrites can occur via the hydrolysis and oxidation of ferrous salts (Guin et al. 2005; Chen et al. 2008), or via neutralization of mixed metal hydroxides (Sue et al. 2011; Velinov et al. 2016). The particle size is controlled mainly through the rate processes of nucleation and grain growth, that will depend upon the reaction temperature. Nucleation might be faster than grain growth at higher temperatures and results in a decrease in particle size. On the other hand, prolonging the reaction time would favour grain growth. The elevated pressures generated in hydrothermal synthesis introduce safety concerns that are not present in atmospheric-pressure methods, especially as the reaction scale increases (Qiao and Swihart 2017).

Thermal Decomposition

The thermal decomposition of iron organic precursors using organic solvents and surfactants has been used to synthesize iron oxide nanoparticles with narrow particle size distribution and uniform in shape. Examples of precursors used include iron complexes such as iron cupferronate, iron pentacarbonyl and iron(III) acetylacetonate. Surfactants such as oleic acid and hexadecylamine provide colloidal stability to the particles. Typically a mixture composed of precursors, surfactants and high boiling solvents are heated from room temperature to high temperature, leading to the generation and accumulation of monomers, followed by burst nucleation (Ling et al. 2015) (Fig. 9.3). In some processes the organometallic precursor is injected into a hot surfactant solution, inducing the simultaneous formation of many nuclei (Mendoza-Garcia and Sun 2016). In both strategies, after the nucleation event the preformed nuclei start to grow from the remaining monomers, without additional nucleation events, leading to the formation of monodispersed nanoparticles (Laurent et al. 2008; Ling et al. 2015; Mendoza-Garcia and Sun 2016). The size and

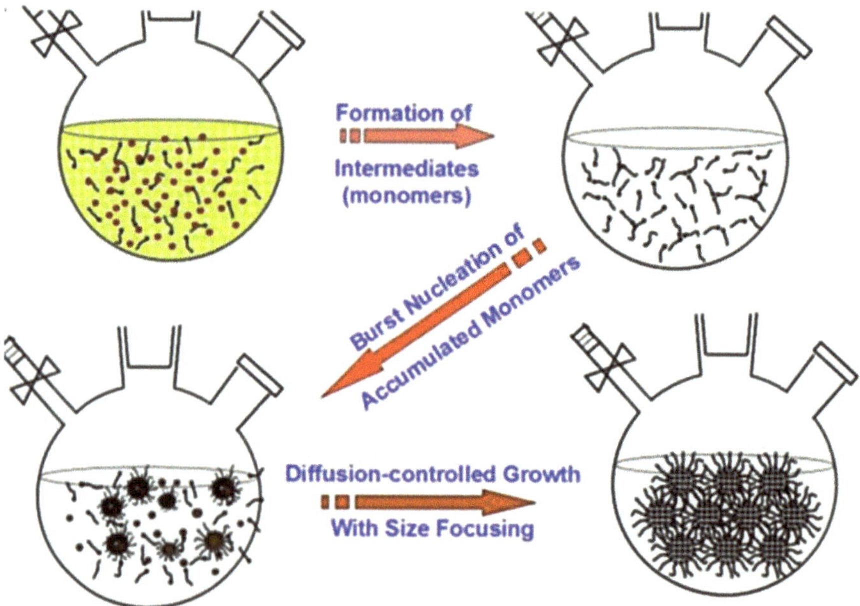

Fig. 9.3 Schematic illustration of the thermal decomposition method for the synthesis of uniformly sized iron oxide nanoparticles. Typically a mixture composed of precursors, surfactants and high boiling solvent are heated from room temperature to high temperature, leading to the generation and accumulation of monomers, followed by burst nucleation. The last step is the diffusion controlled growth of the nanoparticles. (Adapted with permission from Ling et al. 2015)

morphology of the nanoparticles can be controlled by adjusting the reaction times and the temperature but also the concentration and ratios of the reactants, namely the relative amounts of precursor and surfactants. The synthesis using thermal decomposition methods yields iron oxides nanoparticles compatible with organic solvents. For several applications, water-compatibility is required and thus surface modification is necessary using post synthetic methods such as ligand exchange reactions and encapsulation with amphiphilic polymers (Ling et al. 2015).

9.3 Surface Modification of Magnetic Nanoparticles with Biopolymers

Herein we will draw our attention on the synthetic procedures for the production of magnetic biosorbents, via the surface modification of magnetic nanoparticles with biopolymers. Preparative methods to obtain magnetic biosorbents in the particulate form, including composite nanoparticles, microparticles and hydrogel beads will be addressed.

Biopolymers extracted from natural sources present the advantages of biodegradability, reduced toxicity and low cost. Coating magnetic nanoparticles with biopolymers improves their stability against oxidation and provides functional groups to capture target pollutants from water (Avérous and Pollet 2012; Wu et al. 2015; Bohara et al. 2016). In addition, biopolymers improve the colloidal stability of the magnetic nanoparticles in aqueous media and prevent the formation of magnetic aggregates, which otherwise could contribute to diminish the available surface area and sorption capacity (Wu et al. 2008; Mehta et al. 2015). Colloidal stability is improved either due to steric shielding caused by biopolymer chains, or due to electrostatic repulsions between charged moieties present in the biopolymer.

Polysaccharides are among the most commonly used biopolymers for preparing magnetic biosorbents. Examples of polysaccharides are chitosan, starch, carrageenan, alginate, cellulose and derivatives, dextran and natural gum (Table 9.1). They can provide distinct ionic character such as neutral, anionic or cationic, and variable chemical functionalities and physical properties to the magnetic biosorbents. Surface modification with polysaccharides allows to enhance and to tune the chemical affinity of magnetic biosorbents surfaces towards specific target pollutants.

Although an exhaustive description of the properties of these polysaccharides is out of the scope of this chapter and can be found in specific literature (Zafar et al. 2016; Pattanashetti et al. 2017; Saba et al. 2017; Park et al. 2017; Dickinson 2017; Divya and Jisha 2017), it is worth mentioning some properties of relevance for sorption applications for selected polysaccharides.

Chitosan is obtained by deacetylation of chitin, the most abundant natural polymer after cellulose. Its structure contains amine and hydroxyl groups than can interact with metal cations in neutral or alkaline conditions. In acidic environment the protonated amine groups promote the sorption of anions (Choi et al. 2016; Bano et al. 2017; Muxika et al. 2017). Furthermore chitosan can be chemically modified with distinct functionalities to improve the adsorption capacities as well as the mechanical and physical characteristics of the resulting biosorbents (Yong et al. 2013; Vakili et al. 2014; Boamah et al. 2015).

Alginate is extracted mainly from brown seaweeds and can form hydrogels. Its structure contains abundant carboxylate groups that can interact with multivalent metal cations, namely with calcium ions to form the so-called “egg-box structure” through a sol-gel transition (Schnepp 2013). Carrageenan comprises a family of anionic sulfated linear polysaccharides extracted from red seaweeds. Carrageenan undergoes gelation in the presence of monovalent and divalent cations due to the formation of a double helical configuration and helix aggregation (Piculell 2006). Besides hydroxyl groups, carrageenan possess anionic ester sulfate groups that can interact with cationic pollutants (Gholami et al. 2016; Mahdavinia et al. 2016; Fernandes et al. 2017; Soares et al. 2017b). The polysaccharides chitosan, alginate and κ-carrageenan form hydrogels under facile ionotropic gelation and can be easily combined with magnetic nanoparticles and moulded in the form of magnetic beads

Table 9.1 Main characteristics of the polysaccharides and derivatives commonly used for coating magnetic nanoparticles, according to their ionic character, source and functional groups, in the context of environmental applications

Ionic character	Polysaccharide	Source	Functional groups	Target pollutants	References
Neutral	Cellulose	Vascular plants	−OH	Heavy metal ions, resorcinol	Anirudhan and Shainy (2015), Luo et al. (2016) and Ding et al. (2017)
	Dextran	Fermentation of sucrose	−OH	Aromatic hydrocarbons	Cho et al. (2015) and Kumar and Jiang (2017)
	Starch	Green plants	−OH	Cu, dyes	Mahdavinia et al. (2015) and Yang et al. (2016b)
Anionic	Alginate	Cell walls of brown algae	−OH, $-COO^-$	NaF, Sr, dyes	Hong et al. (2016), Li et al. (2016c) and Zhang et al. (2016b)
	Carrageenan	Red seaweeds	−OH, $-OSO_3^-$	Dyes, heavy metal ions, pharmaceuticals, herbicides	Gholami et al. (2016), Soares et al. (2016) and Fernandes et al. (2017)
	Natural gum	Microorganisms	−OH, $-COO^-$	Dyes, heavy metal ions	Sahraei et al. (2017)
Cationic	Chitosan	Shells of shrimp/crustaceans	−OH, $-NH_3^+$	Dyes, heavy metal ions, pharmaceuticals, oils	Chen et al. (2017a), Xiao et al. (2017) and Fan et al. (2017)

for sorption applications (Zhang et al. 2015a; Li et al. 2016c; Bée et al. 2017). Cellulose is the most abundant biopolymer and is an important structural component of the primary cell wall of green plants. Its structure contains several hydroxyl groups that enables cellulose to be relatively stable (Ding et al. 2017). However, the adsorption ability of cellulose which is not modified is insufficient, and the adsorption stability is poor. Cellulose can be chemically modified through several chemical processes to improve the adsorption capability (Lu et al. 2016a; Ding et al. 2017; Daneshfozoun et al. 2017).

9.3.1 *Strategies for Surface Functionalization*

The surface of magnetic nanoparticles (MNPs), typically magnetic iron oxide nanocrystals containing surface hydroxyl groups, can react with different functional groups. Owing to this broad reactivity there is a range of strategies for the surface modification of MNP with biopolymers, which can be carried out either *in situ* during the MNP synthesis, or using post-synthesis routes, i.e. *ex situ* (Bohara et al. 2016; Su 2017). The main interactions that can be present in the adsorption mechanism of biopolymers onto MNP are electrostatic interactions, hydrophobic interactions and hydrogen bonding (Avérous and Pollet 2012; Bohara et al. 2016). As a result of the many strategies available for surface modification, several distinct structures of magnetic bionanocomposites can be obtained, including core–shell structure, multicores or matrix dispersed structure, shell–core–shell, and Janus-type hetero-structures (Wu et al. 2015; Bohara et al. 2016) (Fig. 9.4).

Core-shell structures are obtained when the magnetic core is encapsulated within the biopolymeric material that involves the whole particle. In particular, the core-shell can be called yolk-shell structure when the magnetic core is not located at the center of the functional coating. In the inverse core–shell structure, the magnetic material coats the surface of the nonmagnetic functional material. Multicore structures include two distinct structures: mosaic or matrix-dispersed structure and shell–

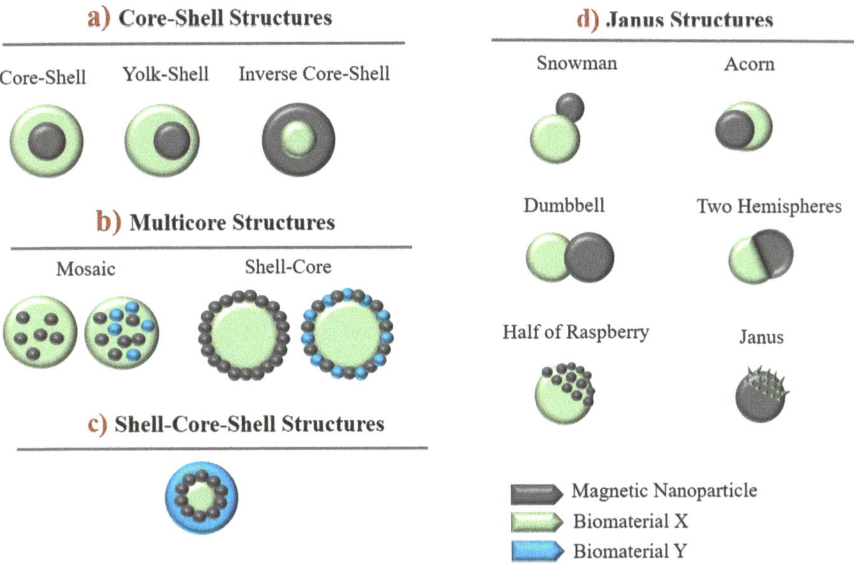

Fig. 9.4 Distinct structures of magnetic bionanocomposites: (**a**) core-shell structures include simple core-shell, yolk-shell and inverse core-shell structures; (**b**) multicore structures include mosaic and shell-core structures; (**c**) shell-core-shell structures; and (**d**) Janus structures comprise several types of arrangements such as snowman, acorn, dumbbell, two hemispheres, half of raspberry and Janus structures

core. The mosaic structure comprises a shell layer made of biopolymer molecules coated to any uniform MNP, i.e. MNPs are dispersed in a continuous biopolymer matrix. In a shell–core structure, MNPs decorate the surface of a core particle of biopolymer. In shell–core–shell structures, the location of MNPs is between the two biopolymeric materials. Magnetic Janus-type structures are hetero-structures where one of the spatial region provides magnetic features while the other compartment provides distinct functionality. Magnetic biopolymer-based Janus structures were barely explored for water treatment purposes. Nevertheless they offer huge potential for sensing applications, namely owing to the possibility of combining distinct analytical functions in a single particle that otherwise would interfere with each other (Fateixa et al. 2015; Pinheiro et al. 2018; Yi et al. 2016). In the context of sensing, Janus-compartmental alginate magnetic microbeads have been reported for the selective and convenient colorimetric detection of Pb(II) in water (Kang et al. 2014). The Pb(II) detection was possible owing to polydiacetylene liposomes embedded in the alginate matrix in the sensory compartment. The magnetic compartment provided convenient collection of the particles by applying magnetic field, after stirring for enhanced sensitivity and fast detection, while the alginate matrix provided additional feature of lead(II) removal.

9.3.1.1 *In situ* Surface Functionalization

This is a one-pot synthesis strategy where both the synthesis of magnetic nanoparticles (MNPs) and their surface functionalization are carried out in a single step. The biopolymer and the precursor of MNPs are added simultaneously to the reaction mixture and the coating process starts as soon as nucleation occurs, preventing further growth of the particles (Bohara et al. 2016). This method benefits from the ability of the biopolymer to interact with metal ions from the precursors of MNPs and from MNPs surface. The metal ion-biopolymer interaction may occur through different modes including complexation and hydrogen bonding (Boury and Plumejeau 2015). Hence, owing to these interactions the biopolymer can play a key role either on the nucleation or particle growth steps. For example, starch was employed to control and tune the size of Fe_3O_4 nanoparticles prepared by oxidation–precipitation of ferrous hydroxide (Tancredi et al. 2015). The size of the MNPs was tuned from 15 to 100 nm by changing the time period for starch addition to the reaction mixture. Starch acted as a kinetically control agent that affected both the size, size distribution and aggregation state of the MNPs. Starch-coated Fe_3O_4 NPs were water-dispersible, presenting good colloidal stability. In our group it has been observed that the particle size and the stability toward oxidation of Fe_3O_4 nanoparticles generated by co-precipitation of iron salts in the presence of carrageenan, strongly depended on the type and concentration of carrageenan used (Daniel-da-Silva et al. 2007). Overall the particle size decreased with increasing carrageenan concentration but the stability to oxidation followed distinct trend. The rheological state (hydrosol or hydrogel) of the final magnetic nanocomposite was found to be a relevant parameter. Considering that increasing biopolymer concentration resulted in smaller Fe_3O_4 nanoparticles and stronger gels and that magnetite oxidation follows a size-dependent diffusion mechanism, it was suggested that small particle sizes accelerate oxidation while strong gels have the opposite effect, as oxygen diffusion within the gel becomes more difficult.

The *in-situ* MNPs synthesis is typically performed using wet chemical routes that require aqueous environment and mild conditions of temperature, compatible with the presence of the biopolymer. Among *in situ* strategies, the co-precipitation of ferric and ferrous ions under alkaline conditions is the most commonly used route to prepare biopolymer coated magnetite (Fe_3O_4) nanoparticles (Lee et al. 1996; Kim et al. 2014). Generally, nanocomposite structures obtained are core-shell structures or mosaic (matrix dispersed) structures (Wu et al. 2015; Bagheri and Julkapli 2016). Nevertheless the morphology and the thickness of the polymer shell are difficult to control using this methodology. Owing to these limitations and even though the one-step synthesis is less time-consuming, magnetic biosorbents prepared by *in situ* procedures have been less reported. Some recent examples are described below that illustrate the usefulness of *in situ* strategy for the functionalization of MNPs aiming applications in the removal of pollutants from water.

Magnetic iron oxide nanoparticles were prepared by co-precipitation of iron ions in the presence of chitosan, followed by the addition of the crosslinker glutaraldehyde (Yang et al. 2016a; Azari et al. 2017) to impart mechanical robustness to the biosorbent. Small nanoparticles, with an average size of 5 nm (Yang et al. 2016a) and 50 nm (Azari et al. 2017) and narrow particle size distribution were obtained. The nanoparticles were homogeneously dispersed and embedded in the biopolymer matrix, leading to the formation of magnetic nanocomposites of undefined morphology. These biosorbents showed good adsorption of mercury species (Azari et al. 2017) even in the presence of competitive cations. In alternative to glutaraldehyde addition, more environmental friendly crosslinking strategies are also being explored. For example, using *in-situ* co-precipitation κ-carrageenan coated magnetite nanoparticles with average size of 4 nm were prepared and then crosslinked with chitosan (Mahdavinia and Mosallanezhad 2016). The electrostatic interactions between positively charged amine groups on chitosan and negatively charged sulfate groups on κ-carrageenan could produce stable complexes, with affinity for cationic dyes such as methylene blue.

Magnetite nanoparticles coated with carboxylated cellulose were prepared using the co-precipitation method, wherein carboxylated cellulose acted as a template and stabiliser for the *in-situ* prepared Fe_3O_4 nanoparticles, preventing its aggregation (Lu et al. 2016a). The magnetic material showed high adsorption capacity of Pb (II) from water owing to carboxylic acid groups of carboxylated cellulose. The biosorbent could be regenerated and reused for at least 5 cycles. Nevertheless the removal capacity decreased with increasing cycle numbers, indicating poor stability of the coating. In another study (Ding et al. 2017) the co-precipitation method was used to prepare cellulose coated Fe_3O_4 nanoparticles that were subsequently reacted with polydopamine, providing a magnetic biosorbent whose surface was enriched in hydroxyl and amino functionalities. The biosorbent was tested in the removal of resorcinol from water showing better adsorption capacity at low pH. The biosorbent was reusable for 5 cycles, with small loss of adsorption capacity, after regeneration in alkaline conditions.

Several examples could be found concerning biopolymer coated magnetic nanophases obtained by in situ co-precipitation that were subsequently modified in post-synthetic steps. In those systems the biopolymer serves as springboard for the addition of reactive groups aiming to improve the adsorption capacity. For instance amino-acids and diethylenetriamine were grafted at the surface of chitosan coated magnetite nanoparticles (size 10–50 nm) prepared via co-precipitation, using epi-chlorohydrin as crosslinker (Galhoum et al. 2015a, 2017). The materials showed binding affinity for uranyl and Dy(III) species present in aqueous solutions. The highest uranyl sorption capacity was obtained with diethylenetriamine functionalized biosorbents, probably due to the increased density of reactive poly-amine groups.

Li et al. (2017) prepared magnetic gelatin by *in situ* co-precipitation of magnetic iron oxide phases with an average size of 15 nm (Fig. 9.5a) and the gelatin was further cross-linked by transglutaminase to maintain its mechanical stability. Compared with other cross-linkers, transglutaminase had the advantage of its specificity towards reaction between lysine and glutamine residues of gelatin, without affecting many of amino groups that remain available for further modification and removal of heavy metals in water. After modification with a dicarboxylic acid, chitosan/polyethyleneimine (PEI) cationic copolymers were grafted on the surface of the nanoparticles, using carbodiimide chemistry, yielding a sheet structure with a mean diameter of 500 nm (Fig. 9.5b). Grafting of PEI greatly increased the number of

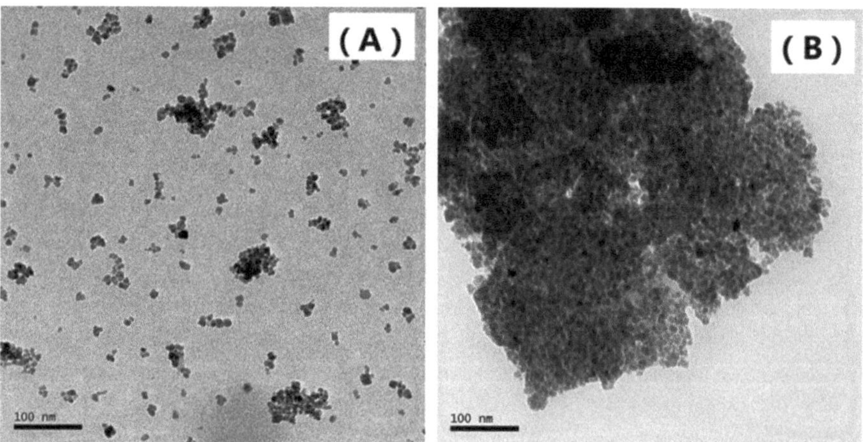

Fig. 9.5 TEM images of (**a**) magnetic gelatin particles with an average size of 15 nm and (**b**) chitosan/PEI-grafted magnetic gelatin sheet structure with a mean diameter of 500 nm. (Adapted with permission from ref. Li et al. (2017). *TEM* transmission electron microscopy, *PEI* polyethyleneimine)

surface amino groups and the adsorption capacity of lead and cadmium cations from water, as well as the stability of the biosorbent in strong alkaline or acidic conditions.

Using a simple one-pot polyol method, magnetic carboxymethylchitosan spherical composite particles with very uniform particle size have been prepared (Charpentier et al. 2016). The average diameter was nearly 500 nm, and each sub-micrometer sized particle was a cluster formed by the aggregation of many small Fe_3O_4 nanocrystals with sizes of 10–15 nm (Fig. 9.6). In this method the polyol (ethylene glycol) acts as high boiling point polar solvent and simultaneously reduces Fe(III) ions from the precursor, to form Fe_3O_4 nanocrystals (Cheng et al. 2009). The primary Fe_3O_4 nanocrystals aggregate into larger secondary particles. The resulting composite nanoparticles possessed high magnetic content and surfaces functionalized with carboxylic moieties and were used for the rapid removal of metal ions (Pb^{2+}, Cu^{2+} and Zn^{2+}) from aqueous solutions.

9.3.1.2 *Ex situ* Surface Functionalization

Ex situ or post surface functionalization strategies are carried out using magnetic nanoparticles (MNPs) previously synthesized. The procedure is divided into two distinct stages, synthesis and surface modification, that allows a better control of each stage individually. It can involve the simple coating of MNPs with biopolymers, physically or chemically crosslinked or the covalent attachment of the biopolymer chains to the surface of MNPs. The latter usually requires the use of a linker ligand such as functional alkoxysilanes (Laurent et al. 2008; Sun et al. 2008; Begin-Colin and Felder-Flesch 2012).

Ionic Crosslinking of Biopolymer Coating

Several natural polyelectrolytes have the ability to undergo ionotropic gelation, i.e. to crosslink in the presence of counter ions to form hydrogels (Jiang et al. 2014b; Nie et al. 2016; Valle et al. 2017). This ability combined with extrusion or emulsification techniques has been widely explored for the encapsulation of magnetic nanoparticles, to form magnetic composite particles with different size and shapes.

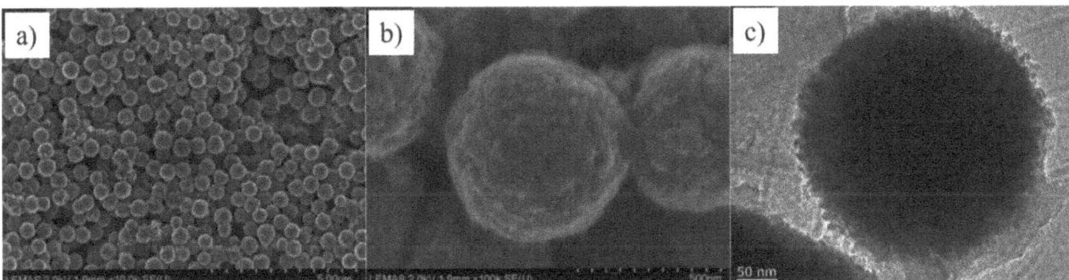

Fig. 9.6 SEM images of carboxymethylchitosan magnetic nanoparticles at (**a**) low and (**b**) high resolution, and (**c**) TEM image. The average size of the particles was 500 nm. (Adapted with permission from ref. Charpentier et al. (2016). *SEM* scanning electron microscopy, *TEM* transmission electron microscopy)

Magnetic alginate core-shell type particles were fabricated using a method of electro-coextrusion, and employed as an biosorbent for separation of fluoride from aqueous solution (Zhang et al. 2016b). In this method the solutions of alginate and Fe_3O_4 nanoparticles were injected simultaneously through a concentric nozzle using an electrostatic spinning machine and droped into a solution of $CaCl_2$ for alginate crosslinking. The resulting particles comprised a core of aggregated Fe_3O_4 nanoparticles coated by a shell of alginate. Because trivalent lanthanum (Lewis hard acid) shows high chemical affinity to fluoride anions (Lewis hard base), the Ca^{2+} ions of the alginate were then replaced by La^{3+} ions, through cation exchange. The resulting particles were able to uptake F^- ions from solution owing to Lewis acid-based interactions. In another study magnetic alginate beads containing cobalt ferrite nanoparticles were prepared using Ca^{2+} crosslinking as well (Li et al. 2016c). The $CoFe_2O_4$ nanoparticles were firstly coated with polydopamine, which is a polymer that provides very good adhesion either to organic and inorganic substrates. Polydopamine coated ferrites were added to an alginate solution and the magnetic beads were obtained by injection into a $CaCl_2$ solution, to induce sol-gel transition. The beads were successfully used in the uptake of organic dyes from water. The adsorption was ascribed to negatively charged groups of the beads (carboxylate from alginate, catechol and amine groups from polydopamine) that could interact with cationic dyes through electrostatic interactions, hydrogen bonding, van der Waals interactions and π-π interactions between aromatic rings of catechol and organic dyes.

Martínez-Cabanas and co-workers (Martínez-Cabanas et al. 2016) have prepared magnetic chitosan beads containing magnetic iron oxide nanoparticles. The MNPs were dispersed in an acidic solution of chitosan and dripped into an alkaline solution under stirring to promote the formation of the beads. The materials were effective in the removal of arsenic from water with short equilibrium times and good adsorption capacity.

Magnetite nanoparticles were coated with κ-carrageenan and used as magnetic biosorbents for the removal of methylene blue from aqueous solutions (Salgueiro et al. 2013). Coating of MNPs (size around 12 nm) was obtained by simple dispersion of the nanoparticles in κ-carrageenan solution, addition of K^+ ions to promote sol-gel transition by physical crosslinking, followed by particle separation. These biosorbents show high methylene blue adsorption capacity due to electrostatic interaction of the cationic dye MB with the ester sulfate moieties of carrageenan. Nevertheless marked loss of adsorption capacity after regeneration and reuse was observed for these biosorbents. This loss was ascribed to possible leaching of the external layer of κ-carrageenan molecules during adsorption/desorption stages, owing to lack of stability of the carrageenan layer that was physically adsorbed at the surface of MNPs.

Covalent Crosslinking of Biopolymer Coating

The mechanical and chemical stability of the biosorbent is an important aspect to consider, namely to ensure successful recycling and reuse of the biosorbents without loss of adsorption capacity. Covalent crosslinking is a versatile method to improve the robustness of the biosorbents. It results in the enhancement of the mechanical properties and insolubility of the biopolymer coating, as the chains are tied together by strong covalent linkages (Maitra and Shukla 2014). Some recent examples of covalently crosslinked magnetic biosorbents are described below. It is worth noting that glutaraldehyde is still one of the most frequently used crosslinkers, in spite of its known ecotoxicological issues (Leung 2001; Hu et al. 2017; Christen et al. 2017). Alternative and much less toxic crosslinkers such as genipin (Muzzarelli 2009; Pujana et al. 2014) have been barely investigated for applications in water remediation (Laus and de Fávere 2011; Mondal et al. 2015).

Magnetic crosslinked-chitosan microparticles were prepared using a reverse-phase suspension cross-linking technique (Funes et al. 2017). The process encompassed the preparation of a water-in-oil emulsion, the aqueous phase comprised an acidic solution of chitosan containing Fe_3O_4 nanoparticles (280 nm in size) dispersed. Glutaraldehyde was added to crosslink the biopolymer and the particles were gathered using a magnet. The microparticles were spherical in shape and polydisperse in size, with a mean diameter of 4.8 μm (Fig. 9.7a). The cross-sectioned particles revealed a homogeneous distribution of Fe_3O_4 nanoparticles within the chitosan microspheres indicating that the embedding process resulted in little Fe_3O_4 aggregation (Fig. 9.7b). These biosorbents were tested in the removal of phosphorous in the form of phosphates from water.

Magnetic biosorbents of a quaternary chitosan, N-(2-hydroxyl)propyl-3-trimethyl ammonium chitosan chloride were prepared using a similar emulsion technique (Li et al. 2016a; Song et al. 2017). Overall, quaternary chitosans provide enhanced adsorption affinity towards anionic pollutants in comparison to unmodified chitosan. The biopolymer coating was covalently crosslinked using glutaraldehyde (Li et al. 2016a) or a combination of formaldehyde and glutaraldehyde (Song et al. 2017). The magnetic biosorbents were used for the removal of anionic pollutants from water such as As(III) (Song et al. 2017) and Cr(VI) (Li et al. 2016a) anionic species and anionic dyes (Li et al. 2016a).

Aiming to prepare chitosan based biosorbents with high specific surface area, Xiao et al. used a distinct approach and produced sub-micron-sized polyethylenimine modified polystyrene/Fe_3O_4/chitosan magnetic composite particles for the uptake of Cu(II) ions from water (Xiao et al. 2017). Polystyrene particles (~300 nm) whose surface was decorated with small Fe_3O_4 nanoparticles (~10 nm) were firstly synthesized. Magnetic polystyrene particles were then coated with chitosan thin-film which was crosslinked by glutaraldehyde, followed by grafting of polyethylenimine (Fig. 9.8). Owing to small particle size and a surface highly enriched in amine groups, these biosorbents exhibited very high Cu(II) adsorption capacity when compared with other magnetic biosorbents, reaching the equilibrium within very short time (15 min).

Besides chitosan and chitosan derivatives, glutaraldehyde was also used to prepare magnetic biosorbents derived from other biopolymers such gum tragacath (Sahraei et al. 2017).

a

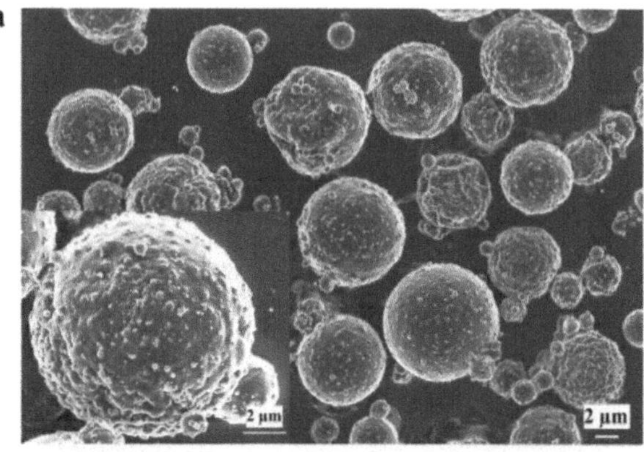

b

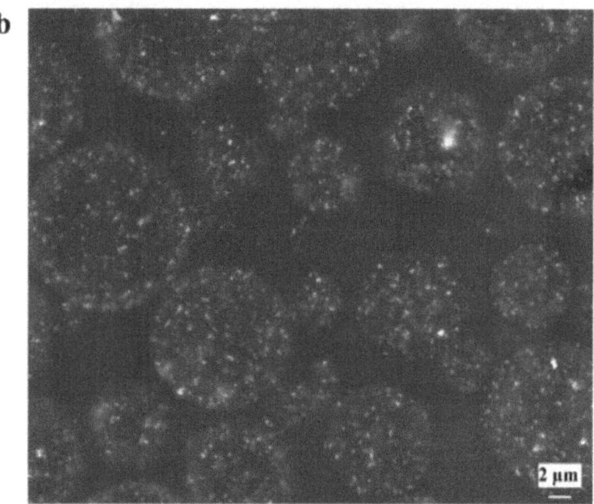

Fig. 9.7 SEM images of (**a**) magnetic chitosan microparticles with an average size of 280 nm and (**b**) cross-sectioned magnetic chitosan microparticles showing a homogeneous distribution of the magentic nanoparticles within the chitosan microspheres. (Adapted with permission from ref. Funes et al. (2017). *SEM* scanning electron microscopy)

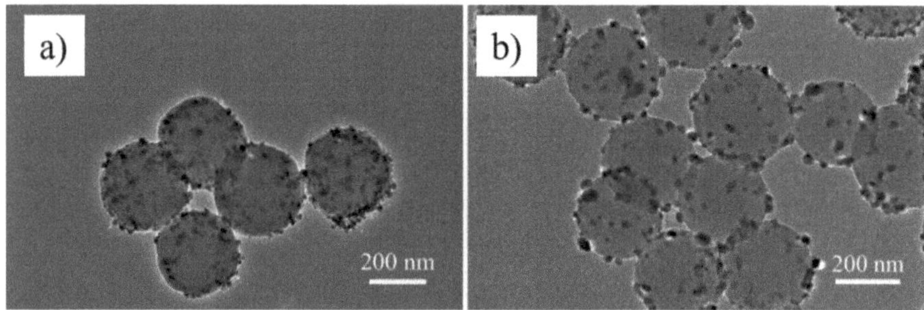

Fig. 9.8 TEM images of (**a**) PS/Fe_3O4/CS and (**b**) PS/Fe_3O_4/CS-PEI microspheres. The average size was 300 nm and 10 nm for PS and Fe_3O_4 particles, respectively. (Adapted with permission from ref. Xiao et al. (2017). *TEM* transmission electron microscopy, *PS* polystyrene, *CS* chitosan, *PEI* polyethylenimine)

Coating by Complexation of Polyelectrolytes

Coating of magnetic nanoparticles can be also performed by assembling polyelectrolytes oppositely charged. Most of biopolymers are polyelectrolytes i.e., macromolecules that are either charged or, under suitable conditions, can become charged. Hence biopolymers can undergo complexation in the presence of other oppositely charged biopolymers (Luo and Wang 2014) or synthetic polyelectrolytes (Gentile et al. 2015). The assembly of oppositely charged polyelectrolytes and MNPs is mainly governed by strong but reversible electrostatic interactions, as well as hydrogen bonds. The layer-by-layer (LbL) assembly technique is a simple and versatile method for the fabrication of polymer-based coatings and has been widely used to modify spherical and planar inorganic substrates (Srivastava and Kotov 2008). This method involves the sequential adsorption of oppositely charged biopolymers onto the surface of the nanoparticles. The thickness of the polymer coating can be tuned by varying the number of layers deposited and the properties of the polymer solutions. Most of magnetic bionanocomposites prepared by LbL technique were developed envisaging biomedical application (Xiong et al. 2013; Luo and Wang 2014; Gentile et al. 2015). Nevertheless examples of magnetic biosorbents for water treatments have been also reported (Mu et al. 2013; Shi et al. 2016). For example polyelectrolyte-coated magnetic composites were prepared by LbL self-assembly of chitosan and cysteine modified β-cyclodextrin on the surface of Fe_3O_4 nanoparticle-decorated attapulgite, and investigated for the adsorption of precious metal ions (Mu et al. 2013).

Using a distinct approach Liang et al. have prepared magnetic chitosan/carrageenan ampholytic microspheres that showed highly efficient adsorption capacity towards both cationic and anionic dyes and heavy metal ions in wastewater (Liang et al. 2017). The microspheres were fabricated via emulsification procedure from the homogeneous chitosan/carrageenan solution in LiOH/KOH/urea aqueous system, showing homogeneous network structure. In alternative to the use of acidic solutions, chitosan was dissolved in the alkali/urea solution via the freezing- thawing process. The dissolution was caused by destruction of the intermolecular hydrogen bonds rather than protonation of the amino groups, and as a result chitosan displayed the neutral feature and no flocculation occurred in chitosan/carrageenan blend solution. The ampholytic chitosan/carrageenan composite matrix played an important role in adsorption of pollutants, and the Fe_3O_4 nanoparticles uniformly embedded in the matrix afforded sensitive magnetic responsiveness.

Grafting of Biopolymers onto Magnetic Particles Surface

Another strategy explored for the preparation of magnetic biosorbents is the grafting of the biopolymer to the surface of magnetic nanoparticles. The biopolymer can be directly grafted onto the surface of the iron oxide particle (Badruddoza et al. 2011; Lu et al. 2016b) or after surface functionalization with ligands such as functional alkoxysilanes that will subsequently link to the biopolymer (Bini et al. 2012; Rodriguez et al. 2017). This strategy might also involve the coating of MNPs with a thin layer of amorphous silica (SiO_2). The method for the encapsulation with SiO_2 shells is well established and involves the hydrolysis of alkoxysilanes (e.g.

tetraethylorthosilicate-TEOS) and subsequent condensation of silica oligomers in the presence of an acid or base catalyst. The amorphous SiO_2 shells protect magnetic iron oxides from oxidation and ion leaching, and provide a suitable surface for further chemical modification aiming the grafting of the biopolymers onto its surface. For example carboxymethylated carrageenans were grafted onto the surface of amino-functionalized silica coated magnetite nanoparticles, via carbodiimide chemistry (Daniel-da-Silva et al. 2015). The biosorbents showed good adsorption of methylene blue due to electrostatic interactions between the sulfonate groups of carrageenan and the dye. Owing to the covalent immobilization of the carrageenan to the magnetic support, the particles prepared with κ-carrageenan were reusable for six adsorption cycles without significant loss of the adsorption capacity.

Kim et al. prepared carboxymethyl chitosan-modified magnetic-cored dendrimers with enhanced adsorption capacity and magnetic properties (Kim et al. 2016). Magnetite nanoparticles were first functionalized with amine groups using an amine functionalized alkoxysilane. Amine terminated dendrimers were then built at the surface of the MNPs. Finally carboxymethyl chitosan was grafted to the dendritic structure upon the reaction between –COOH and $–NH_2$ groups, using the carbodiimide reaction. The resulting material contained carboxylic acid groups and amino groups as terminal sites and showed satisfactory adsorption capacity of both cationic and anionic organic dyes.

Bio-hybrid Coatings

In alternative to the grafting of biopolymers at the surfaces of silica coated nanoparticles, the biopolymer can be introduced in the silica network, yielding an organic-inorganic hybrid material enriched in biopolymer, herein designated by bio-hybrid material. Our group has developed a method for the encapsulation of Fe_3O_4 nanoparticles with bio-hybrid siliceous shells comprising a polysaccharide covalently grafted to the siliceous network (Soares et al. 2016; Fernandes et al. 2017; Soares et al. 2017a). The encapsulation was performed through a one-step procedure, involving the hydrolysis and condensation of a mixture of tetraethyl orthosilicate and an alkoxysilane covalently bound to the biopolymer, in the presence of the magnetic particles and a base that acts as catalyst. Core-shell composite particles comprising a magnetic core of Fe_3O_4 uniformly coated by a thin shell of bio-hybrid siliceous material were obtained (Fig. 9.9). This method was successfully used for the encapsulation using polysaccharides having distinct chemical nature and ionic character, namely κ-carrageenan (Soares et al. 2016; Fernandes et al. 2017; Soares et al. 2017b), chitosan (Soares et al. 2017a) and starch (Fernandes et al. 2017). In comparison to the post encapsulation grafting, this method allowed to obtain surfaces highly enriched with biopolymer functional groups. The resulting magnetic biosorbents were highly effective in the removal of the organic pollutants Paraquat, Methylene Blue and Metoprolol from aqueous solutions (Soares et al. 2016, 2017b; Fernandes et al. 2017). Magnetic hybrid particles prepared from chitosan were tested for the uptake of non-polar organic solvents from water (Soares et al. 2017a).

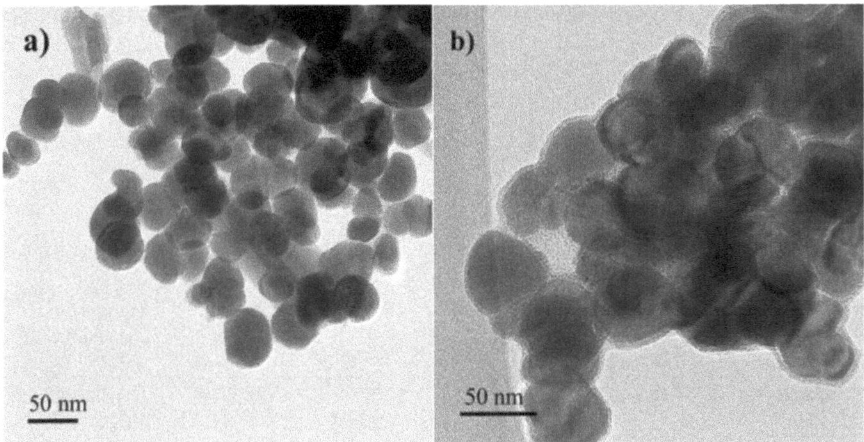

Fig. 9.9 TEM images of (**a**) Fe_3O_4 particles with an average size of 47 nm and (**b**) Fe_3O_4 coated with κ-carrageenan hybrid siliceous shell with a shell thickness of 6 nm. (Adapted with permission from ref. Soares et al. (2016). *TEM* transmission electron microscopy)

9.4 Adsorptive Applications of Magnetic Bionanocomposites in Water Treatment

In this section, the most recent magnetic bionanocomposites for the clean-up of emerging pollutants from water is reviewed taking into consideration adsorptive technologies. The Table 9.2 provides an overview of several materials that have been successfully used for the removal of inorganic and organic pollutants.

9.4.1 Removal of Heavy Metal Species

Although there is no clear classification, heavy metals can be defined as metallic elements that possess a high density (higher than 5 g/cm^3). Herein, metalloids, such as arsenic, will be included in this discussion given their chemical nature and ability to also induce similar toxic effects even at low concentrations (Tchounwou et al. 2012; Chen et al. 2015). Heavy metals have been used for many human applications for at least 5000 years, including for agricultural, mining, industrial or domestic operations. Their accumulation in the environment, i.e. plant tissues, fish, mammals, soil, air and water, poses a serious threat to human health. Even though heavy metals are naturally occurring elements, their unintended environmental accumulation is originated from the uncontrolled disposal of these elements arising from anthropogenic activity. Although some heavy metals have an important biological role (e.g. Cu(II) as cofactor), the excessive exposure to such elements or a simple change on their oxidation state can result in detrimental effects to the ecosystem and human

Table 9.2 Recently reported magnetic bionanocomposites for the uptake for several inorganic and organic pollutants

Biosorbent	Pollutant	Adsorption capacity	pH	References
Magnetic chitosan-graphene oxide	Acid red 17 (AR17) and bromophenol blue (BPB)	8.1 and 9.5 $mg.g^{-1}$	2, 6	Sohni et al. (2017)
Imprinted Fe_3O_4-quaternay ammonium chitosan	As(III)	12 $mg.g^{-1}$	6	Song et al. (2017)
Fe_3O_4-chitosan beads	As(V)	147 $\mu g.g^{-1}$	6–8	Martínez-Cabanas et al. (2016)
Fe_3O_4-chitosan-PAC-clay	Atenolol, ciprofloxacin and gemfibrozil	15.6, 39.1 and 24.8 $mg.g^{-1}$	10, 7, 4	Arya and Philip (2016)
8-hydroxyquinoline anchored imprinted γ-Fe_2O_3@chitosan	Co(II)	100 $mg.g^{-1}$	8	Hossein et al. (2016)
Pyridinium-diethylenetriamine magnetic chitosan	Cr(VI)	176 $mg.g^{-1}$	3	Sakti et al. (2015)
Fe_3O_4-quaternary ammonium chitosan	Cr(VI) and methyl orange	3.25 and 2.5 $mmol.g^{-1}$	2	Li et al. (2016a)
Fe_3O_4-κ-carrageenan-g-poly(methacrylic acid) hydrogel	Crystal violet	28.24 $mg.g^{-1}$	7	Gholami et al. (2016)
Polyethylenimine(PEI)-modified polystyrene/Fe_3O_4/chitosan	Cu(II)	204 $mg.g^{-1}$	6	Xiao et al. (2017)
Starch-g-polyacrylonitrile/montmorillonite/Fe_3O_4	Cu(II)	164 $mg.g^{-1}$	4.7	Mahdavinia et al. (2015)
Fe_3O_4-chitosan	Cu(II)	236 $mg.g^{-1}$	6	Neeraj et al. (2016)
Fe_3O_4@ aminopropyl-functionalized silica-chitosan	Demulsification	n/a	n/a	Lü et al. (2017)
Magnetic cellulose ionomer/layered double hydroxide	Diclofenac	268 $mg.g^{-1}$	9	Hossein et al. (2016)
Gelatin based magnetic beads	Direct red 80 and methylene blue	380 and 465 $mg.g^{-1}$	n/a	Saber-Samandari et al. (2017)
Fe_3O_4-alanine or –serine or –cysteine grafted-chitosan	Dy(III)	14.8, 8.9, and 17.6 $mg.g^{-1}$	5	Galhoum et al. (2015a)
Magnetic chitosan modified with glutaraldehyde	Hg(II)	96 $mg.g^{-1}$	5	Azari et al. (2017)

(continued)

Table 9.2 (continued)

Biosorbent	Pollutant	Adsorption capacity	pH	References
Itaconic acid-grafted-magnetite nanocellulose	Hg(II)	240 $mg.g^{-1}$	8	Anirudhan and Shainy (2015)
Fe_3O_4-cysteine-chitosan	La(III), Nd(III), and Yb(III)	17, 17, 18 $mg.g^{-1}$	5	Galhoum et al. (2015b)
Chitosan/Al_2O_3/Fe_3O_4 microspheres	Methyl orange	419 $mg.g^{-1}$	6	Tanhaei et al. (2016)
Glutaraldehyde cross-linked chitosan-coated Fe_3O_4 nanoparticles	Methyl orange	758 $mg.g^{-1}$	6–10	Yang et al. (2016a)
Fe_3O_4@SiO_2-κ-carrageenan	Methylene blue	530 $mg.g^{-1}$	9	Soares et al. (2017b)
Fe_3O_4-OMWCNT-κ-carrageenan	Methylene blue	1.24×10^{-4} $mol.g^{-1}$	6.5	Duman et al. (2016)
Fe_3O_4-κ-carragenaan crosslinked with chitosan	Methylene blue	123 $mg.g^{-1}$	2–12	Mahdavinia and Mosallanezhad (2016)
Magnetic chitosan/clay beads	Methylene blue	82 $mg.g^{-1}$	> 9	Bée et al. (2017)
Carboxymethyl chitosan-modified magnetic-cored dendrimers	Methylene blue and methyl orange	20.85 and 96.31 $mg.g^{1}$	3; 11	Kim et al. (2016)
Magnetic ampholytic poly-electrolyte microspheres	Methylene blue, Congo red, Cu (II) and Cr(III)	124, 212, 20 and 12 $mg.g^{-1}$	>9, <5 (org. Only)	Liang et al. (2017)
$CoFe_2O_4$ alginate beads	Methylene blue, crystal violet and malachite green	466, 456, 248 $mg.g^{-1}$	5	Li et al. (2016c)
Fe_3O_4@SiO_2-κ-carrageenan	Metoprolol	447 $mg.g^{-1}$	7	Soares et al. (2016)
Ion-imprinted magnetic chitosan/poly(vinyl alcohol)	Ni(II)	500 $mg.g^{-1}$	5.5	Zhang et al. (2015a)
Fe_3O_4-corn stalk	NO^{3-}	102 $mg.g^{-1}$	6–9	Song et al. (2016a)
Fe_3O_4@SiO_2-κ-carrageenan	Paraquat	257 $mg.g^{-1}$	7.3	Fernandes et al. (2017)
Magnetic cellulose beads	Pb(II)	5 $mg.g^{-1}$	2–3	Luo et al. (2016)
Magnetic graphene oxide coated with chitosan	Pb(II)	79 $mg.g^{-1}$	5	Wang et al. (2016)
Chitosan/PEI-grafted magnetic gelatin	Pb(II) and Cd(II)	341 and 321 $mg.g^{-1}$	6–7	Li et al. (2017)
Fe_3O_4-chitosan and -caboxymethylchitosan	Pb(II), Cu(II) and Zn(II)	243, 232, 131 $mg.g^{-1}$	5.2	Charpentier et al. (2016)

(continued)

Table 9.2 (continued)

Biosorbent	Pollutant	Adsorption capacity	pH	References
Magnetic hydrogel beads with gum tragacanth	Pb(II), Cu(II), crystal violet and Congo red	81, 69, 101, 94 mg.g^{-1}	2–6 (M^+) 2–8 (CV) 5–8 (CR)	Sahraei et al. (2017)
Fe_3O_4-β-cyclodextrin-bearing dextran	Phenanthrene and pyrene	K_d: 6095.5, 21,965 l.kg^{-1}	n/a	Cho et al. (2015)
Magnetic chitosan microspheres	Phosphorous	4.84 mg.g^{-1}	7	Funes et al. (2017)
Fe_3O_4 –NH_2 modified cellulose	Reactive Brilliant red K-2BP, methyl Orange and acid red 18	101, 222 and 99 mg.g^{-1}	2–3, 6, 6, >8,	Song et al. (2016b)
Cellulose functionalized with poly(dopamine)	Resorcinol	258 mg.g^{-1}	3	Ding et al. (2017)
Carbon nanotubes–C@Fe–chitosan	Tetracycline	104 mg.g^{-1}	6	Ma et al. (2015)
Carbon disulfide-modified magnetic ion-imprinted chitosan-Fe(II)	Tetracycline and Cd(II)	516 and 194 mg.g^{-1}	7–8	Chen et al. (2017a)
Fe_3O_4-alanine or –serine grafted chitosan	U(VI)	85, 116 mg.g^{-1}	3.6	Galhoum et al. (2015c)

health. In addition, many heavy metals are considered non-essential elements and have been linked to several diseases in humans (Järup 2003; Khan et al. 2008; Tchounwou et al. 2012; Chen et al. 2015; Tóth et al. 2016; Wasana et al. 2017).

Since water is one of the main routes for human exposure to excessive levels of heavy metals, in this section, a review is provided regarding the most recent magnetic bionanocomposites for their water remediation. Due to their well-known toxicity, heavy metals such as lead, cadmium, arsenic, mercury and chromium have been considered priority pollutants since they pose the greatest concern regarding human exposure.

A wide variety of magnetic bionanocomposites have been proposed for the effective removal of Pb(II) from water under several operating conditions (Charpentier et al. 2016; Luo et al. 2016; Wang et al. 2016; Li et al. 2017; Sahraei et al. 2017; Sengupta et al. 2017; Chen et al. 2017b). It is now well documented that Pb(II) can have a serious impact on human health mainly due to increased oxidative stress, with a highly detrimental effect on the hematopoietic, renal, reproductive and central nervous system (Flora et al. 2012). Moreover, even at very low environmental concentrations (<7.5 μg.dL^{-1}), lead can induce neuro–behavioural problems in children (Järup 2003; Lanphear et al. 2005). Despite the limited evidence on humans, lead has been classified as possible human carcinogen (Rousseau et al. 2007). Since that lead poisoning has been considered a major public health risk, several biosorbents have been recently presented for water remediation. Recently, a chitosan/PEI–grafted magnetic composite has been reported with an impressive capacity for the removal of Pb(II) and Cd(II) with a maximum adsorption capacity of 341 and 321 $mg.g^{-1}$ respectively (Li et al. 2017). With an optimum performance in pH range 6–7, the isotherm studies indicated a good agreement with the Langmuir isotherm for both metals. In addition, under the temperature range of 297 to 303 K, the chitosan/polyethylenimine–grafted magnetic composite exhibited negative ΔG° and positive ΔH° values indicating that the adsorption process is thermodynamically favourable and endothermic. It was also found that this material possessed a pseudo-second order kinetics, suggesting that chemisorption is the rate-limiting step. It was proposed that the good removal efficiency could be attributed to presence of biopolymer at the surface of the particles, which resulted in a higher number of amines groups with superior uptake efficiency. Besides the good performance, the chitosan/polyethylenimine–grafted magnetic composite could be effectively regenerated for 5 cycles while keeping the good performance and stability.

Magnetic hydrogel beads containing gum tragacanth in its composition, have also been reported for the effective clean-up of Pb(II) and Cu(II) from water (Sahraei et al. 2017), with a maximum adsorption capacity of 81 and 69 $mg.g^{-1}$ respectively, at pH 6. The good removal capacity of the particles was attributed to the presence of several chelating sulfonic acid, carboxylic acid, hydroxyl and amine groups. The isotherm experiments suggested a good agreement with the Langmuir model and the kinetic studies were more precisely defined by a pseudo-second order kinetics. Besides of the interesting performance, these beads could be regenerated for 3 cycles, maintaining their good performance and stability.

Magnetic carboxymethylchitosan (CMC) nanoparticles have also been proposed for the clean-up of Pb(II), Cu(II) and Zn(II). By preparing the materials through a simple one-step chemical precipitation method, their adsorption efficiency was determined with values of 243, 232 and 131 $mg.g^{-1}$ at pH 5.2 (Charpentier et al. 2016). The good sorption capacity of magnetic carboxymethylchitosan nanoparticles was attributed to the availability of a large number of carboxyl groups which are able to coordinate to metal ions. The isotherm and kinetic experiments revealed that these bionanocomposites follow the Freundlich isotherm and a pseudo-second order kinetics. In addition, the interaction between the metal ions and the biopolymer could be reversed under mild conditions for at least 3 cycles without any loss in their activity or stability.

It is well-known that Cd(II) is a highly toxic metal, where its exposure even at very low concentrations has been linked to several diseases in humans (Satarug et al. 2009). Cadmium is also classified as a human carcinogen (class I) (McMurray and Tainer 2003; Järup 2003; Joseph 2009; Yu et al. 2016). Recently, a novel composite of carbon disulphide-modified magnetic ion-imprinted chitosan-Fe(III) has been reported for the simultaneous clean-up of Cd(II) and tetracycline from water. With an impressive maximum adsorption capacity of 194 and 516 $mg.g^{-1}$ for Cd(II) and tetracycline, respectively, the optimal uptake efficiency was observed on the pH range of 7–8 (Chen et al. 2017a). The presence of pores in the composites along with the functional groups of the biopolymer resulted in good uptake efficiency due to large number of binding sites in the structure. The isotherm experiments indicated that the Langmuir model provides the best fit and that the kinetic profile is characterized by a pseudo-2nd order kinetics. After magnetic clean-up, the composite could be easily regenerated for at least 5 cycles while keeping its optimal adsorption capacity and magnetic behavior.

Arsenic has a wide distribution in nature, either in rocks, soil or water arising from natural and anthropogenic activity. For example, many natural occurring organic arsenic compounds (e.g. arsenobetaine) can be widely found in fish without significant toxic effects. In addition, several organic and inorganic forms of arsenic exist with distinct oxidation states, however the forms with As(III) (arsenite) and As (V) (arsenate) are the ones that raise the greatest concern. It is well-documented that these forms of arsenic are toxic for aquatic systems (Kumari et al. 2017) and are related with several diseases in humans and with well-known carcinogenic effects. Millions of people worldwide are exposed to one of these forms of arsenic due to the contamination of drinking water sources above the established permissible levels (WHO 2001; Hughes 2002; Järup 2003; Shi et al. 2004; Sharma and Sohn 2009).

An As(III) imprinted magnetic Fe_3O_4-N-(2-hydroxy)propyl-3-trimethyl ammonium chitosan (HTCC) composite has been presented for the removal of As(III) from water (Song et al. 2017). The use of HTCC resulted in improved solubility of the composite in water and delivered a high number of cationic biding sites that promoted the removal of As(III) species under optimal pH conditions (pH 6). The isotherm experiments agreed well with the Langmuir-Freundlich model, indicating a maximum adsorption capacity of 12 $mg.g^{-1}$. With a kinetic profile characteristic of pseudo-2nd order kinetics, this biosorbent was recycled for 10 cycles without any significant loss in performance.

Magnetic-chitosan beads have also been proposed for the removal of As(V) from water (Martínez-Cabanas et al. 2016). The isotherm experiment indicated a good agreement with the Langmuir-Freundlich model and a maximum adsorption capacity of 147 $\mu g.g^{-1}$. The material could also be recycled for 4 cycles while maintaining the performance.

Mercury is a heavy metal with a well-known toxicity. It is a natural element with a wide distribution in nature, however a great portion of mercury bioaccumulation is originated from anthropogenic activity. This metal exists in several inorganic (e.g. metallic mercury or mercuric salts (Hg^{2+})) and organometallic forms (e.g. methylmercury). The toxicity of mercury in humans is variable depending on the mercury form, where several forms are known to have a severe impact on human health. Mercury and its several forms have been identified as priority hazardous substance by the European Union (directive 2008/105/EC) and EPA (Clean Water Act). Water contamination of mercury is a serious threat to human health and aquatic ecosystem. For example, the toxic form methylmercury is a major source of human exposure to mercury through the ingestion of contaminated water or fish. This form of mercury is originated from the microbiological conversion of inorganic mercury in aquatic environments, which can then easily accumulate on fish. Moreover, this lipophilic form of mercury is of great concern, given its long retention in the human body (half-life of 70 days) (Zahir et al. 2005; Bernhoft 2012). Since water contamination with mercury is a serious problem, several nanomaterials have been purposed for its remediation. A itaconic acid-grafted-magnetite nanocellulose composite has been purposed for the removal of mercury from water (Anirudhan and Shainy 2015). The maximum adsorption capacity obtained was 240 $mg.g^{-1}$, where the experimental data fitted well with the Freundlich isotherm model. The material also exhibited a good reuse capacity, being recycled for 6 cycles. Recently, magnetic chitosan modified with glutaraldehyde was also purposed for the Hg(II) water remediation (Azari et al. 2017). Showing a maximum adsorption capacity of 96 $mg.g^{-1}$ (pH 5) and a good agreement with the Langmuir isotherm, the material could be recycled for 12 cycles while maintaining the good performance. In this case, glutaraldehyde was important to promote the fixation of chitosan, resulting in superior stability and, consequently, in a large number of strong chelating sites.

Chromium is another natural occurring heavy metal, being Cr(III) the most stable form with a wide distribution in nature. Cr(VI) does not occur naturally, being its bioaccumulation associated from anthropogenic activity. Chromium(III) is an essential nutrient playing a crucial role in several metabolic processes in the human body. Humans and other animals are capable to convert inorganic inactive Cr(III) into biologically active forms. Even though Cr(III) has an important biological role, it is now accepted that the exposure to high levels have been linked to adverse health effects. It is difficult to differentiate the toxic effects of Cr(III) and Cr(VI) since that Cr(VI) is mainly reduced into Cr(III) by the cells. The presence of Cr(VI) in drinking water is of serious concern, since that this form of chromium can easily enter the cells when compared with Cr(III) (Barceloux and Barceloux 1999; Wilbur et al. 2012; Ni et al. 2014). A pyridinium-diethylenetriamine magnetic chitosan (PDFMC) was proposed for the efficient removal of Cr(VI) from water (Sakti et al. 2015). The experimental data agreed well with the Langmuir isotherm, indicating maximum adsorption capacities of 176 $mg.L^{-1}$ (pH 3) and 124 $mg.L^{-1}$ (pH 6). The use of pyridinium units along with chitosan delivered several functional groups (amines and quaternary amines) that promoted the electrostatic interaction with the metal. Moreover, this material could be recycled for 5 cycles without any significant loss in performance.

Several magnetic bionanocomposites have also been proposed for the clean-up of other types of heavy metals where its bioaccumulation may result on detrimental effects to the human health or the ecosystem. The most recent reported materials are summarily presented on Table 9.2.

9.4.2 Removal of Organic Compounds

The impressive diversity of organic compounds commercially available result in their widespread occurrence where some of them possess a long-term persistence, high-bioaccumulation and a significant impact on human health and ecosystems. With the advent of advanced analytical techniques, there was an increase in the scientific public awareness regarding the contamination of water sources of not only hydrophobic persistent pollutants but also of polar contaminants that can deeply affect the water quality criteria. Every day, different classes of organic pollutants are discharged with the potential to contaminate drinking water sources. Some examples include dyes, pharmaceuticals, pesticides, solvents or many other organic by-products originated from industrial manufacturing (Dsikowitzky and Schwarzbauer 2014; Cizmas et al. 2015). Moreover, with the increased limitation of available potable water, there is an added need to implement closed-water cycles involving the reuse of treated wastewater. Unfortunately, many organic pollutants and their by-products are not monitored at water treatment plants. There are now great evidences that a large diversity of hazardous organic pollutants are present in drinking water sources, with an alarming negative impact on human health or aquatic ecosystem. To address this problem, different types of bionanocomposites have been purposed for the water remediation of organic pollutants.

In the last decade, a greater attention has been given to the remediation of organic dyes due to their widespread use and uncontrolled discharge. Some dyes or their by-products have been found to be toxic, mutagenic or carcinogenic (Rai et al. 2005). Furthermore, many bionanocomposites have been explored for the uptake of several common less toxic dyes, given their prevalent use as molecular models for preliminary assessment of material performance. A well-known example of such case is methylene blue. As detailed in Table 9.2, several works have investigated the performance of bionanocomposites for the uptake of this dye from water. Recently, our group explored the ability of hybrid magnetic biosorbents containing a siliceous shell with covalently liked κ-carrageenan for the uptake of methylene blue from water (Soares et al. 2017b). The experimental results indicated a Z-type isotherm, with a maximum adsorption capacity of 530 $mg.g^{-1}$. In addition, the material could be recycled for 6 cycles without loss in performance or stability. It was also revealed that the biopolymer had an important role in the removal of methylene blue by providing functional groups (anionic ester sulfate groups) that could establish electrostatic interactions with the cationic dye under optimal pH conditions.

A gelatin-based magnetic nanocomposite comprising carboxylic acid functionalized carbon nanotube has also been reported for the uptake of methylene blue and direct red 80 (Saber-Samandari et al. 2017). The experimental results indicated a good agreement with the Freundlich isotherm and a maximum adsorption capacity of 380 and 465 $mg.g^{-1}$ for methylene blue and direct red 80 respectively. The authors proposed that the gelatin (type A) delivered cationic functional groups that can remove the organic pollutants from water through electrostatic interactions.

Alginate beads containing dispersed polydopamine coated $CoFe_2O_4$ particles were reported for the uptake methylene blue, crystal violet and malachite green (Li et al. 2016c). The experimental data indicated a good agreement with the Freundlich (methylene blue and crystal violet) and Langmuir (malachite green) isotherm models with maximum adsorption capacities of 466, 456 and 248 $mg.g^{-1}$. It was shown that the presence of carboxylate, catechol and amine groups in these beads, provided important binding sites for the removal of the organic dyes through electrostatic interactions.

Glutaraldehyde cross-linked chitosan-coated Fe_3O_4 nanocomposites were used to efficiently uptake methyl orange (Yang et al. 2016a). Exhibiting a maximum adsorption capacity of 758 $mg.g^{-1}$, the experimental data fitted well with the Freundlich isotherm. The material could be reused for 6 cycles without loss in performance.

In addition to dyes, other organic pollutants have also been investigated for water remediation, given the rising concerns regarding their impact on human health and the ecosystems. For example, our group explored the ability of hybrid magnetic nanoparticles containing κ-carrageenan for the uptake of metoprolol (MP, beta-blocker) (Soares et al. 2016) and paraquat (PQ, herbicide) (Fernandes et al. 2017). The good performance of the material, 447 (MP) and 257 $mg.g^{-1}$ (PQ), and recycling capacity (4+ cycles), make it as a good candidate for the water remediation of these pollutants. It was shown that the biopolymer had an important role in the removal of these pollutants by establishing electrostatic interactions at optimal pH conditions.

Magnetic cellulose ionomer/layered double hydroxide has been successfully used for the removal of diclofenac from water (Hossein et al. 2016). The material exhibited a maximum adsorption capacity of 268 $mg.g^{-1}$, with a good reuse capacity for 3 cycles. The pyridinium rings of cellulose ionomer provided several binding sites for the removal of diclofenac. Different adsorption mechanisms were proposed by the authors, including π-π interaction, hydrogen bonding as well as anion exchange and electrostatic interactions.

Cellulose functionalized with magnetic poly(dopamine) was reported for the uptake of resorcinol from water (Ding et al. 2017). The adsorption process fitted well the Freundlich isotherm, indicating a maximum performance of 258 $mg.g^{-1}$. The material showed a good reusability for 5 consecutive cycles. The cellulose-poly (dopamine) units provided abundant functional groups (hydroxyl and amino groups) that were important for the successful uptake of resorcinol through electrostatic interactions.

A composite material based on chitosan and magnetic carbon nanotubes has been employed for the removal of tetracycline from water (Ma et al. 2015). The adsorption profile fitted well with the Freundlich isotherm model with a maximum performance of 104 $mg.g^{-1}$. In addition, the biosorbent could be recycled for 10 cycles while keeping the performance above 99.3 $mg.g^{-1}$. The kinetic studies along with the FTIR characterization indicated that the chemisorption was the rate-limiting step.

9.5 Fate of Magnetic Biosorbents

The effective recovery of nanosorbents used in from treated water is essential to ensure the absence of nanosized particles in the purified water and in the waste streams of the treatment process (Simeonidis et al. 2016). The use of magnetic biosorbents presents clearly an advantage in comparison to non-magnetic biosorbents since magnetic nanoparticles can be simply and quickly separated from aqueous dispersions using high-gradient fields (Moeser et al. 2004; Yavuz et al. 2006). However in the accidental case of release of the sorbents, the use of magnetic nanoparticles in water treatment may introduce unknown health effects to humans and other living organisms. The effects of the magnetic iron oxide nanoparticles in terms of ecotoxicity and cytotoxicity have been investigated in the last years (Handy et al. 2008; Su 2017). However studies addressing the effect of the magnetite nanoparticles on plants are still scarce (Zhu et al. 2008; Wang et al. 2011). Also few studies have addressed the effects of iron oxide nanoparticles on aquatic living organism (Zhang et al. 2015b; Blinova et al. 2017). More attention has been devoted to potential human exposure and a large number of works investigates the cytotoxicity and genotoxicity of magnetic iron oxide nanoparticles (Singh et al. 2010; Liu et al. 2013; Shen et al. 2015; Zhang et al. 2016a). Parameters such as the size, shape, surface coating and surface charge of the magnetic nanoparticles affect the interaction with cells and living organisms. Generally, neutral surfaces are the most biocompatible, whereas cationic surfaces, which have higher affinity to phospholipid in cell membranes, are more likely to induce undesirable hemolysis (Liu et al. 2013).

Upon release of the magnetic biosorbent particles into the aquatic environment, their potential impact on the environment and human health will largely depend on the particle colloidal stability that is determined by particle surface charge. Environmental factors such as the pH, the salinity and the presence of organic matter will also affect (Handy et al. 2008). For example humic acid, that is a major organic constituent of soil and aquatic environments, if present in high levels, is capable to coat Fe_3O_4 nanoparticles giving rise to particle zeta potential values similar to its own and therefore much more stable nanoparticle suspensions (Hu et al. 2010).

Another concern is the fate of nanoparticles recovered after water treatment. Although magnetic biosorbents are reusable after adequate regeneration treatment, generally after multiple cycles lose adsorptive abilities, generating solid waste that needs to be handled (Gómez-Pastora et al. 2014). Most recently alternative strategies for the effective utilization of pollutant loaded post-sorbents have been reported (Harikishore et al. 2017) and include applications in catalysis, as fertilizer or feed additives. For example, biochar decorated with magnetic nanophases developed for phosphate recovery, at its end-life improved plant growth and the application as fertilizer was proposed (Li et al. 2016b). Nevertheless before implementation of any post-sorption valorization strategy it is critical to investigate the leaching behavior of captured pollutants following experimental protocols that characterize solid wastes as inert, non-hazardous or toxic (Simeonidis et al. 2016).

9.6 Conclusions

The application of magnetic bionanocomposite particles for the uptake of pollutants in water treatment has emerged as an interesting alternative to conventional sorbents. These biosorbents offer clear advantages of magnetic separation and tuned affinity for a variety of pollutants. The rational design of the surface of these materials is essential, in order to attain robust and reusable biosorbents with high adsorption capacities. Thus, this chapter dedicated a specific section to the most relevant chemical strategies for the surface modification of magnetic nanoparticles with biopolymers, aiming the production of magnetic biosorbents optimized and specialized in the targeted pollutants. A direct outcome of the use of optimized biosorbent particles is the reduction of quantities of solids needed in the water treatment, which apart from bringing down the costs of water remediation, also reduces the potential environmental effects of the exhausted sorbents disposed. The successful implementation of magnetic biosorbents based water remediation technology needs of the evaluation of the impact of these materials on the environment and human health. However the use of natural occurring biopolymers anticipates potential bio- and eco-compatibility of the magnetic biosorbents.

Acknowledgements This work was developed in the scope of the exploratory project IF/00405/2014 and the project CICECO-Aveiro Institute of Materials, POCI-01-0145-FEDER-007679 (FCT Ref. UID/CTM/50011/2013), financed by national funds through the FCT/MEC, and when appropriate cofinanced by the European Regional Development Fund (FEDER) under the PT2020 Partnership Agreement. S. F. Soares thanks the Fundação para a Ciência e Tecnologia (FCT) for the PhD grant SFRH/BD/121366/2016. T. Fernandes thanks FCT for the PhD grant SFRH/BD/130934/2017. A. L. D.-d.-S. acknowledges FCT for the research contract under the Program 'Investigador FCT' 2014.

References

Adeleye AS, Conway JR, Garner K, Huang Y, Su Y, Keller AA (2016) Engineered nanomaterials for water treatment and remediation: costs, benefits, and applicability. Chem Eng J 286:640–662. https://doi.org/10.1016/j.cej.2015.10.105

Anirudhan TS, Shainy F (2015) Effective removal of mercury(II) ions from chlor-alkali industrial wastewater using 2-mercaptobenzamide modified itaconic acid-grafted-magnetite nanocellulose composite. J Colloid Interf Sci 456:22–31. https://doi.org/10.1016/j.jcis.2015.05.052

Arya V, Philip L (2016) Adsorption of pharmaceuticals in water using Fe_3O_4 coated polymer clay composite. Microporous Mesoporous Mater 232:273–280. https://doi.org/10.1016/j.micromeso.2016.06.033

Avérous L, Pollet E (2012) Environmental silicate nano-biocomposites, 1st edn. Springer, London

Azari A, Gharibi H, Kakavandi B, Ghanizadeh G, Javid A, Mahvi AH, Sharafi K, Khosravia T (2017) Magnetic adsorption separation process: an alternative method of mercury extracting from aqueous solution using modified chitosan coated Fe_3O_4 nanocomposites. J Chem Technol Biotechnol 92:188–200. https://doi.org/10.1002/jctb.4990

Badruddoza AZM, Tay ASH, Tan PY, Hidajat K, Uddin MS (2011) Carboxymethyl-β-cyclodextrin conjugated magnetic nanoparticles as nano-adsorbents for removal of copper ions: synthesis and adsorption studies. J Hazard Mater 185:1177–1186. https://doi.org/10.1016/j.jhazmat.2010.10.029

Bagheri S, Julkapli NM (2016) Modified iron oxide nanomaterials: functionalization and application. J Magn Magn Mater 416:117–133. https://doi.org/10.1016/j.jmmm.2016.05.042

Bano I, Arshad M, Yasin T, Ghauri MA, Younus M (2017) Chitosan: a potential biopolymer for wound management. Int J Biol Macromol 102:380–383. https://doi.org/10.1016/j.ijbiomac.2017.04.047

Barceloux DG, Barceloux D (1999) Chromium. J Toxicol Clin Toxicol 37:173–194. https://doi.org/10.1081/CLT-100102418

Bartůněk V, Průcha D, Švecová M, Ulbrich P, Huber Š, Sedmidubský D, Jankovský O (2016) Ultrafine ferromagnetic iron oxide nanoparticles: facile synthesis by low temperature decomposition of iron glycerolate. Mater Chem Phys 180:272–278. https://doi.org/10.1016/j.matchemphys.2016.06.007

Bee A, Massart R, Neveu S (1995) Synthesis of very fine maghemite particles. J Magn Magn Mater 149:6–9. https://doi.org/10.1016/0304-8853(95)00317-7

Bée A, Obeid L, Mbolantenaina R, Welschbillig M, Talbot D (2017) Magnetic chitosan/clay beads: a magsorbent for the removal of cationic dye from water. J Magn Magn Mater 421:59–64. https://doi.org/10.1016/j.jmmm.2016.07.022

Begin-Colin S, Felder-Flesch D (2012) Functionalisation of magnetic iron oxide nanoparticles. In: NTK T (ed) Magnetic nanoparticles: from fabrication to clinical applications. CRC Press, Boca Raton, pp 151–192

Bernhoft RA (2012) Mercury toxicity and treatment: a review of the literature. J Environ Public Health 2012:1–10. https://doi.org/10.1155/2012/460508

Bhavani P, Rajababu CH, Arif MD, Reddy IVS, Reddy NR (2017) Synthesis of high saturation magnetic iron oxide nanomaterials via low temperature hydrothermal method. J Magn Magn Mater 426:459–466. https://doi.org/10.1016/j.jmmm.2016.09.049

Bini RA, Marques RFC, Santos FJ, Chaker JA, Jafelicci M (2012) Synthesis and functionalization of magnetite nanoparticles with different amino-functional alkoxysilanes. J Magn Magn Mater 324:534–539. https://doi.org/10.1016/j.jmmm.2011.08.035

Blinova I, Kanarbik L, Irha N, Kahru A (2017) Ecotoxicity of nanosized magnetite to crustacean Daphnia magna and duckweed Lemna minor. Hydrobiologia 798:141–149. https://doi.org/10.1007/s10750-015-2540-6

Boamah PO, Huang Y, Hua M, Zhang Q, Wu J, Onumah J, Sam-Amoah LK, Boamah PO (2015) Sorption of heavy metal ions onto carboxylate chitosan derivatives – a mini-review. Ecotoxicol Environ Saf 116:113–120. https://doi.org/10.1016/j.ecoenv.2015.01.012

Bohara RA, Thorat ND, Pawar SH (2016) Role of functionalization: strategies to explore potential nano-bio applications of magnetic nanoparticles. RSC Adv 6:43989–44012. https://doi.org/10.1039/C6RA02129H

Boury B, Plumejeau S (2015) Metal oxides and polysaccharides: an efficient hybrid association for materials chemistry. Green Chem 17:72–88. https://doi.org/10.1039/C4GC00957F

Carpenter AW, de Lannoy C-F, Wiesner MR (2015) Cellulose nanomaterials in water treatment technologies. Environ Sci Technol 49:5277–5287. https://doi.org/10.1021/es506351r

Carvalho RS, Daniel-da-Silva AL, Trindade T (2016) Uptake of europium(III) from water using magnetite nanoparticles. Part Part Syst Charact 33:150–157. https://doi.org/10.1002/ppsc.201500170

Charpentier TVJ, Neville A, Lanigan JL, Barker R, Smith MJ, Richardson T (2016) Preparation of magnetic carboxymethylchitosan nanoparticles for adsorption of heavy metal ions. ACS Omega 1:77–83. https://doi.org/10.1021/acsomega.6b00035

Chen D, Yu Y-Z, Zhu H-J, Liu Z-Z, Xu Y-F, Liu Q, Qian G-R (2008) Ferrite process of electroplating sludge and enrichment of copper by hydrothermal reaction. Sep Purif Technol 62:297–303. https://doi.org/10.1016/j.seppur.2008.01.003

Chen G, Shi H, Tao J, Chen L, Liu Y, Lei G, Liu X, Smol JP (2015) Industrial arsenic contamination causes catastrophic changes in freshwater ecosystems. Sci Rep 5:17419. https://doi.org/10.1038/srep17419

Chen A, Shang C, Shao J, Lin Y, Luo S, Zhang J, Huang H, Lei M, Zeng Q (2017a) Carbon disulfide-modified magnetic ion-imprinted chitosan-Fe(III): a novel adsorbent for simultaneous removal of tetracycline and cadmium. Carbohydr Polym 155:19–27. https://doi.org/10.1016/j.carbpol.2016.08.038

Chen K, He J, Li Y, Cai X, Zhang K, Liu T, Hu Y, Lin D, Kong L, Liu J (2017b) Removal of cadmium and lead ions from water by sulfonated magnetic nanoparticle adsorbents. J Colloid Interf Sci 494:307–316. https://doi.org/10.1016/j.jcis.2017.01.082

Cheng C, Wen Y, Xu X, Gu H (2009) Tunable synthesis of carboxyl-functionalized magnetite nanocrystal clusters with uniform size. J Mater Chem 19:8782. https://doi.org/10.1039/b910832g

Cheng W, Xu X, Wu F, Li J (2016) Synthesis of cavity-containing iron oxide nanoparticles by hydrothermal treatment of colloidal dispersion. Mater Lett 164:210–212. https://doi.org/10.1016/j.matlet.2015.10.170

Cho E, Tahir MN, Choi JM, Kim H, Yu JH, Jung S (2015) Novel magnetic nanoparticles coated by benzene- and beta-cyclodextrin-bearing dextran, and the sorption of polycyclic aromatic hydrocarbon. Carbohydr Polym 133:221–228. https://doi.org/10.1016/j.carbpol.2015.06.089

Choi C, Nam J-P, Nah J-W (2016) Application of chitosan and chitosan derivatives as biomaterials. J Ind Eng Chem 33:1–10. https://doi.org/10.1016/j.jiec.2015.10.028

Christen V, Faltermann S, Brun NR, Kunz PY, Fent K (2017) Cytotoxicity and molecular effects of biocidal disinfectants (quaternary ammonia, glutaraldehyde, poly(hexamethylene biguanide) hydrochloride PHMB) and their mixtures in vitro and in zebrafish eleuthero-embryos. Sci Total Environ 586:1204–1218. https://doi.org/10.1016/j.scitotenv.2017.02.114

Cizmas L, Sharma VK, Gray CM, McDonald TJ (2015) Pharmaceuticals and personal care products in waters: occurrence, toxicity, and risk. Environ Chem Lett 13:381–394. https://doi.org/10.1007/s10311-015-0524-4

Crini G (2005) Recent developments in polysaccharide-based materials used as adsorbents in wastewater treatment. Prog Polym Sci 30:38–70. https://doi.org/10.1016/j.progpolymsci.2004.11.002

Daneshfozoun S, Abdullah MA, Abdullah B (2017) Preparation and characterization of magnetic biosorbent based on oil palm empty fruit bunch fibers, cellulose and Ceiba pentandra for heavy metal ions removal. Ind Crop Prod 105:93–103. https://doi.org/10.1016/j.indcrop.2017.05.011

Daniel-da-Silva AL, Trindade T, Goodfellow BJ, Costa BFO, Correia RN, Gil AM (2007) In situ synthesis of magnetite nanoparticles in carrageenan gels. Biomacromolecules 8:2350–2357. https://doi.org/10.1021/bm070096q

Daniel-da-Silva A, Carvalho R, Trindade T (2013) Magnetic hydrogel nanocomposites and composite nanoparticles – a review of recent patented works. Recent Pat Nanotechnol 7:153–166. https://doi.org/10.2174/1872210511307990008

Daniel-da-Silva AL, Salgueiro AM, Creaney B, Oliveira-Silva R, Silva NJO, Trindade T (2015) Carrageenan-grafted magnetite nanoparticles as recyclable sorbents for dye removal. J Nanopart Res 17:302. https://doi.org/10.1007/s11051-015-3108-0

Dehabadi L, Wilson LD (2014) Polysaccharide-based materials and their adsorption properties in aqueous solution. Carbohydr Polym 113:471–479. https://doi.org/10.1016/j.carbpol.2014.06.083

Dickinson E (2017) Biopolymer-based particles as stabilizing agents for emulsions and foams. Food Hydrocoll 68:219–231. https://doi.org/10.1016/j.foodhyd.2016.06.024

Ding C, Sun Y, Wang Y, Li J, Lin Y, Sun W, Luo C (2017) Adsorbent for resorcinol removal based on cellulose functionalized with magnetic poly(dopamine). Int J Biol Macromol 99:578–585. https://doi.org/10.1016/j.ijbiomac.2017.03.018

Divya K, Jisha MS (2017) Chitosan nanoparticles preparation and applications. Environ Chem Lett 16:1–12. https://doi.org/10.1007/s10311-017-0670-y

Dsikowitzky L, Schwarzbauer J (2014) Industrial organic contaminants: identification, toxicity and fate in the environment. Environ Chem Lett 12:371–386. https://doi.org/10.1007/s10311-014-0467-1

Duman O, Tunç S, Polat TG, Bozoğlan BK (2016) Synthesis of magnetic oxidized multiwalled carbon nanotube κ carrageenan Fe_3O_4 nanocomposite adsorbent and its application in cationic methylene blue dye adsorption. Carbohydr Polym 147:79–88. https://doi.org/10.1016/j.carbpol.2016.03.099

Fan C, Li K, Li J, Ying D, Wang Y, Jia J (2017) Comparative and competitive adsorption of Pb(II) and Cu(II) using tetraethylenepentamine modified chitosan/$CoFe_2O_4$ particles. J Hazard Mater 326:211–220. https://doi.org/10.1016/j.jhazmat.2016.12.036

Fateixa S, Nogueira HIS, Trindade T (2015) Hybrid nanostructures for SERS: Materials development and chemical detection. Phys Chem Chem Phys 17:21046–21071. https://doi.org/10.1039/C5CP01032B

Fernandes T, Soares S, Trindade T, Daniel-da-Silva A (2017) Magnetic hybrid nanosorbents for the uptake of paraquat from water. Nano 7:68. https://doi.org/10.3390/nano7030068

Flora G, Gupta D, Tiwari A (2012) Toxicity of lead: a review with recent updates. Interdiscip Toxicol 5:2. https://doi.org/10.2478/v10102-012-0009-2

Funes A, de Vicente J, de Vicente I (2017) Synthesis and characterization of magnetic chitosan microspheres as low-density and low-biotoxicity adsorbents for lake restoration. Chemosphere 171:571–579. https://doi.org/10.1016/j.chemosphere.2016.12.101

Galhoum AA, Atia AA, Mahfouz MG, Abdel-Rehem ST, Gomaa NA, Vincent T, Guibal E (2015a) Dy(III) recovery from dilute solutions using magnetic-chitosan nano-based particles grafted with amino acids. J Mater Sci 50:2832–2848. https://doi.org/10.1007/s10853-015-8845-z

Galhoum AA, Mafhouz MG, Abdel-Rehem ST, Gomaa NA, Atia AA, Vincent T, Guibal E (2015b) Cysteine-functionalized chitosan magnetic nano-based particles for the recovery of light and heavy rare earth metals: uptake kinetics and sorption isotherms. Nano 5:154–179. https://doi.org/10.3390/nano5010154

Galhoum AA, Mahfouz MG, Atia AA, Abdel-Rehem ST, Gomaa NA, Vincent T, Guibal E (2015c) Amino acid functionalized chitosan magnetic nanobased particles for uranyl sorption. Ind Eng Chem Res 54:12374–12385. https://doi.org/10.1021/acs.iecr.5b03331

Galhoum AA, Mahfouz MG, Gomaa NM, Vincent T, Guibal E (2017) Chemical modifications of chitosan nano-based magnetic particles for enhanced uranyl sorption. Hydrometallurgy 168:127–134. https://doi.org/10.1016/j.hydromet.2016.08.011

Gentile P, Carmagnola I, Nardo T, Chiono V (2015) Layer-by-layer assembly for biomedical applications in the last decade. Nanotechnology 26:422001. https://doi.org/10.1088/0957-4484/26/42/422001

Gholami M, Vardini MT, Mahdavinia GR (2016) Investigation of the effect of magnetic particles on the crystal violet adsorption onto a novel nanocomposite based on κ-carrageenan-g-poly (methacrylic acid). Carbohydr Polym 136:772–781. https://doi.org/10.1016/j.carbpol.2015.09.044

Girginova PI, Daniel-da-Silva AL, Lopes CB, Figueira P, Otero M, Amaral VS, Pereira E, Trindade T (2010) Silica coated magnetite particles for magnetic removal of Hg^{2+} from water. J Colloid Interf Sci 345:234–240. https://doi.org/10.1016/j.jcis.2010.01.087

Glasgow W, Fellows B, Qi B, Darroudi T, Kitchens C, Ye L, Crawford TM, Mefford OT (2016) Continuous synthesis of iron oxide (Fe_3O_4) nanoparticles via thermal decomposition. Particuology 26:47–53. https://doi.org/10.1016/j.partic.2015.09.011

Gómez-Pastora J, Bringas E, Ortiz I (2014) Recent progress and future challenges on the use of high performance magnetic nano-adsorbents in environmental applications. Chem Eng J 256:187–204. https://doi.org/10.1016/j.cej.2014.06.119

Guin D, Baruwati B, Manorama SV (2005) A simple chemical synthesis of nanocrystalline AFe_2O_4 (A=Fe, Ni, Zn): an efficient catalyst for selective oxidation of styrene. J Mol Catal A Chem 242:26–31. https://doi.org/10.1016/j.molcata.2005.07.021

Gyergyek S, Makovec D, Jagodič M, Drofenik M, Schenk K, Jordan O, Kovač J, Dražič G, Hofmann H (2017) Hydrothermal growth of iron oxide NPs with a uniform size distribution for magnetically induced hyperthermia: structural, colloidal and magnetic properties. J Alloys Compd 694:261–271. https://doi.org/10.1016/j.jallcom.2016.09.238

Handy RD, von der Kammer F, Lead JR, Hassellöv M, Owen R, Crane M (2008) The ecotoxicology and chemistry of manufactured nanoparticles. Ecotoxicology 17:287–314. https://doi.org/10.1007/s10646-008-0199-8

Harikishore KRD, Vijayaraghavan K, Kim JA, Yun Y-S (2017) Valorisation of post-sorption materials: opportunities, strategies, and challenges. Adv Colloid Interf Sci 242:35–58. https://doi.org/10.1016/j.cis.2016.12.002

Hong H-J, Jeong HS, Kim B-G, Hong J, Park I-S, Ryu T, Chung K-S, Kim H, Ryu J (2016) Highly stable and magnetically separable alginate/Fe_3O_4 composite for the removal of strontium (Sr) from seawater. Chemosphere 165:231–238. https://doi.org/10.1016/j.chemosphere.2016.09.034

Hossein BM, Shemirani F, Shirkhodaie M (2016) Aqueous co(II) adsorption using 8-hydroxyquinoline anchored γ-Fe_2O_3@chitosan with co(II) as imprinted ions. Int J Biol Macromol 87:375–384. https://doi.org/10.1016/j.ijbiomac.2016.02.077

Hu J-D, Zevi Y, Kou X-M, Xiao J, Wang X-J, Jin Y (2010) Effect of dissolved organic matter on the stability of magnetite nanoparticles under different pH and ionic strength conditions. Sci Total Environ 408:3477–3489. https://doi.org/10.1016/j.scitotenv.2010.03.033

Hu W, Murata K, Zhang D (2017) Applicability of LIVE/DEAD BacLight stain with glutaraldehyde fixation for the measurement of bacterial abundance and viability in rainwater. J Environ Sci 51:202–213. https://doi.org/10.1016/j.jes.2016.05.030

Hughes MF (2002) Arsenic toxicity and potential mechanisms of action. Toxicol Lett 133:1–16. https://doi.org/10.1016/S0378-4274(02)00084-X

Järup L (2003) Hazards of heavy metal contamination. Br Med Bull 68:167–182. https://doi.org/10.1093/bmb/ldg032

Jiang F, Li X, Zhu Y, Tang Z (2014a) Synthesis and magnetic characterizations of uniform iron oxide nanoparticles. Phys B Condens Matter 443:1–5. https://doi.org/10.1016/j.physb.2014.03.009

Jiang X-S, Mathew MP, Du J (2014b) Polyelectrolyte hydrogels: thermodynamics. In: Visakh PM, Bayraktar O, Picó GA (eds) Polyelectrolytes: thermodynamics and rheology. Springer, Cham, pp 183–214

Joseph P (2009) Mechanisms of cadmium carcinogenesis. Toxicol Appl Pharmacol 238:272–279. https://doi.org/10.1016/j.taap.2009.01.011

Kang DH, Jung H-S, Ahn N, Yang SM, Seo S, Suh K-Y, Chang P-S, Jeon NL, Kim J, Kim K (2014) Janus-compartmental alginate microbeads having polydiacetylene liposomes and magnetic nanoparticles for visual lead(II) detection. ACS Appl Mater Interfaces 6:10631–10637. https://doi.org/10.1021/am502319m

Kaur R, Hasan A, Iqbal N, Alam S, Saini MK, Raza SK (2014) Synthesis and surface engineering of magnetic nanoparticles for environmental cleanup and pesticide residue analysis: a review. J Sep Sci 37:1805–1825. https://doi.org/10.1002/jssc.201400256

Khan S, Cao Q, Zheng YM, Huang YZ, Zhu YG (2008) Health risks of heavy metals in contaminated soils and food crops irrigated with wastewater in Beijing, China. Environ Pollut 152:686–692. https://doi.org/10.1016/j.envpol.2007.06.056

Kim J-H, Kim S-M, Kim Y-I (2014) Properties of magnetic nanoparticles prepared by co-precipitation. J Nanosci Nanotechnol 14:8739–8744. https://doi.org/10.1166/jnn.2014.9993

Kim H-R, Jang J-W, Park J-W (2016) Carboxymethyl chitosan-modified magnetic-cored dendrimer as an amphoteric adsorbent. J Hazard Mater 317:608–616. https://doi.org/10.1016/j.jhazmat.2016.06.025

Kumar ASK, Jiang S-J (2017) Synthesis of magnetically separable and recyclable magnetic nanoparticles decorated with β-cyclodextrin functionalized graphene oxide an excellent adsorption of as(V)/(III). J Mol Liq 237:387–401. https://doi.org/10.1016/j.molliq.2017.04.093

Kumari B, Kumar V, Sinha AK, Ahsan J, Ghosh AK, Wang H, DeBoeck G (2017) Toxicology of arsenic in fish and aquatic systems. Environ Chem Lett 15:43–64. https://doi.org/10.1007/s10311-016-0588-9

Lanphear BP, Hornung R, Khoury J, Yolton K, Baghurst P, Bellinger DC, Canfield RL, Dietrich KN, Bornschein R, Greene T, Rothenberg SJ, Needleman HL, Schnaas L, Wasserman G, Graziano J, Roberts R (2005) Low-level environmental lead exposure and children's intellectual function: an international pooled analysis. Environ Health Perspect 113:894–899. https://doi.org/10.1289/ehp.7688

Laurent S, Forge D, Port M, Roch A, Robic C, Vander Elst L, Muller RN (2008) Magnetic iron oxide nanoparticles: synthesis, stabilization, vectorization, physicochemical characterizations, and biological applications. Chem Rev 108:2064–2110. https://doi.org/10.1021/cr068445e

Laus R, de Fávere VT (2011) Competitive adsorption of Cu(II) and Cd(II) ions by chitosan crosslinked with epichlorohydrin–triphosphate. Bioresour Technol 102:8769–8776. https://doi.org/10.1016/j.biortech.2011.07.057

Lee J, Isobe T, Senna M (1996) Magnetic properties of ultrafine magnetite particles and their slurries prepared via *in-situ* precipitation. Colloids Surfaces A Physicochem Eng Asp 109:121–112. https://doi.org/10.1016/0927-7757(95)03479-X

Leung H-W (2001) Ecotoxicology of glutaraldehyde: review of environmental fate and effects studies. Ecotoxicol Environ Saf 49:26–39. https://doi.org/10.1006/eesa.2000.2031

Li K, Li P, Cai J, Xiao S, Yang H, Li A (2016a) Efficient adsorption of both methyl orange and chromium from their aqueous mixtures using a quaternary ammonium salt modified chitosan magnetic composite adsorbent. Chemosphere 154:310–318. https://doi.org/10.1016/j.chemosphere.2016.03.100

Li R, Wang JJ, Zhou B, Awasthi MK, Ali A, Zhang Z, Lahori AH, Mahar A (2016b) Recovery of phosphate from aqueous solution by magnesium oxide decorated magnetic biochar and its potential as phosphate-based fertilizer substitute. Bioresour Technol 215:209–214. https://doi.org/10.1016/j.biortech.2016.02.125

Li X, Lu H, Zhang Y, He F, Jing L, He X (2016c) Fabrication of magnetic alginate beads with uniform dispersion of $CoFe_2O_4$ by the polydopamine surface functionalization for organic pollutants removal. Appl Surf Sci 389:567–577. https://doi.org/10.1016/j.apsusc.2016.07.162

Li B, Zhou F, Huang K, Wang Y, Mei S, Zhou Y, Jing T (2017) Environmentally friendly chitosan/PEI-grafted magnetic gelatin for the highly effective removal of heavy metals from drinking water. Sci Rep 7:43082. https://doi.org/10.1038/srep43082

Liang X, Duan J, Xu Q, Wei X, Lu A, Zhang L (2017) Ampholytic microspheres constructed from chitosan and carrageenan in alkali/urea aqueous solution for purification of various wastewater. Chem Eng J 317:766–776. https://doi.org/10.1016/j.cej.2017.02.089

Lin S, Lin K, Lu D, Liu Z (2017) Preparation of uniform magnetic iron oxide nanoparticles by co-precipitation in a helical module microchannel reactor. J Environ Chem Eng 5:303–309. https://doi.org/10.1016/j.jece.2016.12.011

Ling D, Lee N, Hyeon T (2015) Chemical synthesis and assembly of uniformly sized iron oxide nanoparticles for medical applications. Acc Chem Res 48:1276–1285. https://doi.org/10.1021/acs.accounts.5b00038

Liu G, Gao J, Ai H, Chen X (2013) Applications and potential toxicity of magnetic iron oxide nanoparticles. Small 9:1533–1545. https://doi.org/10.1002/smll.201201531

Lu J, Jin R-N, Liu C, Wang Y-F, Ouyang X (2016a) Magnetic carboxylated cellulose nanocrystals as adsorbent for the removal of Pb(II) from aqueous solution. Int J Biol Macromol 93:547–556. https://doi.org/10.1016/j.ijbiomac.2016.09.004

Lu S, Li H, Zhang F, Du N, Hou W (2016b) Sorption of Pb(II) on carboxymethyl chitosan-conjugated magnetite nanoparticles: application of sorbent dosage-dependent isotherms. Colloid Polym Sci 294:1369–1379. https://doi.org/10.1007/s00396-016-3893-8

Lü T, Chen Y, Qi D, Cao Z, Zhang D, Zhao H (2017) Treatment of emulsified oil wastewaters by using chitosan grafted magnetic nanoparticles. J Alloys Compd 696:1205–1212. https://doi.org/10.1016/j.jallcom.2016.12.118

Luo Y, Wang Q (2014) Recent development of chitosan-based polyelectrolyte complexes with natural polysaccharides for drug delivery. Int J Biol Macromol 64:353–367. https://doi.org/10.1016/j.ijbiomac.2013.12.017

Luo X, Lei X, Xie X, Yu B, Cai N, Yu F (2016) Adsorptive removal of lead from water by the effective and reusable magnetic cellulose nanocomposite beads entrapping activated bentonite. Carbohydr Polym 151:640–664. https://doi.org/10.1016/j.carbpol.2016.06.003

Ma J, Zhuang Y, Yu F (2015) Facile method for the synthesis of a magnetic CNTs-C@Fe-chitosan composite and its application in tetracycline removal from aqueous solutions. Phys Chem Chem Phys 17:15936–15944. https://doi.org/10.1039/C5CP02542G

Mahdavinia GR, Mosallanezhad A (2016) Facile and green rout to prepare magnetic and chitosan-crosslinked κ-carrageenan bionanocomposites for removal of methylene blue. J Water Proces Eng 10:143–155. https://doi.org/10.1016/j.jwpe.2016.02.010

Mahdavinia GR, Hasanpour S, Behrouzi L, Sheykhloie H (2015) Study on adsorption of Cu (II) on magnetic starch-*g*-polyamidoxime/montmorillonite/Fe_3O_4 nanocomposites as novel chelating ligands. Starch-Stärke 68:188–199. https://doi.org/10.1002/star.201400255

Mahdavinia GR, Rahmani Z, Mosallanezhad A, Karami S, Shahriari M (2016) Effect of magnetic laponite RD on swelling and dye adsorption behaviors of κ-carrageenan-based nanocomposite hydrogels. Desalin Water Treat 57:20582–20596. https://doi.org/10.1080/19443994.2015.1111808

Maitra J, Shukla VK (2014) Cross-linking in hydrogels – A Review. Am J Polym Sci 4:25–31. https://doi.org/10.5923/j.ajps.20140402.01

Martínez-Cabanas M, López-García M, Barriada JL, Herrero R, Sastre de Vicente ME (2016) Green synthesis of iron oxide nanoparticles. Development of magnetic hybrid materials for efficient as (V) removal. Chem Eng J 301:83–91. https://doi.org/10.1016/j.cej.2016.04.149

McMurray CT, Tainer JA (2003) Cancer, cadmium and genome integrity. Nat Genet 34:239–241. https://doi.org/10.1038/ng0703-239

Mehta D, Mazumdar S, Singh SK (2015) Magnetic adsorbents for the treatment of water/wastewater—a review. J Water Proces Eng 7:244–265. https://doi.org/10.1016/j.jwpe.2015.07.001

Mendoza-Garcia A, Sun S (2016) Recent advances in the high-temperature chemical synthesis of magnetic nanoparticles. Adv Funct Mater 26:3809–3817. https://doi.org/10.1002/adfm.201504172

Moeser GD, Roach KA, Green WH, Alan Hatton T, Laibinis PE (2004) High-gradient magnetic separation of coated magnetic nanoparticles. AICHE J 50:2835–2848. https://doi.org/10.1002/aic.10270

Mondal S, Li C, Wang K (2015) Bovine serum albumin adsorption on gluteraldehyde cross-linked chitosan hydrogels. J Chem Eng Data 60:2356–2362. https://doi.org/10.1021/acs.jced.5b00264

Mu B, Kang Y, Wang A (2013) Preparation of a polyelectrolyte-coated magnetic attapulgite composite for the adsorption of precious metals. J Mater Chem A 1:4804. https://doi.org/10.1039/c3ta01620j

Muxika A, Etxabide A, Uranga J, Guerrero P, de la Caba K (2017) Chitosan as a bioactive polymer: processing, properties and applications. Int J Biol Macromol 105:1358–1368. https://doi.org/10.1016/j.ijbiomac.2017.07.087

Muzzarelli RAA (2009) Genipin-crosslinked chitosan hydrogels as biomedical and pharmaceutical aids. Carbohydr Polym 77:1–9. https://doi.org/10.1016/j.carbpol.2009.01.016

Nair NR, Sekhar VC, Nampoothiri KM, Pandey A (2017) Biodegradation of biopolymers. In: Pandey A, Negi S, Soccol CR (eds) Current bevelopments in biotechnology and bioengineering. Elsevier, Amsterdam, pp 739–755

Neeraj G, Krishnan S, Kumar PS, Shriaishvarya KR, Kumar VV (2016) Performance study on sequestration of copper ions from contaminated water using newly synthesized high effective chitosan coated magnetic nanoparticles. J Mol Liq 214:335–346. https://doi.org/10.1016/j.molliq.2015.11.051

Ni W, Huang Y, Wang X, Zhang J, Wu K (2014) Associations of neonatal lead, cadmium, chromium and nickel co-exposure with DNA oxidative damage in an electronic waste recycling town. Sci Total Environ 472:351–362. https://doi.org/10.1016/j.scitotenv.2013.11.032

Nie J, Wang Z, Hu Q (2016) Chitosan hydrogel structure modulated by metal ions. Sci Rep 6:36005. https://doi.org/10.1038/srep36005

Pankhurst QA, Connolly J, Jones SK, Dobson J (2003) Applications of magnetic nanoparticles in biomedicine. J Phys D Appl Phys 36:R167–R181. https://doi.org/10.1088/0022-3727/36/13/201

Park S-B, Lih E, Park K-S, Joung YK, Han DK (2017) Biopolymer-based functional composites for medical applications. Prog Polym Sci 68:77–105. https://doi.org/10.1016/j.progpolymsci.2016.12.003

Pattanashetti NA, Heggannavar GB, Kariduraganavar MY (2017) Smart biopolymers and their biomedical applications. Procedia Manuf 12:263–279. https://doi.org/10.1016/j.promfg.2017.08.030

Piculell L (2006) Gelling carrageenans. In: Alistair M, Stephen GOP (eds) Food polysaccharides and their applications. CRC Press, Hoboken, pp 239–287

Pinheiro PC, Daniel-da-Silva AL, Nogueira HIS, Trindade T (2018) Functionalized inorganic nanoparticles for magnetic separation and SERS detection of water pollutants. Eur J Inorg Chem https://doi.org/10.1002/ejic.201800132

Pinheiro PC, Tavares DS, Daniel-da-Silva AL, Lopes CB, Pereira E, Araújo JP, Sousa CT, Trindade T (2014) Ferromagnetic sorbents based on nickel nanowires for efficient uptake of mercury from water. ACS Appl Mater Interfaces 6:8274–8280. https://doi.org/10.1021/am5010865

Pujana MA, Pérez-Álvarez L, Iturbe LCC, Katime I (2014) Water soluble folate-chitosan nanogels crosslinked by genipin. Carbohydr Polym 101:113–120. https://doi.org/10.1016/j.carbpol.2013.09.014

Pušnik K, Goršak T, Drofenik M, Makovec D (2016) Synthesis of aqueous suspensions of magnetic nanoparticles with the co-precipitation of iron ions in the presence of aspartic acid. J Magn Magn Mater 413:65–75. https://doi.org/10.1016/j.jmmm.2016.04.032

Qiao L, Swihart MT (2017) Solution-phase synthesis of transition metal oxide nanocrystals: morphologies, formulae, and mechanisms. Adv Colloid Interf Sci 244:199–266. https://doi.org/10.1016/j.cis.2016.01.005

Rai HS, Bhattacharyya MS, Singh J, Bansal TK, Vats P, Banerjee UC (2005) Removal of dyes from the effluent of textile and dyestuff manufacturing industry: a review of emerging techniques with reference to biological treatment. Crit Rev Environ Sci Technol 35:219–238. https://doi.org/10.1080/10643380590917932

Rebelo R, Fernandes M, Fangueiro R (2017) Biopolymers in medical implants: a brief review. Procedia Eng 200:236–243. https://doi.org/10.1016/j.proeng.2017.07.034

Reddy DHK, Yun Y-S (2016) Spinel ferrite magnetic adsorbents: alternative future materials for water purification? Coord Chem Rev 315:90–111. https://doi.org/10.1016/j.ccr.2016.01.012

Reguyal F, Sarmah AK, Gao W (2017) Synthesis of magnetic biochar from pine sawdust via oxidative hydrolysis of $FeCl_2$ for the removal sulfamethoxazole from aqueous solution. J Hazard Mater 321:868–878. https://doi.org/10.1016/j.jhazmat.2016.10.006

Resch-Fauster K, Klein A, Blees E, Feuchter M (2017) Mechanical recyclability of technical biopolymers: potential and limits. Polym Test 64:287–295. https://doi.org/10.1016/j.polymertesting.2017.10.017

Rodriguez AFR, Costa TP, Bini RA, Faria FSEDV, Azevedo RB, Jafelicci M, Coaquira JAH, Martínez MAR, Mantilla JC, Marques RFC, Morais PC (2017) Surface functionalization of magnetite nanoparticle: a new approach using condensation of alkoxysilanes. Phys B Condens Matter 521:141–147. https://doi.org/10.1016/j.physb.2017.06.043

Roth H-C, Schwaminger SP, Schindler M, Wagner FE, Berensmeier S (2015) Influencing factors in the CO-precipitation process of superparamagnetic iron oxide nano particles: a model based study. J Magn Magn Mater 377:81–89. https://doi.org/10.1016/j.jmmm.2014.10.074

Rousseau M-C, Parent M-E, Nadon L, Latreille B, Siemiatycki J (2007) Occupational exposure to lead compounds and risk of cancer among men: a population-based case-control study. Am J Epidemiol 166:1005–1014. https://doi.org/10.1093/aje/kwm183

Saba N, Jawaid M, Sultan MTH, Alothman OY (2017) Green biocomposites for structural applications. In: Jawaid M, Salit MS, Alothman OY (eds) Green biocomposites: design and applications. Springer, Cham, pp 1–27

Saber-Samandari S, Saber-Samandari S, Joneidi-Yekta H, Mohseni M (2017) Adsorption of anionic and cationic dyes from aqueous solution using gelatin-based magnetic nanocomposite beads comprising carboxylic acid functionalized carbon nanotube. Chem Eng J 308:1133–1144. https://doi.org/10.1016/j.cej.2016.10.017

Sahraei R, Sekhavat Pour Z, Ghaemy M (2017) Novel magnetic bio-sorbent hydrogel beads based on modified gum tragacanth/graphene oxide: removal of heavy metals and dyes from water. J Clean Prod 142(Part):2973–2984. https://doi.org/10.1016/j.jclepro.2016.10.170

Sakti SCW, Narita Y, Sasaki T, Nuryono TS (2015) A novel pyridinium functionalized magnetic chitosan with pH-independent and rapid adsorption kinetics for magnetic separation of Cr(VI). J Environ Chem Eng 3:1953–1961. https://doi.org/10.1016/j.jece.2015.05.004

Salgueiro AM, Daniel-da-Silva AL, Girão AV, Pinheiro PC, Trindade T (2013) Unusual dye adsorption behavior of κ-carrageenan coated superparamagnetic nanoparticles. Chem Eng J 229:276–284. https://doi.org/10.1016/j.cej.2013.06.015

Satarug S, Garrett SH, Sens MA, Sens DA (2009) Cadmium, environmental exposure, and health outcomes. Environ Health Perspect 118:182–190. https://doi.org/10.1289/ehp.0901234

Schnepp Z (2013) Biopolymers as a flexible resource for nanochemistry. Angew Chemie Int Ed 52:1096–1108. https://doi.org/10.1002/anie.201206943

Sengupta A, Rao R, Bahadur D (2017) Zn^{2+}–silica modified cobalt ferrite magnetic nanostructured composite for efficient adsorption of cationic pollutants from water. ACS Sustain Chem Eng 5:1280–1286. https://doi.org/10.1021/acssuschemeng.6b01186

Sharma VK, Sohn M (2009) Aquatic arsenic: toxicity, speciation, transformations, and remediation. Environ Int 35:743–759. https://doi.org/10.1016/j.envint.2009.01.005

Shen Y, Huang Z, Liu X, Qian J, Xu J, Yang X, Sun A, Ge J (2015) Iron-induced myocardial injury: an alarming side effect of superparamagnetic iron oxide nanoparticles. J Cell Mol Med 19:2032–2035. https://doi.org/10.1111/jcmm.12582

Shi H, Shi X, Liu KJ (2004) Oxidative mechanism of arsenic toxicity and carcinogenesis. Mol Cell Biochem 255:67–78. https://doi.org/10.1023/B:MCBI.0000007262.26044.e8

Shi H, Yang J, Zhu L, Yang Y, Yuan H, Yang Y, Liu X (2016) Removal of Pb^{2+}, Hg^{2+}, and Cu^{2+} by chain-like Fe_3O_4@SiO_2@chitosan magnetic nanoparticles. J Nanosci Nanotechnol 16:1871–1882. https://doi.org/10.1166/jnn.2016.10712

Simeonidis K, Mourdikoudis S, Kaprara E, Mitrakas M, Polavarapu L (2016) Inorganic engineered nanoparticles in drinking water treatment: a critical review. Environ Sci Water Res Technol 2:43–70. https://doi.org/10.1039/C5EW00152H

Singh N, Jenkins GJS, Asadi R, Doak SH (2010) Potential toxicity of superparamagnetic iron oxide nanoparticles (SPION). Nano Rev 1:5358. https://doi.org/10.3402/nano.v1i0.5358

Soares SF, Simões TR, António M, Trindade T, Daniel-da-Silva AL (2016) Hybrid nanoadsorbents for the magnetically assisted removal of metoprolol from water. Chem Eng J 302:560–569. https://doi.org/10.1016/j.cej.2016.05.079

Soares SF, Rodrigues MI, Trindade T, Daniel-da-Silva AL (2017a) Chitosan-silica hybrid nanosorbents for oil removal from water. Colloids Surf A Physicochem Eng Asp 532:305–313. https://doi.org/10.1016/j.colsurfa.2017.04.076

Soares SF, Simões TR, Trindade T, Daniel-da-Silva AL (2017b) Highly efficient removal of dye from water using magnetic carrageenan/silica hybrid nano-adsorbents. Water Air Soil Pollut 228:87. https://doi.org/10.1007/s11270-017-3281-0

Sohni S, Gul K, Ahmad F, Ahmad I, Khan A, Khan N, Bahadar Khan S (2017) Highly efficient removal of acid red-17 and bromophenol blue dyes from industrial wastewater using graphene oxide functionalized magnetic chitosan composite. Polym Compos. https://doi.org/10.1002/pc.24349

Song W, Gao B, Xu X, Wang F, Xue N, Sun S, Song W, Jia R (2016a) Adsorption of nitrate from aqueous solution by magnetic amine-crosslinked biopolymer based corn stalk and its chemical regeneration property. J Hazard Mater 304:280–290. https://doi.org/10.1016/j.jhazmat.2015.10.073

Song W, Gao B, Xu X, Xing L, Han S, Duan P, Song W, Jia R (2016b) Adsorption–desorption behavior of magnetic amine/Fe_3O_4 functionalized biopolymer resin towards anionic dyes from wastewater. Bioresour Technol 210:123–130. https://doi.org/10.1016/j.biortech.2016.01.078

Song X, Li L, Geng Z, Zhou L, Ji L (2017) Effective and selective adsorption of as(III) via imprinted magnetic Fe_3O_4/HTCC composite nanoparticles. J Environ Chem Eng 5:16–25. https://doi.org/10.1016/j.jece.2016.11.016

Sousa FL, Daniel-da-Silva AL, Silva NJO, Trindade T (2015) Bionanocomposites for magnetic removal of water pollutants. In: Thakur VK (ed) Eco-friendly polymer nanocomposites: chemistry and applications. Springer, New Delhi, pp 279–310

Srivastava S, Kotov NA (2008) Composite layer-by-layer (LBL) assembly with inorganic nanoparticles and nanowires. Acc Chem Res 41:1831–1841. https://doi.org/10.1021/ar8001377

Su C (2017) Environmental implications and applications of engineered nanoscale magnetite and its hybrid nanocomposites: a review of recent literature. J Hazard Mater 322:48–84. https://doi.org/10.1016/j.jhazmat.2016.06.060

Sue K, Aoki M, Sato T, Nishio-Hamane D, Kawasaki S, Hakuta Y, Takebayashi Y, Yoda S, Furuya T, Sato T, Hiaki T (2011) Continuous hydrothermal synthesis of nickel ferrite nanoparticles using a central collision-type micromixer: effects of temperature, residence time, metal salt molality, and NaOH addition on conversion, particle size, and crystal phase. Ind Eng Chem Res 50:9625–9631. https://doi.org/10.1021/ie200036m

Sugimoto T, Matijević E (1980) Formation of uniform spherical magnetite particles by crystallization from ferrous hydroxide gels. J Colloid Interf Sci 74:227–243. https://doi.org/10.1016/0021-9797(80)90187-3

Sun C, Lee J, Zhang M (2008) Magnetic nanoparticles in MR imaging and drug delivery. Adv Drug Deliv Rev 60:1252–1265. https://doi.org/10.1016/j.addr.2008.03.018

Tancredi P, Botasini S, Moscoso-Londoño O, Méndez E, Socolovsky L (2015) Polymer-assisted size control of water-dispersible iron oxide nanoparticles in range between 15 and 100nm. Colloids Surf A Physicochem Eng Asp 464:46–51. https://doi.org/10.1016/j.colsurfa.2014.10.001

Tang SCN, Lo IMC (2013) Magnetic nanoparticles: essential factors for sustainable environmental applications. Water Res 47:2613–2632. https://doi.org/10.1016/j.watres.2013.02.039

Tanhaei B, Ayati A, Lahtinen M, Mahmoodzadeh Vaziri B, Sillanpää M (2016) A magnetic mesoporous chitosan based core-shells biopolymer for anionic dye adsorption: kinetic and isothermal study and application of ANN. J Appl Polym Sci 133:22. https://doi.org/10.1002/app.43466

Tavares DS, Lopes CB, Daniel-da-Silva AL, Vale C, Trindade T, Pereira ME (2016) Mercury in river, estuarine and seawaters – is it possible to decrease realist environmental concentrations in order to achieve environmental quality standards? Water Res 106:439–449. https://doi.org/10.1016/j.watres.2016.10.031

Tchounwou PB, Yedjou CG, Patlolla AK, Sutton DJ (2012) Heavy metal toxicity and the environment. EXS 101:133–164. https://doi.org/10.1007/978-3-7643-8340-4_6

Tóth G, Hermann T, Da Silva MR, Montanarella L (2016) Heavy metals in agricultural soils of the European Union with implications for food safety. Environ Int 88:299–309. https://doi.org/10.1016/j.envint.2015.12.017

Vakili M, Rafatullah M, Salamatinia B, Abdullah AZ, Ibrahim MH, Tan KB, Gholami Z, Amouzgar P (2014) Application of chitosan and its derivatives as adsorbents for dye removal from water and wastewater: a review. Carbohydr Polym 113:115–130. https://doi.org/10.1016/j.carbpol.2014.07.007

Valle LJ, Díaz A, Puiggalí J (2017) Hydrogels for biomedical applications: cellulose, chitosan, and protein/peptide derivatives. Gels 3:27. https://doi.org/10.3390/gels3030027

Vandenbossche M, Jimenez M, Casetta M, Traisnel M (2015) Remediation of heavy metals by biomolecules: a review. Crit Rev Environ Sci Technol 45:1644–1704. https://doi.org/10.1080/10643389.2014.966425

Velinov N, Petrova T, Tsoncheva T, Genova I, Koleva K, Kovacheva D, Mitov I (2016) Auto-combustion synthesis, Mössbauer study and catalytic properties of copper-manganese ferrites. Hyperfine Interact 237:24. https://doi.org/10.1007/s10751-016-1222-8

Wang H, Kou X, Pei Z, Xiao JQ, Shan X, Xing B (2011) Physiological effects of magnetite (Fe_3O_4) nanoparticles on perennial ryegrass (*Lolium perenne* L.) and pumpkin (*Cucurbita mixta*) plants. Nanotoxicology 5:30–42. https://doi.org/10.3109/17435390.2010.489206

Wang Y, Li L, Luo C, Wang X, Duan H (2016) Removal of Pb^{2+} from water environment using a novel magnetic chitosan/graphene oxide imprinted Pb^{2+}. Int J Biol Macromol 86:505–511. https://doi.org/10.1016/j.ijbiomac.2016.01.035

Wasana HMS, Perera GDRK, Gunawardena PDS, Fernando PS, Bandara J (2017) WHO water quality standards vs synergic effect(s) of fluoride, heavy metals and hardness in drinking water on kidney tissues. Sci Rep 7:42516. https://doi.org/10.1038/srep42516

WHO (2001) Environmental health criteria 224: arsenic and arsenic compounds, Geneva

Wilbur S, Abadin H, Fay M, Yu D, Tencza B, Ingerman L, Klotzbach J, James S (2012) Toxicological profile for chromium. U.S. Department of health and human services, Atlanta

Wu W, He Q, Jiang C (2008) Magnetic iron oxide nanoparticles: synthesis and surface functionalization strategies. Nanoscale Res Lett 3:397–415. https://doi.org/10.1007/s11671-008-9174-9

Wu W, Wu Z, Yu T, Jiang C, Kim W-S (2015) Recent progress on magnetic iron oxide nanoparticles: synthesis, surface functional strategies and biomedical applications. Sci Technol Adv Mater 16:23501. https://doi.org/10.1088/1468-6996/16/2/023501

Xiao C, Liu X, Mao S, Zhang L, Lu J (2017) Sub-micron-sized polyethylenimine-modified polystyrene/Fe_3O_4/chitosan magnetic composites for the efficient and recyclable adsorption of Cu(II) ions. Appl Surf Sci 394:378–385. https://doi.org/10.1016/j.apsusc.2016.10.116

Xiong Z, Qin H, Wan H, Huang G, Zhang Z, Dong J, Zhang L, Zhang W, Zou H (2013) Layer-by-layer assembly of multilayer polysaccharide coated magnetic nanoparticles for the selective enrichment of glycopeptides. Chem Commun 49:9284–9286. https://doi.org/10.1039/c3cc45008b

Xu P, Zeng GM, Huang DL, Feng CL, Hu S, Zhao MH, Lai C, Wei Z, Huang C, Xie GX, Liu ZF (2012) Use of iron oxide nanomaterials in wastewater treatment: a review. Sci Total Environ 424:1–10. https://doi.org/10.1016/j.scitotenv.2012.02.023

Yang D, Qiu L, Yang Y (2016a) Efficient adsorption of methyl orange using a modified chitosan magnetic composite adsorbent. J Chem Eng Data 61:3933–3940. https://doi.org/10.1021/acs.jced.6b00706

Yang X, Jin D, Zhang M, Wu P, Jin H, Li J, Wang X, Ge H, Wang Z, Lou H (2016b) Fabrication and application of magnetic starch-based activated hierarchical porous carbon spheres for the efficient removal of dyes from water. Mater Chem Phys 174:179–186. https://doi.org/10.1016/j.matchemphys.2016.02.073

Yavuz CT, Mayo JT, Yu WW, Prakash A, Falkner JC, Yean S, Cong L, Shipley HJ, Kan A, Tomson M, Natelson D, Colvin VL (2006) Low-field magnetic separation of monodisperse Fe_3O_4 nanocrystals. Science 314:964–967. https://doi.org/10.1126/science.1131475

Yi Y, Sanchez L, Gao Y, Yu Y (2016) Janus particles for biological imaging and sensing. Analyst 141:3526–3539. https://doi.org/10.1039/C6AN00325G

Yong SK, Bolan NS, Lombi E, Skinner W, Guibal E (2013) Sulfur-containing chitin and chitosan derivatives as trace metal adsorbents: a review. Crit Rev Environ Sci Technol 43:1741–1794. https://doi.org/10.1080/10643389.2012.671734

Yu W, Ma Y, Srivastava R, Shankar S (2016) Abstract 4065: mechanistic role of heavy metal cadmium exposure in the etiology of pancreatic cancer. Cancer Res 76:4065–4065. https://doi.org/10.1158/1538-7445.AM2016-4065

Zafar R, Zia KM, Tabasum S, Jabeen F, Noreen A, Zuber M (2016) Polysaccharide based bionanocomposites, properties and applications: a review. Int J Biol Macromol 92:1012–1024. https://doi.org/10.1016/j.ijbiomac.2016.07.102

Zahir F, Rizwi SJ, Haq SK, Khan RH (2005) Low dose mercury toxicity and human health. Environ Toxicol Pharmacol 20:351–360. https://doi.org/10.1016/j.etap.2005.03.007

Zhang Y, Ren Z, Fu Y, Yuan X, Zhai Y, Huang H, Zhai H (2009) An investigation on the behavior of fine-grained magnetite particles as a function of size and surface modification. J Phys Chem Solids 70:505–509. https://doi.org/10.1016/j.jpcs.2008.11.017

Zhang L, Zhong L, Yang S, Liu D, Wang Y, Wang S, Han X, Zhang X (2015a) Adsorption of Ni (II) ion on Ni(II) ion-imprinted magnetic chitosan/poly(vinyl alcohol) composite. Colloid Polym Sci 293:2497–2506. https://doi.org/10.1007/s00396-015-3626-4

Zhang Y, Zhu L, Zhou Y, Chen J (2015b) Accumulation and elimination of iron oxide nanomaterials in zebrafish (Danio rerio) upon chronic aqueous exposure. J Environ Sci 30:223–230. https://doi.org/10.1016/j.jes.2014.08.024

Zhang L, Wang X, Miao Y, Chen Z, Qiang P, Cui L, Jing H, Guo Y (2016a) Magnetic ferroferric oxide nanoparticles induce vascular endothelial cell dysfunction and inflammation by disturbing autophagy. J Hazard Mater 304:186–195. https://doi.org/10.1016/j.jhazmat.2015.10.041

Zhang Y, Lin X, Zhou Q, Luo X (2016b) Fluoride adsorption from aqueous solution by magnetic core-shell Fe_3O_4@alginate-La particles fabricated via electro-coextrusion. Appl Surf Sci 389:34–45. https://doi.org/10.1016/j.apsusc.2016.07.087

Zhu H, Han J, Xiao JQ, Jin Y (2008) Uptake, translocation, and accumulation of manufactured iron oxide nanoparticles by pumpkin plants. J Environ Monit 10:713–717. https://doi.org/10.1039/b805998e

Chapter 10
Remediation of Polycyclic Aromatic Hydrocarbons Using Nanomaterials

Manviri Rani and Uma Shanker

Contents

M. Rani · U. Shanker (✉)
Department of Chemistry, Dr. B. R. Ambedkar National Institute of Technology, Jalandhar, Punjab, India

G. Crini, E. Lichtfouse (eds.), *Green Adsorbents for Pollutant Removal*, Environmental Chemistry for a Sustainable World 18,
https://doi.org/10.1007/978-3-319-92111-2_10

Abstract Polycyclic aromatic hydrocarbons are major contaminants in environmental bodies due to ubiquitous occurrence, toxicity and potential to bioaccumulation. Increased population, rapid industrialization and extensive use of oil fuels are one of the major cause of pollution by polycyclic aromatic hydrocarbons. Here, we review the issues related to polycyclic aromatic hydrocarbons (PAHs) and their removal techniques using nanoparticles through adsorption, photocatalytic and redox degradation. Among the dye removal techniques, adsorption was found best in terms of its efficiency and economy. For that, traditional techniques such as microbial, photolysis and conventional adsorbents such as commercial activated carbon, agricultural and natural waste are highly employed. Lately, low cost nanomaterials with high surface-area come out as most economic, rapid and effective green adsorbent cum photocatalyst under UV and sun-light irradiation. Green synthesized nanomaterials with advanced characteristics of adsorbent and photocatalysts are gaining importance in degradation of various organic-pollutants due to low cost of production and mediated effect of biogenic sources. We also discuss the use of TiO_2, ZnO and metal hexacyanoferrate to remove polycyclic aromatic hydrocarbons pollution.

10.1 Introduction

Currently, our environment is facing lots of challenges due to several types of pollution which ultimately harming biota on Earth. Therefore, in order to have sustainable and better life, it is imperative to follow the rules of environmental management. In this direction, advanced treatment technologies like use of nanomaterials are being developed to eradicate such problems from the environment. Green adsorbent and photocatalysts are the low-cost materials with satisfactory adsorption properties and environmentally-friendly nature, which reduces the procedural cost and tackle out the problem of economic and environmental crisis.

One of the emerging and greatest problems worldwide is the water pollution caused by polycyclic aromatic hydrocarbons (PAHs). They are being discharged annually estimated to about 80,000 tons from the unwanted leakage/spilling of oil (Wright and Welbourn 2002; Guo et al. 2007). PAHs are the first recognized ubiquitously present environmental carcinogens and mutagens (IARC 1983) and exposed to humans via several routes like air, food, water and contact with soil and dust (Menzie et al. 1992). The pictorial representation of the biological transfer of pesticides to man has been depicted in Fig. 10.1. Their persistence and toxicity increases with molecular weight (Haritash and Kaushik 2016), mainly due to low water solubility (Cerniglia 1992).

The potential sources of PAHs are natural and anthropogenic including forest fires volcanic eruptions, and petroleum discharge (Kaushik and Haritash 2006). They also originated during thermal decomposition of organic molecules at high temperature like 500–800 °C and as well as low 100–300 °C temperature (Patnaik

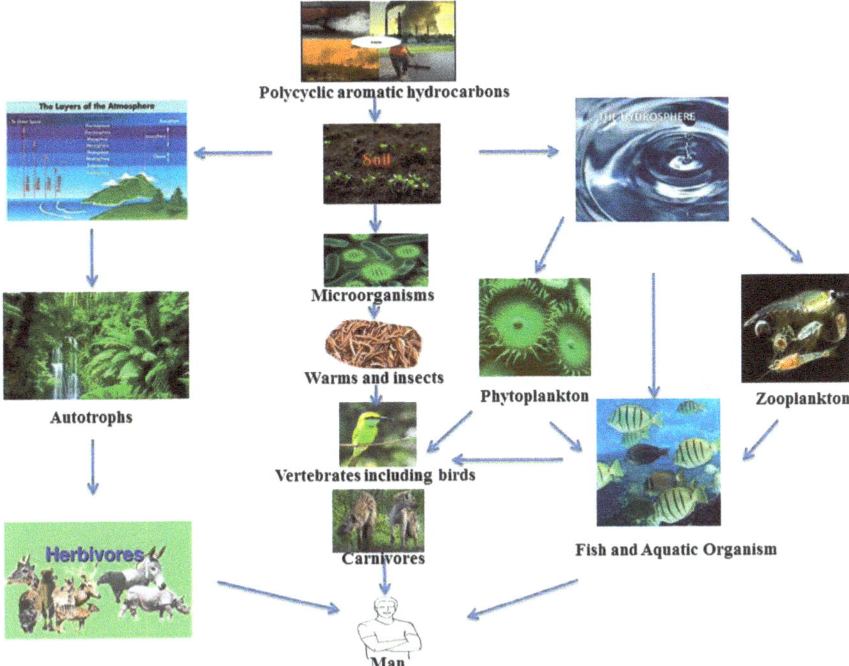

Fig. 10.1 Biological transfer of polycyclic aromatic hydrocarbons to man via food chain and food web. Polycyclic aromatic hydrocarbons reach to atmosphere, soil and water bodies through release from natural and anthropogenic sources such as forest fires, volcanic eruptions, industries and petroleum discharge. The bioaccumulation of carcinogenic PAHs has an environmental concern Herbivores: cow, buffalo, pig and goat; Carnivores: dogs, wildcats and snake; Autotrophs: plants that synthesize food in presence of sunlight; Phytoplankton: Small plants such as dinoflagellates and diatoms at the surface of water bodies. They provide food for a wide range of sea creatures including whales, shrimp, snails, and jellyfish; Zooplankton: small organism sizes including small protozoans such as Amoeba and Paramecium

1999). In urbanized areas, motor vehicle emission contributes to 46–90% of the mass of individual PAHs (Dubowsky et al. 1999; Tonne et al. 2004). PAHs level due to indoor emission accounts to about 16% in United States, 29% in Sweden and 33% in Poland (Fromme et al. 1998; Maliszewska-Kordybach 1999; Li and Ro 2000; Lung et al. 2003). People spend 80–93% of their time indoor and exposed to PAH indoor sources via inhalation (Brunekreef et al. 2005).

Over the last few decades, the levels of PAHs have been increased tremendously and detected in all environmental segments such as air, soil, sediments, water, oil, tars and foodstuffs (Freeman and Cattell 1990; Lijinsky 1991; Latimer and Zheng 2003; Baklanov et al. 2007). United States Environmental Protection Agency (US-EPA) defined 17 unsubstituted PAHs (Fig. 10.2a) as priority pollutants owe to their wide contamination of at least 600 sites out of 1408, resistance towards biodegradation and potential of bioaccumulation (Agency 1995).

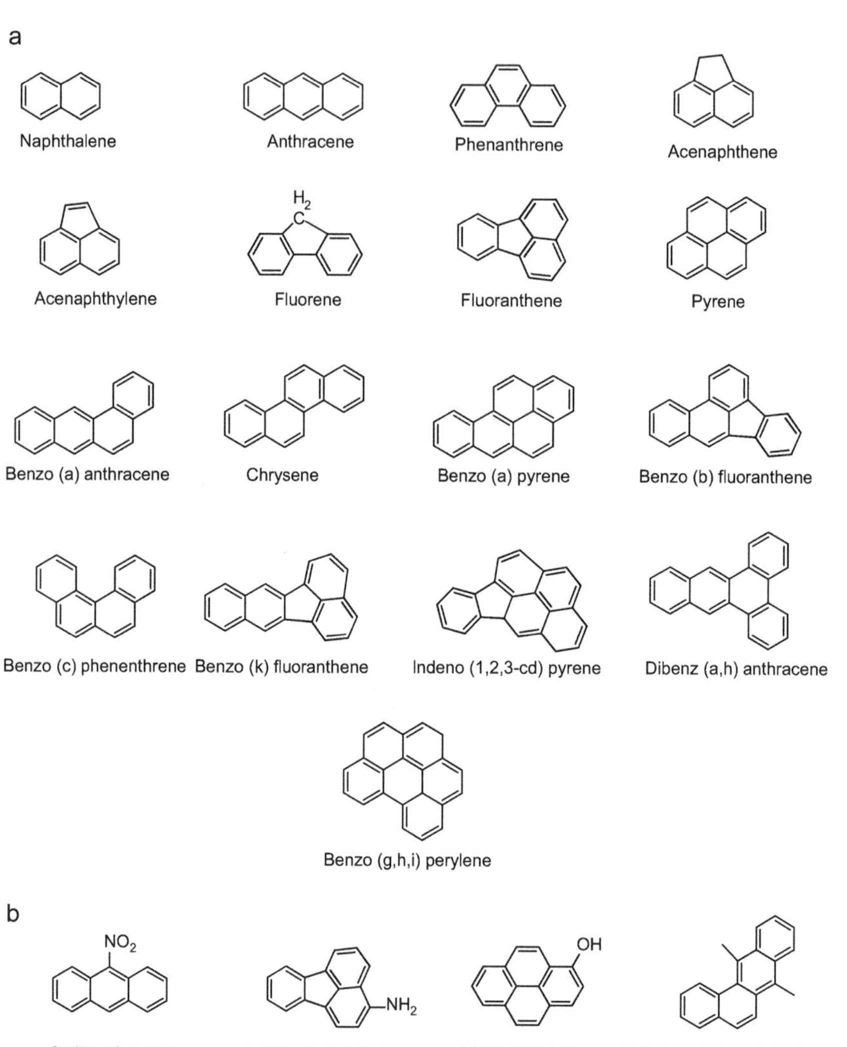

Fig. 10.2 (**a**) Chemical structures of the various polycyclic aromatic hydrocarbons present in environment. (**b**) Chemical structures of some substituted polycyclic aromatic hydrocarbons

PAHs can be oxidized in presence of UV irradiation and air contaminants such as OH• radicals, ozone or nitrogen oxides. Consequently, their substituted derivatives such as hydroxylated, nitrated, methylated and aminated PAHs are formed (Fig. 10.2b). Recent studies reveal that some of these substituted PAHs may be even more toxic than the parent PAHs (Fu et al. 2012). Like other persistent organic pollutants, they may also transformed into more toxic byproducts (Gupta et al. 2011, 2012a, b, c; Rani 2012; Rani et al. 2017a; Rani and Shanker 2017a, b, Shanker et al. 2017a).

Due to toxicity and prevalence of PAHs and their metabolites, there is an urgent need to detect and develop low-cost, easy-to-handle, green and effective treatment techniques to remove those perilous contaminants. A review on PAHs source, environmental impact, effect on human health and remediation indicated that PAHs may undergo adsorption, leaching, volatilization, photolysis, and chemical degradation, however, microbial or biodegradation is the major eco-friendly alternative degradation process (Abdel-Shafy and Mohamed-Mansour 2016). Moreover, high production cost of commercially available adsorbents (activated carbon and chitosan) led to development and commercialization of efficient and economical alternative for environmental remediation (Jassal et al. 2015a, b, 2016a, b, c; Shanker et al. 2016a; Rani et al. 2017b; Rani and Shanker 2017b). Ukiwe et al. (2013) reviewed briefly the several efficient and cost effective remediation techniques like chemical, bio, phyto, combined, photo-and chemical oxidation for PAHs. Biodegradation faced the limitations of slow pathways, unclear biochemical mechanism and formation of some toxic intermediates and consequent lesser use on industrial scale (Wammer and Peters 2005; Haritash and Kaushik 2009). Catalytic reduction chemically converts the PAHs into less toxic compounds (Castaño et al. 2006; Nelkenbaum et al. 2007). However, such methods are not economical and often require elevated temperatures and pressures. Hence, in view of these shortcomings, present chapter deals exclusively with the remediation of emerging contaminants PAHs by different nanomaterials itself and doped or coated that operated through adsorption, and photocatalytic mechanism. The possible degradation pathways of several PAHs and their by-products have been widely discussed and evaluated. In addition, role of green synthesized nanoparticles in degradation of PAHs and other various organic pollutants have been highlighted. Physiochemical properties of PAHs are presented in Table 10.1.

10.2 Major Polycyclic Aromatic Hydrocarbons

EPA has listed some PAHs as priority pollutants based on their toxicity and increasing level in the environment. The concentrations of these PAHs in the ambient air are even dangerous at few ng/m^3 that calls for their immediate removal from the environment (Ramadahl et al. 1982).

10.2.1 Naphthalene

Naphthalene has been identified as carcinogenic, mutagenic, and teratogenic substance (Ge et al. 2015). In the water-soluble fraction of petroleum, this bicyclic aromatic compound along with its methylated derivatives is considered to be one of the most noxious compounds (Heitkamp et al. 1987). It is highly persistent in nature and has low water solubility (Ncibi et al. 2007).

Table 10.1 Physiochemical properties of various polycyclic aromatic hydrocarbons indicated that the compounds with increasing ring or molecular weight have more persistence and are much toxic

S. No.	Name	Ring	Mol. wt.	Cas No.	M. Pt (°C)	B. Pt. (°C)	Vapor Pressure (kPa)	Solubility (mg L^{-1})	Half-life ($t_{1/2}$)[a] (d)	Env. concentration	
										Soil (ng/g)	Crude oil (ng/m^3)
1.	Naphthalene	2	123	91-20-3	80.2	218	1.1 × 10–2	30	3–8 h	273	30,000
2.	Acenaphthylene	3	152	208-96-8	90–93	265–280	3.9 × 10–3	Insoluble	12	160	–
3.	Acenaphthene	3	154	83-32-9	90–96	278–279	2.1 × 10–3	0.04	<24 h	265	–
4.	Fluorene	3	166	86-73-7	116–118	293–295	8.7 × 10–5	1.992	2–64	238	–
5.	Anthracene	3	178	120-12-7	215–219	340	3.6 × 10–5	2.0 × 10–4	17 h	230	–
6.	Phenanthrene	3	178	85-01-8	96–101	339–340	2.3 × 10–5	6.8 × 10–4	47 h	392	–
7.	Fluoranthene	4	202	206-44-0	107–111	375–393	6.5 × 10–7	6.0 × 10–6	5	503	–
8.	Pyrene	4	202	129-00-0	150–156	360–404	3.1 × 10–6	6.8 × 10–7	40 h	525	–
9.	Benz (a) anthracene	4	228	56-55-3	157–167	435	1.5 × 10–8	5.0 × 10–9	199–252	235	6400
10.	Chrysene	4	229		252–256	441–448	5.7 × 10–10	6.3 × 10–7	371–381	245	6400
11.	Benzo (b) fluoranthene	5	252	205-99-2	167–168	481	6.7 × 10–8	–	294	224	6400
12.	Benzo (k) fluoranthene	5	252	207-08-9	198–217	471–480	2.1 × 10–8	–	2 y	197	6400
13.	Benzo (a) pyrene	5	252	50-32-8	177–179	493–496	7.3 × 10–10	5.0 × 10–7	5–10 y	273	640
14.	Benzo (g,h,i) perylene	6	276	191-24-2	275–278	525	1.3 × 10–11	1.0 × 10–10	600–650	176	–
15.	Indeno (1,2,3-cd) pyrene	6	276	193-39-5	162–163	NA	ca × 10–11	1.0 × 10–10	139	193	6400
16.	Dibenzo (a,h) anthracene	5	278	53-70-3	266–270	524	1.3 × 10–11	1.0 × 10–10	18–21	95	580
17.	Coronene	7	300	191-07-1	438–440	525	~1.0 × 10–4	4.0 × 10–4	16 y	–	–

[a]y = years; w = week; d = days; h = hour; s = seconds

10.2.2 Anthracene

Anthracene is a toxic polycyclic compound used in dyes, insecticides, wood preservatives and coal tar (Juhasz et al. 2000) and commonly found at the sites of gas factory pollution. It is sparingly soluble in water, exhibits high potential to bio-accumulate, highly recalcitrant to nucleophilic attack and hence, resistant to biodegradation (Guieysse et al. 2001). It may cause irritation to eyes and is a possible inducer of tumors (Das et al. 2008).

10.2.3 Phenanthrene

Phenanthrene is commonly detected, three fused benzene rings compound highly resistant to photodegradation (Wang et al. 1995; Wen et al. 2002; Jia et al. 2012) and gets photo-oxidized by absorbing light in the UV regions (Kou et al. 2008). It is known to be an inducer of sister chromatid exchanger, a human skin photosensitizer, and a potent inhibitor of gap-junctional intercellular communication (Samanta et al. 2002).

10.2.4 Fluorene

Fluorene is a tricyclic compound with two benzene rings fused to a cyclopentane ring. This xenobiotic along with its derivatives is a major environmental concern associated with petroleum and oil spills, industrial discharge and waste incineration (Lu and Zhu 2007). In addition, they are also being used as thermo and light sensitizers (Yuanfu et al. 2007).

10.2.5 Pyrene

Because of hydrophobic, recalcitrant, and bioaccumulating characteristics, pyrene is strongly adsorbed on suspended particulate matters and biota, thus, leading to its accumulation in soil and sediment. It is listed as a priority pollutant by the US EPA (Agency 1995). Pyrene is an important mutagen ubiquitously present in contaminated soils (Schneider et al. 1996).

10.2.6 Chrysene

Chrysene is a colorless solid also known as benzo (*a*) phenanthrene mostly found as the by-product from the incomplete combustion of organic matter. Due to its mutagenic and carcinogenic nature it has gained high environmental concerns worldwide. Chrysene can be obtained from the catalytic cracking in the high boiling fraction during petroleum refining or by the pyrolysis of fats and oils (Harvey 1997).

10.2.7 Benzo (a) Pyrene

Benzo (a) pyrene is a five-ring compound and classified by the US Environmental Protection Agency (USEPA) as a priority pollutant due to its high carcinogenicity, genotoxiciy (in wide range of prokaryotic and mammalian cell assays) and acute toxicity.

10.2.8 Acenaphthene

Acenaphthene consists of naphthalene with an ethylene bridge connecting positions 1 and 8. Nearly 0.3% of acenaphthene is present in coal tar (Connelly and Geiger 1996). It is used on a large scale to prepare naphthalene dicarboxylic anhydride, a precursor used in making dyes and optical brighteners (Griesbaum et al. 2002). This PAH possesses high recalcitrance in the environment and because of its long half-life in the soil of 270 days to 5.2 years, it is highly toxic (Juhasz and Naidu 2000).

10.2.9 Acenaphthylene

Acenaphthylene is a low molecular weight and bicyclic EPA priority pollutant (Irwin 1997a). It is a component of crude oil, coal tar and a product of combustion which may be produced and released to the environment during natural fires. Emissions from petroleum refining and coal tar distillation are major contributors of acenaphthylene to the environment. A calculated Koc range of 950 to 3315 indicates acenaphthylene will have a low to slight mobility in soil (Irwin 1997a).

10.2.10 Fluoranthene

Fluoranthene is a high molecular weight and 4-ring PAH. It has high recalcitrance in the environment and because of its half-life in the soil of 4–5 days, it is toxic in nature. Fluoranthene when released into air and water through combustion, gets rapidly adsorbed to sediment and particulate matter in the water column, and bioconcentrate into aquatic organisms (Irwin 1997b).

10.2.11 Benzo (a) *Anthracene*

Benzo (a) anthracene is a tetracyclic PAH found to be genotoxic and carcinogenic in animal studies with development of tumours in the liver, skin and lungs of young mice after ingestion, dermal application and subcutaneous and intraperitoneal injection (The MAK-Collection 2012). Of all estimated environmental releases of BAA, 94% are to air (Irwin 1997c). In addition to this, BAA has been found at 62 out of 1177 sites on the National Priorities List (NPL) of hazardous waste sites in the United States (Irwin 1997c).

10.2.12 Benzo (b) *Fluoranthene*

Benzo (b) fluornthene is a 5-ring compound present in coal, coke oven emissions and petroleum products. BBF is a probable carcinogen in humans as well as in other animals and may even cause reproductive damage in humans (Hazardous substances fact sheet 2001). Its photolysis and photo-oxidation is expected to occur but the adsorbed BFF resists these processes and remains as such (Irwin, 1997d).

10.2.13 Benzo (k) *Fluoranthene*

Benzo (k) fluoranthene is a 5-ring polycyclic compound with ubiquitous occurrence of half-life of 2 years in products of incomplete combustion (IARC 1983) and in soils, groundwater, and surface waters at hazardous waste sites (ATSDR 1990). BKFs presence in distant places indicates that it is reasonably stable in the atmosphere and capable of long-distant transport (Irwin 1997e).

10.2.14 Benzo (c) Phenanthrene

Benzo (*c*) phenanthrene is an organic compound having four benzene rings fused together. This compound is of mainly theoretical interest but it is also environmentally occurring (Imma et al. 2004). X-ray analysis shows that it is not planar and, therefore, has a dipole moment. The concentration of BCP in urban air is reported as 9.8 μg/1.500 m^3 (Wagoner 1976). BCP and eight of its alkyl derivatives are strongly carcinogenic in nature and may even produce tumors in mice (Christensen and Fairchild 1976).

10.2.15 Indeno (1,2,3-cd) Pyrene

Indeno (1,2,3-cd) pyrene is a five ring polycyclic compound. This compound often occurs together with other PAHs, and is carcinogenic. IP is released to the soil and will get adsorbed strongly with estimated $K_{oc} = 20{,}146$ and, hence, is not expected to leach. The primary route of human exposure to IP will probably be through ingestion of contaminated food. Other exposure to IP may be from contaminated drinking water and breathing air (Irwin 1997f).

10.2.16 Dibenz (a,h) Anthracene

Dibenz (*a,h*) anthracene, being heavier than the other PAHs, it is more persistent in the environment. Based on an estimated BCF of 51,000, DBA is expected to bioconcentrate in aquatic organisms (Irwin 1997g). Its presence in remote areas demonstrates the potential of long range transport as well as considerable stability in the air (Irwin 1997g).

10.2.17 Benzo (g,h,i) Perylene

Benzo (g,h,i) perylene is six-ring and high molecular weight toxic PAH. Emissions from petroleum refining, coal tar distillation, and the combustion of wood, coal, oil, propane, gasoline and diesel fuels are major contributors of BP to the environment. The reported biodegradation half-lives and Koc value for BP in aerobic soil ranges from 600 to 650 days and 9×10^4 to 4×10^5, respectively (Irwin 1997h).

10.3 Environmental Concern of Polycyclic Aromatic Hydrocarbons

These are the most widespread natural and anthropogenic organic contaminants in soils and water (Puglisi et al. 2007). Due to low solubility in water, they have the ability of getting adsorbed on the surface of solid particulate matters, especially soil (Tremblay et al. 2005). Due to their toxic, mutagenic effects on the living organisms, these compounds have been identified as priority pollutants by different State and Central Pollution Control Boards. Diesel emissions are reported to contain excessive amount of PAHs, thus causing damage to the ecosphere (Rybak and Olejniczak 2014). Phenanthrene due to its photo sensitizing and mild allergic nature has been reported to affect human skin (Fawell and Hunt 1988). In water PAHs accumulate in fish, bivalves (Kasiotis and Emmanouil 2015) and other marine organisms and in soil (Villanneau et al. 2013) they get easily adsorbed on the sediments rich in organic content. The highest levels of PAHs i.e., 104 ± 78 ng g^{-1}d.w were detected in river sediment downstream from an industrial discharge in south central Chile (Barra et al. 2009). From these, PAHs gets transferred to humans via food consumption (Meador et al. 1995). Due to ship breaking activities, the sediment of the coastal region of Alang-Sosiya ship breaking yard, Gujarat, India is polluted with enormous amount of PAHs (Reddy et al. 2003, 2005). Their accidental use, spillage or misuse damaged practically all the segments of the environment.

Oil-spilling is one of the major reasons contributing to PAH pollution and death of several aquatic species. Some important PAHs related accidents due to oil spills have been listed in Table 10.2.

The Deepwater Horizon drilling rig explosion on April 20, 2010 in the Macondo field killed 11 life's and caused largest oil spill in the US history. As a consequence, several aquatic species died due to the toxic PAHs present in the oil spill (Suzanne and Terry 2012). Due to extensive petroleum trade, the Cotonou coastal zones of

Table 10.2 Some important polycyclic aromatic hydrocarbons related accidents

Oil spill	Place	Year	Causes
PAHs (Lakeview Gusher)	California, United States	1910–1911	1200 tonnes of crude oil released
PAHs (Kuwaiti oil lakes)	Kuwait	1991	3409–6818 tonnes of crude oil released
PAHs (Kuwaiti oil fires)	Kuwait	1991	136,000 tonnes of crude oil released
PAHs (MT *Hebei Spirit* oil spill)	South Korea	2007	10,800 tonnes of crude oil released
PAHs (*Deep Water Horizon* oil spill)	United States	2010	11 deaths, instant killing of aquatic species, accumulation
PAHs (Sundarbans oil spill)	Bangladesh	2014	Captain of the ship died, 1,000,000,000 bangladeshi taka (currency) was lost.
PAHs (Ennore oil spill)	Chennai	2017	More than 70 tonnes of crude oil released

Benin are contaminated by large amount of 25 ± 1450 ng g^{-1} of petrogenic PAHs, whereas the Aquitaine sediment of France are contaminated by pyrolytic PAHs of 4 ± 855 ng g^{-1} due to waste oils from mechanics shops (Soclo et al. 2000). From the above discussion it is clear that toxic and highly persistent PAHs have their residues in environment. This needs initiative to promote use of various nanoparticles for complete degradation as such methods are efficient, cheap and environmental friendly.

Several studies reported that the uptake of PAHs through diet (foodstuffs, vegetables and crops) is much higher than the intake through inhalation of air (Menzie et al. 1992). The excretion is mainly by feces, and sometimes urine depending upon the animal (Baird et al. 2005; Xue and Warshawski 2005; Yusuf et al. 2007). Since PAHs are ubiquitously present in the environment (WHO 1998), they have also been detected in some body tissues and eggs as well. Oiled birds transfer oil from their feathers to the eggshell while incubating and also ingest oil when preening. The PAHs present in eggs are not only because of maternal transfer but also due to direct application to the egg shell by feet of oiled birds (Altenburger et al. 2003). Large amounts PAHs have been detected in avian species, such as herring gulls such as Larus argentatus from Ontario, Canada, contained various PAHs of lipid like, anthracene 0.15 μg/kg, pyrene 0.076 μg/kg, fluoranthene 0.082 μg/kg, fluorene 0.044 μg/kg, naphthalene 0.05 μg/kg, benzo[a]pyrene 0.038 μg/kg and acenaphthene 0.038 μg/kg (Hallett and Brecher 1984). Since PAHs are lipophilic, they have more chances to be present in eggs (Naf et al. 1992; Stronkhorst et al. 1993). Thus, eggs are potentially good biomonitors of PAH contamination in birds.

They are discharged into natural waters through both point as well as non-point sources (Manoli and Samara 1999). Since PAH exist abundantly in surface water as well as in sediment (Guo et al. 2007), therefore, they are vital pollution source having potential harmful effect on aquatic organisms. PAHs are also reported to have toxic effects on the zooplankton community by modifying their structure and function, thus affecting the lake food web (Hanazato 2001).

Oil spilling is popular source of toxicity to aquatic ecosystem including animals. Oil contains all carcinogenic PAHs that lead to a number of deaths immediately after spilling, e.g. after the Deep Horizon Oil Spill, 11 people went missing and were never found. Also, thousands of marine species died on the spot (Campbell and Clifford 2010; Rick and Alan 2010). This calls for the immediate need for the development of effective PAHs removal techniques after an oil spill. Several methods are available for the degradation of crude oil. They consume exposure to lamps or direct photooxidation under sunlight. Since such approaches are not feasible everywhere, therefore, there is an immediate need for developing efficient techniques for overcoming the serious issue of oil spill. As discussed above, nanomaterials have been successfully employed for the degradation of PAHs. They are cheap, easily synthesized (Jassal et al. 2015a, 2016a, b, c), eco-friendly and highly efficient catalysts in removing PAHs. Moreover, the amount of nanomaterials required for the degradation studies is very small and can be recycled to many runs (few mg), therefore, no toxicity is contributed to the environment by these adsorbents.

10.4 Need to Degrade Polycyclic Aromatic Hydrocarbons from Water

PAHs are commonly detected carcinogen, mutagens, and potent immune-suppressants with high persistence in air, soil and water and lesser degradation i.e., high stability towards heat, light, oxidizing agents and microbial attack (Latimer and Zheng 2003; Baklanov et al. 2007). Toxicity impact of PAHs depend upon their release rate from several sources. Cooking accounts for almost 32.8% of total indoor PAHs emissions (Zhu et al. 2009). The total emission factors for 15 PAHs from coal stoves in North China ranged from 53 to 1435 mg/kg. Oanh et al. (1999) reported that burning of wood released the highest amount of 17 PAHs. Chen et al. (2007) studied that from a scrap tyre plant, 42 g/day of PAHs were released via pyrolysis with an emission factors of 4 mg/kg. Thermal decompositions of organic materials contributed to emission factors in the range of 0.4 ± 0.13 mg/g and 9.0 ± 0.5 mg/g for cellulose and tyre, respectively. Emission factors from industrial stacks ranged from 0.1 to 4 mg/kg feedstock (Fabbri and Vassura 2006). Combustion of heavy oils led to the highest emission factors of total PAHs (Yang et al. 1998). Abrantes et al. (2009) observed that light motor vehicles using ethanol fuel released lesser total emission of toxic PAHs of 11.7 to 27.4 μg/km than those using gasohol of 41.9 to 612 μg/km. Low molecular weight polycyclic aromatic hydrocarbons are mainly emitted from helicopters, which accounts for nearly 97.5% of the total polycyclic aromatic hydrocarbons emissions (Chen et al. 2006). Biomass burning emitted 80–90% of low molecular weight polycyclic aromatic hydrocarbons such as naphthalene, flouranthene, acenaphthylene, phenanthrene and pyrene. On a global basis, 56.7% of total 17 priority PAHs are released from the combustion of biofuel (Zhang and Tao 2009).

Increasing rates of release of polycyclic aromatic hydrocarbons in the environment have led to the deposition of large amount of these thermally stable carcinogens. Sometimes, their metabolites are more toxic than parent compounds such as oxons. Owing to these adverse effects of extensive use of PAHs as well as their toxic metabolites formed in environmental bodies, their degradation is highly imperative. To overcome this serious problem several methods such as biological treatment, oxidation have been reported for the removal of PAHs. The recent advancement in nanotechnology led to an extensive research using nanoparticles which resulted in sharp increase in the removal of PAHs from environment. They act as excellent adsorbents, so can be effectively used in solving this problem. No doubt the use of nanomaterials for degrading PAHs is increasing nowadays, but it still needs further exploration. Therefore, it was considered worthwhile to write this review in order to summarize and address the advancements in this field of removal of PAHs using nanoparticles. One of the main driving forces in the writing of this chapter is to organize all the scattered information in this particular field.

10.5 Green Adsorption and Photocatalysis

PAHs can be removed by various low-cost conventional adsorbents like activated carbon, plants, biodegradation, advanced oxidation processes such as solar ultraviolet radiation, direct photolysis, heterogeneous photocatalysis, ultrasound frequency, and combined degradation techniques. However, sonochemical degradation, photocatalytic degradation as well as current density enhanced degradation is novel approaches that have been tested successfully for the compound's removal. Some of these conventional clean-up methods are not only environmental friendly; they also present a novel approach in reducing the ability of PAHs to cause prospective risk to humans and the ecosystem. The adsorption technique has become more popular in recent years for wastewater treatment owing to its efficiency in the removal of pollutants too stable for biological methods. Adsorption is influenced by many factors such as adsorbate or adsorbent interaction, adsorbent's surface area, particle size, temperature, pH and contact time. The main advantage of adsorption recently became the use of low-cost materials, which reduces the procedure cost i.e., "green adsorption". The selection for the most appropriate adsorbent would be based on some major characteristic properties such as: (i) the low-cost along with the satisfactory adsorption properties like capacity, reuse, industrial-scale use and (ii) the environmentally-friendly nature of each adsorbent from abundant natural sources, biodegradable, non-toxic. Green adsorbents are either the agricultural waste or materials which can work in very small amount at room temperature or sunlight and hence reduce the cost of process. Having unique characteristics, nanomaterials or their modified forms like doped or coated degradaed the PAHs and other pollutants via adsorption, and photocatalytic mechanism. Green synthesized nanoparticles acted same as other nanomaterials but showed enhanced efficiency due to introduction of phytoconstituents in crystal lattice.

10.6 Polycyclic Aromatic Hydrocarbons Degradation Techniques

10.6.1 Conventional Techniques

Chemical or oxidative photochemical degradation is chiefly initiated by oxidants like peroxides, H_2O_2, hydroxyl radicals and ozone. They convert PAHs into reasonably complex mixtures of other highly resistant PAHs because loss of aromatic character requires large amount of energy (Mallakin et al. 2000). The dead-end product thus formed gets build-up in the environment and become a cause for further contamination (Freeman and Harris 1995). Fenton's reagent (Nam et al. 2009), ozone (Legrini et al. 1993) and electron donors such as acetate, lactate and pyruvate (Bach et al. 2005) are also promising chemicals used to degrade PAHs. These methods have limitations of synthesis of reagents, low aqueous solubility and vapor pressure of PAHs, production of numerous intermediates and mixture of stable organic compounds (Prak and Pritchard 2002). Thus, such processes cannot be employed at large scale. Biodegradation by microorganisms have considerable potential in treatment of oil-contaminated sediments and removal of chrysene from aqueous solution (Ramsay et al. 2000; Tam et al. 2002). High-molecular-weight PAHs like BaP and chrysene are difficult to be biodegraded whereas low-molecular-weight ones such as phenanthrene and naphthalene can be efficiently degraded (Yamada et al. 2003). The details of the microorganism used for degradation of PAHs have been summarized in Table 10.3. Degradation of naphthalene by aerobic bacterium *Bacillus thermoleovorans* (Annweiler et al. 2000), fluorene by *Sphingomonas* sp. LB126 (Wattiau et al. 2001), chrysene by *Polyporus* sp. S133 (Hadibarata et al. 2009) and BaP by *Mycobacterium* sp. of strain RJGII-135 (Schneider et al. 1996) have been shown in Figs. 10.3, 10.4, 10.5, and 10.6, respectively.

Moreover, microbes are expensive and difficult to isolate. Combined degradation are more efficient and cost effective as no end products are left behind e.g., chemical pre-oxidation and bioremediation, anaerobic digestion and ozonation, biodegradation and Fenton's reagent, biological, chemical and electrochemical treatment, as well as Fenton reagent versus ozonation (Nam et al. 2001; Goi and Trapido 2004; Bernal-Martínez et al. 2005; Zheng et al. 2007). Main disadvantage of this method is the longer time required from several hours to days for degrading PAHs. This demands for the development of advanced techniques which can completely remove whole of the PAH in a very short span of time.

10.6.2 Nanomaterials Mediated Degradation

The unique properties of nanomaterials offer great advantages in removal of hazardous contaminants over the other adsorbents like activated carbon (Sharma et al. 2009; Keng et al. 2014; Baruah et al. 2016). The use of nanofiltration membranes, photocatalytic nanomaterials, adsorption nanomaterials and reducing nanomaterials is increasing day by day. Carbon nanotubes, magnetic nanoparticles, noble metal nanomaterials and quantum dots are being highly used for the detection of trace pollutants and pathogens (Xue et al. 2017). Liu (2012) reviewed the importance of magnetic semiconductor nano-photocatalysts for the degradation of organic pollutants in order to take advantage of the high activity and enable the semiconductor nano-photocatalysts to be reused.

The details of nanomaterial mediated degradation are presented in Table 10.4.

The time consuming and expensive chemical/microorganisms assisted degradation is now getting replaced with heterogeneous photocatalysis that involving the generation of highly reactive free radicals species from metal based nanomaterials. These active radicals convert organic contaminants or PAHs into mineral salts and

Table 10.3 Polycyclic aromatic hydrocarbons degraded using microorganisms

S. No.	PAH	Microorganisms	Remarks	References
1.	BaP	*P. chrysosporium*	3–15% mineralization within 24 h to 30 d	Sanglard et al. (1986)
2.	BaP	*Pleurotus* sp. Florida	Mineralization over a 15-w; solid-state fermentation:	Wolter et al. (1997)
3.	Chrysene	*Polyporus* sp. S133	86% degradation in crude oil contaminated soil.	Hadibarata and Tachibana (2009)
4.	Chrysene	*Pseudoxanthomonas* sp. PNK-04	A catabolic degradation into three novel metabolites, hydroxyphenanthroic acid, 1-hydroxy-2-naphthoic acid and salicylic acid; microorganism isolated from coal	Nayak et al. (2011)
5.	Chrysene	*Polyporus* sp. S133	Degradation rate: agitation (65%) > non-agitated culture (24%) in 30 d	Hadibarata et al. (2009)
6.	Pyrene	*Rhodococcus*	Mineralization (72%) within 2 weeks	Walter et al. (1991)
7	Pyrene	Mycobacterium sp. strain KR2	Metabolised (60%) into protocatechuic acid within 8 d at 20 °C	Rehmann et al. (1998)
8.	Fluorene	Fungal stains	More than 90% removal in 288 h	Garon et al. (2004)
9.	Fluorene	*Agrocybe* sp. CU-43	99% degradation into less toxic intermediates within 6 d	Chupungars et al. (2009)
10.	Anthracene (300 mg L^{-1})	*Bacillus licheniformis*	>95% degradation into non-toxic by-products within 22 d	Swaathy et al. (2014)
11.	Anthracene	Brachybacterium paraconglomeratum	70.32% degradation within 10 d	Lily et al. (2013)
12.	Anthracene	*Bacillus badius*	Successful degradation in 60 h	Ahmed et al. (2012)
13.	Naphthalene	*Pseudomonas gessardii* strain LZ-E	77% of degradation in 48 h from contaminated discharge waste-water of petrochemical corporations.	Huang et al. (2016)
14.	Naphthalene	Bacterial consortium (DV-AL)	Complete degradation of a mix-ture of pollutants within 24 h	Patel et al. (2012a)
15.	Naphthalene (2000 mg L^{-1})	*Pseudomonas* sp. HOB1	Complete degradation in 24 h under alkaline pH at 35–37 °C.	Pathak et al. (2009)
16.	Phenanthrene	Bacterial strains	95% of degradation into phthalic acid in 120 h under alkaline pH at 37 °C	Patel et al. (2012b)

(continued)

Table 10.3 (continued)

S. No.	PAH	Microorganisms	Remarks	References
17.	Benzo (a) anthracene	Mycobacterium sp. strain RJGII-135	Successful degradation into two dihydrodiols, namely, 5,6-BAA-dihydrodiol and 10,11-BAA-dihydrodiol.	Schneider et al. (1996)
18.	Benzo (a) anthracene	*Burkholderia cepacia* strains VUN 10 003	Catabolized around 9.5% within 7–10 days.	Juhasz et al. (1997)

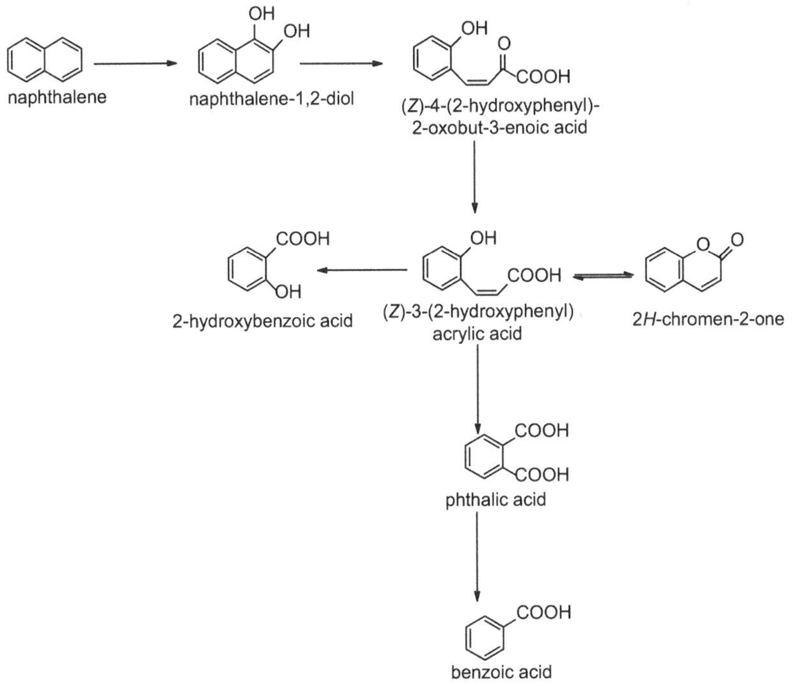

Fig. 10.3 Proposed degradation pathway of naphthalene by aerobic bacterium Bacillus thermoleovorans (Annweiler et al. 2000)

nontoxic byproducts (Anipsitakis and Dionysiou 2003; Bandala et al. 2002, 2006). Free radicals works by hydrogen abstraction or by electrophilic addition to double bonds and generate new $R^{•}$ free radicals and peroxyradicals that lead to mineralization of the organics via oxidative chain reactions as shown: RH + $HO^{•}$ or $SO_4^{•-}$ to $HR^{•}$ + H_2O (Zhang and Noshaka, 2014). Among, TiO_2 and ZnO are extensively studied for removal of PAHs and other organic pollutants. TiO_2 is inexpensive, abundant, photostable, non-toxic, highly reactive and easily form highly active •OH on UV or visible irradiation due to low band gap energy (Oncescu et al., 2010).

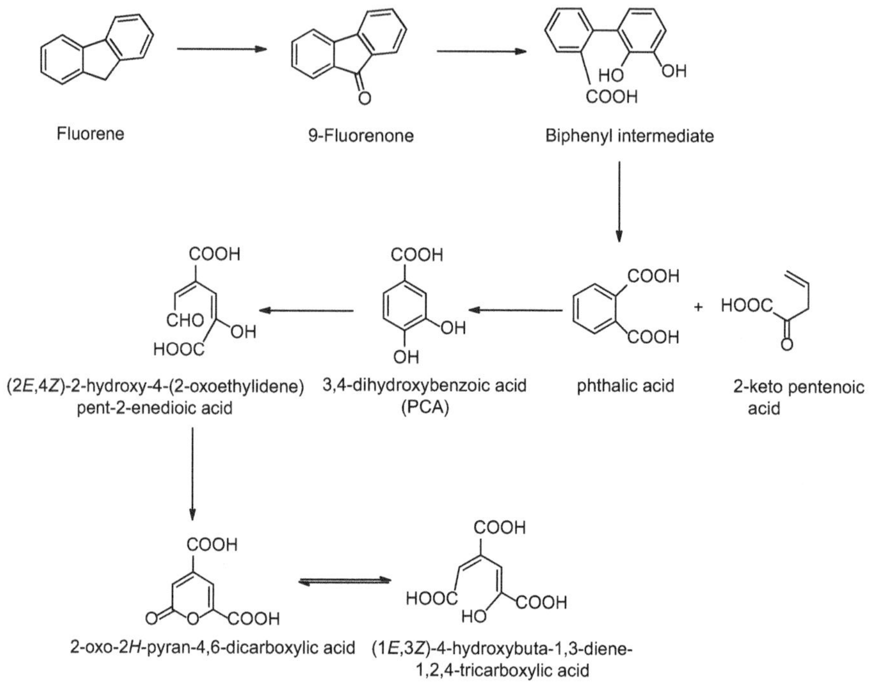

Fig. 10.4 Microbial degradation of fluorene by *Sphingomonas* sp. LB126 into small metabolites like phthalic acid and 3,4-dihydroxybenzoic acid. (With permission through rightslink Wattiau et al. 2001)

Using TiO_2, anthracene was degraded under UV light into non-toxic 9, 10-anthraquinone (Karam et al. 2014). Efficiency of the TiO_2 photocatalyst was further improved by extending its absorption spectra to the visible region. Coupling of TiO_2 with a narrow band gap semiconductor such as bismuth oxide enhanced charge separation (Ayekoe et al. 2016). Anatase form of TiO_2 reduced soil half-life of phenanthrene from 45.90 to 31.36 h under UV-light within 25 h (Gu et al. 2012). La and N co-doped TiO_2 nanoparticles supported on activated carbon could degrade 93.5% of naphthalene within short period of 120 min (Liu et al. 2016). By doping nitrogen with more electron rich medium, the chemical statuse of TiO_2 changed into O–Ti–N, and, hence, binding energy of Ti can be considerably reduced (Chen and Burda, 2008). The probable fact behind that is less-electronegativity of nitrogen than oxygen (Sathish et al. 2007).

Xia et al. (2015) fabricated Ti/ZnO–Cr_2O_3 composite for degrading 90.2% of naphthalene within 240 min via photo reaction mechanism. Aqueous solution of TiO_2 containing nonionic surfactant micelles was utilized for degrading phenanthrene under UV light and alkaline conditions (Zhang et al. 2011) (Fig. 10.7).

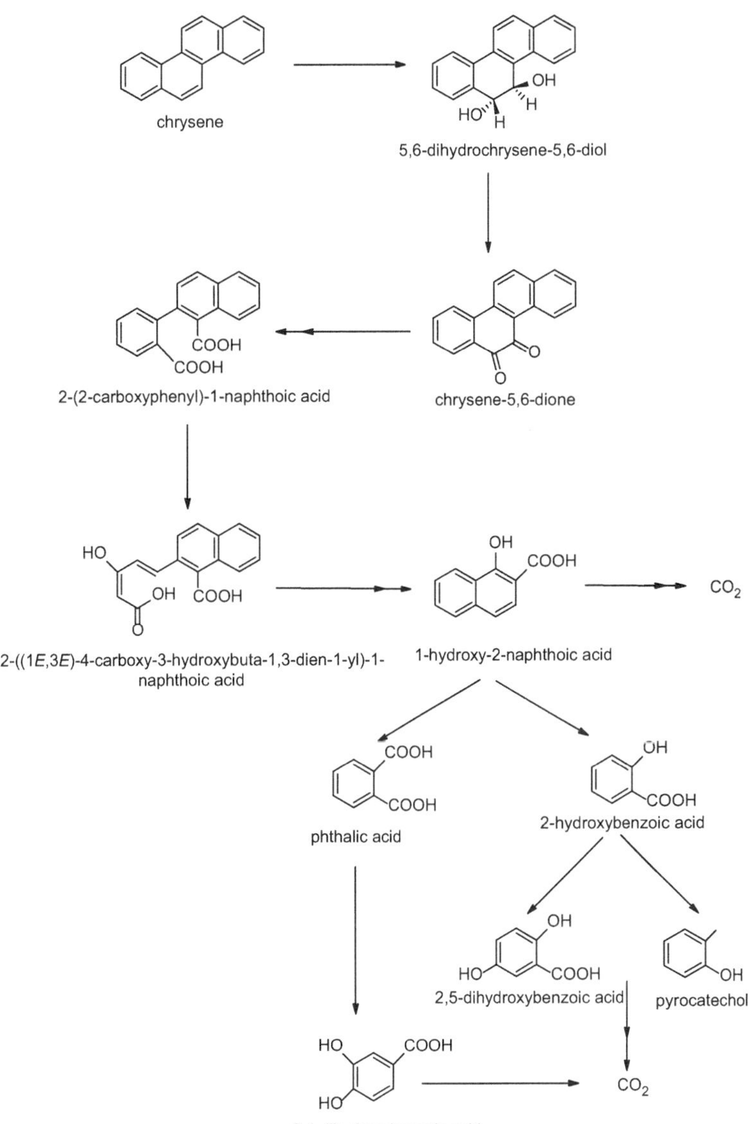

Fig. 10.5 Proposed pathway for the microbial degradation of chrysene by Polyporus sp. S133 into small products like pyrocatechol, 2,5-dihydroxybenzoic acid, or 3,4-dihydroxybenzoic acid and ultimately mineralization. (With permission through rightslink; Hadibarata et al. 2009)

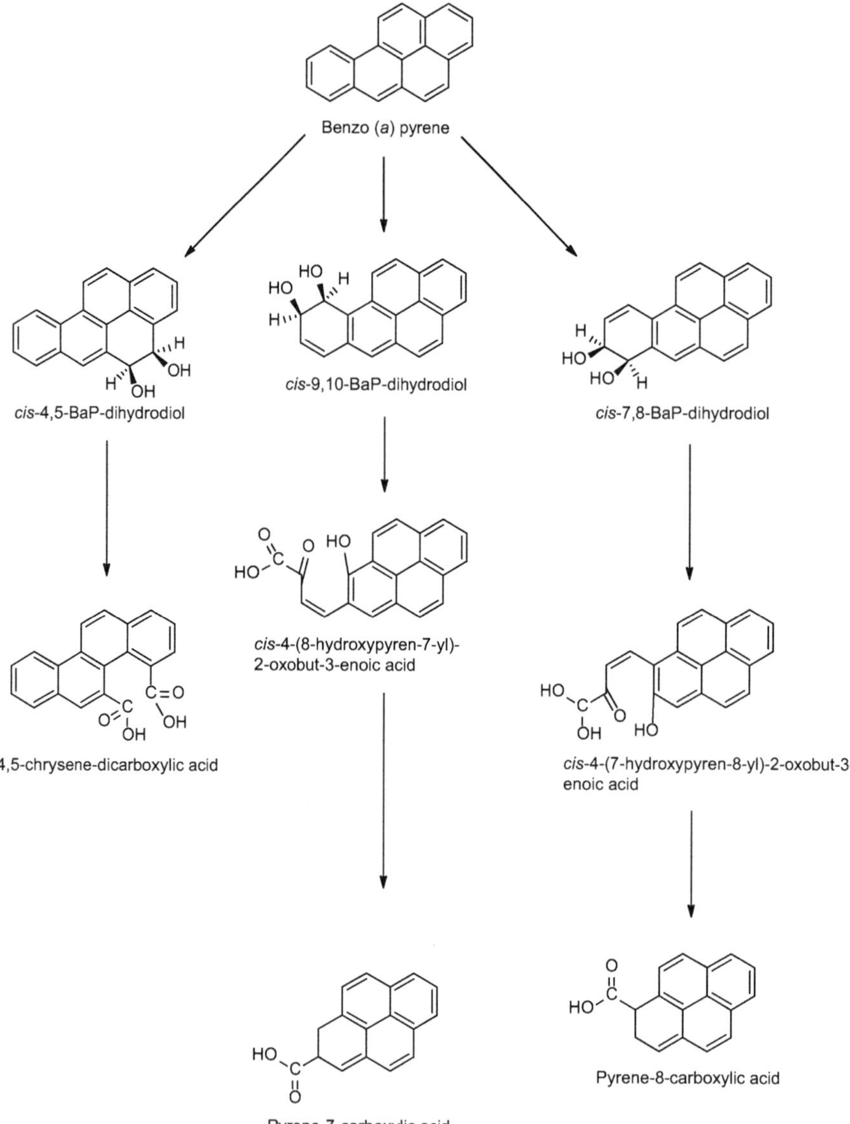

Fig. 10.6 Microbial degradation of large structure of BaP by *Mycobacterium* sp. strain RJGII-135 into toxic metabolites like pyrene-7-carboxylic acid and pyrene-8-carboxylic acid. (With permission through rightslink; Schneider et al. 1996)

Table 10.4 List of various polycyclic aromatic hydrocarbons degraded using nanoparticles; chemically and green synthesized

S. No.	PAH	Nanoparticles	Remarks	Mechanism	References
1.	BaP	Fe_3O_4 coated with fatty acids	Complete adsorption within 5 min at room temperature from water.	Adsorption	Liao et al. (2015)
2.	Chrysene	Curcumin conjugated ZnO	100% removal in 2.2 h	Photodegradation	Moussawi and Patra (2016)
3.	Pyrene	TiO_2 (anatase)	43.5 ± 1.1% of degradation from soil within 25 h via photocatalytically under UV irradiation	Photodegradation	Dong et al. (2010a)
4.	Pyrene	TiO_2 (rutile)	Photocatalytically decomposed completely within 25 h	Photodegradation	Dong et al. (2010b)
5.	Naphthalene	La and N co-doped TiO_2 supported on activated carbon	Efficient degradation (93.5%) within 120 min	Photodegradation	Liu et al. (2016)
6.	Phenanthrene	TiO_2 (anatase)	$t_{1/2}$ decreased from 46 to 31 h under UV-light.	Photodegradation	Gu et al. (2012)
7.	Phenanthrene	Co-doped titanate nanotubes	98.6% removal in 12 h	Photodegradation	Zhao et al. (2016)
8.	Anthracene	ZnO	Photodegradation (96%) into non-toxic 9, 10-anthraquinone.	Photodegradation	Hassan et al. (2015)
9.	Anthracene, Phenanthrene chrysene, Fluorene, and benzo (a) pyrene	FeHCF and KZnHCF	Anthracene (90%) phenanthrene (80%), fluorene, chrysene and benzo (*a*) pyrene were ~70–80%.	Photodegradation (natural sunlight)	Shanker et al. (2017b, c)

Complete mineralization of phenanthrene was obtained using highly potential hybrid gel-derived ZrO_2-acetylacetonate in aqueous solution and in the absence of light at 30 °C (Sannino et al. 2014). Fe^{3+}-modified montmorillonite degraded PAHs into quinones, ring-opening products and benzene derivatives under visible light and transformation rate followed the order of benzo[a]pyrene ≈ anthracene > benzo[a] anthracene > phenanthrene (Zhao et al. 2017). Quantum simulation results confirmed the crucial role of "cation-π" interaction between Fe^{3+} and PAHs on their

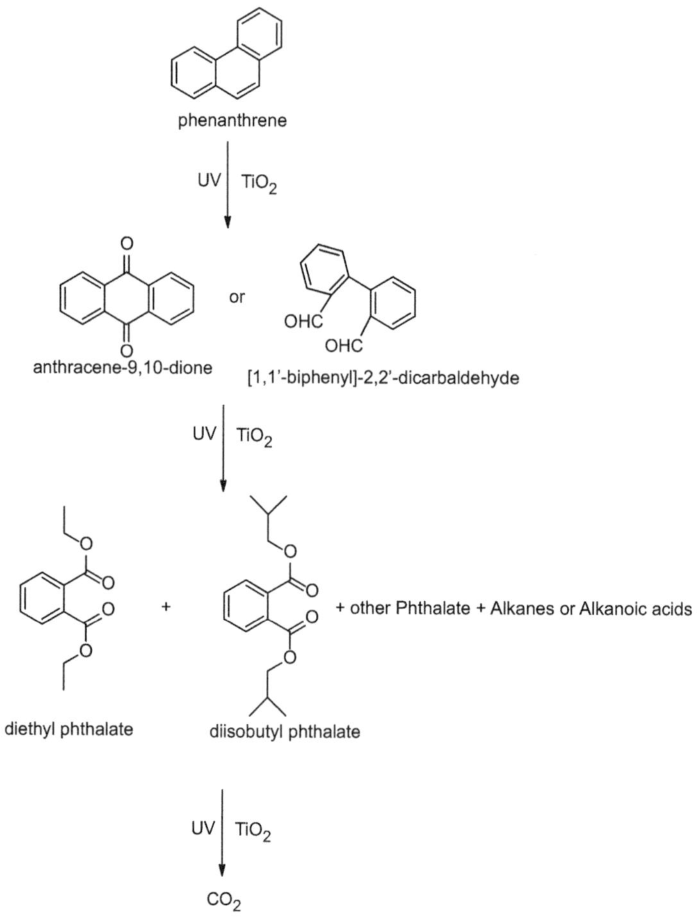

phenanthrene

UV TiO_2

anthracene-9,10-dione or [1,1'-biphenyl]-2,2'-dicarbaldehyde

UV TiO_2

diethyl phthalate + diisobutyl phthalate + other Phthalate + Alkanes or Alkanoic acids

UV TiO_2

CO_2

Fig. 10.7 Photodegradation of phenanthrene by TiO_2 under UV irradiation. •OH generated by electronic excitement of TiO_2 converted phenenthrene into anthracene-9,10-dione or [1,1′-biphenyl]-2,2′-dicarbaaldehyde followed by diethyl phthalate and other smaller products that will finally mineralized to H_2O and CO_2. (With permission through rightslink; Zhang et al. 2011)

transformation kinetics. Photolysis of PAHs on clay surface firstly occurs by electron transfer from PAHs to Fe^{3+}-montmorillonite, followed by degradation involving photo-induced ROS such as ·OH and $\cdot O_2^-$.

Co-doped titanate nanotubes offer 10 times higher photodegradation rate over conventional TiO_2 with 98.6% removal of phenanthrene in 12 h (Zhao et al. 2016). Using TiO_2 as a catalyst, fluorene was also degraded under artificial or sunlight radiation sources (Dass et al. 1994). TiO_2 nanoparticles successfully photocatalytically degraded 43.5 ± 1.1% of pyrene on soil surface for 25 h under UV irradiation (Dong et al. 2010a). The same group also utilized other form of TiO_2 nanoparticles i.e. rutile form for photocatalytically decomposing pyrene within 25 h (Dong et al. 2010b). This limited research and the work on TiO_2 facilitated the need for exploring new pathways such as use of nanoadsorbents like metals, metal oxides, mixed metals, doped metals. Recently, Moussawi and Patra (2016) used curcumin conjugated ZnO nanostructures for removing chrysene in 2.2 h. Gupta and Gupta (2015) utilized different iron oxides such as α-FeOOH, β-FeOOH, γ-Fe_2O_3, α-Fe_2O_3 and Fe_3O_4 for the photodegradation of BaP in different soil pH, wavelength, catalyst and oxalic acid dose. α-FeOOH degraded BaP into smaller pyrene molecule in 120 h (Fig. 10.8). Complete decomposition of BaP in 20 h was obtained on the surface of titanium dioxide using UV/photocatalytic oxidation process (Asano et al. 2012). Addition of TiO_2 and Fe_2O_3 decreased the half-life photodegradation of BaP from 363.22 h to 89.34 and 99.85 h, respectively, suggesting the effectiveness of using semi-conducting materials (Zhang et al. 2006). Gold nanoparticles immobilized onto multifunctional homopolymer vesicle i.e. poly[2-hydroxy-3-(naphthalen-1-ylamino) propyl methacrylate] (PHNA) were applied for the removal of pyrene with adsorption efficiency of 1.09×10^{-3} g pyrene/g PHNA in 1 h (Zhu et al. 2014).

A multifunctional vesicle based on a statistical copolymer, poly[2-hydroxy-3-(naphthalen-1-ylamino) propyl methacrylate]-stat-[2-(diethylamino)ethyl methacrylate]] [P(HNA_{21}-stat-DEA_{35})] was employed for immobilizing gold nanoparticles. This was then employed for removing various PAHs like naphthalene, anthracene and pyrene from wastewater with adsorption capacity of mg PAHs/g vesicles of 4.0, 9.3 and 8.0, respectively (Xiao and Du 2016). Very scarce data is available for the degradation of BaP using nanoparticles. Fatty acids coated Fe_3O_4 nanoparticles completely adsorbed BaP from water sample within 5 min at room temperature (Liao et al. 2015). This reveals the high potential of nanomaterials to act as excellent adsorbents for the removal of BaP in a very short time. Since good results are obtained by using conventional materials like α-FeOOH or TiO_2, therefore, their nanoscale counterparts must be explored for better results.

10.7 Utilization of Green Synthesized Nanomaterials in Remediation for Environment

Nanomaterials generated *via* green route are cheap. Currently, a lot of researchers worldwide are working on synthesis of nanoparticles using by employing sunlight, or plant-based surfactants or microorganisms and water or combination of those. For example, silver nanoparticles were biosynthesized extracellularly by bacteria *Bacillus cereus* with high activity against bacteria (Prakash et al. 2011). Shukla and Iravani (2017) reviewed green strategies for the preparation of metallic nanoparticles applied for water purification. The use of enzymes, vitamins, monosaccharides, polysaccharides and biodegradable polymers including microwave-assisted synthesis were discussed in detail. These biosynthesis methods are economical and eco-friendly, thus, can be used on large-scale. Application of these bioactive nanoparticles in the degradation of harmful PAHs will help in reconstructing the polluted environment. Green process not only control the morphology but are also

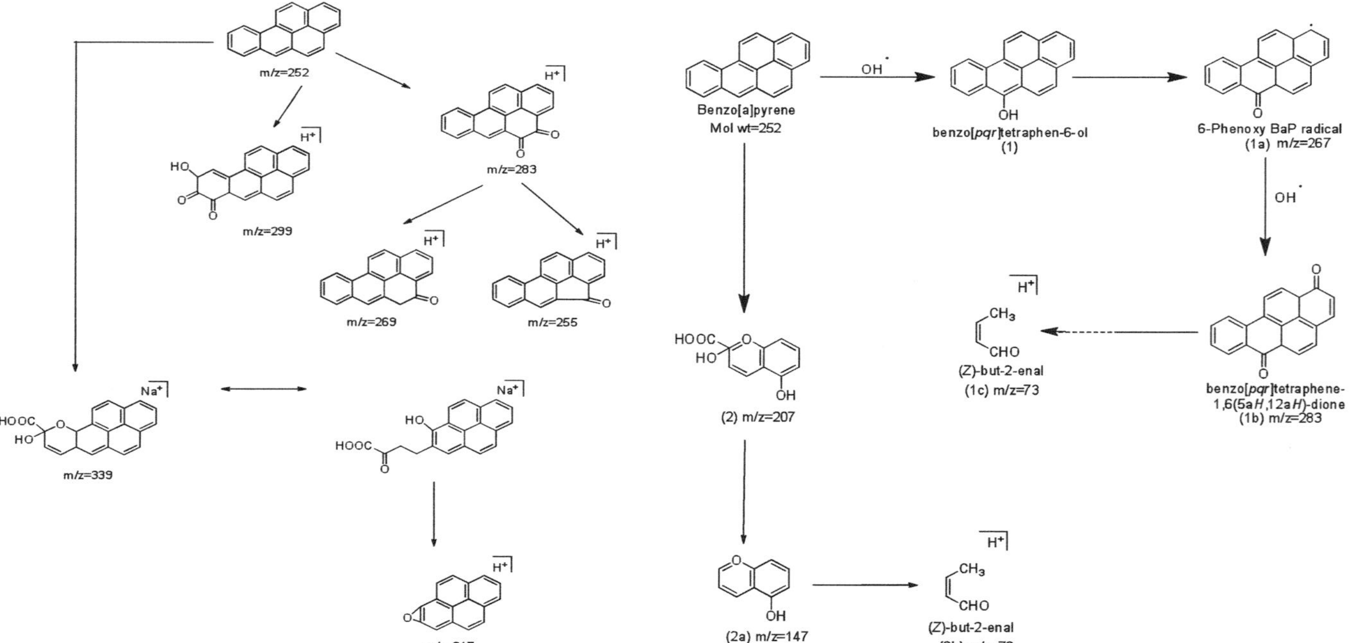

Fig. 10.8 Possible pathway for photodegradation of benzopyrene using iron oxides and metal hexacyanoferrates nanoparticles
Metal hexacyanoferrates nanoaprticles were better than iron oxides because of complete transformation of benzopyrene into smaller products and non-toxic byproducts. (With permission through rightslink; Gupta and Gupta 2015; Shanker et al. 2017c)

efficient than chemically synthesized nanoparticles. Shahwan et al. 2011 synthesized iron nanopartciles such as iron oxide and oxyhydroxide using green tea leaf extract and evaluated it as Fenton-like catalyst in degradation of cationic and anionic dyes such as methylene blue and methyl orange dyes. They observed that green tea-iron iron nanoparticles were a better catalyst than Fe nanoparticles produced by borohydride reduction. As we already discussed there are lack of reports on degradation or removal of PAHs by nanomaterials, only few studies are recently reported on use of green synthesized nanoparticles implemented for treatment of PAHs. Biosynthesized nanoparticles ZnO showed good potential in photodegradation of anthracene 96% for the into non-toxic 9,10-anthraquinone (Fig. 10.9) (Hassan et al. 2015).

Shanker and Co-workers (Rani and Shanker 2017a; Shanker et al. 2017b, c) and Rani et al. synthesized metal hexacyanoferrates via green route using *sapindus mukorossi* as natural surfactant. *Sapindus mukorossi*, a bio-surfactant that belongs to plant family Sapindeae contains saponins which are responsible for the surfactant action (Ghagi et al. 2002) (Fig. 10.10).

Highly crystalline sharp potassium zinc hexacyanoferrate nanocubes of ~100 nm and hexagonal, rod and spherical shaped iron hexacyanoferrate nanoparticles, size range: 10–60 nm were characterized (Figs. 10.11 and 10.12). The potential of these green synthesized nanoparticles was tested for the treatment of toxic PAHs, namely, anthracene, phenanthrene chrysene, fluorene, and benzo (a) pyrene in water as well as in soil under dark exposure, UV-and sunlight (Fig. 10.13). The selection of PAHs was based on different rings. Using metal hexacyanoferrate nanoparticles different toxic PAHs like anthracene > phenanthrene > fluorene > chrysene > BaP were degraded to good extent in water (70–93%) as well as in soil (68–84%) at neutral pH under sunlight exposure. The order of degradation was found to be dependent on the molecular weight, size, structure and aromaticity of the PAHs. Small and non-toxic byproducts like malealdehyde, 4-oxobut-2-enoic acid and*o*-xylene were identified. This indicates the positive aspects of using synthesized nanoparticles in removal of PAHs. Working mechanism of metal hexacyanoferrate nanoparticles indicated the initial adsorption followed by photo-degradation process occurring over their surface (Fig. 10.14). Metal hexacyanoferrate nanoparticles like KCuHCF > KNiHCF > KCoHCF synthesized using *Aegle marmelos* (Fig. 10.10) could able to degrade 90–95% of dye.

There are some examples highlighting the use of other green synthesized nanomaterials in removal of various organic pollutants and provide the feasibility in treatment of pesticides in wastewater. Recently, sunlight assisted green synthesis of various transition metal oxide such as ZnO, CuO, Co_3O_4, NiO and Cr_2O_3 nanoparticles and their catalytic application in removal of organic colorants were studied (Shanker et al. 2016b). Using *Morinda tinctoria* leaf extract silver nanoparticles were synthesized and effectively degraded 95% of dye in 72 h under sunlight irradiation (Vanaja et al. 2014). In addition to this, *Polygonum Hydropiper*

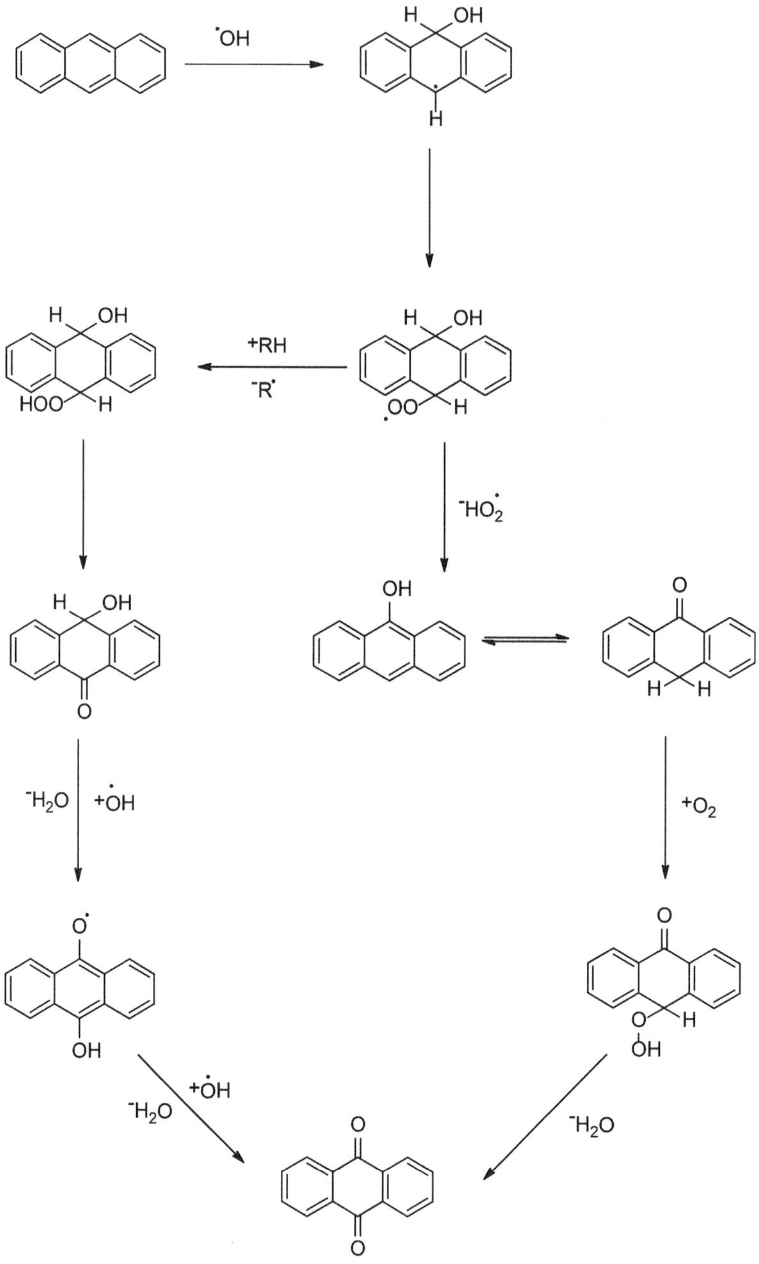

Fig. 10.9 Pathway for the 96% photodegradation of anthracene by green synthesized ZnO nanoparticles into non-toxic compound 9,10-anthraquinone via generation of OH radical under sunlight. (With permission through rightslink; Hassan et al. 2015)

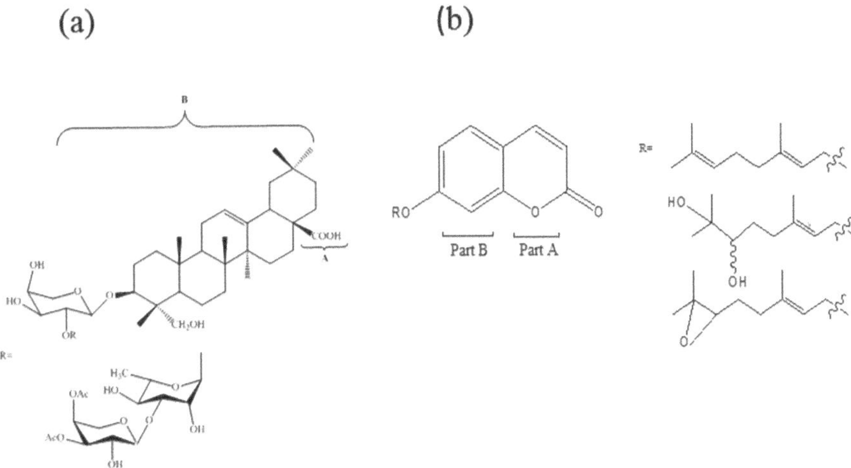

Fig. 10.10 Structure of various natural products present in the plant based surfactant (**a**) *Sapindus mukorossi* (**b**) *Aegle marmelos*. Just like synthetic surfactants they lower the interfacial tension and promotes the formation of nano-emulsions that will control the aggregation and growth of the particles

was also used as a biogenic source for synthesizing silver nanoparticles with high catalytic efficiency in degrading MB completely within 13 min (Bonnia et al. 2016). A high degradation of 98% for Orange II was reported using green synthesized bimetallic Fe/Pd nanoparticles, whereas, only 16% of the dye was degraded using Fe nanoparticles (Luo et al. 2016). Green synthesized copper oxide nanoparticles and ZnO/Graphene Oxide (GO) nanocomposite were also used for decolorization of the dye (Lellala et al. 2016).

10.8 Conclusion

There is lack of reports on degradation of PAHs using nanoparticles. Most of the studies are confined to microorganisms mediated degradation. Such methodologies need more time for complete metabolism from several days to month and are also very expensive, consequently, cannot be employed at industrial scale. The limited research facilitates the need for exploring new pathways, especially the use of nanomaterials as they have a unique property of high surface-to-volume ratio, which makes them excellent adsorbents. Moreover, green synthesized nanoparticles have been showing potential role in making environment green. Nanoadsorbents like TiO_2 or ZnO as such or doped are playing an excellent role in complete

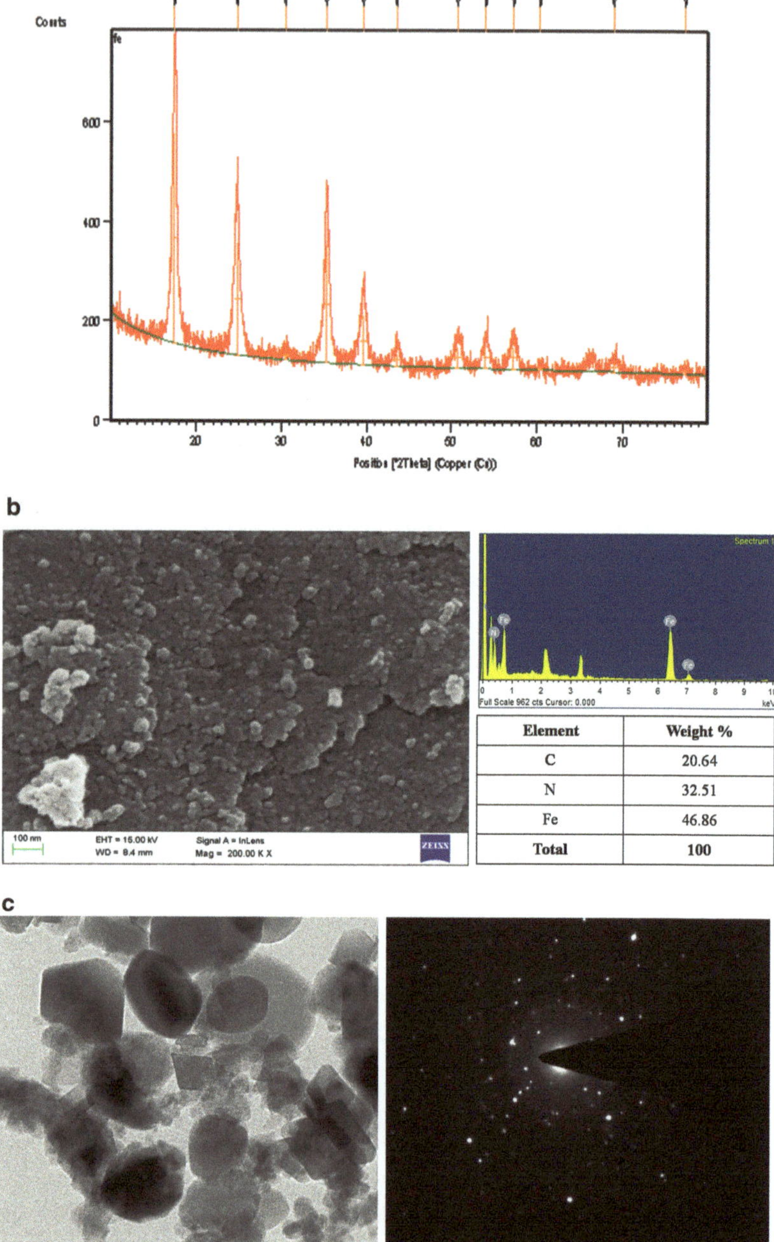

Element	Weight %
C	20.64
N	32.51
Fe	46.86
Total	100

Fig. 10.11 (**a**) Powder X-ray diffraction pattern of iron hexacyanoferrates nanoparticles showing the good intensity of peaks that reflects the crystalline nature of particles (**b**) Field Emmision-Scanning electron microscopic image shows that the particles are below 100 nm size with little agglomeration and energy dispersive x-ray analysis shows the elements presents in the iron hexacyanoferrates (**c**) Transmission electron microscopic image shows inner morphology with more resolved image and selected area diffraction pattern of iron hexacyanoferrates nanoparticles which supports the facts that these are highly crystalline material (Shanker et al. 2017b)

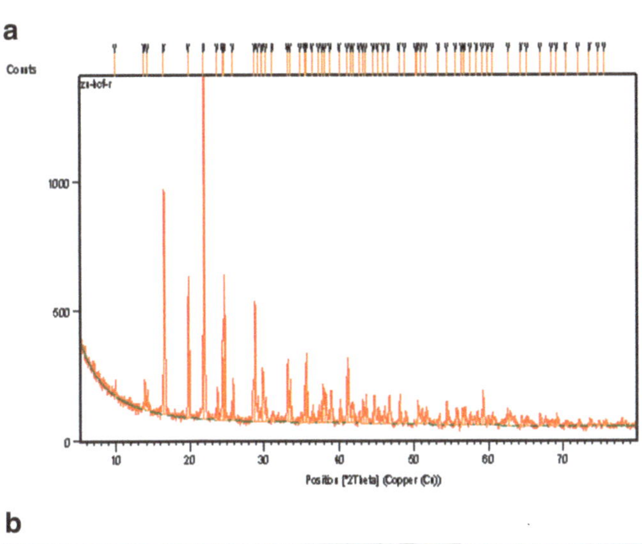

Element	Weight %
Zn	25.72
C	22.67
N	27.17
K	9.04
Fe	15.39
Total	100

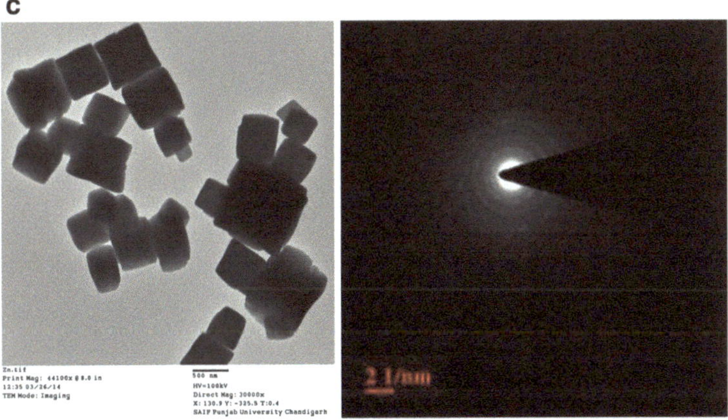

Fig. 10.12 (**a**) Powder X-ray diffraction pattern of potassium zinc hexacyanoferrates nanoparticles showing the good intensity of peaks that reflects the crystalline nature of particles (**b**) Field Emmision-Scanning electron microscopic image shows that the particles are below 100 nm size with cubic surface morphology and energy dispersive x-ray analysis shows the elements presents in

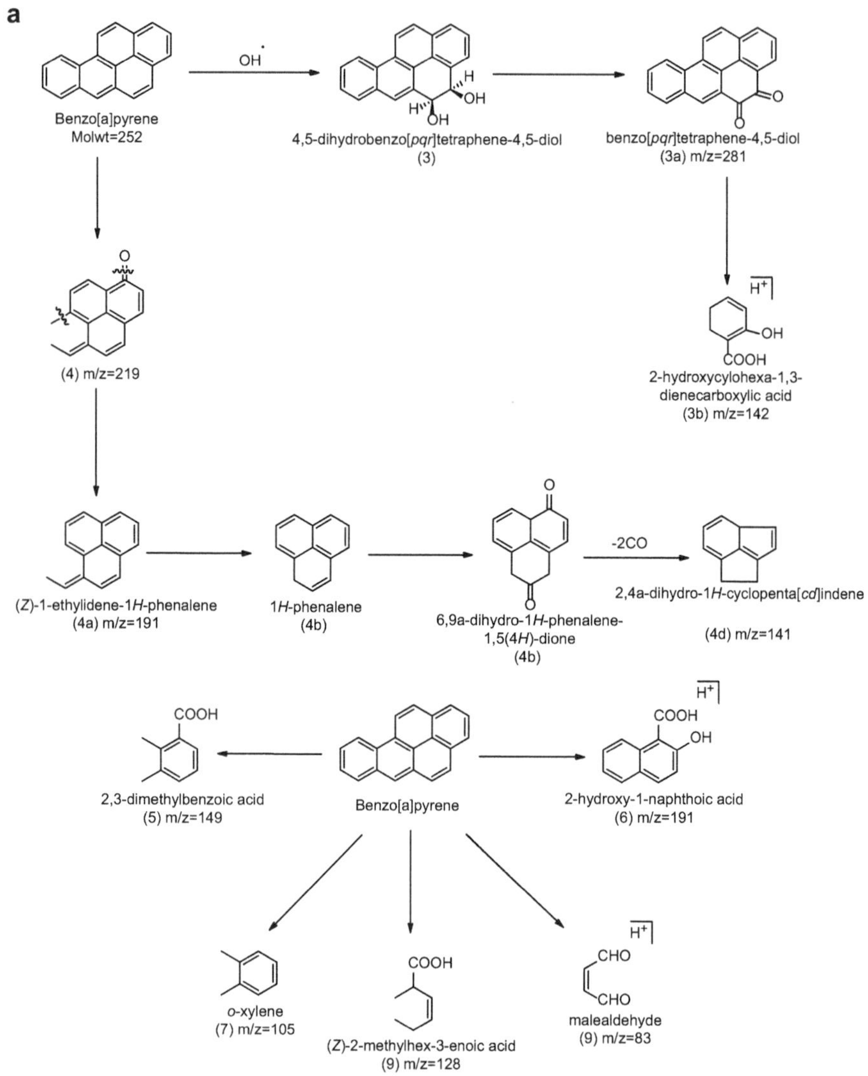

Fig. 10.13 Proposed photodegradation pathway of (**a**) BaP (**b**) chrysene (**c**) fluorene (**d**) phenanthrene (**e**) anthracene by metal hexacyanoferrates nanostructures. Metal hexacyanoferrates being low band gap chemicals can be easily excited under the sunlight. OH radical is main species that converted polycyclic aromatic hydrocarbons to the small and non-toxic metabolites. (With permission through rightslink; Shanker et al. 2017c)

Fig. 10.12 (continued) the potassium zinc hexacyanoferrates (**c**) Transmission electron microscopic image shows inner morphology with more resolved image and selected area diffraction pattern of potassium zinc hexacyanoferrates nanoparticles which supports the facts that these are highly crystalline material (Shanker et al. 2017c)

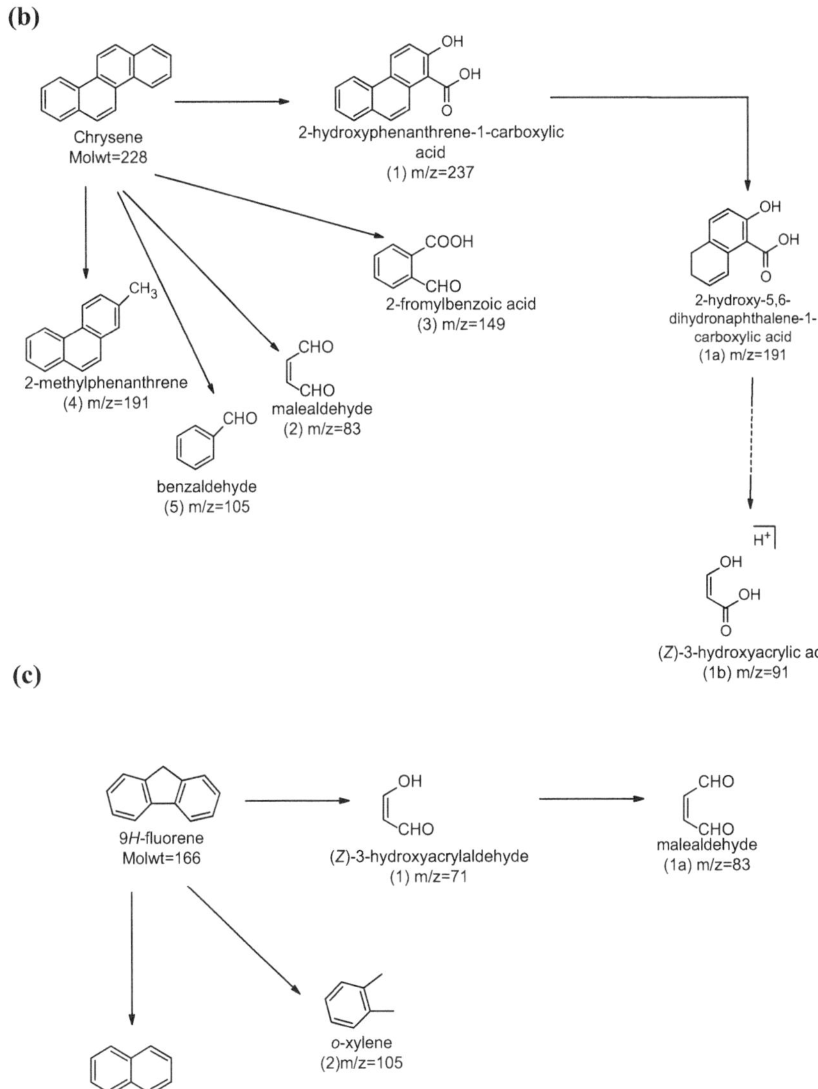

Fig. 10.13 (continued)

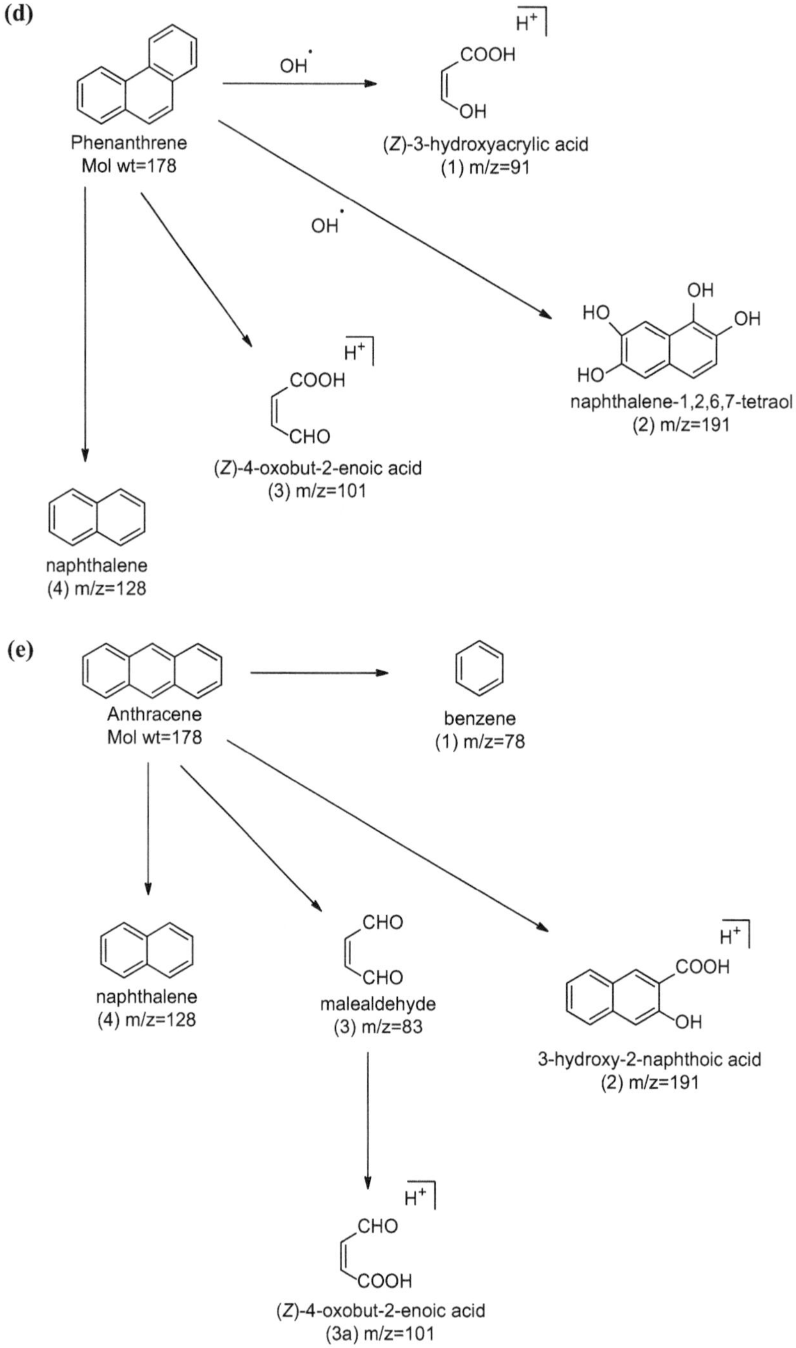

(d)

Phenanthrene
Mol wt=178

$OH^{\cdot}$

COOH
OH
H^+
(*Z*)-3-hydroxyacrylic acid
(1) m/z=91

$OH^{\cdot}$

OH
HO
OH
HO
naphthalene-1,2,6,7-tetraol
(2) m/z=191

COOH
CHO
H^+
(*Z*)-4-oxobut-2-enoic acid
(3) m/z=101

naphthalene
(4) m/z=128

(e)

Anthracene
Mol wt=178

benzene
(1) m/z=78

naphthalene
(4) m/z=128

CHO
CHO
malealdehyde
(3) m/z=83

COOH
OH
H^+
3-hydroxy-2-naphthoic acid
(2) m/z=191

CHO
COOH
H^+
(*Z*)-4-oxobut-2-enoic acid
(3a) m/z=101

Fig. 10.13 (continued)

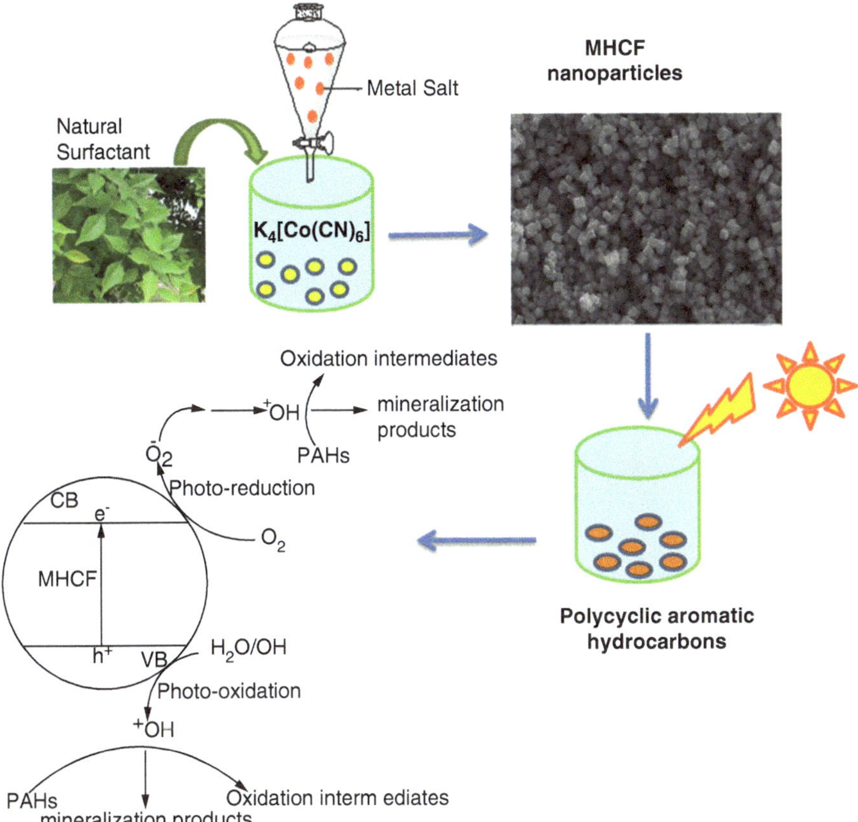

Fig. 10.14 Schematic representation of the green synthesis of metal hexacyanoferrates nanoparticles utilizing a plant based natural surfactant followed by the degradation of individual polycyclic aromatic hydrocarbons under the sunlight. OH radical generated either by the photooxidation or photo-reduction is responsible for the photo-degradation process occurring over the surface of metal hexacyanoferrates nanoparticles

mineralization of PAHs than microorganisms in a time as short as few minutes (Fig. 10.15). Furthermore, these advanced nanomaterials are observed to be highly potential candidates as revealed by either absence of toxic by-products or transient metabolites which get quickly mineralized into non-toxic smaller by-products. Photooxidation mechanisms of PAHs could be operated by singlet oxygen mechanism or radical chain mechanism (Calvert et al. 2002) (Fig. 10.16).

Use of nanoparticles for the degradation of high molecular weight PAHs like Dibenzo (*a,h*) anthracene, Indeno (1,2,3-*cd*) pyrene, Benzo (*g,h,i*) perylene, Benzo

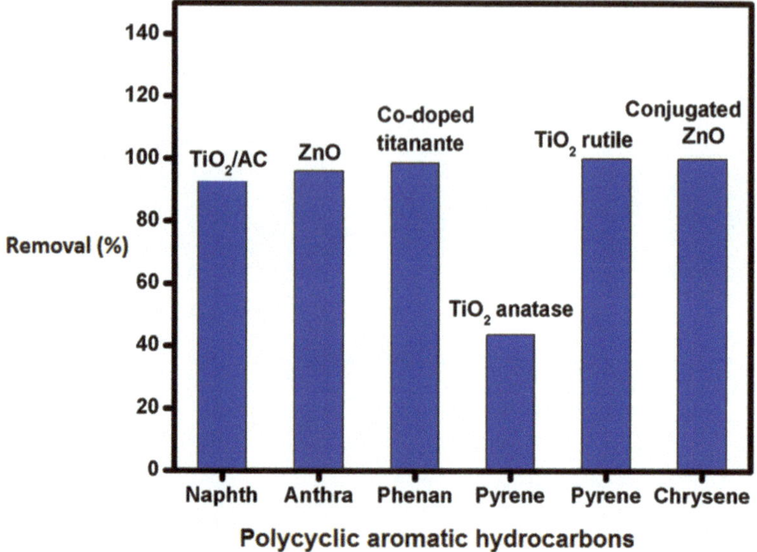

Fig. 10.15 Graphical representation of removal (%) of polycyclic aromatic hydrocarbons using various photocatalysts e.g., ZnO, TiO_2 both anatase and rutile form or their doped like TiO_2 with activated carbon, cobalt doped TiO_2 or conjugated ZnO
where, Naphtha = Naphthalene; Anthra = Anthracene; Phenan = Phenanthrene

(*k*) fluoranthene is still awaited. Due to high aromaticity, these compounds are very stable and do no undergo biodegradation. Hence, must be degraded using effective nanoadsorbents due to their long half-lives from days to years and high persistence in nature.

10.9 Future Scope

This chapter covered the encouraging performances of a variety of nanomaterials and nanocomposites as adsorbent, heterogeneous redox and photo-catalyst for the effective degradation of hydrophobic, recalcitrant and bioaccumulating PAHs. Extensively used microbial degradation is associated with several drawbacks like high cost, long time requirements, difficulty in handling and selectivity. Nanomaterials not only adsorb the toxic PAHs, but also transform them into into less hazardous forms in an eco-friendly manner with minimum requirement of energy, time and cost. It was also observed that nanomaterials may be utilized up to many cycles. Being inexpensive, commercially available, non-toxic in nature, and of high efficiency, TiO_2 and ZnO nanomaterials as such or doped have proven to be effective adsorbents in degrading PAHs completely in short period of few minutes to

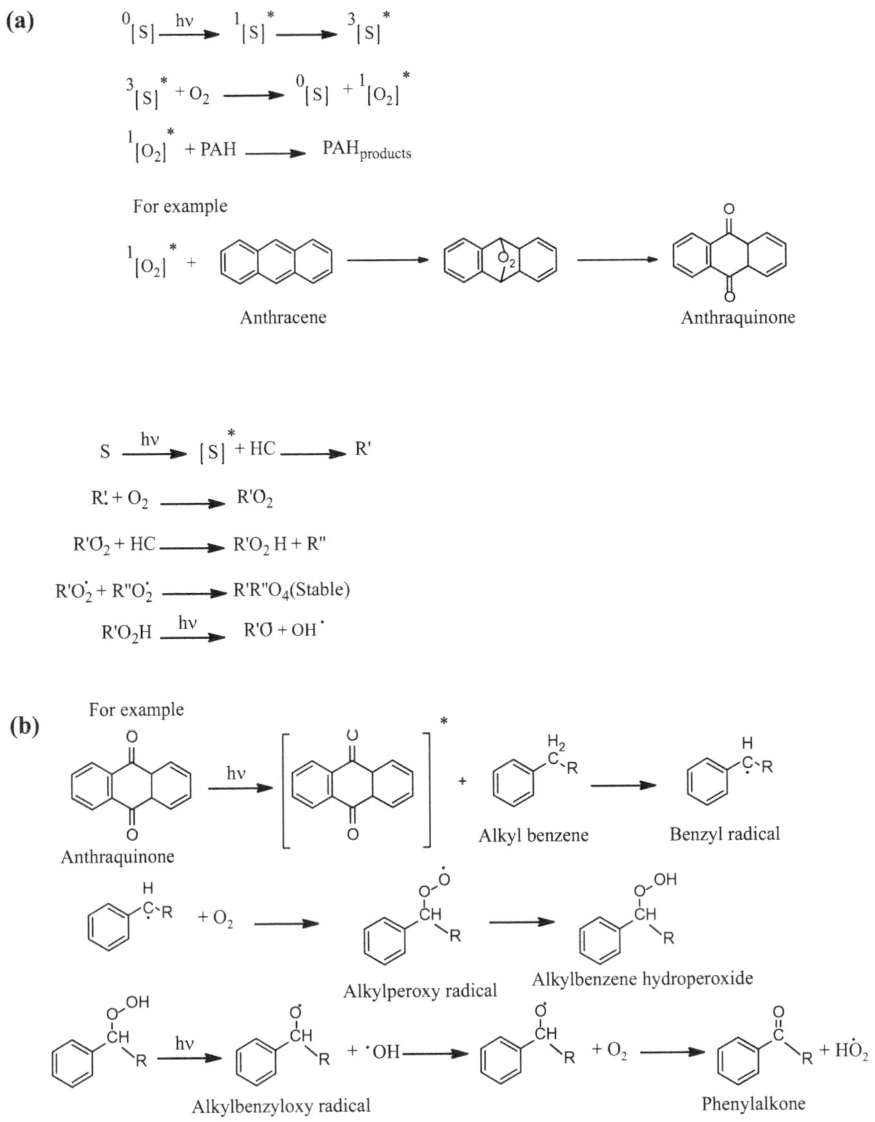

(a)

$$^{0}[S] \xrightarrow{h\nu} {}^{1}[S]^{*} \longrightarrow {}^{3}[S]^{*}$$

$$^{3}[S]^{*} + O_2 \longrightarrow {}^{0}[S] + {}^{1}[O_2]^{*}$$

$$^{1}[O_2]^{*} + PAH \longrightarrow PAH_{products}$$

For example

$$^{1}[O_2]^{*} + \text{Anthracene} \longrightarrow (O_2 \text{ adduct}) \longrightarrow \text{Anthraquinone}$$

Anthracene

Anthraquinone

$$S \xrightarrow{h\nu} [S]^{*} + HC \longrightarrow R'$$

$$R'^{\cdot} + O_2 \longrightarrow R'O_2$$

$$R'O_2^{\cdot} + HC \longrightarrow R'O_2H + R''$$

$$R'O_2^{\cdot} + R''O_2^{\cdot} \longrightarrow R'R''O_4(\text{Stable})$$

$$R'O_2H \xrightarrow{h\nu} R'O^{\cdot} + OH^{\cdot}$$

(b)

For example

Anthraquinone $\xrightarrow{h\nu}$ [Anthraquinone]* + Alkyl benzene $\longrightarrow$ Benzyl radical

Benzyl radical + O_2 $\longrightarrow$ Alkylperoxy radical $\longrightarrow$ Alkylbenzene hydroperoxide

Alkylbenzene hydroperoxide $\xrightarrow{h\nu}$ Alkylbenzyloxy radical + $^{\cdot}OH$ $\longrightarrow$ Alkylbenzyloxy radical + O_2 $\longrightarrow$ Phenylalkone + $H\dot{O}_2$

Fig. 10.16 Photo-oxidation mechanisms (**a**) singlet oxygen mechanism (**b**) radical chain mechanism for the degradation of various polycyclic aromatic hydrocarbons. (With permission through rightslink; Calvert et al. 2002)
where, HC = Hydrocarbon, S = light absorbing molecule

hours with UV-Visible irradiation. Majority of the nanoadsorbents are able to degrade more than 70 to 80% of the PAHs and various even achieved complete degradation in a short span of time. Further, the band gap of TiO_2 can be altered via the insertion of other metal oxides or electron rich elements. Doping could increase the catalytic potential by manifolds of the nanoparticles. Addition of H_2O_2 enhanced the oxidative efficiency of nanomaterials because of production of more •OH radical. Keeping in view the future aspects, nanomaterials with high degrading efficiencies can be used as commercialized adsorbents on industrial scale. Furthermore, degradation of high molecular weight PAHs like Dibenzo (*a,h*) anthracene, Indeno (1,2,3-*cd*) pyrene, Benzo (*g,h,i*) perylene, Benzo (*k*) fluoranthene is still not explored extensively yet using nanomaterials. These PAHs can cause severe damage to the environment due to their toxicity and needs immediate treatment. Finally chapter ends with suggested future scope of using green synthesized nanomaterials that would be cheaper and effective catalysts. For sustainable developments, biopolymers-based nano-biocomposites should be developed. The emission of PAHs containing waste should be avoided and its recovery should be promoted. Control at the source could drastically reduce the levels of PAHs in the environment.

References

Abdel-Shafy H, Mohamed-Mansour MS (2016) A review on polycyclic aromatic hydrocarbons: source, environmental impact, effect on human health and remediation. Egypt J Petrol 25:107–123. https://doi.org/10.1016/j.ejpe.2015.03.011

Abrantes R, Assunção JV, Pesquero CR, Bruns RE, Nóbrega RB (2009) Emission of polycyclic aromatic hydrocarbons from gasohol and ethanol vehicles. Atmos Environ 43:648–654. https://doi.org/10.1016/j.atmosenv.2008.10.014

Agency for Toxic Substances and Disease Registry (1995) Toxicological profile for polycyclic aromatic hydrocarbons (PAHs). U.S. Department of Health and Human Services, Washington, DC

Ahmed T, Ahmed MA, Othman DV, Sarwade KRG (2012) Degradation of Anthracene by Alkaliphilic Bacteria Bacillus badius. Environ Poll 1:97–104

Altenburger R, Segner H, van der Oost R (2003) Biomarkers and PAHs – prospects for the assessment of exposure and effects in aquatic systems. In: PAHs: an ecotoxicological perspective. Wiley, Chichester, pp 297–328. https://doi.org/10.1002/0470867132.ch16

Anipsitakis GP, Dionysiou ESDD (2003) Radical generation by the interaction of transition metals with common oxidants. J Phys Chem B 109:13052–13055. https://doi.org/10.1021/es035121o

Annweiler E, Richnow HH, Antranikian G, Hebenbrock S, Garms C, Franke S, Francke W, Michaelis W (2000) *Naphthalene degradation and incorporation of naphthalene-derived carbon* into biomass by the thermophile *Bacillus thermoleovorans*. Appl Environ Microbiol 66:518–523

Asano M, Sumino S, Jiku F (2012) Decomposition of benzo (a) pyrene on artificial sea water using of UV/photocatalytic oxidation process. J Environ Sci Eng A 18:195–199

ATSDR (Agency for Toxic Substances and Disease Registry) (1990) Toxicological profile for polycyclic aromatic hydrocarbons. Acenaphthene, Acenaphthylene, Anthracene, Benzo(a) anthracene, Benzo(a)pyrene, Benzo(b)fluoranthene, Benzo(g,i,h)perylene, Benzo(k) fluoranthene, Chrysene, Dibenzo(a,h)anthracene, Fluoranthene, Fluorene, Indeno(1,2,3-c,d) pyrene, Phenanthrene, Pyrene. Prepared by Clement International Corporation, under contract no. 205-88-0608. ATSDR/TP-90-20

Ayekoe PY, Robert D, Goné DL (2016) Preparation of effective TiO_2/Bi_2O_3 photocatalysts for water treatment. Environ Chem Lett 14:387. https://doi.org/10.1007/s10311-016-0565-3

Bach QD, Kim SJ, Choi SC, Oh YS (2005) Enhancing the intrinsic bioremediation of PAHs contaminated anoxic estuarine sediments with biostimulating agents. J Microbiol 43:319–324

Baird WM, Hooven LA, Mahadevan B (2005) Carcinogenic polycyclic aromatic hydrocarbon-DNA adducts and mechanism of action. Environ Mol Mutagen 4:106–114. https://doi.org/10.1002/em.20095

Baklanov A, Hänninen O, Slørdal LH, Kukkonen J, Bjergene N, Fay B (2007) Integrated systems for forecasting urban meterology, air pollution and population exposure. Atmos Chem Phys 7:855–874. https://doi.org/10.5194/acp-7-855-2007

Bandala ER, Gelover S, Leal MT, Arancibia-Bulnes C, Jimenez A, Estrada CA (2002) Solar photocatalytic degradation of Aldrin. Catal Today 76:189–199. https://doi.org/10.1016/S0920-5861(02)00218-3

Bandala ER, Andres-Octaviano J, Pastrana P, Torres LG (2006) Removal of aldrin, dieldrin, heptachlor, and heptachlor epoxide using activated carbon and/or pseudomonas fluorescens free cell cultures. J Environ Sci Health B 41:553–569. https://doi.org/10.1080/03601230600701700

Barra R, Quiroz R, Saez K et al (2009) Sources of polycyclic aromatic hydrocarbons (PAHs) in sediments of the Biobio River in south Central Chile. Environ Chem Lett 7:133. https://doi.org/10.1007/s10311-008-0148-z

Baruah S, Najam Khan M, Dutta J (2016) Perspectives and applications of nanotechnology in water treatment. Environ Chem Lett 14:1. https://doi.org/10.1007/s10311-015-0542-2

Bernal-Martínez A, Carrière H, Patureau D, Delgenè JP (2005) Combining anaerobic digestion and ozonation to remove PAH from urban sludge. Process Biochem 40:3244–3250. https://doi.org/10.1016/j.procbio.2005.03.028

Bonnia NN, Kamaruddin MS, Nawawi MH, Ratimd S, Azlinac HN, Ali ES (2016) Green biosynthesis of silver nanoparticles using 'polygonum hydropiper' and study its catalytic degradation of methylene blue. Procedia Chem 19:594–602. https://doi.org/10.1016/j.proche.2016.03.058

Brunekreef B, Janssen NA, de Hartog JJ, Oldenwening M, Meliefste K, Hoek G, Lanki T, Timonen KL, Vallius M, Pekkanen J, Van Grieken R (2005) Personal, indoor and outdoor exposures of PM_{25} and its components for groups of cardiovascular patients in Amsterdam and Helsinki. Res Rep Health Eff Inst 127:1–70

Calvert JG, Atkinson R, Becker KH, Kamesns RM, Seinfeld JH, Wallington TH, Yarwood G (2002) The mechanisms of atmospheric oxidation of the aromatic hydrocarbons. Oxford University Press, New York, pp 370–383

Campbell R, Clifford K (2010) Gulf spill is the largest of its kind, scientists say. The New York Times

Castaño P, Pawelec B, Fierro JLG, Arandes JM, Bilbao J (2006) Aromatics reduction of pyrolysis gasoline (PyGas) over HY-supported transition metal catalysts. Appl Catal A 315:101–113. https://doi.org/10.1016/j.apcata.2006.09.009

Cerniglia CE (1992) Biodegradation of polycyclic aromatic hydrocarbons. Biodegradation 3:351–368. https://doi.org/10.1007/BF00129093

Chen X, Burda C (2008) The electronic origin of the visible-light absorption properties of C-, N- and S-doped TiO_2 nanomaterials. J Am Chem Soc 130:5018–5019. https://doi.org/10.1021/ja711023z

Chen YC, Lee WJ, Uang SN, Lee SH, Tsai PJ (2006) Characteristics of polycyclic aromatic hydrocarbon (PAH) emissions from a UH-1H helicopter engine and its impact on the ambient environment. Atmos Environ 40:7589–7597. https://doi.org/10.1016/j.atmosenv.2006.06.054

Chen S, Su B, Chang JE, Lee WJ, Huang KL, Hsieh LT, Huang JC, Lin WJ, Lin CC (2007) Emissions of polycyclic aromatic hydrocarbons (PAHs) from the pyrolysis of scrap tires. Atmos Environ 41:1209–1220. https://doi.org/10.1016/j.atmosenv.2006.09.041
Christensen HE, Fairchild EJ (1976) Registry of toxic effects of chemical substances. Prepared by Tracor Jitco Inc., Rockville, National Institute of Occupational Safety and Health, HEW Publication No. (NIOSH) 76–191
Chupungars K, Rerngsamran P, Thaniyavarn S (2009) Polycyclic aromatic hydrocarbons degradation by Agrocybe sp. CU-43 and its fluorene transformation. Int Biodeterior Biodegrad 63:93–99
Connelly NG, Geiger WE (1996) Chemical redox agents for organometallic chemistry. Chem Rev 96:877–910. https://doi.org/10.1021/cr940053x
Das P, Mukherjee S, Sen R (2008) Improved bioavailability and biodegradation of a model polyaromatic hydrocarbon by a biosurfactant producing bacterium of marine origin. Chemosphere 72:1229–1234. https://doi.org/10.1016/j.chemosphere.2008.05.015
Dass S, Muneer M, Gopidas KR (1994) Photocatalytic degradation of wastewater pollutants. Titanium-dioxide-mediated oxidation of polynuclear aromatic hydrocarbons. J Photochem Photobiol A Chem 77:83–88. https://doi.org/10.1016/1010-6030(94)80011-1
Dong D, Li P, Li X, Zhao Q, Zhang Y, Jia C, Li P (2010a) Investigation on the photocatalytic degradation of pyrene on soil surfaces using nanometer anatase TiO_2 under UV irradiation. J Hazard Mater 174:859–863. https://doi.org/10.1016/j.jhazmat.2009.09.132
Dong D, Li P, Li X, Xu C, Gong D, Zhang Y, Zhao Q, Li P (2010b) Photocatalytic degradation of phenanthrene and pyrene on soil surfaces in the presence of nanometer rutile TiO_2 under UV-irradiation. Chem Eng J 158:378–383. https://doi.org/10.1016/j.cej.2009.12.046
Dubowsky SD, Wallace LA, Buckley TJ (1999) The contribution of traffic to indoor concentrations of polycyclic aromatic hydrocarbons. J Expo Anal Env Epid 9:312–321. https://doi.org/10.1038/sj.jea.7500034
Fabbri D, Vassura I (2006) Evaluating emission levels of polycyclic aromatic hydrocarbons from organic materials by analytical pyrolysis. J Anal Appl Pyrolysis 75:150–158
Fawell JK, Hunt S (1988) The polyaromatic hydrocarbons. Enviormental toxicology: organic pollutants. Ellis Harwood, Chichester, pp 241–269
Freeman DJ, Cattell FCR (1990) Woodburning as a source of atmospheric polycyclic aromatic hydrocarbons. Environ Sci Technol 24:1581–1585. https://doi.org/10.1021/es00080a019
Freeman HM, Harris EF (1995) Hazardous waste remediation: innovative treatment technologies, 3rd edn. Technomic Publishing Company, Pennsylvania, p 463
Fromme H, Oddoy A, Piloty M, Krause M, Lahrz T (1998) Polycyclic aromatic hydrocarbons (PAH) and diesel engine emission (elemental carbon) inside a car and a subway train. Sci Total Environ 217:165–173. https://doi.org/10.1016/S0048-9697(98)00189-2
Fu PP, Xia Q, Sun X, Yu H (2012) Phototoxicity and environmental transformation of polycyclic aromatic hydrocarbons (PAHs)-light induced reactive oxygen species, lipid peroxidation, and DNA damage. J Environ Sci Health Part C 30:1–41. https://doi.org/10.1080/10590501.2012.653887
Garon D, Sage L, Murandi FS (2004) Effects of fungal bioaugmentation and cyclodextrin amendment on fluorene degradation in soil slurry. Biodegradation 15:1–8
Ge XY, Tian F, Wu ZL, Yan YJ, Cravotto G, Wu ZS (2015) Adsorption of naphthalene from aqueous solution on coal-based activated carbon modified by microwave induction: microwave power effects. Chem Eng Process 91:67–77. https://doi.org/10.1016/j.cep.2015.03.019
Ghagi RK, Satpute SK, Chopade BA, Banpurkar AG (2002) Study of functional properties of Sapindusmukorossi as a potential biosurfactant. Indian J Sci Technol 4:530–533. https://doi.org/10.17485/ijst/2011/v4i5/30055
Goi A, Trapido M (2004) Degradation of polycyclic aromatic hydrocarbons in soil: the Fenton reagent versus ozonation. Environ Technol 25:155–164. https://doi.org/10.1080/09593330409355448

Griesbaum K, Behr A, Biedenkapp D, Voges HW, Garbe D, Paetz C, Collin G, Mayer D, Höke H (2002) Hydrocarbons. In: Ullmann's encyclopedia of industrial chemistry. Wiley-VCH, Weinheim. https://doi.org/10.1002/14356007.a13_227

Gu J, Dong D, Kong L, Zheng Y, Li X (2012) Photocatalytic degradation of phenanthrene on soil surfaces in the presence of nanometer anatase TiO_2 under UV-light. J Environ Sci 24:2122–2126. https://doi.org/10.1016/S1001-0742(11)61063-2

Guieysse B, Cirne MD, Mattiasson B (2001) Microbial degradation of phenanthrene and pyrene in a two-liquid phase partitioning bioreactor. Appl Microbiol Biotechnol 56:796–802. https://doi.org/10.1007/s002530100706

Guo W, He M, Yang Z, Lin C, Quan X, Wang H (2007) Distribution of polycyclic aromatic hydrocarbons in water, suspended particulate matter and sediment from Daliao River watershed. Chemosphere 68:93–104. https://doi.org/10.1016/j.chemosphere.2006.12.072

Gupta H, Gupta B (2015) Photocatalytic degradation of polycyclic aromatic hydrocarbonbenzo[a] pyrene by iron oxides and identification of degradation products. Chemosphere 138:924–931. https://doi.org/10.1016/j.chemosphere.2014.12.028

Gupta B, Rani M, Kumar R, Dureja P (2011) Decay profile and metabolic pathways of quinalphos in water, soil and plants. Chemosphere 85:710–716. https://doi.org/10.1016/j.chemosphere.2011.05.059

Gupta B, Rani M, Kumar R (2012a) Degradation of thiraminwater, soil and plants: a study by high-performance liquid chromatography. Biomed Chromatogr 26(1):69–75. https://doi.org/10.1002/bmc.1627

Gupta B, Rani M, Kumar R, Dureja P (2012b) Identification of degradation products of thiram in water, soil and plants using LC-MS technique. J Environ Sci Health Part B 47:823–831. https://doi.org/10.1080/03601234.2012.676487

Gupta B, Rani M, Salunke R, Kumar R (2012c) In vitro and in vivo studies on degradation of quinalphos in rats. J Hazard Mater 213–214:285–291. https://doi.org/10.1016/j.jhazmat.2012.01.089

Hadibarata T, Tachibana S (2009) Enhanced chrysene biodegradation in presence of a synthetic surfactant. Obayashi Y, Isobe T, Subramanian A, Suzuki S, Tanabe S. Interdisciplinary studies on environmental chemistry-environmental research in Asia. 2:301–308

Hadibarata T, Tachibana S, Itoh K (2009) Biodegradation of chrysene, an aromatic hydrocarbon by *Polyporus*sp.S133 in liquid medium. J Hazard Mater 164:911–917. https://doi.org/10.1016/j.jhazmat.2008.08.081

Hallett DJ, Brecher RW (1984) Cycling of polynuclear aromatic hydrocarbons in the Great Lakes ecosystem. In: Niagru JO, Simmons MS (eds) Toxic contaminants in the Great Lakes. Wiley, New York, pp 213–238

Hanazato T (2001) Pesticide effects on freshwater zooplankton: an ecological perspective. Environ Pollut 112:1–10. https://doi.org/10.1016/S0269-7491(00)00110-X

Haritash AK, Kaushik CPJ (2009) Biodegradation aspects of polycyclic aromatic hydrocarbons (PAHs): a review. J Hazard Mater 169:1–15. https://doi.org/10.1016/j.jhazmat.2009.03.137

Haritash AK, Kaushik CP (2016) Degradation of low molecular weight polycyclic aromatic hydrocarbons by microorganisms isolated from contaminated soil. Int J Environ Sci 6:472–482. https://doi.org/10.6088/ijes.6053

Harvey RG (1997) Polycyclic aromatic hydrocarbons, chemistry and carcinogenicity. Cambridge University Press, Cambridge

Hassan SSM, Azab WIME, Ali HR, Mansour MSM (2015) Green synthesis and characterization of ZnO nanoparticles for photocatalytic degradation of anthracene. Adv Nat Sci Nanosci Nanotechnol 6:1–11. https://doi.org/10.1088/2043-6262/6/4/045012

Hazardous substances fact sheet, July (2001) New Jersey Department of Health and Senior Services. In: 1–6 http://nj.gov/health/eoh/rtkweb/documents/fs/0208.pdf

Heitkamp MA, Freeman JP, Cerniglia CE (1987) Naphthalene biodegradation in environmental microcosms: estimates of degradation rates and characterization of metabolites. Appl Environ Microbiol 53:129–136

Huang H, Wu K, Khan A, Jiang Y, Ling Z, Liu P, Chen Y, Tao X, Li X (2016) A novel Pseudomonas gessardii strain LZ-E simultaneously degrades naphthalene and reduces hexavalent chromium. Bioresour Technol 207:370–378

Imma T, Stephen de M, Sheikholeslami Reza M, Jean-Pierre V, Jean B, Chantal C (2004) Aliphatic and aromatic hydrocarbons in coastal Caspian Sea sediments. Mar Pollut Bull 48:44–60. https://doi.org/10.1016/S0025-326X(03)00255-8

International Agency for Research on Cancer (IARC) (1983) Benzo[a]pyrene, polynuclear aromatic compounds. Part 1, Chemical, environmental and experimental data. Monographs on the Evaluation of the Carcinogenic Risk of Chemicals to Humans 32:211–224

Irwin RJ (1997a–h) Environmental contaminants encyclopedia a: acenaphthylene; b: fluoranthene; c: benzo (*a*) anthracene; d: benzo (*b*) fluoranthene; e: benzo (*k*) anthracene; f: indeno (1,2,3-c,d) pyrene; g: dibenz (*a,h*) anthracene [dibenzo (*a,h*) anthracene]; h: diabenzo (*g,h,i*) perylene. National park service water resources divisions, Water operations branch 1201 Oakridge drive, Suite 250 Fort Collins, Colorado 80525

Jassal V, Shanker U, Shankar S (2015a) Synthesis characterization and applications of nano-structured metal hexacyanoferrates: a review. J Environ Anal Chem 2:1000128–1000141. https://doi.org/10.4172/2380-2391.1000128

Jassal V, Shanker U, Kaith BS, Shankar S (2015b) Green synthesis of potassium zinc hexacyanoferrate nanocubes and their potential application in photocatalytic degradation of organic dyes. RSC Adv 5:26141–26149. https://doi.org/10.1039/C5RA03266K

Jassal V, Shanker U, Kaith BS (2016a) *Aegle marmelos* mediated green synthesis of different nanostructured metal hexacyanoferrates: activity against photodegradation of harmful organic dyes. Scientifica 2016:1–13. https://doi.org/10.1155/2016/2715026

Jassal V, Shanker U, Gahlot U (2016b) Green synthesis of some iron oxide nanoparticles and their interaction with 2-amino, 3-amino and 4-aminopyridines. Mater Today Proc 3:1874–1882. https://doi.org/10.1016/j.matpr.2016.04.087

Jassal V, Shanker U, Gahlot S, Kaith BS, Kamaluddin IMA, Samuel P (2016c) Sapindus mukorossi mediated green synthesis of some manganese oxide nanoparticles interaction with aromatic amines. Appl Phys A Mater Sci Process 122:271–282. https://doi.org/10.1007/s00339-016-9777-4

Jia H, Zhao J, Fan X, Dilimulati K, Wang C (2012) Photodegradation of phenanthrene on cation-modified clays under visible light. Appl Catal B 123–124:43–51. https://doi.org/10.1016/j.apcatb.2012.04.017

Juhasz L, Naidu R (2000) Bioremediation of high molecular weight polycyclic aromatic hydrocarbons: a review of the microbial degradation of benzo[a]pyrene. Int Biodeterior Biodegrad 45:57–88. https://doi.org/10.1016/S0964-8305(00)00052-4

Juhasz AL, Britz ML, Stanley GA (1997) Degradation of fluoranthene, pyrene, benz[a]anthracene and dbenz[a,h]anthracene by Burkholderiacepacia. J Appl Microbiol 83:189–198

Juhasz AL et al (2000) Microbial degradation and detoxification of high molecular weight polycyclic aromatic hydrocarbons by Stenotrophomonas maltophilia strain VUN 10,003. Lett Appl Microbiol 30:396–401

Karam FF, Hussein FH, Baqir SJ, Halbus AF, Dillert R, Bahnemann D (2014) Photocatalytic degradation of anthracene in closed system reactor. Int J Photoenergy 2014:1–6. https://doi.org/10.1155/2014/503825

Kasiotis KM, Emmanouil C (2015) Advanced PAH pollution monitoring by bivalves. Environ Chem Lett 13:395. https://doi.org/10.1007/s10311-015-0525-3

Kaushik CP, Haritash AK (2006) Polycyclic aromatic hydrocarbons (PAHs) and environmental health. Our Earth 3:1–7. https://doi.org/10.1016/j.jhazmat.2009.03.137

Keng PS, Lee SL, Ha ST et al (2014) Removal of hazardous heavy metals from aqueous environment by low-cost adsorption materials. Environ Chem Lett 12:15. https://doi.org/10.1007/s10311-013-0427-1

Kou J, Zhang H, Yuan Y, Li Z, Wang Y, Yu T, Zou Z (2008) Efficient photodegradation of phenanthrene under visible light irradiation via photosensitized electron transfer. J Phys Chem C 112:4291–4296. https://doi.org/10.1021/jp7111022
Latimer J, Zheng J (2003) The sources, transport, and fate of PAH in the marine environment. In: Douben PET (ed) PAHs: an ecotoxicological perspective. Wiley, New York
Legrini O, Oliveros E, Braun AM (1993) Photochemical processes for water treatment. Chem Rev 93:671–698. https://doi.org/10.1021/cr00018a003
Lellala K, Namratha K, Byrappa K (2016) Microwave assisted synthesis and characterization of nanostructure zinc oxide-graphene oxide and photo degradation of brilliant blue. Mater Today Proc 3:74–83
Li CS, Ro YS (2000) Indoor characteristics of polycyclic aromatic hydrocarbons in the urban atmosphere of Taipei. Atmos Environ 34:611–620. https://doi.org/10.1016/S1352-2310(99)00171-5
Liao W, Ma Y, Chen A, Yang Y (2015) Preparation of fatty acids coated Fe_3O_4 nanoparticles for adsorption and determination of benzo(*a*)pyrene in environmental water samples. Chem Eng J 271:232–239. https://doi.org/10.1016/j.cej.2015.03.010
Lijinsky W (1991) The formation and occurrence of polynuclear aromatic hydrocarbons associated with food. Mutat Res 259:251–262. https://doi.org/10.1016/0165-1218(91)90121-2
Lily MK, Bahuguna A, Bhatt KK, Dangwal K (2013) Degradation of Anthracene by a novel strain Brachybacterium paraconglomeratum BMIT637C (MTCC 9445) Int. J Environ Sci 3:1242–1252
Liu SQ (2012) Magnetic semiconductor nano-photocatalysts for the degradation of organic pollutants. Environ Chem Lett 10:209. https://doi.org/10.1007/s10311-011-0348-9
Liu D, Wu Z, Tian F, Ye BC, Tong Y (2016) Synthesis of N and La co-doped TiO_2/AC photocatalyst by microwave irradiation for the photocatalytic degradation of naphthalene. J Alloys Compd 676:489–498. https://doi.org/10.1016/j.jallcom.2016.03.124
Lu H, Zhu L (2007) Pollution patterns of polycyclic aromatic hydrocarbons in tobacco smoke. J Hazard Mater 139:193–198. https://doi.org/10.1016/j.jhazmat.2006.06.011
Lung SC, Kao MC, Hu SC (2003) Contribution of incense burning to indoor PM_{10} and particle-bound polycyclic aromatic hydrocarbons under two ventilation conditions. Indoor Air 13:194–199. https://doi.org/10.1034/j.1600-0668.2003.00197.x
Luo F, Yang D, Chen Z, Megharaj M, Naidu R (2016) One-step green synthesis of bimetallic Fe/Pd nanoparticles used to degrade orange II. J Hazard Mater 303:145–153
Maliszewska-Kordybach B (1999) Sources, concentrations, fate and effects of polycyclic aromatic hydrocarbons (PAHs) in the environment. Part A: PAHs in air. Pol J Environ Stud 8:131–136
Mallakin A, Dixon DG, Greenberg BM (2000) Pathway of anthracene modification under simulated solar radiation. Chemosphere 40:1435–1441. https://doi.org/10.1016/S0045-6535(99)00331-8
Manoli E, Samara C (1999) Polycyclic aromatic hydrocarbons in natural waters: sources, occurrence and analysis. Trends Anal Chem 18:417–428. https://doi.org/10.1016/S0165-9936(99)00111-9
Meador JP, Stein JE, Reichert WL (1995) Bioaccumulation of polyaromatic hydrocarbon by marine organisms. Rev Environ Contam T 79:143–145
Menzie CA, Potocki BB, Santodonato J (1992) Exposure to carcinogenic PAHs in the environment. Environ Sci Technol 26:1278–1284. https://doi.org/10.1021/es00031a002
Moussawi RN, Patra D (2016) Nanoparticle self-assembled grain like curcumin conjugated zno: curcumin conjugation enhances removal of perylene, fluoranthene, and chrysene by ZnO. Sci Rep 6:24565. https://doi.org/10.1038/srep24565
Naf C, Broman D, Brunstrom B (1992) Distribution and metabolism of polycyclic aromatic hydrocarbons (PAHs) injected into eggs of chicken (*Gallus domesticus*) and common eider duck (Somateriamollissima). Environ Toxicol Chem 11:1653–1660. https://doi.org/10.1002/etc.5620111114

Nam K, Rodriguez W, Kukor JJ (2001) Enhanced degradation of polycyclic aromatic hydrocarbons by biodegradation combined with a modified Fenton reaction. Chemosphere 45:11–20. https://doi.org/10.1016/S0045-6535(01)00051-0

Nam JJ, Sweetman AJ, Jones KC (2009) Polynuclear aromatic hydrocarbons (PAHs) in global background soils. J Environ Monit 11:45–48. https://doi.org/10.1039/b813841a

Nayak A, Sanganal SK, Mudde SK, Karegoudar TB (2011) A catabolic pathway for the degradation of chrysene by Pseudoxanthomonas sp. PNK-04. FEMS Microbiol Lett 320(2):128–134

Ncibi MC, Mahjoub B, Gourdon R (2007) Effects of aging on the extractability of naphthalene and phenanthrene from Mediterranean soils. J Hazard Mat 146:378–384. https://doi.org/10.1016/j.jhazmat.2006.12.032

Nelkenbaum E, Dror I, Berkowitz B (2007) Reductive hydrogenation of polycyclic aromatic hydrocarbons catalyzed by metalloporphyrins. Chemosphere 68:210–217. https://doi.org/10.1016/j.chemosphere.2007.01.034

Oanh NTK, Reutergardh LB, Dung NT (1999) Emission of polycyclic aromatic hydrocarbons and particulate matter from domestic combustion of selected fuels. Environ Sci Technol 33:2703–2709. https://doi.org/10.1021/es980853f

Oncescu T, Stefan MI, Oancea P (2010) Photocatalytic degradation of dichlorvos in aqueous TiO_2 suspensions. Environ Sci Pollut Res 17:1158–1166. https://doi.org/10.1007/s11356-009-0292-4

Patel V, Jain S, Madamwar D (2012a) Naphthalene degradation by bacterial consortium (DV-AL) developed from Alang-Sosiya ship breaking yard, Gujarat, India. Bioresour Technol 107:122–130

Patel V, Cheturvedula S, Madamwar D (2012b) Phenanthrene degradation by Pseudoxanthomonas sp. DMVP2 isolated from hydrocarbon contaminated sediment of Amlakhadi canal, Gujarat. India J Hazard Mat 201–202:43–51

Pathak H, Kantharia D, Malpani A, Madamwar D (2009) Naphthalene degradation by Pseudomonas sp. HOB1: in vitro studies and assessment of naphthalene degradation efficiency in simulated microcosms. J Hazard Mat 166:1466–1473

Patnaik P (1999) A comprehensive guide to the properties of hazardous chemical substances, 2nd edn. Wiley, New York

Prak DJ, Pritchard PH (2002) Solubilization of polycyclic aromatic hydrocarbons mixtures in micellular non-ionic surfactant solution. Water Res 36:3463–3472. https://doi.org/10.1016/S0043-1354(02)00070-2

Prakash A, Sharma S, Ahmad N, Ghosh A, Sinha P (2011) Synthesis of Agnpsbybacilus cereus bacteria and their antimicrobial potential. J Biomater Nanobiotechnol 2:156–162. https://doi.org/10.4236/jbnb.2011.22020

Puglisi E, Cappa F, Fragoulis G, Trevisan M, Re AAMD (2007) Bioavailability and degradation of phenanthrene in compost amended soils. Chemosphere 67:548–556. https://doi.org/10.1016/j.chemosphere.2006.09.058

Ramadahl T, Alfheim I, Rustad S, Olsen T (1982) Chemical and biological characterisation of emissions from small residential stoves burning wood and charcoal. Chemosphere 11:601–611. https://doi.org/10.1016/0045-6535(82)90205-3

Ramsay MA, Swannell RPJ, Shipton WA, Duke NC, Hill RT (2000) Effect of bioremediation community inoiled mangrove sediments. Mar Pollut Bull 20:413–419. https://doi.org/10.1016/S0025-326X(00)00137-5

Rani M(2012) Studies on decay profiles of quinalphos and thiram pesticides. Ph.D thesis, Indian Institute of Technology Roorkee, Roorkee, Uttarakhand, India, Chapter 1, 5

Rani M, Shanker U (2017a) Degradation of traditional and new emerging pesticides in water by nanomaterials: recent trends and future recommendations. Int J Environ Sci Technol. https://doi.org/10.1007/s13762-017-1512-y

Rani M, Shanker U (2017b) Removal of carcinogenic aromatic amines by metal hexacyanoferrates nanocubes synthesized via green process. J Environ Chem Engg 5:5298–53112017. https://doi.org/10.1016/j.jece.2017.10.028

Rani M, Shanker U, Chaurasia A (2017a) Catalytic potential of laccase immobilized on transition metal oxides nanomaterials: degradation of alizarin red S dye. J Environ Chem Engg 5:2730. https://doi.org/10.1016/j.jece.2017.05.026

Rani M, Shanker U, Jassal V (2017b) Recent strategies for removal and degradation of persistent & toxic organochlorine pesticides using nanoparticles: a review. J Environ Manag 190:208–222. https://doi.org/10.1016/j.jenvman.2016.12.068

Reddy SM, Shaik B, Kumar VGS, Joshi HV, Ghosh PK (2003) Quantification and classification of ship scrapping waste at Alang-Sosiya, India. Mar Pollut Bull 46:1609–1614. https://doi.org/10.1016/S0025-326X(03)00329-1

Reddy MS, Basha S, Joshi HV, Ramachandraiah G (2005) Seasonal distribution and contamination levels of total PHCs, PAHs and heavy metals in coastal waters of the Alang-Sosiya ship scrapping yard, Gulf of Cambay, India. Chemosphere 61:1587–1593. https://doi.org/10.1016/j.chemosphere.2005.04.093

Rehmann K, Noll HP, Steinberg CE, Kettrup AA (1998) Pyrene degradation by Mycobacterium sp. strain KR2. Chemosphere 36(14):2977–2992. https://doi.org/10.1016/S0045-6535(97)10240-5

Rick J, Alan L (2010) Obama, in Gulf, pledges to push on stopping leak. USA Today. Associated Press. Retrieved 3 March 2013

Rybak J, Olejniczak T (2014) Accumulation of polycyclic aromatic hydrocarbons (PAHs) on the spider webs in the vicinity of road traffic emissions. Environ Sci Pollut Res Int 21(3):2313–2324. https://doi.org/10.1007/s11356-013-2092-0

Samanta SK, Singh OV, Jain RK (2002) Polycyclic aromatic hydrocarbons: environmental pollution and bioremediation. Trends Biotechnol 20:243–248. https://doi.org/10.1016/S0167-7799(02)01943-1

Sanglard D, Leisola MSA, Fiechter A (1986) Role of extracellular liginases in biodegradation of benzo[a]pyrene by Phanerochaetechrysoporium. Enzym Microb Technol 8:209–212

Sannino F, Pirozzi D, Vitiello G, D'errico G, Aronne A, Fanelli E, Pernice P (2014) Oxidative degradation of phenanthrene in the absence of light irradiatiion by hybrid ZrO_2-acetylacetonate gel-derived catalyst. Appl Catal B 156–157:101–107. https://doi.org/10.1016/j.apcatb.2014.03.006

Sathish M, Viswanathan B, Viswanath RP (2007) Characterization and photocatalytic activity of N-doped TiO_2 prepared by thermal decomposition of Ti–melamine complex. Appl Catal B Environ 74:307–312. https://doi.org/10.1016/j.apcatb.2007.03.003

Schneider J, Grosser R, Jayasimhulu K, Xue W, Warshawsky D (1996) Degradation of pyrene, benzo[*a*]anthrancene, and benzo[*a*]pyrene by mycobacterium sp. strain RJGII-135, isolated from a former coal gasification site. Appl Environ Microbiol 13:13–19

Shahwan T, Abu Sirriah S, Nairat M et al (2011) Green synthesis of iron nanoparticles and their application as a Fenton-like catalyst for the degradation of aqueous cationic and anionic dyes. Chem Eng J 172:258–266. https://doi.org/10.1016/j.cej.2011.05.103

Shanker U, Jassal V, Rani M, Kaith BS (2016a) Towards green synthesis of nanoparticles: from bio-assisted sources to benign solvents. A review. Int J Environ Anal Chem 96:801–835. https://doi.org/10.1080/03067319.2016.1209663

Shanker U, Jassal V, Rani M (2016b) Catalytic removal of organic colorants from water using some transition metal oxide nanoparticles synthesized under sunlight. RSC Adv 6:94989–94999. https://doi.org/10.1039/C6RA17555D

Shanker U, Rani M, Jassal V (2017a) Degradation of hazardous organic dyes in water by nanomaterials. Environ Chem Lett 15:1–20. https://doi.org/10.1007/s10311-017-0650-2

Shanker U, Jassal V, Rani M (2017b) Green synthesis of iron hexacyanoferrate nanoparticles: potential candidate for the degradation of toxic PAHs. J Environ Chem Engg 5:4108–4120. https://doi.org/10.1016/j.jece.2017.07.042

Shanker U, Jassal V, Rani M (2017c) Degradation of toxic PAHs in water and soil using potassium zinc hexacyanoferrate nanocubes. J Environ Manag 204:337–348. https://doi.org/10.1016/j.jenvman.2017.09.015

Sharma YC, Srivastava V, Singh VK, Kaul SN, Weng CH (2009) Nano-adsorbents for the removal of metallic pollutants from water and wastewater. Environ Technol 30:583–609. https://doi.org/10.1080/09593330902838080
Shukla AK, Iravani S (2017) Metallic nanoparticles: green synthesis and spectroscopic characterization. Environ Chem Lett 15:223. https://doi.org/10.1007/s10311-017-0618-2
Soclo HH, Garriguesa PH, Ewald M (2000) Origin of polycyclic aromatic hydrocarbons (PAHs) in coastal marine sediments: case studies in Cotonou (Benin) and Aquitaine (France) areas. Mar Pollut Bull 40:387–396. https://doi.org/10.1016/S0025-326X(99)00200-3
Stronkhorst J, Ysebaert TJ, Smedes F, Meininger PL, Dirksen S (1993) Contaminants in eggs of some waterbird species from Scheldt Estuary, SW Netherlands. Mar Pollut Bull 26:572–578
Suzanne G, Terry MA (2012) BP suspended from new US federal contracts over deepwater disaster. The Guardian, London
Swaathy S, Kavitha V, Pravin AS, Mandal AB, Gnanamani A (2014) Microbial surfactant mediated degradation of anthracene in aqueous phase by marine Bacillus licheniformis MTCC 5514. Biotechnol Rep (Amst) 4:161–170
Tam NFY, Guo CL, Yau WY, Wong YS (2002) Preliminary study on biodegradation of phenanthrene by bacteria isolated from mangrove sediments in Hong Kong. Mar Pollut Bull 45:316–324. https://doi.org/10.1016/S0025-326X(02)00108-X
The MAK-Collection Part I (2012) Occupational Toxicants, Deutsche Forschungsgemeinschaft@ 2012. Wiley-VCH Verlag GmbH & Co. KGaA, Weinheim, p 27
Tonne CC, Whyatt RM, Camann DE, Perera FP, Kinney PL (2004) Predictors of personal polycyclic aromatic hydrocarbon exposures among pregnant minority women in New York City. Environ Health Perspect 112:754–759. https://doi.org/10.1289/ehp.5955
Tremblay L, Kohl SD, Rice JA, Gagne JP (2005) Effects of temperature, salinity, and dissolved humic substances on the sorption of polycyclic aromatic hydrocarbons to estuarine particles. Mar Chem 96:21–34. https://doi.org/10.1016/j.marchem.2004.10.004
Ukiwe LN, Egereonu UU, Njoku PC, Nwoko CIA (2013) Combined chemical and water hyacinth (Eichhorniacrassipes) treatment of PAHs contaminated soil. IJSER 4:1–12
Vanaja M, Paulkumar K, Baburaja M, Rajeshkumar S, Gnanajobitha G, Malarkodi C, Sivakavinesan M, Annadurai G (2014) Degradation of methylene blue using biologically synthesized silver nanoparticles. Bioinorg Chem Appl 2014:1–8
Villanneau EJ, Saby NPA, Orton TG et al (2013) First evidence of large-scale PAH trends in French soils. Environ Chem Lett 11:99. https://doi.org/10.1007/s10311-013-0401-y
Wagoner D (1976) Compilation of ambient trace substances. Draft of report prepared by research triangle institute under contract no. 68-02-1325 for US Environmental Protection Agency. Available from W.G. Tucker, Project officer, IERL-EPA, Project Triangle Park, N.C
Walter U, Beyer M, Klein J, Rehm HJ (1991) Degradation of pyrene by Rhodococcus sp. UW1. Appl Microbiol Biotechnol 34:671–676
Wammer KH, Peters CA (2005) Polycyclic aromatic hydrocarbon biodegradation rates: a structure based study. Environ Sci Technol 39:2571–2578
Wang C, Yediler A, Peng A, Kettrup A (1995) Photodegradation of phenanthrene by N-doped TiO_2 photocatalyst. Chemosphere 30:501–510. https://doi.org/10.1016/0045-6535(94)00413-O
Wattiau P, Bastiaens L, Herwijnen RV, Daal L, Parsonsc JR, Renard ME, Springael D, Cornelis GR (2001) Fluorene degradation by Sphingomonas sp. LB126 proceeds through protocatechuic acid: a genetic analysis. Res Microbiol 152:861–872. https://doi.org/10.1016/S0923-2508(01)01269-4
Wen S, Zhao J, Sheng G, Fu J, Peng PA (2002) Photocatalytic reactions of phenanthrene at TiO_2/water interfaces. Chemosphere 46:871–877. https://doi.org/10.1016/S0045-6535(01)00149-7
WHO (1998) Environmental health criteria 202: selected non-heterocyclic polycyclic aromatic hydrocarbons. World Health Organization, Geneva, p 883
Wolter M, Zadrazil F, Martens R, Bahadir M (1997) Degradation of eight highly condensed polycyclic aromatic hydrocarbons by Pleurotus sp. Florida in solid wheat straw substrate. Appl Microbiol Biotechnol 48:398–404

Wright DA, Welbourn P (2002) Environmental toxicology. Cambridge University Press, Cambridge

Xia S, Zhang L, Zhou X, Shao M, Pan G, Ni Z (2015) Fabrication of highly dispersed Ti/ZnO–Cr_2O_3 composite as highly efficient photocatalyst for naphthalene degradation. Appl Catal B 176–177:266–277. https://doi.org/10.1016/j.apcatb.2015.04.008

Xiao J, Du J (2016) A multifunctional statistical copolymer vesicle for water remediation. Polym Chem 7:4647–4653. https://doi.org/10.1039/C6PY00763E

Xue D, Warshawski D (2005) Metabolic activation of polycyclic and heterocyclic aromatic hydrocarbons and DNA damage: a review. Toxicol Appl Pharmacol 206:73–93. https://doi.org/10.1016/j.taap.2004.11.006

Xue X, Cheng R, Shi L, Ma Z, Zheng X (2017) Nanomaterials for water pollution monitoring and remediation. Environ Chem Lett 15:23. https://doi.org/10.1007/s10311-016-0595-x

Yamada M, Takada H, Toyoda K, Yoshida A, Shibata A, Nomura H, Wada M, Nishimura M, Okamoto K, Ohwada K (2003) Study on the fate of petroleum-derived polycyclic aromatic hydrocarbons (PAHs) and the effect of chemical dispersant using an enclosed ecosystem, mesocosm. Mar Pollut Bull 47:105–113. https://doi.org/10.1016/S0025-326X(03)00102-4

Yang HH, Lee WJ, Chen SJ, Lai SO (1998) PAH emission from various industrial stacks. J Hazard Mater 60:159–174. https://doi.org/10.1016/S0304-3894(98)00089-2

Yuanfu P, Otake M, Vacha M, Sato H (2007) Synthesis and characterization of a novel electroluminescent polymer based on phenoxazine and fluorene derivatives. React Funct Polym 67:1211–1217. https://doi.org/10.1016/j.reactfunctpolym.2007.07.011

Yusuf N, Timares L, Seibert MD, Xu H, Elmets CA (2007) Acquired and innate immunity to polyaromatic hydrocarbons. Toxicol Appl Pharmacol 224:308–312. https://doi.org/10.1016/j.taap.2006.12.009

Zhang J, Noshaka Y (2014) Mechanism of the OH radical generation in photocatalysis with TiO_2 of different crystalline types. J Phys Chem C 118(20):10824–10832. https://doi.org/10.1021/jp501214m

Zhang Y, Tao X (2009) Global atmospheric emission inventory of polycyclic aromatic hydrocarbons (PAHs) for 2004. Atmos Environ 43:812–819. https://doi.org/10.1016/j.atmosenv.2008.10.050

Zhang LH, Li PJ, Gong ZQ, Oni AA (2006) Photochemical behaviour of benzo [a] pyrene on soil surface under UV light irradiation. J Environ Sci 18:1226–1232. https://doi.org/10.1016/S1001-0742(06)60067-3

Zhang Y, Wong JWC, Liu P, Yuan M (2011) Heterogeneous photocatalytic degradation of phenanthrene in surfactant solution containing TiO_2 particles. J Hazard Mater 191:136–114. https://doi.org/10.1016/j.jhazmat.2011.04.059

Zhao X, Cai Z, Wang T, O'Reilly SE, Liu W, Zhao D (2016) A new type of cobalt-deposited titanate nanotubes for enhanced photocatalytic degradation of phenanthrene. Appl Catal B 187:134–143. https://doi.org/10.1016/j.apcatb.2016.01.010

Zhao S, Jia H Nulaji G, Gao H, Wang F, Wang C (2017) Photolysis of polycyclic aromatic hydrocarbons (PAHs) on Fe^{3+}–montmorillonite surface under visible light: degradation kinetics, mechanism, and toxicity assessments. Chemosphere 84:1346–1354. https://doi.org/10.1016/j.chemosphere.2017.06.106

Zheng XJ, Blais JF, Mercier G, Bergeron M, Drogui P (2007) PAH removal from spiked municipal wastewater sewage sludge using biological, chemical and electrochemical treatments. Chemosphere 68:1143–1152. https://doi.org/10.1016/j.chemosphere.2007.01.052

Zhu L, Lu H, Chen S, Amagai T (2009) Pollution level phase distribution and source analysis of polycyclic aromatic hydrocarbons in residential air in Hangzhou, China J Hazard Mater 162:1165–1170. https://doi.org/10.1016/j.jhazmat.2008.05.150

Zhu Y, Fan L, Yang B, Du J (2014) Multifunctional Homopolymer vesicles for facile immobilization of gold nanoparticles and effective water remediation. ACS Nano 8:5022–5031. https://doi.org/10.1021/nn5010974

Index

G. Crini, E. Lichtfouse (eds.), *Green Adsorbents for Pollutant Removal*, Environmental Chemistry for a Sustainable World 18,
https://doi.org/10.1007/978-3-319-92111-2

Zeitfracht Medien GmbH
Ferdinand-Jühlke-Straße 7
99095 Erfurt, Deutschland
produktsicherheit@kolibri360.de